Mechanisms and Machine Science

Volume 81

Series Editor

Marco Ceccarelli, Department of Industrial Engineering, University of Rome Tor Vergata, Roma, Italy

Editorial Board

Alfonso Hernandez, Mechanical Engineering, University of the Basque Country, Bilbao, Vizcaya, Spain

Tian Huang, Department of Mechatronical Engineering, Tianjin University, Tianjin, China

Yukio Takeda, Mechanical Engineering, Tokyo Institute of Technology, Tokyo, Japan

Burkhard Corves, Institute of Mechanism Theory, Machine Dynamics and Robotics, RWTH Aachen University, Aachen, Nordrhein-Westfalen, Germany

Sunil Agrawal, Department of Mechanical Engineering, Columbia University, New York, NY, USA

This book series establishes a well-defined forum for monographs, edited Books, and proceedings on mechanical engineering with particular emphasis on MMS (Mechanism and Machine Science). The final goal is the publication of research that shows the development of mechanical engineering and particularly MMS in all technical aspects, even in very recent assessments. Published works share an approach by which technical details and formulation are discussed, and discuss modern formalisms with the aim to circulate research and technical achievements for use in professional, research, academic, and teaching activities.

This technical approach is an essential characteristic of the series. By discussing technical details and formulations in terms of modern formalisms, the possibility is created not only to show technical developments but also to explain achievements for technical teaching and research activity today and for the future.

The book series is intended to collect technical views on developments of the broad field of MMS in a unique frame that can be seen in its totality as an Encyclopaedia of MMS but with the additional purpose of archiving and teaching MMS achievements. Therefore, the book series will be of use not only for researchers and teachers in Mechanical Engineering but also for professionals and students for their formation and future work.

The series is promoted under the auspices of International Federation for the Promotion of Mechanism and Machine Science (IFToMM).

Prospective authors and editors can contact Mr. Pierpaolo Riva (publishing editor, Springer) at: pierpaolo.riva@springer.com

Indexed by SCOPUS and Google Scholar.

More information about this series at http://www.springer.com/series/8779

Veniamin Goldfarb · Evgenii Trubachev ·
Natalya Barmina

Editors

New Approaches to Gear Design and Production

 Springer

Editors
Veniamin Goldfarb (1941–2019)
Kalashnikov Izhevsk State Technical
University
Izhevsk, The Udmurt Republic, Russia

Natalya Barmina
Institute of Mechanics
Kalashnikov Izhevsk State Technical
University
Izhevsk, The Udmurt Republic, Russia

Evgenii Trubachev
Institute of Mechanics
Kalashnikov Izhevsk State Technical
University
Izhevsk, The Udmurt Republic, Russia

.

ISSN 2211-0984   ISSN 2211-0992   (electronic)
Mechanisms and Machine Science
ISBN 978-3-030-34947-9   ISBN 978-3-030-34945-5   (eBook)
https://doi.org/10.1007/978-3-030-34945-5

Chapters 1, 6, 11, 12, 18, 19, 20, 21 were translated from Russian into English by Natalya Barmina, Ph.D.,
the other chapters were written in English by their authors.

This Springer imprint is published by the registered company Springer Nature Switzerland AG
The registered company address is: Gewerbestrasse 11, 6330 Cham, Switzerland

Foreword

This book deals with the technology of gearing systems by looking at theory, research, and practice with recent updates and results from activity within the IFToMM community. The gearing systems are important design in an important area of MMS (Mechanism and Machine Science) both in design and performance of machines, even in the most sophisticated solutions with modern mechatronic structures.

IFToMM is the International Federation for the Promotion of Mechanism and Machine Science (www.iftomm.net) and among its Technical Committees one is specifically addressed to Gearing Systems to work out specific IFToMM activities (in collaboration and dissemination of achievements and trends) in this specific domain.

This third volume is organized as an update and enlargement of the previous books as coming from experts within the communities of several IFToMM member organizations from all over the world. Therefore, I am sure that readers will find interesting discussions and research updates that can reinforce and indeed stimulate their attention and activity in the field of both research and applications.

I congratulate the guest editors for the successful result of their efforts and time they have spent in coordinating and preparing this volume. I thank the authors for their valuable contributions that show clearly that gearing systems are of fundamental importance in further developing mechanical and mechatronic systems. This volume can be also considered as a source of inspiration for better and better consideration of gearing systems also within challenges of modern developments of science and technology within activity for teaching, design, and research.

Roma, Italy   Marco Ceccarelli
March 2019   IFToMM President 2016–2019

Introduction

The present issue-related contributed volume is the third in the series that started in 2016 [1] and continued in 2018 [2], and it includes publications in the field of gears and transmissions.

The theme of manuscripts covers a wide range of issues related to research, design, and production of gears and gearbox systems. In particular, we present works devoted to:

- reviews of the history and contemporary state of the theory and practice of gears and gearbox systems in Russia and Bulgaria;
- development of basic concepts and approaches to geometrical and kinematic analysis of gearing;
- synthesis of schemes differing by interesting kinematic properties of gears;
- load tooth contact analysis of gears;
- methods and systems of the integrated design (including the thorough analysis and basic production preparation) of different types of gears;
- simulation, research, and methods of synthesis of specific types of gears: spur and helical (including those having an approximate meshing), cylindrical-bevel, worm and complex cases of planetary gears;
- proposal of new schemes of tooth generation;
- improvement of methods of tooth machining and techniques of gear production.

The contributed volume also includes manuscripts related to the rational design of transmissions and their units that cover:

- invariant methods of computer-aided design of gearbox systems;
- methods of calculation and experimental evaluation of bearing units;
- methods of evaluation of transmission vibration activity.

A good deal of published manuscripts integrates two or several parts of the gear technology (for instance, methods of gear calculation with the corresponding CAD systems, gear design with production, tooth control/measurement with gearing design/analysis) that reflects the up-to-date requirements to the science and engineering.

The book can be interesting and useful to scientific researchers and engineers that deal with the issues of gear design, research, and production; and to young researchers and students that choose new promising areas of development and application of progressive gears and methods of their synthesis, analysis, and production.

The editors welcome and thank both authors that are regularly publishing their manuscripts in this series (which is already becoming the regular one) and new authors with the strong confidence that the participation in this series is beneficial and interesting; and we wish all the authors and the volume as a whole to be useful and highly demanded for the readers and mechanical engineers in general.

By tradition, we want to express our deep gratitude to Springer publisher, especially to the chief editor of the MMS series Prof. Marco Ceccarelli for the constant support of this series of books and also to Dr. Eng. Ph.D. Sergey Lagutin for his sustained assistance and advice in gear terminology issues.

[1] Goldfarb V., Barmina N. (eds.): Theory and Practice of Gearing and Transmissions: In Honor of Professor Faydor L. Litvin, vol. 34, p. 450. Springer International Publishing, AG Switzerland (2016). ISBN: 978-3-3 19-19739-5, DOI 10.1007/978-3-319-19740-1. https://www.springer.com/gp/book/9783319197395

[2] Goldfarb V., Trubachev E., Barmina N. (eds.): Advanced Gear Engineering, vol. 51, p. 497. Springer International Publishing, AG Switzerland (2018). ISBN: 978-3-319-60398-8, DOI 10.1007/978-3-319-60399-5. https://www.springer.com/gp/book/9783319603988

Contents

About the Editors

Veniamin Goldfarb (1941–2019) D.Sc. in Engineering, Full Professor, Honored Scientist of Russia and Udmurt Republic, Director of Scientific Department "Institute of Mechanics" of Kalashnikov Izhevsk State Technical University, Vice-president of International Federation for the Promotion of Mechanism and Machine Science (IFToMM) in 2012–2015. Prof. Veniamin Goldfarb was the initiator and the chief editor of the previous two contributed volumes on gears. He took an active part in editing and preparation of this third book, but, to our great regret, he did not live till its publication.

Evgenii Trubachev D.Sc. in Engineering, Full Professor, Honored Scientist of Udmurt Republic, Vice-Director of Scientific Department "Institute of Mechanics", Professor of Mechanical Engineering Department of Kalashnikov Izhevsk State Technical University. E-mail: truba@istu.ru

Natalya Barmina Ph.D. in Engineering, Senior Researcher of Scientific Department "Institute of Mechanics", Associate Professor of English Department of Kalashnikov Izhevsk State Technical University, Member of PC CPA of International Federation for the Promotion of Mechanism and Machine Science (IFToMM). E-mail: barmina-nat@mail.ru

Chapter 1
Russian School of the Theory and Geometry of Gearing. Part 2. Development of the Classical Theory of Gearing and Establishment of the Theory of Real Gearing in 1976–2000

Dmitry T. Babichev, Sergey A. Lagutin and Natalya A. Barmina

Abstract Authors of this paper initiated the work on Review of the Russian school in the theory and geometry of gearing in the second half of the 20th century. The Review comprise the analysis of the carried out works with an application of tables that allow to obviously see who of the experts published their papers with the date and the theme of these publications. Essays are provided for the experts who contributed most into the development of the theory of gearing. Part 1 of the Review under name "Origin of the theory of gearing and its golden period 1935–1975" has been published in 2016. The presented part 2 considers about 230 publications in Russian and 30 publications in English made by the Russian gear experts in the period 1976–2000. The important feature of this period is that gear design and production techniques started to be regarded as the united complex issue which requires the agreed solution of many interrelated tasks. Note here, that, publication reviews usually indicate which problems have been solved in the cited references, but they do not comprise the methods for their solution; the original source should be addressed for this purpose. In order to facilitate this point for the readers, especially beginners in the theory of gearing (students, engineers, post-graduates), we supplemented the review with the statement of methods for the solution of the most important issues.

D. T. Babichev
Tyumen Industrial University, Volodarskogo Str. 38, 62500, Tyumen, Russia
e-mail: babichevdt@rambler.ru

S. A. Lagutin
Electrostal Heavy Engineering Works, JSC, 144000, Krasnaya Str. 19, Electrostal city, Russia
e-mail: lagutin@eztm.ru

N. A. Barmina (✉)
Institute of Mechanics, Kalashnikov Izhevsk State Technical University, Studencheskaya Str. 7, 426069, Izhevsk, Russia
e-mail: barmina-nat@mail.ru

© Springer Nature Switzerland AG 2020
V. Goldfarb et al. (eds.), *New Approaches to Gear Design and Production*,
Mechanisms and Machine Science 81,
https://doi.org/10.1007/978-3-030-34945-5_1

Keywords Theory of gearing (TG) · Gear geometry · Theory of generation ·
Review of publications in Russian

1.1 Introduction

During the second half of the 20th century, Soviet scientists were among the world
leaders in the gear geometry and applied theory, that are the theoretical fundamentals
of designing gears and gear-cutting tools.

The authors of the current paper have considered works related to the development
of the theory of gearing (TG) published in Russia from the mid-1930s to the present.
Overall, the review covers more than 400 publications of nearly 200 authors (among
more than 1000 studied works).

The first part of this review was published under name "Russian school of the
theory and geometry of gearing: Its origin and golden period (1935–1975)" in J. Front.
Mech. Eng. Higher Education Press and Springer-Verlag Berlin Heidelberg, 2016,
11(1), pp. 44–59. Having examined 165 publications of more than 70 researchers,
the authors talked about the authoritative scientific schools that emerged and worked
actively during this period in Leningrad, Moscow, Saratov, Novocherkassk, Gorky,
Kurgan, Izhevsk. The main scientific and practical achievements of these schools
were considered. Among them:

– the kinematic, matrix and other methods of analysis and synthesis of gearing,
– the technological synthesis of spiral bevel and hypoid gears.
– the mastering of new types of gears, including the W-N gears, globoid and spiroid
  gears etc.

The present, second part relates on development of Russian school of the theory
of gearing in 1976–2000 years. This period has a number of features provided by
historical reasons. First, the USSR industry revealed the first obvious symptoms of
stagnation at the late 1970s. Second, the number of post-graduates decreased, since
universities generally solved the problem of teaching stuff that was urgent in the
early 1960s because of the abrupt increase of the number of engineer students.

However, within these hard 15–20 years preceding the USSR breakup, certain
old scientific schools continued functioning in the Soviet Union and new centers
developing the theory and practice of gearing arose.

The list of publication involves the most profound works of the leading scientists
and the most original works of less famous experts with the advanced ideas and
results. To make this review interesting for a wide audience, it involves a certain
number of works on designing tools, on gear strength and accuracy.

1.2   Review of Works Published in Russian in 1976–2000

By the mid 1970s the theory of gearing as an individual science has reached its completeness: basic and priority problems of gearing analysis and synthesis have been solved. The number of novel theoretical works has declined. The structure of scientific theses has been mainly unified: construction of design formulas for a specific gearing → algorithmization → computer-aided analysis → production → experiment → implementation certificates. Investigations came closer to the production; and the theory of gearing became, first of all, the tooling for solving specific engineering problems.

The main results of the review of works dated the last quarter of the 20th century [1–227] are introduced in two tables. The table in Fig. 1.1 represents 5 groups including 26 scientific problems that comprised the main results of theoretical investigations. The table in Fig. 1.2 enumerates 30 types of gears, mechanisms and their parts, and also 3 areas of activity that implement the practical results of works in 1976–2000.

Each of the tables contains: (a) names of the solved problems (investigated objects, spheres of their application), (b) family names of main authors who solved these problems (with indication of the work number in the reference list and publication date), (c) connection lines that relate the authors and their solved problems. Family names of the authors who contributed most to the theory of gearing (to our opinion) and the most essential works are highlighted in bold in the tables. The tables allow for tracking:

• themes that are considered to be the most popular and important at that time;
• researchers working on the theme of your interest;
• issues considered by the researchers and the scope of the subject matter for each of them.

The tables cover both directions of development of the theory of gearing: "*world-wide*"—IGD, and "*Russian*"—theory of real gearing. Features of the "*Russian*" direction is mainly considered. As far as the "*world-wide*" direction is concerned, it comprises the following technique for solving the problems of analysis and synthesis for in Russia and abroad:

• methods of the classical theory of gearing (at computer-aided design of geometry of any type of gears);
• non-differential methods of analysis of gear generating processes when implementing both versatile and special applied software,
• numerical methods of optimization (at synthesis of geometry of tooth operating flanks and transient surfaces);
• the expanded list of quality features of gearing which includes the criteria for assessing the terms of oil wedge generation and other;
• methods of boundary and finite elements at analysis of contact interaction of work surfaces and of the tooth deformation mode;

Let us consider the features of the "*Russian*" approach in four paragraphs:

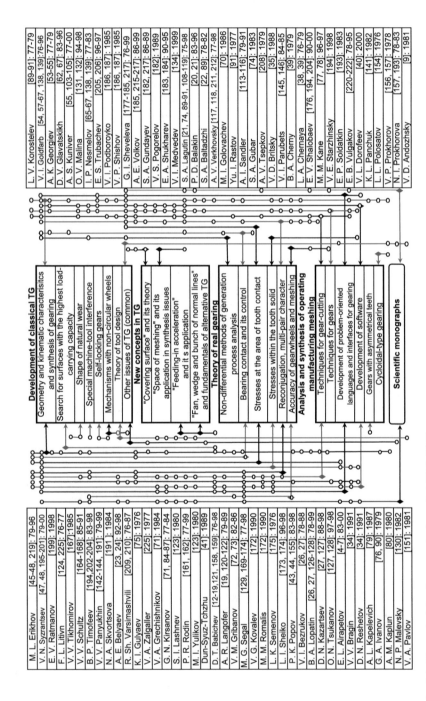

Fig. 1.1 Theoretical developments in the period 1976–2000

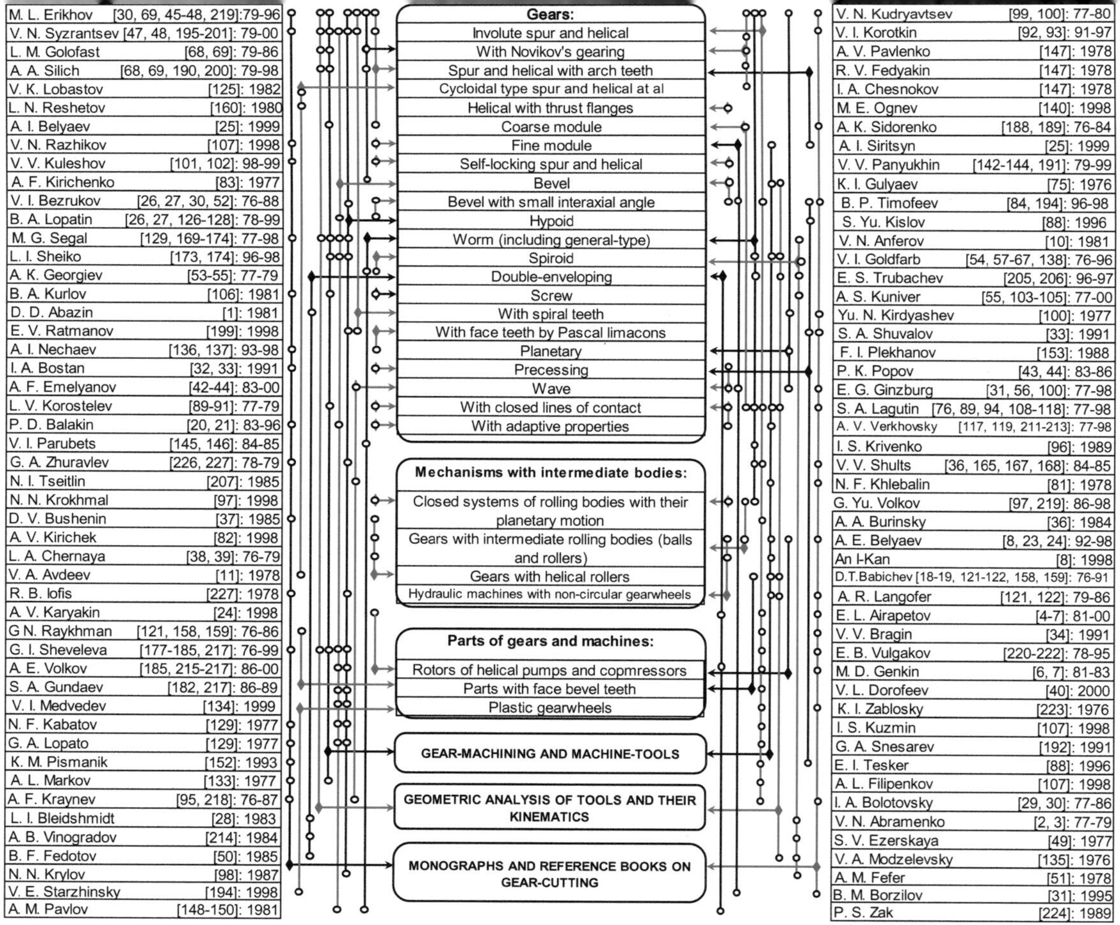

Gears:
- Involute spur and helical
- With Novikov's gearing
- Spur and helical with arch teeth
- Cycloidal type spur and helical at al
- Helical with thrust flanges
- Coarse module
- Fine module
- Self-locking spur and helical
- Bevel
- Bevel with small interaxial angle
- Hypoid
- Worm (including general-type)
- Spiroid
- Double-enveloping
- Screw
- With spiral teeth
- With face teeth by Pascal limacons
- Planetary
- Precessing
- Wave
- With closed lines of contact
- With adaptive properties

Mechanisms with intermediate bodies:
- Closed systems of rolling bodies with their planetary motion
- Gears with intermediate rolling bodies (balls and rollers)
- Gears with helical rollers
- Hydraulic machines with non-circular gearwheels

Parts of gears and machines:
- Rotors of helical pumps and copmressors
- Parts with face bevel teeth
- Plastic gearwheels

GEAR-MACHINING AND MACHINE-TOOLS

GEOMETRIC ANALYSIS OF TOOLS AND THEIR KINEMATICS

MONOGRAPHS AND REFERENCE BOOKS ON GEAR-CUTTING

V. N. Kudryavtsev [99, 100]: 77-80
V. I. Korotkin [92, 93]: 91-97
A. V. Pavlenko [147]: 1978
R. V. Fedyakin [147]: 1978
I. A. Chesnokov [147]: 1978
M. E. Ognev [140]: 1998
A. K. Sidorenko [188, 189]: 76-84
A. I. Siritsyn [25]: 1999
V. V. Panyukhin [142-144, 191]: 79-99
K. I. Gulyaev [75]: 1976
B. P. Timofeev [84, 194]: 96-98
S. Yu. Kislov [88]: 1996
V. N. Anferov [10]: 1981
V. I. Goldfarb [54, 57-67, 138]: 76-96
E. S. Trubachev [205, 206]: 96-97
A. S. Kuniver [55, 103-105]: 77-00
Yu. N. Kirdyashev [100]: 1977
S. A. Shuvalov [33]: 1991
F. I. Plekhanov [153]: 1988
P. K. Popov [43, 44]: 83-86
E. G. Ginzburg [31, 56, 100]: 77-98
S. A. Lagutin [76, 89, 94, 108-118]: 77-98
A. V. Verkhovsky [117, 119, 211-213]: 77-98
I. S. Krivenko [96]: 1989
V. V. Shults [36, 165, 167, 168]: 84-85
N. F. Khlebalin [81]: 1978
G. Yu. Volkov [97, 219]: 86-98
A. A. Burinsky [36]: 1984
A. E. Belyaev [8, 23, 24]: 92-98
An I-Kan [8]: 1998
D. T. Babichev [18-19, 121-122, 158, 159]: 76-91
A. R. Langofer [121, 122]: 79-86
E. L. Airapetov [4-7]: 81-00
V. V. Bragin [34]: 1991
E. B. Vulgakov [220-222]: 78-95
M. D. Genkin [6, 7]: 81-83
V. L. Dorofeev [40]: 2000
K. I. Zablosky [223]: 1976
I. S. Kuzmin [107]: 1998
G. A. Snesarev [192]: 1991
E. I. Tesker [88]: 1996
I. A. Filipenkov [107]: 1998
I. A. Bolotovsky [29, 30]: 77-86
V. N. Abramenko [2, 3]: 77-79
S. V. Ezerskaya [49]: 1977
V. A. Modzelevsky [135]: 1976
A. M. Fefer [51]: 1978
B. M. Borzilov [31]: 1995
P. S. Zak [224]: 1989

M. L. Erikhov [30, 69, 45-48, 219]79-96
V. N. Syzrantsev [47, 48, 195-201]: 79-00
L. M. Golofast [68, 69]: 79-86
A. A. Silich [68, 69, 190, 200]: 79-98
V. K. Lobastov [125]: 1982
L. N. Reshetov [160]: 1980
A. I. Belyaev [25]: 1999
V. N. Razhikov [107]: 1998
V. V. Kuleshov [101, 102]: 98-99
A. F. Kirichenko [83]: 1977
V. I. Bezrukov [26, 27, 30, 52]: 76-88
B. A. Lopatin [26, 27, 126-128]: 78-99
M. G. Segal [129, 169-174]: 77-98
L. I. Sheiko [173, 174]: 96-98
A. K. Georgiev [53-55]: 77-79
B. A. Kurlov [106]: 1981
D. D. Abazin [1]: 1981
E. V. Ratmanov [199]: 1998
A. I. Nechaev [136, 137]: 93-98
I. A. Bostan [32, 33]: 1991
A. F. Emelyanov [42-44]: 83-00
L. V. Korostelev [89-91]: 77-79
P. D. Balakin [20, 21]: 83-96
V. I. Parubets [145, 146]: 84-85
G. A. Zhuravlev [226, 227]: 78-79
N. I. Tseitlin [207]: 1985
N. N. Krokhmal [97]: 1998
D. V. Bushenin [37]: 1985
A. V. Kirichek [82]: 1998
L. A. Chernaya [38, 39]: 76-79
V. A. Avdeev [11]: 1978
R. B. Iofis [227]: 1978
A. V. Karyakin [24]: 1998
G N. Raykhman [121, 158, 159]: 76-86
G. I. Sheveleva [177-185, 215-217]: 76-99
A. E. Volkov [185, 215-217]: 86-00
S. A. Gundaev [182, 217]: 86-89
V. I. Medvedev [134]: 1999
N. F. Kabatov [129]: 1977
G. A. Lopato [129]: 1977
K. M. Pismanik [152]: 1993
A. L. Markov [133]: 1977
A. F. Kraynev [95, 218]: 76-87
L. I. Bleidshmidt [28]: 1983
A. B. Vinogradov [214]: 1984
B. F. Fedotov [50]: 1985
N. N. Krylov [98]: 1987
V. E. Starzhinsky [194]: 1998
A. M. Pavlov [148-150]: 1981

Fig. 1.2 Practical results in the period 1976–2000

2.1. new geometric concepts in the theory of gearing;
2.2. development of methods of optimization synthesis;
2.3. development of methods of generating process analysis;
2.4. software: features of development and implementation.

1.2.1 New Geometric and Geometry and Kinematic Concepts

The crucial contribution to the development of the theory of gearing after 1970s was the introduction and application of new concepts:
(1) *covering surface* (Sheveleva [180]); (2) *fan, wedge, bunch of normal lines* (Babichev [14]); (3) *feeding-in acceleration* [121]; (4) *space of meshing* (Lagutin [108]); Let us consider these innovations.

1.2.1.1   Enveloping and Covering Surfaces

In the theory of surface generation by moving solids two following types of the *obtained surfaces* should be distinguished: undulated covering line Σ_1 and smooth enveloping line Σ_2—see Fig. 1.3.

The covering surface Σ_1 is the set of a discrete family of fragments of basic tool surfaces (BTS); the enveloping surface Σ_2 is the surface contacting all elements of this discrete set. The enveloping surface Σ_2 is usually a smooth surface, while the covering surface Σ_1 is always the faceted surface. The concept "covering" was proposed by G. I. Sheveleva and effectively applied at study of bevel and hypoid gears [180]. Note, that the term "*covering*" is not perfect when translated into English. In practice it means a real, wavy or rough surface.

The surface Σ_2 is usually determined as the enveloping for one- or two-parametric family of generating surfaces. The covering surface Σ_1 is determined by direct tracking of the position of points of the generating surface Σ_0 with respect to the blank; it is the method of *direct digital modeling* (DDM) sometimes called the "*method of direct enveloping*". Methods of DDM are more reliable than kinematic ones, so it is preferable to determine the covering surface Σ_1 rather than the enveloping surface

Fig. 1.3 Covering Σ_1 and enveloping surface Σ_2 of the generating surface Σ_0

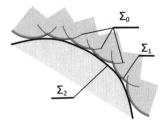

Fig. 1.4 The band type of roughnesses

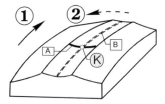

Σ_2 in many CAD and CAE systems. Although, in order to determine Σ_1 it is required to perform hundreds of calculations more than for Σ_2.

Note, that covering surfaces Σ_1 can consist of stripes (bands), for instance, at line grinding of teeth—see Fig. 1.4, each of the "bands" is the surface obtained by enveloping process (line A is the contact line at one-parametric enveloping). Each stripe is contacting the two-parametric Σ_2 enveloping surface along the line (this line passes along the stripe). When the feed is reduced, the "bands" become narrower. When the feed value tends to zero, the generated surface will be approaching the enveloping surface of a family of "bands". It will be exactly the surface Σ_2 generated by *two-parametric enveloping by the surface Σ_0.*

The covering Σ_1 can consist of individual "flakes", for example, when cutting the gearwheels by module hobs or diagonal gear grinding. Figure 1.5 shows the deviations of the flake covering surface Σ_1 from the enveloping Σ_2. "Flakes" are contacting with Σ_2 at the centre of the flake. The shape of "flakes" is in general case hexagonal that can degenerate into parallelogram.

The curvature of band and flake covering surfaces Σ_1 crucially and essentially differs from the curvature of the enveloping surface Σ_2. Thus, the radius of curvature for the involute is smoothly increasing from its base circle. And the radius of curvature for the covering surface generated by the rectilinear rack is $\rho = \infty$ at all points (and at jogs it is $\rho = 0$), since the covering profile is the broken line contacting the involute by each of its rectilinear segments.

Using of the concept of the covering surface instead of the enveloping one creates a number of difficulties when solving the contact problem by numerical methods: errors of strains and contact stress analysis increase, boundaries of instant contact areas lose their smoothness. The common means of "approaching" the covering surface with the enveloping one is the issue of small feeds at computer-aided modeling of the generating process. It reduces the roughness of the surface obtained at modeling and

Fig. 1.5 The shape of roughness is hexagonal "flakes"

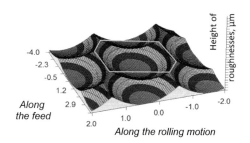

decreases the errors at solving the contact problem. But it does not solve the problem of analysis of the enveloping curvature, since the curvature of the covering surface does not depend on the feed value.

1.2.1.2   Jogs on Tooth Profiles and Flanks

Jogs (lines of surface intersection or points of profile intersection) are present practically on all gear parts and tools. The concept of jogs as of specific features of generating surfaces comes from Rodin [161]. He wrote: «the surface of a part consisting of a number of adjacent areas can be *considered as a single surface*. The point of jog of the profile located at the boundary of adjacent areas can be considered as an arc of a circumference with the radius tending to zero». Figure 1.6 (left) shows three possible types of jogs which give rise to three types of normal lines families: a fan (at intersection of two profiles), a wedge (at intersection of two surfaces) and a bunch (at intersection of three surfaces). Figure 1.6 (right) specifies, why the point B at the jog of the rack is the contacting one, that is, it lies on the generated transient curve of the tooth fillet; and the point C does not belong to the transient curve. The reason is that one of the normals of the fan at the point B passes through the pitch point, while there is no such a normal line at the fan of the jog C. Note, that the classic theory of gearing does not answer the question why the point B is the contact one and the point C is not. Terms "fan, wedge, bunch" have been later replaced by terms "sector, prism, pyramid" – see chap. 6 by Babichev and Barmina in this book.

Figure 1.7 shows that the jog and its adjacent segments of surfaces are the *single surface with the continuous curvilinear coordinate v*. In this figure: **1** is **v**-line; **2** are brushes of normal lines; **3** is fan of normal lines; $\mathbf{n_A}$, $\mathbf{n_B}$ are main unit vectors.

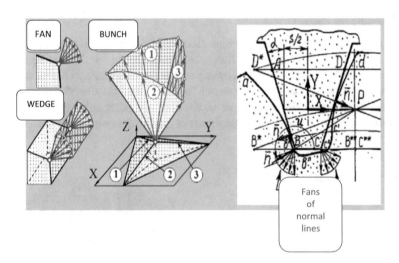

Fig. 1.6 Types of jogs and contact at the point B

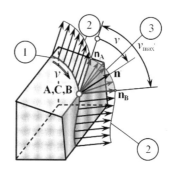

Fig. 1.7 Parameters of the edge jog: 1—v-line; 2—brushes of normal lines; 3—fan of normal lines; n_A, n_B—main unit vectors of the fan of normal lines

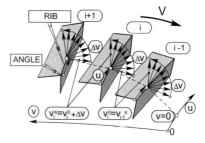

Fig. 1.8 Curvilinear coordinates on the real BTS for hobs and broaches

Publications [12–19] are devoted to the development of the kinematic method for investigation of generating processes with higher reliability as compared to the classic method. It applies both jogs and curvilinear coordinates that are single for the whole tooth or the gear rim (and even along the whole BTS for all edge tools: hobs, mills, broaches, shavers [13]—see Fig. 1.8).

Figure 1.8 shows several neighboring teeth of the edge tool with pointing the curvilinear coordinates u (along the cutting edge) and v (across it). The coordinate v is related here with the tooth number i.

From physical point of view, the real BTS is the set of cutting edges with adjoining front and back planes (surfaces). From mathematical point of view, the real BTS is the smooth surface with two continuous curvilinear coordinates on it.

For the real BTS we should know all information related to the geometry that influences the generating and cutting processes. Computational models for such tools are to be found in [17].

Summary: jogs are the useful means when analyzing the generation processes.

1.2.1.3 Feeding-in Parameters of the Generating Surface to the Solid of the Generated Part: Velocities and Acceleration of Feeding-in, etc.

One of the main issues of the theory of gearing is to determine the tooth surface Σ_2 of the second element when knowing the tooth surface Σ_1 of the first element.

When solving this problem one should consider the motion of the **surface** of the *first element* solid in the coordinate system of the second one, that is, in the three-dimensional space of the *second element* (Cartesian). It is logical to assume that the motion parameters of **the solid 1 surface** in the **space of the solid 2** should be used. However, in classical theory of gearing there is no common system of concepts and characteristics specifying the interaction of the **solid surface with the space**. And in order to describe this interaction the parameters are used that specify the spatial motion of only the *solid as the set of points* rather than the *solid limited by the surface*. In order to eliminate the misunderstanding of the essence of the said above, let us explain the difference of any point A of the solid from the point K lying on surface Σ_1 of this solid. At the point K there is the normal line to the solid surface Σ_1 (and the tangent surface and curvature parameters), while at the point A (inside the solid) there is nothing similar to it. For the generation process of the solid surface Σ_2 of the second element it is important to know the motion in space of the second element of points of the surface Σ_1 projected on its normal **N** at these points. V. A. Shishkov paid attention on it at the end of 1940s, proposed the *local parameter of interaction of the solid surface with the space* and called it the "velocity of feeding-in". It is obtained as the scalar product:

$$V_N = \mathbf{V}_{12} \cdot \mathbf{n} = V_{12x} \cdot n_x + V_{12y} \cdot n_y + V_{12z} \cdot n_z \tag{1.1}$$

where **n** is the unit vector of the normal **N** directed outwards from the solid 1, $\mathbf{V}_{12}$ is the vector of the relative velocity at this point: $\mathbf{V}_{12} = \mathbf{V}_1 - \mathbf{V}_2$.

Velocity V_N is also called the velocity of mutual approach (for $V_N > 0$) and removal (for $V_N < 0$). If $V_N = 0$, surfaces Σ_1 and Σ_2 are contacting at this time instant and at this point. V_N is the important but not the only parameter of feeding-in motion. Other parameters should be considered: acceleration of feeding-in a_N, as the first time derivative of the velocity V_N; and its higher time derivatives.

Applying V_N and a_N, it became possible to develop the generalized method of study of load for cutting edges of a wide range of enveloping gear cutting tools (hobs and grinding wheels) for rough and final machining. It also became possible to implement this method by the software that allowed for investigating the operation of a large group of different types of tools [120–122].

Let us show examples of investigations carried out by means of this software.

Figure 1.9 shows the example of investigating the cutting process for arch teeth of cylindrical gears by spiral disk mills proposed by M. L. Erikhov. On the left in this figure the cutting scheme is shown; on the right the boundaries of the cutting zone and thickness of the layers to be cut (in mm when the mill is fed along its axis S = 1 mm/rev) are shown. Parameters of the machine-tool meshing are: z = 20, m = 5, mill radius $r_m = 100$ mm, the number of racks (chaser) k = 8, the ratio of angular velocities of the mill ω_1 and the blank ω_2 is equal to $i_{12} = \omega_1/\omega_2 = z$.

Results of computer-aided modeling of spiral disk mills operation show that:

(1) two teeth of each three-teeth chaser are constantly participating in cutting;
(2) the most loaded teeth are the middle teeth of all chasers;

Fig. 1.9 Scheme for cutting arch teeth and boundaries of the cut layer on the developed view of the cutting edge trace

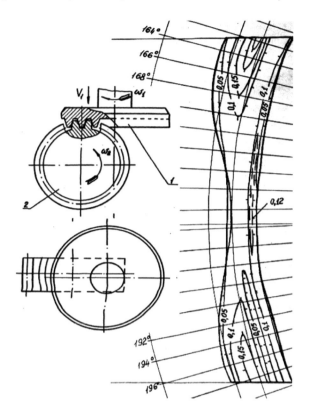

(3) outer cutting edges of all chasers practically do not participate in operation;
(4) the maximum thickness of the cut is at the top edge of the fifth rack, it is S_{max} = 0.35 mm since the cutting start;
(5) the maximum thickness S_{max} of the cut layers in the course of the tool cutting-in is transmitted from the top and the transient segments to the lateral ones;
(6) the double increase in the radius r_m of the tool widens the cutting layer by 8%, but it practically does not influence the value of S_{max};
(7) the decrease in the number of chasers doubles the value of S_{max};
(8) the double decrease in the radial feed decreases the value of S_{max} in two times with simultaneous narrowing the cut by 30%.

Figure 1.10 shows the cutting scheme and cutting zone for face bevel teeth which applied as half-couplings in systems of control and as gearwheels in wave gears.

Flanks of face bevel teeth are the enveloping lines of a two-parametric family of lines (cutting edges of the disk hob in Fig. 1.10). These surfaces were investigated by Ya. S. Davydov, V. I. Bezrukov, L. Ya. Liburkin; their works are described in the first part of our review. But only in [158, 159] the problem was solved for determining the curvature radii of surfaces generated by two-parametric motion of the generating line. The difficulty of this problem solution is that the surface circumscribed by the

Fig. 1.10 Scheme for
cutting the face bevel teeth
and boundaries of cutting
zones: dotted
line—computer-aided
analysis; solid lines—wax
model experiment

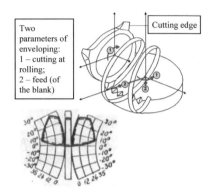

cutting edge when running-in 1 (Fig. 1.10) is the "spiral surface on the bagel". And
the spiral pitch is decreased when the tool is approaching the axis of the blank rotation
due to the feed motion. That is, the tooth surface on the item is the envelope of the
single-parametric family of a "spiral surface on the bagel" *deformed* at motion (due
to feed 2).

Further in Figs. 1.11, 1.12, and 1.13 the investigation results for cutting the spur
involute gears by shaping cutters are shown. Note, that Figs. 1.11 and 1.12 are

Fig. 1.11 Thicknesses of
layers cut by the involute
shaper with $z_0 = 20$, $x_0 = 0.5$
when cutting the gearwheels
with $z_1 = 20$ and $x_1 = 0$

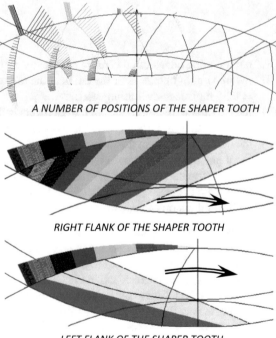

A NUMBER OF POSITIONS OF THE SHAPER TOOTH

RIGHT FLANK OF THE SHAPER TOOTH

LEFT FLANK OF THE SHAPER TOOTH

Fig. 1.12 Cutting zones at machining of the gearwheel with internal teeth with $z_1 = 30$ and $x_1 = 0.5$ by the shaper with $z_0 = 20$ and $x_0 = 0.5$

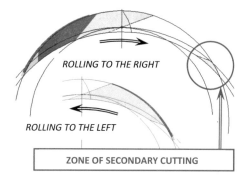

Fig. 1.13 Cutting zone and lines of equal thicknesses of cut layers for the finishing pass of the involute shaper

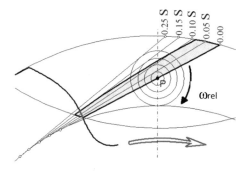

based on the concept of the feeding-in velocity (as by V. A. Shishkov). The role of the feeding-in acceleration a_N is lower: to increase the computational accuracy for thicknesses of cuts and the boundaries of cutting zones (including the zones of secondary cutting) when analyzing the finishing passes of the tool.

In planar meshing the following theorem allows for assessing the load of the tool:

Theorem If the point of the tool profile is in the machining allowance, then the tool load is proportional to the moment developed by the normal line to the profile with respect to the pitch point (Fig. 1.13).

Figure 1.13 shows the example of the finishing gear shaping of the involute gearwheel. The cutting zone with the variable width is stretched along the meshing line (as the "oasis" along one river bank). When the direction of running-in is changed, the cutting zone ("oasis") is transmitted to the other side of the meshing line ("river") scarcely changing its width.

The velocity V of the material removal by abrasives and the thickness S of the cut layer are:

$$V = (\omega_0 - \omega_1) \cdot R_\Sigma$$

$$S = S_0 \cdot R_\Sigma \cdot \left| \frac{\omega_0 - \omega_1}{V_P} \right|$$

where ω_0 and ω_1 are angular velocities of running-in motion for the tool and blank ($\omega > 0$ when rotating counterclockwise); R is the arm of the normal line to the tool profile with respect to the pitch point P; V_P is the circumferential velocity at centroids; S_0 is the circumferential feed per lead.

Note: The point at the tool is work loaded if the profile normal line *coming out of the tool solid* creates the moment directed along the relative angular velocity $\omega_0 - \omega_1$ with respect to pitch point P.

In accordance with the theorem, it is seen in Fig. 1.13 that the maximum thickness of the cut is $S = 0.15 \cdot S_0$ (at the beginning of the tool operation).

When determining the thickness of the cut layers, it is rather difficult to find the boundaries of cutting zones, especially at finish gear machining. Acceleration of the cutting-in allowed for solving this problem (though approximately) even for such a complex process as gear hobbing by generating method. These issues are studied in details in the thesis by Langofer [120].

The issues considered above in p. 2.1.3 and shown in Figs. 1.9, 1.10, 1.11, 1.12, and 1.13 reveal only one of possible direction of applying the concepts of cutting-in parameters (velocity, acceleration and higher derivatives) for analysis of tool loading. But application of the concept "acceleration of cutting-in" simplifies the solution of other important problems of gear geometry. For instance, when $a_N > 0$, the envelope is formed inside the solid of the generating element which is physically impossible. When determining Σ_2, the value a_N allows for simultaneously determining the radii of curvature and micro-roughness height along the whole surface formed by generating for the band-type and flaky shapes of roughnesses—see Fig. 1.4 above.

By means of the concept of the feeding-in acceleration the new formula has been obtained to analyze the reduced curvature at any normal section at the point of contact of two moving solids that are contacting along the line:

$$\frac{1}{R_\Sigma} = -\frac{\omega_W^2}{a_N}$$

where ω_W is the angular velocity of solids rolling on the plane of the section where the curvature is analyzed; The fundamental nature of this formula means that according to its generality it is commensurate with the known in geometry Rodrigues and Frenet formulas, but it is intended for surfaces developed in the process of their forming by generating methods.

1.2.1.4 Space of Meshing

This concept was introduced to the scientific circulation by Lagutin [108]. The term implies the physical space obtained at axes of gearwheels of the synthesized gear, each point of the space possessing its set of beams of the screw of relative motion (Fig. 1.14). By considering each of these beams as the probable contact normal and relating it to velocities of elements by vector equations, different isosurfaces can

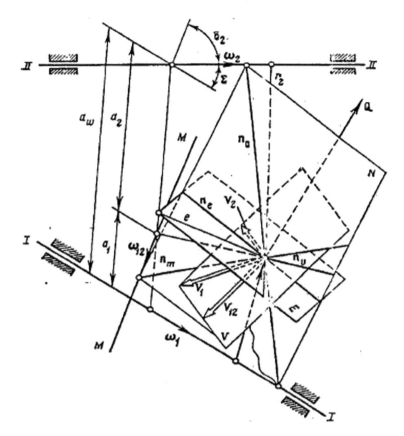

Fig. 1.14 Point and plane of contact normal lines in the space of meshing

be singled out of this space, that is, the geometric loci for points at which the gear will possess specific properties: the assigned pressure angle, specific sliding etc. S. A. Lagutin proposed to apply such isosurfaces as boundaries of the area of gear existence and as potential surfaces of meshing.

Publication [108] enriches and makes more profound the investigation results on synthesis of gears with helical motion of elements [91], on determining analogs of meshing axes in general-type worm gears [89], on gear synthesis by method of loci [90] that were obtained together with L. V. Korostelev and his followers.

Later, the author performed the synthesis of general-type worm gears by means of the meshing space [110, 119] and was looking for equivalent settings for cutting bevel gears with circular teeth. We suppose that the significance of the analysis of meshing space at gear synthesis will be increased in time.

1.2.2 Development of Methods for Synthesis and Optimization of Gearing

The priority practical problem of gear design is to determine tooth profiles or surfaces that provide the maximum load carrying capacity of a gear. Figure 1.15 presents the basic methods for increasing the load carrying capacity of gearing (by example of spur and helical gears). This figure systemizes the methods based on advancement of tooth geometry. The left column (blocks 1–4) represents the ways of increasing the tooth bending strength, that is, procedures of changing the tooth shape aimed at decreasing the stresses inside teeth and, first of all, at the tooth dedendum.

The central column in Fig. 1.15 (blocks 5–8) presents the methods for synthesis of tooth flanks that provide the least contact stresses and thereby the increased lifetime of teeth.

The right column (blocks 9–12) includes the ways of correction and modification of profiles (both for the synthesized and the existing gears, first of all, for the involute ones). Here the word "correction" means the agreed variation of conjugate segments on the pinion and gearwheel at once (block 9), for example, in order to decrease the contact stresses by increasing the reduced radii of curvature in the danger zone.

"Modification" is the deviation of one of the conjugate segments of profiles (usually on the addendum) from mutual enveloping in order to redistribute the forces at multi-pair contact or within the areas of reconjugation (with account of tooth deformation under load). Modification also decreases the forces of tooth impact at the initial moment of tooth contact, thus decreasing the vibration activity of the meshing which is vital for heavy-loaded high-speed gears. The following procedures are distinguished at modification: simple flanking (block 10), when the curve is assigned (segment of the straight line, circumference arc, etc.) on the profile of the generating rack or on the tooth; and complex flanking (block 11), when the profile is synthesized on the modified segment. Such flanking simultaneously eliminates the edge tooth contact at the initial (final) stage of their contact; and it gives the smooth load increase on teeth coming into contact (and the smooth load decrease for teeth coming out of the contact) etc.

The complexity of the problem in blocks 7–9 is that tooth deformations depend on the value of the torque T, that is why, optimal values and the shape of modification lines and flanking are various for different T. The gear should be designed so that it could operate properly at all its loading modes. The situation is worsened by the tendency of heavy-loaded gears to trochoidal interference, that is, untimely coming of teeth into the edge contact at tooth deformation.

Based on Figs. 1.1, 1.2, and 1.15, let us enumerate the works on analysis and optimization of geometry for tooth profiles and flanks for plane and spatial gearings in order to increase the load-carrying capacity of gears:

Block 1 (Fig. 1.15): optimization of the shape of the transient curve in order to decrease the stress at tooth root: works by Vulgakov [220, 221] specifying bending strength of heavy-loaded aviation gears. Investigations of tooth mode of deformation are also presented in works by Airapetov and Genkin [4–7], Reshetov and Bragin

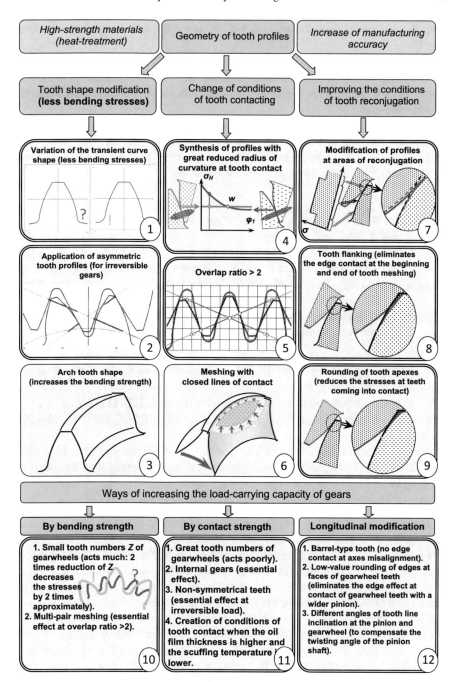

Fig. 1.15 Ways of increasing the gear load-carrying capacity

[34], Syzrantsev [196], Goldfarb with his followers [61, 64], Dorofeyev [40]—see Fig. 1.1.

Block 2 (Fig. 1.15): application of asymmetric tooth profiles: Kapelevich [79] and Vulgakov [221]. They are efficient in many gear drives, in which the tooth load on one flank is significantly higher and is applied for longer periods of time than for the opposite one.

Block 3 (Fig. 1.15): arch tooth shape that increases the tooth bending strength. In spur and helical gears there are works by Erikhov [46, 48] and Syzrantsev [48, 196]; Belyaev and Siritsyn [25]; Reshetov [122]. In bevel and hypoid gears there are up to 20 authors and dozens of publications—see "gears"–>"bevel" and "hypoid" in Fig. 1.2.

Block 4 (Fig. 1.15): synthesis of tooth profiles and flanks for the big reduced radius of curvature R_Σ at tooth contact thus reducing the Hertz contact stresses. There are different methods for such a synthesis:

Method 1. For the assigned type of conjugated profiles (or flanks) it is necessary to determine the segments where $R_\Sigma \to$ max. It is the method for designing involute gears by determining the optimal shift coefficient x_1 and x_2 (for instance, applying the blocking contours): Bolotovsky et al. [29, 30].

Method 2. It is necessary to search for new types of gearing and (or) optimize the geometry of new and existing types of gears. Such gears involve the following types (see Fig. 1.2):

- W-N gears: Kudryavtsev [99], Fedyakin, Chesnokov, Pavlenko [147], Korotkin [92, 93], Kirichenko [83] et al.
- Worm gears: Lagutin, Verkhovsky and Sandler [89, 108–110, 112–119, 211–213], Krivenko [96], Parubets [145, 146], Shultz [164, 165, 167, 168], Pavlov [148–150], Volkov [97, 219].
- Spiroid: Georgiev [53–55], Goldfarb [54, 58, 60–66], Kuniver [55, 103–105], Anferov [10], Trubachev [205], Abramenko [2, 3], Ezerskaya [49], Modzelevsky [135], Fefer [81].
- Gears with face teeth by Pascal limacons: Nechayev [136, 137].

Method 3. It is necessary to apply numerical methods at optimization synthesis of gearing. It can be used both at local synthesis and global one. The global approach is more complex but it gives more reliable results. The versatile method of the global synthesis by criterion of location and dimensions of the bearing pattern is developed and implemented for bevel and hypoid gears by M. G. Segal in the early 1970s. The certain contribution to the development of the global method of synthesis at the same period was made by B. A. Cherny and K. I. Gulyaev. These works were considered in our previously published first part of the review. Note, that from mathematical point of view the optimization synthesis is reduced in general case either to the problem of non-linear programming or to the variation problem. In the first case the *values* of the set of parameters (vector $\mathbf{x}$) are determined for which the target function $F(\mathbf{x})$ reaches its extreme point (under the set of limitations). In the second case the control function $f(u,v)$ (the profile or flank of a tooth with curvilinear coordinates u and v) is determined for which the target function $F(f(u,v))$ reaches its extreme point.

Such approaches started producing the results only after 2000, that is why, we do not consider them in this second part of the review.

Block 5 (Fig. 1.15): application of gears with multi-pair engagement. They are:

(1) *Spur gears with the overlap ratio more than 2* ($\varepsilon > 2$); teeth of such gears are shown in Block 5 in Fig. 1.15. The minimum pinion tooth number in these gears is about 30. At any instant of meshing at least two pairs of teeth are interacting; and the force transmitted by the tooth is reduced almost in two times, thus decreasing contact and bending stresses. Gears require high production accuracy: the error of pitch and profile should be an order less than tooth deformations under load.

These gears were first applied in aviation in 1980 abroad and only in the 21st century in Russia. It became common to replace classical gears with $1 < \varepsilon < 2$ by gears with $\varepsilon > 2$ (for instance, in gearboxes of lorries "Belarus": the casing geometry, shafts and bearings, mounting surfaces are left the same; gearwheel tooth number are increased in 1.5 times with decreasing the module by 1.5 times (with correcting the shift coefficients if necessary); parameters of the basic rack profile are taken to be as follows: $\alpha = 20°$, $h_a{}^* = 1.25$, $h_l{}^* = 2.5$, $c_0{}^* = (0.2–0.25)$. It provides the increase in the transmitted power up to 30...50%.

(2) *Wave gears* (Fig. 1.16) are also related to gears with multi-pair meshing: up to 30...40% of the whole tooth number are in the simultaneous contact, thus providing high load carrying capacity, kinematic accuracy and smoothness of operation. In these gears one of the meshing gearwheels is flexible (3 in Fig. 1.16). It is used to transmit the deformation wave from the element 1 which is called the generator. Gearwheels are always spur and helical, but sometimes there are gears with face bevel teeth. The gear ratio of *all* wave gears is usually i= 40...300 and it is determined by the expression:

$$i = \frac{z_{driven}}{z_{driven} - z_{stationary}}$$

Fig. 1.16 Example of wave gear: 1—driving element (generator); 2—driven element (rigid gearwheel); 3—column (thin-wall elastic barrel with teeth of the *flexible* gearwheel cut on the outside)

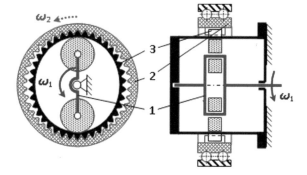

Fig. 1.17 Example of spiroid gear

The tangible contribution to the design and implementation of wave gears was made by Ginzburg [56]; Emelyanov and Popov [44], Kraynev [218] et al.

(3) *Spiroid gears* (Fig. 1.17) can also have multi-pair contact (up to 10 tooth pairs in simultaneous contact). By the end of the 20th century the leader in investigation, design and implementation of spiroid gears in Russia (and in fact in the world) was acknowledged to be Izhevsk Mechanical Institute (now Kalashnikov Izhevsk State Technical University). It was presented by two active groups of scientists headed by A. K. Georgiev and V. I. Goldfarb. Great contribution to development of theoretical investigations and implementation of spiroid gears was made by their followers: D. V. Glavatskikh, A. S. Kuniver, O. V. Malina, I. P. Nesmelov, V. N. Anferov, E. S. Trubachev, V. N. Abramenko, S. V. Ezerskaya, A. M. Fefer, V. A. Modzelevsky et al. The most important works published in 1976–2000 by these experts are presented in tables in Figs. 1.1 and 1.2.

(4) *Planetary type precession gears with small tooth number difference.* It is common that internal spur and helical gears can have small difference of gearwheel tooth numbers (up to $z_2\text{-}z_1 = 1$). The advantage of such gears is low contact stresses because of a high reduced radius of curvature and big number of teeth that are in simultaneous contact. The drawback is the gear ratio $i = z_2/z_1$ close to 1. Similar properties are also specific for bevel gears at the shaft angle Σ close to 180°. The main drawback (closeness of i to 1) can be overcome by making precession planetary gears. Definite layout features are also revealed by bevel gears with the shaft angle Σ close to 0° (Fig. 1.18a). Figure 1.18b–d presents the schemes of precession gears studied thoroughly in Zlatoust by Bezrukov, Lopatin and their followers [26, 27, 52, 126–128]); in Kishinev by Bostan [32, 33]), in Volgograd and Saint-Petersburg by Kislov, Tesker, Timofeev [88]); in Snezhinsk by Nechaev [136, 137]), and lately in Tyumen by Syzrantsev.

Block 6 (Fig. 1.15): gearing with closed lines of contact (CLC). Investigations of *worm* gears with CLC were carried out in the late 1960s in the USSR. Later similar investigations were made in Germany, Czech Republic and Japan. In gears with CLC the closed space is developed between contacting teeth, where the lubrication

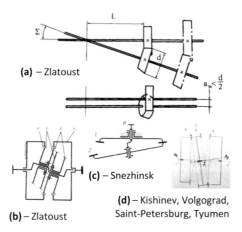

(a) – Zlatoust

(c) – Snezhinsk

(b) – Zlatoust

(d) – Kishinev, Volgograd, Saint-Petersburg, Tyumen

Fig. 1.18 Bevel and precession gears: **a** bevel gear for the boat; **b–d** precession gears with bevel gearwheels

is blocked—see Fig. 1.19. The force from the driving tooth to the driven one is transmitted here not only by their CLC, but through the blocked oil. The volume of the space with oil is decreased at rotation of the gear elements and the pressed out oil provides good lubrication conditions for teeth. It reduces the wear and power losses in gearing and increases the transmitted force. In the late 1980s the issue of organizing a serial production of worm gearboxes with CLC was discussed, though it required high accuracy of meshing (5th–6th degrees of accuracy), the increased rigidity of the whole layout and precise bearings. Investigation results for worm gears with CLC in 1976–2000 are presented in works by Lagutin and Verkhovsky [117–119, 211–213].

Blocks 7–9 (Fig. 1.15): these issues are not considered in this part of the review, since works on these methods of increasing the load carrying capacity were only at the primary stage before 2000 both in Russia and CIS countries. Works according to blocks 10–12 are also not described here since the approaches considered there are of auxiliary importance.

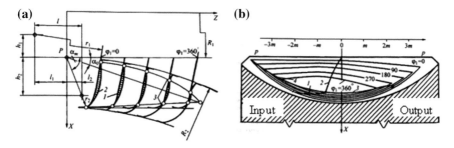

Fig. 1.19 Worm gear with closed lines of contact: **a** section by the mid plane of the gearwheel; **b** projections of contact lines on the plane perpendicular to the worm axis

Finishing the p. 2.2 "Development of synthesis methods…", we refer the reader to the Table in Fig. 1.2. It can be used in order to:

Step (1) choose the interesting gear, mechanism, gear part or theme of investigations;

Step (2) follow the lines coming from the right and left to your chosen theme, family name of authors and number of publications;

Step (3) look through the title of publications in the list of reference and decide the priority of their studying.

1.2.3 Development of Methods for Analysis of Generating Processes and Theory of Gearing

In order to obtain information on this theoretical theme, refer to the Table in Fig. 1.1. By performing all steps described above, get the necessary information.

1.2.4 Software: Development and Application

Software applied at gear design and performance of other engineering analysis can be divided into 4 categories:

Category 1. Commercial software focused on solving the wide range of typical problems. They are CAD/CAM/CAE—systems with English interface, for instance, **Pro/ENGINEER, ANSYS, CATIA, Solid Works** and oth. The special software is intended for gear design: **KISSsoft** (Switzerland); **LTCA** (Gleason, USA); **Kimos** (Klingelnberg, Germany); **PCM** (USA); **PCD** (Japan) and oth.

Category 2. Russian software that tends to be commercial but usually solves a less number of problems. It involves the following systems for design of gears and their elements: (1) **WinMachine** by NTC, Korolev; (2) **CAD KOMPAS** by "Askon" supplemented by software "**Compas#Gears**" on gear analysis.

Category 3. Russian software developed in the leading scientific schools on gears and tools for "internal use". They are:

(1) Software **Reduk 43** for analysis of geometry and strength of involute spur and helical gears with internal and external meshing, planetary gears, bevel gears with straight and circular teeth. The software was developed in 1990s by Department of gear technology of TsNIITMASH (D. E. Goller, A. A. Birbrayer and S. L. Berlin). It implements the techniques of analysis stated in the corresponding standards (GOSTs 16532-70, 1643-81, 21354-87, 19326-74, 1758-81 and other) that were developed at the same Department. This software allows for analyzing the load carrying capacity and durability of both individual gears and multi-stage gearboxes with account of tooth heat treatment, loading cyclogram,

efficiency, power flow, etc. The software is actively applied at heavy engineering plants.

(2) CAM/CAE "**Volga 5**" for geometric and manufacturing analysis of bevel gears with circular teeth developed in 1980s at Saratov SKBZS headed by M. G. Segal.

(3) Software "**SPDIAL+**" for design, optimization and preproduction of spiroid and worm gears (Izhevsk, ISTU, Institute of Mechanics, headed by V. I. Goldfarb, leading developers are E. S. Trubachev, O. V. Malina et al.).

(4) Software "**Expert**" for analysis and synthesis of bevel gearwheels with circular teeth (Moscow, "Mosstankin", headed by G. I. Sheveleva, leading developers are A. E. Volkov and V. I. Medvedev).

Category 4. "Personal" software for solving the problems of analysis and synthesis of gears and tools developed for and gear design at enterprises. Majority of high-skilled experts have such programs. Basics of these programs are standard calculations plus own developments. Let us name only several such software:

(1) Analysis and optimization design of aviation gears (V. L. Dorofeev, Moscow).
(2) Synthesis and optimization of bevel gears (V. N. Syzrantsev, Tyumen).
(3) Design and optimization of parameters of spur and helical, worm and double-enveloping gears (S. A. Lagutin, Electrostal).

Note that during the last third part of the 20th century there were works in the USSR on development of convenient and rational interfaces of programs on analysis and synthesis of gears (Saratov, Izhevsk, Khabarovsk, Leningrad, Tyumen, Minsk and oth.). In particular, an intensive work on development of task-oriented programming language was done in Tyumen [18, 19]. The dialog support system for IBM compatible computer was also worked out. The language involved the means for description of: (1) systems of coordinates and motions; (2) geometrical objects; (3) deviations and deformations; (4) the problem to be solved; (5) types of output data; (6) processing and cataloguing of results; (7) operation with regulating information (tables and formulas). Operations with interval arithmetic were thought out, that is, information (and results) could be represented not only by number, but also by range of numbers (for example, of allowable stresses). Syntax of the developed language was described by Backus-Naur method, since the formal grammar was made to be context-free. Moreover, it was LL(1) grammar which assumed, as known, the efficient methods of syntax analysis and translation by push-down automatons (PDA). When applying PDAs, semantics of the language was implemented by semantic sub-programs. Unfortunately, events of 1990s and rapid development of computer technologies along with the lack of experience and means put an end to that work that seemed to be future-oriented.

1.3 Achievements and Problems of Development of the Theory of Gearing in 1975–2000

1.3.1 Development of the Theory of Gearing

- New concepts were introduced and applied: (**a**) the space of meshing (S. A. Lagutin); (**b**) the covering surface (G. I. Sheveleva); (**c**) the fan, wedge and bunch of normal lines, acceleration of feeding-in (D. T. Babichev).
- Non-differential and kinematic methods for analysis of working and manufacturing meshing gained their development.
- Having applied "the acceleration of feeding-in", new techniques for solving a number of problems were developed: analysis of radii of curvature in meshing, determination of cutting zones and thicknesses of layers removed by cutting edges of tools; analysis of faceting values and other.

1.3.2 Main Theoretical Results on Development of the Theory of Real Gearing

- Methods of gear synthesis were developed with account of many quality characteristics and factors of gear operation: terms of reconjugation, accuracy of production, deformations, vibration activity and other.
- Techniques for investigation of tooth contact interaction and deformation mode of the whole tooth (including multi-pair tooth contact) were developed and implemented. It became possible to control the contact pattern.
- The system of assigning the real ITS (initial tool surfaces) was developed and mathematical models for ITS were created.

1.3.3 The Important Practical Results—A Great Number of Very Adverse Types of Gearing Was Investigated (and Often Implemented: In Gears, in Hydraulic and Pneumatic Machines, in Machine-Tool Gearing)

- Traditional gears and gearing: (**a**) spur and helical: involute (small module, big module, high-speed heavy loaded), with Novikov gearing and other; (**b**) bevel and hypoid; (**c**) spiroid, worm, double-enveloping, screw; (**d**) face gears; and other (Fig. 1.2).
- Non-traditional gears and gearing: (**a**) gears with intermediate rolling bodies; (**b**) devices with closed systems of rolling bodies; (**c**) precessing bevel gears with

shaft angles close to 0° and 180°; (**d**) hydraulic and pneumatic machines: helical (two- and three-rotor) common and planetary type; (**e**) hydraulic and pneumatic machines with non-circular gearwheels; and other.

Appendix 1: Review of Publications of Russian Authors in Foreign Editions in 1976–2000 (Published in English)

During considered period the Soviet science was held in respect in the world; and the leading scientific technical journals ("Vestnik mashinostroyeniya", "Stanki i instrument", "Mashinovedeniye"—since 1980) were fully or partially republished in English. They were published with different titles, in particular, "Machine & Tooling", "Journal of Soviet Machine Science". Later, the journal "Soviet Engineering Research" was issued in the USA in which the featured manuscripts from all three mentioned above journals were published.

Unfortunately, there was "The Iron Curtain" in that period between the USSR and Western countries. It prevented much the cooperation between Russian authors with their foreign colleagues. As a rule, translations of the manuscripts were not authorized by the authors. The authors did not get the proofs of their published papers and even did not know the exact bibliographic data of those publications. That was why, though the majority of papers from the mentioned above journals were translated and published in English, we decided to list the Russian originals in the Reference to the review.

Chronologically the first manuscripts directly published in English in the leading USA editions were the papers by Professor F. L. Litvin and his co-authors. These papers provided the foreign reader with the latest at that time fundamental achievements of the Russian school of the theory of gearing. But the value of those publications was not only in their specific content. The advanced user of the Internet now can not perceive their psychological effect papers on contemporaries due to the breakthrough of the Iron Curtain to the international society.

In 1990s after the breakdown of the Iron Curtain the Russian investigators began actively participating in international conferences, including IX (Milano, 1995) and X (Oulu, 2000) World Congresses on Theory of Machines and Mechanisms, IV World Congress on Gearing and Power Transmission (Paris, 1999) and other.

Professor G. I. Sheveleva and her colleagues were taking an active part in international conferences during the considered period; their presented works were devoted to analysis and synthesis of spiral bevel gears. Papers of S. A. Lagutin considered synthesis and application of general type worm gears. Contributions of Prof. V. N. Syzrantsev and his co-authors discussed the issues of theoretical and experimental research of different types of gears.

A special attention should be paid to the active participation of Professor V. I. Goldfarb in international events. Along with his colleagues he made contributions on investigation and implementation of spiroid gears in practice of mechanical engineering, as well as on general issues of the theory and practice of gearing. Merits of Prof. V. I. Goldfarb in the development of international cooperation of gear experts were noted by his two elections for the position of the Chair of the Gearing Committee of IFToMM, then the Vice-President of this Federation (2011–2015) and finally the Dedicated Service Award in IFToMM (2017).

Being the Chair of the IFToMM Gearing Committee, V. I. Goldfarb was publishing the International Journal "Gearing and Transmissions" twice a year in 1994–2004. It included publications of Russian and foreign experts in Russian and English. Such parallel publication in two languages promoted the increase in the level of mutual understanding between scientists from different countries. As a rule, works of Russian authors published in this journal had the fundamental character and were mentioned above in different paragraphs of this review.

Here is the list of the most fundamental publications in English within the described period:

Airapetov, E. L., Goldfarb, V. I., Novosyolov, V. Yu. (1999). "Analytical and Experimental Assessment of Spiroid Gear Tooth Deflection." Proc. 10th World Congress on TMM. Oulu, 6, 2257–2262.

Goldfarb, V. I., Isakova, N. V. (1995). "Variants of spiroid gearing from pitch realization point of view." *J Gearing and Transmissions*, 1, 25–34.

Goldfarb, V. I., Kuniver, A. S., Koshkin, D. V. (2000). "Investigation of Spiroid Gear Tooth Tangency under Action of Errors." *Proc. Int. Conf. "Gearing, Transmissions & Mechanical Systems"*. Nottingham, 99–108.

Goldfarb, V. I., Russkikh, A. G. (1991). "Skew Axis Gearing Scheme Synthesis." *Proc. MPT'91 Int. Conf. on Motion and Power Transmissions (JSME)*. Japan, Hiroshima, 649–653.

Goldfarb, V. I., Trubachev, E. S. (1997). "Peculiarities of Non-orthogonal Spiroid Gears Parametric Synthesis." *Proc. Int. Conf. on Mechanical Transmissions and Mechanisms.* China, Tjanjin, 613–616.

Goldfarb, V. I. (1995). "The Nondifferential Method of Geometrical Modeling of the Enveloping Process." *Proc. 9th World Congress on the ToMM.* Milan, 1, 424–427.

Goldfarb, V. I. (1994). "The Synthesis of Nontraditional Kind of Skew Axes Gearing." *Proc. Int. Gearing Conference "BGA Transmission Technology"*. Newcastle—London, 513–516.

Goldfarb, V. I. (1995). "Theory of Design and Practice of Development of Spiroid Gearing." *Proc. Congress "Gear Transmissions'95"*. Sofia, 2, 1–5.

Lagutin, S. A. (2000). "Envelope Singularities and Tooth Undercutting in Rack and Worm Gearing." *Proc. Int. Conf. "Gearing, Transmissions & Mechanical Systems"*. Nottingham, 99–108.

Lagutin, S. A. (1999). "Local Synthesis of General Type Worm gearing and its Applications." *Proc. 4th World Congress on Gearing and Power Transmissions.* Paris, 1, 501–506.

Lagutin, S. A. (1999). "Synthesis of Gearings Transmitting a Screw Motion." *Proc. 10th World Congress on the Theory of Machines and Mechanisms.* Oulu, Finland, 6: 2293–2298.

Litvin, F. L., Krylov, N. N., Erikhov, M. L. (1975). "Generation of Tooth Surfaces by Two-Parameter Enveloping," *Chapter in "Mechanism and Machine Theory"*, 10(5), 365–373.

Litvin, F. L., Petrov, K. M., Ganshin, V. A. (1974). "The Effect of Geometrical Parameters of Hypoid and Spiroid Gears on its Quality Characteristics." ASME J of Engineering for Industry, 96, 330–334.

Lopatin, B. A., Tsukanov, O. N. (1999). "Design Cylindrobevel Gears in Generalizing Parameters." *J Gearing and Transmissions*, 2, 24–35.

Sheveleva, G. I., Gundaev, S. A., Volkov, A. E. (1988). "Analysis and Synthesis of Conical Gearing." *Proc. World Congress INTER-GEAR' 88*, China, 1, 1025–1028.

Sheveleva, G. I., Medvedev, V. I., Volkov, A. E. (1995). "Mathematical simulation of spiral bevel gears production and processes with contact and bending stressing." *Proc. 9th World Congress on the ToMM*, Milan, 1, 509–513.

Syzrantsev, V. N., Golofast, S. L., Syzrantseva, K. V. (1999). "Gearing serviceability diagnostic with the help of integral strain gauges." *Proc. 4th World Congress on Gearing and Power Transmission*. Paris, 2, 1845–1850.

Syzrantsev, V. N., Shteen, O. A. (1997). "Research of contact and bending durability of cylindrical gears with arch-shaped teeth with two-point contact." *J Gearings and Transmissions,* 1, 17–29.

Syzrantsev, V. N., Erikhov, M. L., et al. (2000). "Novikov-Wildhaber Gearing. New Methods of Geometrical Analysis, Technology of Manufacturing and Control. Experimental Studies of Load Capacity and Service Life." *J Gearing and Transmissions*, 1, 29–37.

Syzrantsev, V., Kotlikova, K. (2000). "Mathematical and program provision of design of bevel gearing with small shaft angle." *Proc. Int. Conf. on Gearing, Transmissions, and Mechanical Systems*. Nottingham, 13–18.

Syzrantsev, V., Seelich, A. (2000). "Theoretical and experimental research of Novikov gear shaving." *Proc. Int. Conf. on Gearing, Transmissions, and Mechanical Systems.* Nottingham, 143–150.

Syzrantsev, V. N., Golofast, S. L., et al. (1995). "New Methods for Experimental Research of Gear Transmissions." *Proc. of Congress "Gear Transmissions'95"*, Sofia, 1, 71–73.

Volkov, A. E., Gundaev, S. A., Sheveleva, G. I. (1985). "Load-carrying capacity and quality of gear transmissions and reduction unit elements of a CAD system for gear-cutting processes." *J Soviet Engineering Research*, 5(10), 9–12 (translation from *J "Vestnik mashinostroyeniya"*).

Volkov, A. E., Sheveleva, G. I. (1990). "Computer calculation of tooth-broaching heads for machining of straight-tooth bevel gears." *J Soviet Engineering Research*, 10, 11, 97–101 (translation from *J "Machine-tools and Tooling"*).

Appendix 2: Theory of Gearing in Persons

The first part of this review was called "Russian School of the Theory and Geometry of Gearing: Its Origin and Golden Period (1935–1975)". In this part, the authors wrote that after the publication of the first edition monograph "Theory of Gearing" by Professor Litvin in 1960, the 1960s became the period of the formation of this theory as an independent science.

During these years of *Sturm und Drang* a whole galaxy of brilliant researchers built an analytic theory of gearing, collaborating and competing with each other. Along with F. L. Litvin the patriarchs of our science N. I. Kolchin, V. N. Kudryavtsev, and V. A. Gavrilenko actively worked. N. N. Krylov and L. V. Korostelev recently defended their doctoral theses. Ya. S. Davydov, K. M. Pismanik, G. I. Sheveleva, M. L. Erikhov, I. I. Dusev, E. B. Vulgakov, M. G. Segal et al. already began to create their own scientific schools.

Most of them continued to work in those years, which are considered in the second part of our review. Paying tribute to the memory of our predecessors, we present below essays for some of them.

Yakov Samuilovich Davydov (1914–2003)

Yakov Samuilovich Davydov entered the Gorky mechanical engineering (later polytechnic) institute in 1933 after three years of working experience. In 1937 he was assigned to Moscow State Technical University named after Bauman and graduated from in1939. That autumn he was accepted to the post-graduate course at the MMS department.

In the first days of the Great Patriotic War Ya. S. Davydov was enrolled as an engineer at the Armored repair plant in Moscow. There he got acquainted with gears of the company Fellow that comprised the involute spur pinion and its paired face gearwheel. When returned to the post-graduate course in autumn 1945, he made an analytical investigation of these gears under the supervision of V. A. Gavrilenko. After completing postgraduate studies, he came to the Gorky institute of water transport engineers where had worked continuously till 1987 as the assistant, associate and full professor at the department of the theory of mechanisms and machine parts.

Ya. S. Davydov summarized the ideas stated in the Ph.D. thesis as the monograph "Non-involute gearing" (Mashgiz, 1950). This book played a great role in the development of the theory of gearing not only due to a new class of gears investigated and discussed there, but due to the kinematic method proposed there by the author (simultaneously with V. A. Shishkov and F. L. Litvin and independently of them). This method became the main technique for analysis and synthesis of gearing in the second part of the XX century.

A number of papers published at the 1960s had no less importance for the theory of gearing. In these works Ya. S. Davydov showed that in order to generate conjugated gearing with the point contact of active flanks (such as W-N gearing), it is not obligatory to reject Olivier's principles inextricably related to the gear-cutting technique. It is just necessary and enough to spread these principles for application of rigid non-congruent pairs of generating surfaces and curve lines.

Brilliant erudition, tactfulness, carefulness in his work and assessment of other works acquired the great authority of Ya. S. Davydov in the gear expert society. At the end of 1990s he emigrated to the USA where lived till the end of his days.

Kalman Malkielevich PISMANIK (1914–1990)

Kalman Malkielevich Pismanik was born in Saint-Petersburg on August 24, 1914. In 1933 he graduated from the railway college. In 1935 he tested out the exams for the first two courses and was accepted to the third year of Leningrad Polytechnic Institute. In 1939 he graduated with honors from the evening department of this institute with the major "Cold machining of metals". At the end of 1939 he entered the post-graduate course and held the study with the work of the chief designer first at the Kirov plant of lifting equipment, then at the Special Designing Bureau 19. During the Great Patriotic War K. M. Pismanik was working as the head of the designer group at the plant #174 of the Ministry of armored vehicle industry.

In 1948 K. M. Pismanik presented his Ph.D. thesis on "Fundamentals of the kinematic theory of spatial gearing and its application for investigating the technique and meshing of hypoid gears". By this time the applicant had 12-year experience in industrial plants.

The further working activity of K. M. Pismanik was closely related to education. Since 1948 he had worked at Saratov road transport institute as the senior lecturer first, then as the associate professor of "Mechanical engineering" department. Since 1960 he had worked as the associate professor of "Metal cutting machines and tools" department. In 1973 K. M. Pismanik presented his DSc thesis on "Theoretical fundamentals of tooth profiling for bevel and hypoid gears". After awarding him the professor title in 1973 he became the head of "Mechanism and machine science" department of Saratov polytechnic institute.

Professor K. M. Pismanik contributed much to the generation of the Russian gear cutting science. He published 118 scientific papers, including 5 books. His monograph "Hypoid Gears" (1964) remains today the reference book of any engineer engaged in the production of these gears. He supervised 10 Ph.D. theses. A number of very urgent theoretical issues on the geometry of gearing and gear machining technique for hypoid and worm gears had been solved under his leadership. His works with V. N. Kedrinskiy became the foundation for the development of a whole range of Russian machine-tools for machining bevel and hypoid gears at the Saratov plant and the Special Designing Bureau of heavy gear cutting and shaping machine tools. The last book with him as the first author was published in 1993 [152].

Scientific activity of professor K. M. Pismanik was specific by his high integrity, insistence on quality and practical focus of works that gained the worthy tribute and wide popularity among scientists and experts in this field of science and production both in Russia and abroad.

Lev Vasilyevich KOROSTELEV (1923–1978)

As a rule, a fundamental discovery appears in a rather complex and confused way. The founder involuntarily reproduces the thought passing through the labyrinth of accumulated knowledge prior to the light of a new Truth and enriches the statement by the variety of formulas, calculations and details that promoted the obtained conclusions. The author presents all the known facts and proofs as if in front of a trial jury in order to convince them to reach the only right, to his opinion, verdict. The works by Ch. I. Gochman and E. Wildhaber are the visual examples of such an approach. And there is some "period of information settling" during which juridical riots are calming down, random features are vanishing and the truth appears as a perfectly simple statement, transparent as the main law of meshing in the form $\mathbf{n \cdot V} = \mathbf{0}$. The more serious the discovery is, the longer this period is. And usually only representatives of the next generation manage to bring the simplicity and importance of fundamental discoveries to wide recognition.

But there are rare exceptions to this rule. The history of science knows scientists of Kepler's type, capable of intuitively conceive the truth in the last resort at once and state the information so simply and evidently as the verdict is final without appeal. Prof. L. V. Korostelev had this lucky ability to the full extent, being the bright representative of the second generation of scientists who developed the theory of gearing in Russia.

Lev Vasilyevich Korostelev was born on September 15, 1923 in a hereditary intelligent family. During the Great Patriotic War, he was working as a gear cutter at the military plant. In 1946 he entered the Moscow Machine-Tool Institute (Stankin) and devoted all his further life to it. In 1950 he graduated from the machine-tool department and entered the post-graduate course in this specialty. In 1954 he presented the Ph.D. thesis on development and production of bevel gears.

Having worked a little at Izhevsk mechanical institute, he returned to his Alma-Mater as the associate professor of MMS department. In 1964 he presented his DSc thesis on "Geometrical and kinematical characteristics of load-carrying capacity of spatial gearing". In 1968 he was award the professor title. Since 1971 he held the head of MMS department position with the responsibilities of Vice-rector for Research. He died in the road accident on August 9, 1978.

During a short period of his scientific activity, L. V. Korostelev published works on: investigation of curvature of mutually enveloping surfaces, kinematic sensitivity of gears to errors of mutual arrangement of gearwheel axes, synthesis of spatial gearing by means of screw generating wheel and investigation of general-type worm gears. He discovered the analog of the Willis theorem for spatial gearing. The worm gears with closed lines of contact are among his 30 inventions.

He passed away painfully early. His papers and reports are scattered within different journals. He did not have time to write a monograph. He did not publish any of his works abroad. All this does not promote the knowledge of his scientific heritage for new generations of researchers. But he is worth to be studied at least not to apply numerical methods for solving those issues that are evident from general laws determined by him, to find grains of further investigations in his works and, finally, to inherit his simplicity and clearness of the line of thoughts .

Edgar Borisovich Vulgakov (1928–2006)

Education: 1952—Moscow Automobile and Road and Institute, 1963—Ph.D. degree at Moscow State Technical University, Thesis—"Gears with modified basic rack profile", 1974—Dr.Sci degree at the same University, Thesis—"Gears with improved characteristics. Generalized theory and design".

Main employment: 1956–1962 –VNIINMASH (Russian Research Institute of Standardization and Certification in Mechanical Engineering), Research sector chief, 1962–2006—CIAM (Central Institute of Aviation Motors), Gear department chief.

In the 1960s E. B. Vulgakov researched a new basic rack profile that improved gear quality characteristics. He had developed the unique block-contour album and defined the local bending stresses for gears with different basic contours. Later, he created the Gear Theory in Generalized Parameters that describes the involute gear mesh independently of gear fabrication technology and tooling generation profile. This theory is an outstanding contribution by E. B. Vulgakov to involute gearing that breaks the dogma about the necessity of using a preselected basic (or generating) rack profile to define involute gear parameters and gear design. In traditional gear design, the typically standard tooling generating gear rack is primary, and gear parameters and performance characteristics are secondary. In a fact, Vulgakov's theory makes gear drive performance parameters primary—where completely defined gear geometry dictates tooling proportions and dimensions—just like when designing practically any other mechanical parts and components. This approach allows for gear geometry to be fully optimized for gear drive performance for a particular application.

In his theory, Prof. E. B. Vulgakov defined the functions and parameter limits of the involute mesh of cylindrical gears and introduced their areas of existence. He described the generalized rack generating tooling profile, its area of existence and shaping of the gear tooth root fillet by the protuberance hobs. He also extended his theory to asymmetric tooth gears and epicyclic gears.

The Gear Theory in Generalized Parameters was ahead of its time and, unfortunately, despite the many publications of Vulgakov and his followers, it did not find deserved understanding and application in Russia. In the USA it became a foundation for the development of the Direct Gear Design method that is used to design high-performance custom gear drives.

E. B. Vulgakov was an editor of the handbook "Aviation Gearboxes" that became very popular among aerospace gear engineers in Russia. In 1993, E. B. Vulgakov founded the "Conversion in Machine Building" journal and became its chief editor. This magazine published articles about the restructuring and conversion of Russia's defense industries and enterprises for production of civil goods.

(The essay about E. B. Vulgakov was prepared by his follower A. L. Kapelevich).

Galina Ivanovna SHEVELEVA (1929–2005)

Galina Ivanovna Sheveleva graduated with honors from Moscow Machine-Tool Institute (Stankin) in 1952. In 1955 she entered the post-graduate courses of the same institute and in 1957 she was enrolled as the assistant at the department "Theoretical mechanics" where she had been working till her death. There she presented her Ph.D. (1960) and DSc Theses (1969) became the Assistant Professor (1965), Professor (1973) and the holder of the award by the Council of Ministers of the USSR (1985).

Already in the 1960s she proposed new methods for solving the problems of the theory of gearing by decomposing the functions into power series; that allowed carrying out the synthesis of Revacycle bevel gears for Soviet and Polish enterprises. She developed the theory of covering lines; on its basis the algorithm was developed for metal removal from the workpiece by multi-point cutting tools. This allowed for efficiently eliminating such a generating defect as the tooth flank cutting-off by one of the cutters of the broaching mill.

During the age of computerization these methods proved to be high-demanded; and they became the foundation of a new trend in solving the problems of synthesis and analysis of gears by numerical simulation of gear-machining processes and meshing gearwheel pairs. She proposed a new solution for the contact problem of the elasticity theory by consequent loading as applied to gears.

The main practical outcome for scientific studies of G. I. Sheveleva is the software system "Expert" developed under her leadership and at her direct participation. It is intended to master the production of spiral bevel and hypoid gears. "Expert" is competitive with the best world-wide analogs developed by the world leaders of gear machining equipment (Gleason, USA and Klingelnberg, Germany); and it is widely used at aviation and mechanical engineering plants.

Being the eminent and miscellaneous teacher, she had been unchallenged for many years as the best lector of Stankin. Her lectures on theoretical mechanics, computer science and programming captivated and admired the audience, acquainting the students to the flawless logical process developed from the very beginning of narration to the final conclusions.

She created her own scientific school, published the monograph, several textbooks, about 200 scientific and guidance manuscripts. She was the research advisor of successfully presented theses: two DSc and 11 Ph.D., including in Poland. Galina Ivanovna paid much attention to her followers and was proud of them. She was strict and fair. No disingenuous motives and nobody's authority could make her change the exam mark of student's work or the Ph.D. presentation.

Mark Gerasimovich SEGAL (1931–2000)

Mark Gerasimovich Segal was born on December 19, 1931 in Latvia, Rezekna. In 1954 he graduated from the mechanical and mathematical department of Saratov State University on the major "Mechanics". He started his working activity as the teacher of physics at school.

From 1956 till 1984 he was working at Saratov Plant of Heavy Gear-cutting Machines (SPHGM), at the department on development of heavy gear-cutting and gear-grinding machines. He made a career from the senior designing engineer to the Chief Designer and Head of the Department of theoretical investigations. In 1964 M. G. Segal presented the Ph.D. thesis and in 1979 the DSc thesis.

Since 1981 he headed the department of Metal-cutting machine-tools at Saratov Polytechnic Institute. In 1987 he was awarded the title of Professor. He is the author of 120 scientific works and 8 inventions.

The main contribution of Prof. M. G. Segal to the theory of gearing is that unlike his predecessors he introduced the concept of the ease off between the nominal and theoretical surfaces of the gear tooth for the analysis of contact condition of active tooth flanks. The line of the level of reduced clearances that corresponds to the thickness of the paint layers compressed at gear running-in at the testing machine is the contour of the theoretical area of contact. This idea is being commonly used nowadays in the world.

M. G. Segal was also the first to state the issue of determining the value of undercut in bevel and hypoid gears. Before him there were efficient methods for determining the fact of the presence or absence of undercut rather than its value. Segal showed that the requirement for the complete absence of the undercut is often unjustified; and he introduced the concept of the allowable value of undercut by proposing the distance h_Q to the axis of the pinion rotation from the point Q of jog, that is, from the point of intersection of the enveloping and the transient surfaces in the chosen section.

Professor M. G. Segal headed the development of the program system for solving the problems of manufacturing synthesis and analysis of contact conditions, first of all, for bevel teeth with circular teeth. The last version of this system "Volga 5" is being still actively applied at many mechanical engineering plants for analysis of gear-cutting machine-tool settings at the SPHGM enterprise.

At the same time, M. G. Segal developed the mathematical model for the process of tooth generation at CNC machine-tools based on the theory of envelopes. At any assigned laws of motion of CNC machine-tools elements this model allows for analyzing all criteria of the quality of the gear cut at these machine-tools, the criteria characterizing the conjugacy of tooth flanks and their geometrical shapes.

In 1990s M. G. Segal was invited to work in Germany where he became the ideologist and chief developer of the program system "Kimos" for Klingelnberg gear-cutting machine-tools. His results on the choice of optimal schemes of multi-coordinate CNC machine-tools for gearwheel machining were also applied by Klingelnberg Company.

He died on October 29, 2000 in the hospital in Israel and he was buried there.

Maks L'vovich Erikhov (1937–2002)

Maks L'vovich Erikhov was one of few encyclopaedists in the field of gearing theory and practice of gearbox engineering. He had the competency to thoroughly discuss fine mathematical nuances of the theory of envelopes, logics of the structure of a higher educational course on MMS, geometry of double-enveloping gears with ground worms, kinematics of machine-tools for cutting the arch teeth, the system of tolerances for W-N gears, and any issues of the theory and practice of real gearing. Maks L'vovich was born in 1937 in Velikie Luki, Pskov region. He graduated from the mechanical engineering department of LPI—Leningrad Polytechnic Institute (now Saint-Petersburg Polytechnic University—SPU) in 1960. He started his working activity in the same year at the design department of Khabarovsk enterprise named after Gorky, where he was working up to entering the post-graduate course at MMS department of LPI in 1962.

Having presented the Ph.D. thesis, he was working since the end of 1965 to 1975 at Khabarovsk Polytechnic Institute as the senior lecturer of the "Machine parts" department, and since June 1966 he was the head of this department. During this period Prof. M. L. Erikhov managed not only to re-establish the staff of the department by captivating the most talented graduates of the institute by scientific investigations, but to organize the scientific laboratory at the department to carry out the experimental investigation of gears and methods of gear generation. By continuing and developing the investigations started during the post-graduate course, he obtained a number of new results in the theory of gearing that became the subject of DSc thesis presented in 1972 at the age of 35.

In 1975 Prof. M. L. Erikhov along with his few followers moved to work at Kurgan Mechanical Institute (now Kurgan State University) where he was the head of the "Machine parts" department and in 1986–1990 he was the vice-rector of the institute on the scientific work. In Kurgan M. L. Erikhov continued his intense scientific activity. During a short period, he managed to re-organize the work of the department and create laboratories for experimental research where investigation of new types of gears and methods for gear machining were continued and fundamentals of new scientific directions were laid.

Thanks to the authority and organizational efforts of M. L. Erikhov, in 1979 Kurgan became the venue for the 3rd All-Union Symposium "Theory and geometry of spatial gearing". After three symposia "Theory of real gearing" were held there in 1988, 1993 and 1997. The change of the title of the Symposium represented the increase in the number of considered issues and main trends in the development of the gearing science at the advanced level.

Since 1998 and till the end M. L. Erikhov worked as the professor of SPU.

Professor M. L. Erikhov was one of the founders of a new scientific direction in the theory of gearing—the application of the principle of enveloping with two parameters to analysis and synthesis of gearing. Introduction of two-parametric models into the classical theory allowed for reconsidering the whole theory of conjugate gearing generation and to give it the logical structure and completeness.

Original methods of gearing synthesis developed by Erikhov allowed for developing and investigating a number of new types of advanced gears and methods for gear generation, including gears with arch and barrel-shaped teeth.

M. L. Erikhov was one of the first in Russia to understand that further development of methods for gear analysis and optimization was impossible without complex consideration of such factors as manufacturing errors and deformation of teeth and other gear elements. Such an approach allowed him and his followers to state and solve a number of problems of complex investigation of the loaded state of gears with arch teeth, W-N gears, double-enveloping gears and, above all, to refine the design experimental methodology of solving the spatial problems of the theory of real gearing.

It was stimulated by mastering the method for assessment of the loaded state of gear elements by means of an integral type strain gauge—a new direction in experimental mechanics that has been successfully developed by the followers of M. L. Erikhov and the followers of his followers.

M. L. Erikhov did his best to prepare the skilled experts in the field of gears and to assist young scientists. He was the supervisor of more than 30 Ph.D. and two DSc theses. He is the author of more than 130 publications. Papers published by the staff of the scientific school created by M. L. Erikhov are widely known in Russia and abroad.

Maks L'vovich had the natural Saint-Petersburg commitment with respect to people and ability to spread positive emotions around him. This wonderful ability along with the versatility of his knowledge promoted the uniting of the gearing society of our country.

References

1. Abazin, D.D.: Double-enveloping gear with point contact with the worm ground by the plane. J. Izvestiya VUZov. Mashinostroenie **1**, 31–34 (1981)
2. Abramenko, V.N.: Geometry investigation of the worm of one variety of spiroid bevel gear. In: Mechanical Transmissions, pp. 40–44. Izhevsk, ISTU (1977)
3. Abramenko, V.N.: Influence of the basic cone angle of the worm on reduced radii of curvature of mutually enveloping surfaces of bevel spiroid gears. In: Proceedings of All-Union Meeting "Prospects of Development and Application of Spiroid Gears and Gearboxes, pp. 10–13. Izhevsk, ISTU (1979)
4. Airapetov, E.L.: Approximated solution of the contact task at arbitrary geometry of tooth contacting surfaces. In: Proceedings of International Conference on "Theory and Practice of Gears", pp. 23–28. Izhevsk, ISTU (1996)
5. Airapetov, E.L.: State of Art and Prospects of Development of Methods for Analysis of the Loaded State of Gears. Izhevsk-Moscow, Astrel (2000)
6. Airapetov, E.L., Genkin, M.D., Melnikova, T.N.: Statics of Double-Enveloping Gears. Nauka, Moscow (1981)
7. Airapetov, E.L., Genkin, M.D., Ryasnov, YuA: Statics of Gears. Nauka, Moscow (1983)
8. An I-Kan, Belyaev, A.E.: Cutting of non-circular teeth. In: Proceedings of International Conference on "Theory and Practice of Gears", pp. 369–372. Izhevsk, ISTU
9. Andozhsky, V.D., Vasilenok, V.D., Zelenkova, T.M.: Geometry of truncation of the longitudinal edge of the tooth. J. Izvestiya VUZov. Mashinostroenie **3**, 37–41 (1981)
10. Anferov, V.N.: Investigation of wear resistance of spiroid cylindrical gears by means of roller analogy. J. Vestnik Mashinostroeniya **6**, 27–29 (1981)

11. Avdeyev, V.A., Lapin, N.V.: Conditions of existence of helical gears. J. Izvestiya VUZov. Mashinostroenie **7**, 47–50 (1978)
12. Babichev, D.T.: Development of the kinematic method for investigation of meshing. In: Proceedings of 4th All-Union Symposium "Theory of Real Gearing. Part 1. Geometry and CAD of Real Gearing", pp. 28–29. Kurgan, KSU
13. Babichev, D.T.: Unified curvilinear coordinates of gearwheel surfaces. Ibid, pp. 30–31
14. Babichev, D.T.: To basic geometrical primitives of the theory of gearing. In: Proceedings of International Conference on "Theory and Practice of Gearing", pp. 469–474. Izhevsk, ISTU
15. Babichev, D.T.: Application of the upgraded basic geometrical concepts for analysis of operating and machine meshing. In: Proceedings of 6th International Symposium on "Theory of Real Gearing. Informative Materials. Part 1, pp. 58–59. Kurgan, KSU (1997)
16. Babichev, D.T.: Search for conjugated surfaces of teeth having the maximum load carrying capacity. Ibid, pp. 55–58 (1997)
17. Babichev, D.T.: Basic rack surface of edge tools. In: Proceedings of International Conference on "Theory and Practice of Gearing", pp. 412–421. Izhevsk, ISTU (1998)
18. Babichev, D.T., Langofer, A.R.: Fundamentals of the problem oriented language and dialog development system for execution of engineering analysis. In: Proceedings of All-Union Scientific and Technical Conference "Problems of Quality of Gears and Gearboxes", pp. 37–39. Leningrad
19. Babichev, D.T., Langofer, A.R., Dolgushin, V.V.: Development of the problem oriented language and the database for computer-aided solution of analysis and optimization synthesis for gear machining tools and gears. In: Research Report on the Theme F3, Tyumen (1989)
20. Balakin, P.D.: Mechanical Gears with Adaptive Properties. Omsk, OmSTU (1996)
21. Balakin, P.D., Lagutin, S.A.: Generating surface at two-parametrical enveloping. J. Mech. Mach. Mosc. Nauk. **61**, 16–20 (1983)
22. Baltadzhi, S.A.: Optimization synthesis of non-orthogonal worm gears. J. Izvestiya VUZov. Mashinostroenie **6**, 40–43 (1982)
23. Belyaev, A.E.: Mechanical Gears with Ball Intermediate Bodies. Tomsk, TsNTI (1992)
24. Belyaev, A.E., Karyakin, A.V.: New methods for analysis of conjugate profiles. In: Proceedings of International Conference on "Theory and Practice of Gearing", 141–147. Izhevsk, ISTU (1998)
25. Belyaev, A.I., Siritsin, A.I.: Geometrical analysis and technique for cutting the gearwheels with arc-shaped teeth. J. Vestnik Mashinostroeniya **1**, 3–6 (1991)
26. Bezrukov, V.I., Lopatin, B.A.: Determination of kinematic and loading parameters of contact in hyperboloid gears. J. Adv. Mach. Des. Methods Parts Mach. **215**,18–23. Chelyabinsk, ChPI (1978)
27. Bezrukov, V.I., Lopatin, B.A., Kazartsev, D.N.: Choice of geometrical parameters of hyperboloid gears with involute bevel gearwheels. J. Adv. Mach. Des. Methods Parts Mach. 45–49. Chelyabinsk, ChPI (1988)
28. Bleishmidt, L.I., Rubin, M.A., Kisin, I.L.: Determination of load carrying capacity of modified double-enveloping gears. J. Izvestiya VUZov. Mashinostroenie **12**, 31–34 (1983)
29. Bolotovskaya, T.P., Bolotovsky, I.A., et al.: Cylindrical Involute Internal Gear Pairs. Calculation of Geometrical Parameters. Reference book, Moscow, Mashinostroenie (1977)
30. Bolotovsky, I.A., Bezrukov, V.I., Vasilyev, O.F., et al.: Handbook on Geometrical Analysis of Involute Gears and Worm Gears. Mashinostroenie, Moscow (1986)
31. Borzilov, B.M., Ginzburg, E.G., Kadatsky, A.I.: Analysis and Design of Small-Sized Gearboxes. VIKKA n.a. Mozhaisky, Study guide, Leningrad (1995)
32. Bostan, I.A.: Development of high-stresses planetary precession gearboxes of new generation. J. Gearing Transm. **1**, 35–39 (1991)
33. Bostan, I.A.: Precession Gears with Multi Pair Meshing. Kishinev (1991)
34. Bragin, V.V., Reshetov, D.N.: Design of high-stressed spur and helical gears. Mashinostroenie, Moscow (1991)
35. Britsky, V.D.: Invariants of internal geometry of surfaces forming the higher kinematic pair. J. Izvestiya VUZov. Mashinostroenie **6**, 40–43 (1988)

36. Burinsky, A.A., Schultz, V.V.: Analysis of geometry of double involute gears. J. Izvestiya VUZov. Mashinostroenie **11**, 34–38 (1984)
37. Bushenin, D.V.: Non-coaxial Screw-Nut Mechanisms. Mashinostroenie, Moscow (1985)
38. Chernaya, L.A.: Method for synthesis of geometrical parameters of roller planetary gear by contact strength. PhD Thesis, Moscow, MVTU (1976)
39. Chernaya, L.A., Cherny, B.A.: To one method for decomposition of the reverse task of the theory of gearing. J. Izvestiya VUZov. Mashinostroenie **7**, 38–41 (1979)
40. Dorofeyev, V.L.: Fundamentals of computer-aided modeling of fields by methods of complex analytical functions. In: Proceedings of 3rd International Conference on "New Techniques of Controlling the Motion of Technical Objects", vol. 2, pp. 104–109. Novocherkassk (2000)
41. Dun-Syue-Chzhu: To the theory of gear meshing with two degrees of freedom. J. Izvestiya VUZov. Mashinostroenie **8**, 46–49 (1989)
42. Emelyanov, A.F.: Analysis of quality parameters of gears with backlash compensation for high-precision drives. DSc Thesis, Snezhinsk (2000)
43. Emelyanov, A.F., Popov, P.K.: Complex investigation of gears with backlash compensation. J. Izvestiya VUZov. Mashinostroenie 11, 30–37, 12, 26–29 (1986)
44. Emelyanov, A.F., Popov, P.K., Reibakh, YuS: Stand testing of the feed drive of the CNC machine-tool with harmonic gear. J. Vestnik Mashinostroeniya **7**, 13–14 (1983)
45. Erikhov, M.L.: Geometrical and kinematical schemes of machine-tool meshing and principles of their classification. In: Proceedings of 3rd Symposium on "Theory and Practice of Spatial Gearing", pp. 7–9. Kurgan, KSU (1979)
46. Erikhov, M.L.: Cylindrical gears with arc-shaped teeth. Features and possibilities. In: Proceedings of Symposium "Cylindrical Gears with Arc-Shaped Teeth. Analysis, Design, Manufacture", pp. 3–5. Kurgan, KSU (1983)
47. Erikhov, M.L., Syzrantsev, V.N.: "Determination of the range of allowable values of parameters of the grinding wheel at relieving the spiral disk mills. In: Analysis and Synthesis of Mechanisms and Theory of Gears, pp. 37–47. Khabarovsk, KhPI
48. Erikhov, M.L., Syzrantsev, V.N.: Several methods of generation of conjugate surfaces with two-point contact in gearing with arc-shaped teeth. In: Proceedings of International Conference on "Theory and Practice of Gears", pp. 241–246. Izhevsk, ISTU
49. Ezerskaya, S.V., Shpilkin, I.A.: Investigation of meshing surface of spiroid gears with Archimedes worms. In: Mechanical Transmissions, vol. 2, pp. 22–26. Izhevsk, ISTU (1977)
50. Fedotov, B.F.: Foundation of standard geometrical parameters of double-enveloping gears. J. Vestnik Mashinostroeniya **10**, 50–52 (1985)
51. Fefer, A.M., Chekalkin, G.T.: Several issues of geometry of meshing of fine-module spiroid gears. J. Vestnik Mashinostroeniya **5**, 13–16 (1978)
52. Gavrilenko, V.A., Bezrukov, V.I.: Geometrical design of gears comprising involute bevel gearwheels. J. Vestnik Mashinostroeniya **9**, 15–18 (1976)
53. Georgiev, A.K.: Main features, classification and areas of effective application of spiroid gears. In: Prospects for the Development and Application of Spiroid Gears and Gearboxes, pp. 3–9. Izhevsk, IMI (1979)
54. Georgiev, A.K., Goldfarb, V.I.: Preferable combination of directions of rotation for elements of non-orthogonal hyperboloid gear. J. Izvestiya VUZov. Mashinostroenie **6**, 34–37 (1978)
55. Georgiev, A.K., Kuniver, A.S.: Issues of generation and techniques investigation of geometry of wheel teeth with modified flanks for spiroid cylindrical gear. In: Advances in Processes of Metalworking by Cutting, vol. 2, pp. 58–66. Izhevsk, IMI (1977)
56. Ginzburg, E.G.: Practice of application of harmonic gears. In: Proceedings of International Conference on "Theory and Practice of Gears", pp. 308–312. Izhevsk, ISTU (1998)
57. Goldfarb, V.I.: Equation of the ideal pitch surface of the worm. J. Izvestiya VUZov. Mashinostroenie **3**, 52–55 (1976)
58. Goldfarb, V.I.: Shape of the ideal pitch surface of the worm of orthogonal spiroid gears. J. Izvestiya VUZov. Mashinostroenie **11**, 38–41 (1976)
59. Goldfarb, V. I.: Trends in development of CAD of gears. In: Computer-Aided Design of Elements of Transmissions, pp. 5–7. Izhevsk (1987)

60. Goldfarb, V.I.: Problems of computer-aided design of gears. In: Proceedings of 4th All-Union Symposium on "Theory of Real Gearing. Part 1. Geometry and CAD of Real Gearing", pp. 8–17. Kurgan, KSU (1988)
61. Goldfarb, V.I.: Aspects of the computer-aided design of gears and gearboxes. J. Gearing Transm. **1**, 20–24 (1991)
62. Goldfarb, V.I., Glavatskikh, D.V., Voznyuk, R.V.: Instrumental system of modeling the enveloping process. In: Proceedings of International Conference on "Theory and Practice of Gears", pp. 481–484. Izhevsk, ISTU (1996)
63. Goldfarb, V.I., Mardanov, I.I.: Development of the range of new spiroid gear-motors and gearboxes. J. Vestnik Mashinostroeniya **12**, 54–57 (1990)
64. Goldfarb, V.I., Mokretsov, V.N., Spiridonov, V.M., et al.: Practice of development and production of spiroid gearboxes and gear-motors. In: Proceedings of International Conference on "Theory and Practice of Gears", pp. 213–218. Izhevsk, ISTU (1996)
65. Goldfarb, V.I., Nesmelov, I.P.: Application of interpolation methods at investigation of spiroid gearwheel flanks. In: Advances in Processes of Metalworking by Cutting, vol. 2, pp. 28–32. Izhevsk, IMI (1977)
66. Goldfarb, V.I., Nesmelov, I.P.: Choice of geometrical parameters of non-orthogonal spiroid gear. J. Izvestiya VUZov. Mashinostroenie **8**, 48–51 (1981)
67. Goldfarb, V.I., Nesmelov, I.P., Glavatskikh, D.V.: Computer-aided simulation of generation of flanks by enveloping method. In: Proceedings of All-Union Symposium on "Application of Computer-Aided Systems of Layout Design in Mechanical Engineering, pp. 75–77. Moscow
68. Golofast, L.M., Silich, A.A.: Investigation of errors of profiling of worm grinding wheels (for machining of Novikov gears). In: Theory of Machines of Metallurgy and Mining Machinery, pp. 123–131. Sverdlovsk, UPI (1986)
69. Golofast, L.M., Erikhov, M.L., Silich, A.A.: Investigation of machine-tool meshing when cutting Novikov gears by the hob. In: Proceedings of 3rd All-Union Symposium on "Theory and Practice of Spatial Gearing", pp. 11–12. Kurgan, KSU
70. Golovachev, M.I.: Effect of technological localization of contact of arc-shaped teeth. J. Izvestiya VUZov. Mashinostroenie **5**, 54–58 (1986)
71. Grechishnikov, V.A., Kirsanov, G.R., Katayev, A.V.: Computer-Aided Design of Metal Cutting Tools. Mosstankin, Moscow (1984)
72. Gribanov, V.M.: Analytical theory of accuracy of spatial gearing. J. Izvestiya VUZov. Mashinostroenie **4**, 49–52 (1982)
73. Gribanov, V.M.: Multi-criteria synthesis and analysis of basic rack profiles for Novikov gears. J. Izvestiya VUZov. Mashinostroenie **4**, 36–40 (1986)
74. Gubar, S.A., Lagutin, S.A.: Longitudinal localization of contact in gears with arc-shaped teeth. J. Izvestiya VUZov. Mashinostroenie **7**, 24–28 (1983)
75. Gulyaev, K.I.: Theoretical fundamentals of synthesis and finish machining of bevel gears. DSc Thesis, Leningrad (1976)
76. Ivanov, G.A., Lagutin, S.A.: Determination of efficiency of orthogonal gears with cylindrical worms. J. Izvestiya VUZov. Mashinostroenie **5**, 30–35 (1979)
77. Kane, M.M.: Choice of rational accuracy of spur and helical gears at different operations of their machining. J. Vestnik Mashinostroeniya **8**, 3–8 (1996)
78. Kane, M.M.: Controlling the processes of design and production of gears. J. Vestnik Mashinostroeniya **11**, 8–12 (1997)
79. Kapelevich, A.L.: Synthesis of asymmetric involute gearing. J. Mashinovedenie **1**, 62–67 (1987)
80. Kaplun, A.M.: Sliding of teeth of spatial gears. J. Izvestiya VUZov. Mashinostroenie **1**, 40–43 (1980)
81. Khlebalin, N.F.: Cutting of Bevel Gears. Mashinostroenie, Leningrad (1978)
82. Kirichek, A.V.: Stresses state of threads of parts in the vacuum membrane pump. In: Proceedings of International Conference on "Theory and Practice of Gears", pp. 108–113. Izhevsk, ISTU (1998)

83. Kirichenko, A.F.: Geometrical parameters of Novikov cylindrical gears with two points of contact. J. Izvestiya VUZov. Mashinostroenie **5**, 30–33 (1977)
84. Kirsanov, G.N.: Planar methods of displaying of the Ball cylindroid. J. Izvestiya VUZov. Mashinostroenie **9**, 28–33 (1977)
85. Kirsanov, G.N.: Profiling the running-in tools having new design. J. Izvestiya VUZov. Mashinostroenie **12**, 107–111 (1979)
86. Kirsanov, G.N.: Basics of the screw theory of profiling the gear machining tools. In: Mechanics of Machines, vol. 61, pp. 10–16. Moscow, Nauka (1983)
87. Kirsanov, G.N.: Design of Tools, Kinematic Methods. Mashinostroenie, Moscow (1984)
88. Kislov, S. Yu., Tesker, E. I., Timofeev, B. P.: Precessing bevel gears with internal meshing. In: Proceedings of International Conference on "Theory and Practice of Gears", pp. 387–392. Izhevsk, ISTU
89. Korostelev, L.V., Baltadzhi, S.A., Lagutin, S.A.: Conjugate lines of meshing of general-type worm. J. Mashinovedenie **5**, 49–56 (1978)
90. Korostelev, L.V., Ivanov, G.A., Lagutin, S.A.: Synthesis of gearing by means of the method of loci. In: Proceedings of Third All-Union Symposium "Theory and Practice of Spatial gearing", pp. 3–4. Kurgan, KSU (1979)
91. Korostelev, L.V., Lagutin, S.A., Rastov, YuI: Kinematics of machine-tool meshing at machining of gearwheels transmitting the helical motion. J. Mashinovedenie **4**, 68–73 (1977)
92. Korotkin, V.I.: To account of edge effects when analyzing the contact endurance of Novikov gears. J. Vestnik Mashinostroeniya **6**, 8–11 (1997)
93. Korotkin, V.I., Kharitonov, YuD: Novikov Gears. Rostov, RGU (1991)
94. Kovtushenko, A.A., Lagutin, S.A., Yatsin, YuL: Mastering the new types of gears for drives of metallurgy equipment. J. Tyazheloye Mashinostroenie **1**, 18–22 (1993)
95. Kraynev, A.F.: Reference Dictionary on Mechanisms. Mashinostroenie, Moscow (1987)
96. Krivenko, I.S.: To determination of basic parameters of a gear with ZT worm. J. Izvestiya VUZov. Mashinostroenie **11**, 43–46 (1989)
97. Krokhmal', N.N., Volkov, G.Y.: Analysis of schemes of high-speed mechanical gears. In: Proceedings of International Conference on "Theory and Practice of Gears", pp. 236–239. Izhevsk, ISTU (1998)
98. Krylov, N.N., Popov, V.A.: Geometry of contact of non-conjugate surfaces of worm gears. J. Mashinovedenie **5**, 32–36 (1987)
99. Kudryavtsev, V.N.: Machine Parts. Mashinostroenie, Leningrad (1980)
100. Kudryavtsev, V.N., Kirdyashev, Y.N., Ginzburg, E.G., et al.: Planetary Gears. Reference book, Moscow, Mashinostroenie (1977)
101. Kuleshov, V.V.: Elements of mechanical logics of self-locking gearing. In: Proceedings of International Conference on "Theory and Practice of Gears", pp. 283–287. Izhevsk, ISTU (1998)
102. Kuleshov, V.V.: Self-Locking Gears with Parallel Axes. Chelyabinsk, Chelyabinsk Printing House (1999)
103. Kuniver, A.S.: Method for assessment of contact in modified spiroid gears. J. Izvestiya VUZov. Mashinostroenie **1–3**, 39–43 (1998)
104. Kuniver, A.S.: Controlling the parameters of the bearing contact in modified spiroid gears. J. Vestnik Mashinostroeniya **8**, 3–7 (2000)
105. Kuniver, A.S., Kuniver, YuA: Modification of spiroid gearwheel teeth by generating worms with two pitch surfaces. J. Izvestiya VUZov. Mashinostroenie **1–3**, 13–18 (1996)
106. Kurlov, B.A.: Crossed Helical Involute Gears. Reference book, Moscow, Mashinostroenie (1981)
107. Kuzmin, I.S., Razhikov, V.N., Filipenkov, A.L.: Issues of improving the methods for gear analysis. In: Proceedings of International Conference on "Theory and Practice of Gears", pp. 248–250. Izhevsk, ISTU (1998)
108. Lagutin, S.A.: Meshing space and its elements. J. Mashinovedenie **4**, 69–73 (1987) (Reprinted J. Sov. Mach. Sci. **4**, 63–68 (1987))

109. Lagutin, S.A.: Once again to the problem of singularities and tooth undercut. In: Proceedings of International Conference on "Theory and Practice of Gears", pp. 193–199. Izhevsk, ISTU (1998)

110. Lagutin, S.A.: Space of meshing and synthesis of worm gears with localized contact. In: Proceedings of the International Conference on "Theory and Practice of Gears", pp. 185–192. Izhevsk, ISTU (1998)

111. Lagutin, S.A.: Synthesis of spatial gearing by the method of screws. J. Gearing Transm. **2**, 59–70 (1998)

112. Lagutin, S.A., Puzina, V.M.: Profiling the tool for cutting gears with closed lines of contact. J. Izvestiya VUZov. Mashinostroenie **4**, 57–60 (1982)

113. Lagutin, S.A., Sandler, A.I.: Relieving the flanks of teeth of hobs. J. Izvestiya VUZov. Mashinostroenie **1**, 115–120 (1979)

114. Lagutin, S.A., Sandler, A.I.: Providing the back angles of hobs for cutting the multi-thread worm gears. J. Mach.-Tools Tool. **5**, 19–21 (1984)

115. Lagutin, S.A., Sandler, A.I.: Radial axial relieving of hobs. J. Mach.-Tools Tool. **5**, 20–23 (1989)

116. Lagutin, S.A., Sandler, A.I.: Grinding of Helical and Relieved Surfaces. Mashinostroenie, Moscow (1991)

117. Lagutin, S.A., Verkhovsky, A.V.: Accuracy of assembly for worm gears with closed lines of contact. J. Mach.-Tools Tool. **8**, 14–17 (1977)

118. Lagutin, S.A., Verkhovsky, A.V.: Conditions of contact in worm gears with liquid friction. Computer-aided design of transmission elements, pp. 13–20. Izhevsk (1987)

119. Lagutin, S.A., Verkhovsky, A.V., Yatsyn, Y.L.: General-type worm gears for metallurgy equipment. In: Proceedings of International Congress "Gears'95", vol. II, pp. 41–44. Bulgaria, Sofia (1995)

120. Langofer, A.R.: Improving the design and operation of running-in gear-cutting hobs by mathematical modeling of their loading. PhD Thesis, Moscow, Mosstankin (1987)

121. Langofer, A.R., Babichev, D.T., Raikhman, G.N., Shunayev, B.K.: Computer-aided investigation of the load on cutting edges of gear cutting tools. J. Mach.-Tools Tool. **1**, 18–19 (1986)

122. Langofer, A.R., Babichev, D.T., Shunayev, B.K.: Computer-aided investigation of loading the teeth of hobs. In: Proceedings of 3rd All-Union Symposium "Theory and Practice of Spatial Gearing", pp. 30–31. Kurgan, KSU (1979)

123. Lashnev, S.I., Yulikov, M.I.: Computer-Aided Design of the Cutting Part of Tools. Mashinostroenie, Moscow (1980)

124. Litvin, F.L.: Sufficient feature of existence of the enveloping characteristics on the enveloping surface of the gearwheel teeth of the mechanism. In: Theory of Machines and Mechanisms. Moscow, Nauka (1976)

125. Lobastov, V.K.: Machine meshing and machining of gears with off-centroid hypocycloidal gearing. J. Izvestiya VUZov. Mashinostroenie **4**, 52–57 (1982)

126. Lopatin, B.A.: Development of theoretical fundamentals of design, manufacture and testing of cylindrical bevel gears with low interaxial angles. DSc Thesis, Chelyabinsk (1999)

127. Lopatin, B.A., Kazartsev, D.N., Lopatin, D.B., Tzukanov, O.N.: Application of gears with small shaft angle in machine drives. In: Proceedings of International Conference on "Theory and Practice of Gears", pp. 288–293. Izhevsk, ISTU (1998)

128. Lopatin, B.A., Tsukanov, O.N.: Methods of generation of tooth flanks of gears having low interaxial angle. J. Gearing Transm. **2**, 38–49 (1997)

129. Lopato, G.A., Kabatov, N.F., Segal, M.G.: Bevel and Hypoid Gears with Circular Teeth. Mashinostroenie, Moscow (1977)

130. Malevsky, N.P., Smirnov, A.E.: Derivation of the equation for the helical surface conjugated with the gear part having an arbitrary tooth profile. J. Izvestiya VUZov. Mashinostroenie **6**, 18–23 (1982)

131. Malina, O.V.: Methodology for development of the computer-aided system for design of spiroid gearboxes. In: Proceedings of International Symposium on "Progressive Gears", pp. 185–191. Izhevsk (1994)

132. Malina, O.V.: Intellectualization of computer-aided design of spiroid gearboxes. In: Proceedings of International Conference on "Theory and Practice of Gears", pp. 537–542. Izhevsk, ISTU (1998)
133. Markov, A.L.: Measurement of Gearwheels: Tolerances, Methods and Means of Control. Leningrad, Mashinostroenie (1977)
134. Medvedev, V.I.: Synthesis of running-in non-orthogonal bevel and hypoid gear pairs. J. Probl. Mech. Eng. Reliab. Mach. **5**, 3–12 (1999)
135. Modzelevsky, V.A.: Features of the design and manufacturing for cutting the gearwheels of spiroid gears with bevel helicoid worms of the convex-concave profile. In: Inter-University Contributed Volume "Improving the Processes of Metal Working by Cutting, vol. 1, pp. 92–95. Izhevsk, IMI (1976)
136. Nechayev, A.I.: Plane and Spatial Gearing According to Pascal Snails. Krasnoyarsk, KGU, DSP (1993)
137. Nechayev, A.I.: Development, kinematic and strength analyses of gears with meshing of end teeth on Pascal snails. DSc Thesis, Krasnoyarsk (1998)
138. Nesmelov, I.P., Goldfarb, V.I.: Dialog system of computer-aided design of spiroid gears. In: Proceedings of Sc. Tech. Conference on "Computer-Aided Design of Mechanical Gears", pp. 5–7. Izhevsk, ISTU (1982)
139. Nesmelov, I.P., Goldfarb, V.I.: Non-differential approach to solving the task of enveloping. In: Mechanics of Machines, vol. 61, pp. 3–10. Moscow, Nauka (1983)
140. Ognev, M.E., Gromov, D.P., Sedov, A.S.: Practice of application of helical gears with thrust washers in gearboxes of rocking machines. In: Proceedings of International Conference on "Theory and Practice of Gears", pp. 321–323. Izhevsk, ISTU (1998)
141. Panchyuk, K.L.: Geometrical synthesis of plane gearing. J. Izvestiya VUZov. Mashinostroenie **6**, 35–39 (1982)
142. Panyukhin, V.V.: Conditions of self-locking in meshing of mechanical gears. J. Vestnik Mashinostroeniya **2**, 34–37 (1979)
143. Panyukhin, V.V.: Geometrical analysis of self-locking gears having the point contact. J. Izvestiya VUZov. Mashinostroenie **12**, 28–33 (1984)
144. Panyukhin, V.V.: Investigation of self-locking of mechanisms and development of methods for design of highly effective gearing with self-locking profiles. DSc Thesis, Vladimir (1999)
145. Parubets, V.I.: Repeated contact in cylindrical worm gears. J. Vestnik Mashinostroeniya **1**, 15–19 (1984)
146. Parubets, V.I.: Features of the contact in worm gears with involute worms. J. Vestnik Mashinostroeniya **7**, 17–21 (1985)
147. Pavlenko, A.V., Fedyakin, R.V., Chesnokov, V.A.: Gears with Novikov Gearing. Kiev, Technika (1978)
148. Pavlov, A.M.: Worm poly-globoid gear with the offset worm. J. Izvestiya VUZov. Mashinostroenie **12**, 9–13 (1982)
149. Pavlov, A.M.: Method for auxiliary generating surfaces and contact localization in bevel gear pairs. J. Gearing Transm. **1**, 42–45 (1991)
150. Pavlov, A.M., Bogatsky, M.M.: Reduced curvature at the pitch point of worm hyperboloid gears. J. Izvestiya VUZov. Mashinostroenie **2**, 23–26 (1983)
151. Pavlov, V.A.: Approximation of profile of the grinding wheel for hob sharpening. J. Izvestiya VUZov. Mashinostroenie **6**, 142–144 (1981)
152. Pismanik, K.M., Sheiko, L.I., Denisov, V.M.: Machine-Tools for Bevel Gears Machining. Mashinostroenie, Moscow (1993)
153. Plekhanov, F.I.: Synthesis of the approximated internal meshing of planetary gears without a carrier. J. Vestnik Mashinostroeniya **2**, 14–17 (1988)
154. Polosatov, L.P., Tarkhanov, K.S.: Undercutting in non-involute gearing. J. Izvestiya VUZov. Mashinostroenie **2**, 46–49 (1976)
155. Popov, P.K., Shtripling, L.O.: Requirements to the accuracy of gears and power transmissions at the eve of the 21st century. In: Proceedings of International Conference on "Theory and Practice of Gears", pp. 508–511. Izhevsk, ISTU (1998)

156. Prokhorov, V.P., Chernysheva, I.N.: Investigation of quality indicators and choice of constant parameters of cycloidal gearing. J. Izvestiya VUZov. Mashinostroenie **12**, 50–54 (1978)
157. Prokhorov, V.P., Prokhorova, N.I.: To the relative motion of elements of the spatial gearing. J. Mashinovedenie **6**, 36–41 (1978)
158. Raikhman, G.N., Babichev, D.T.: Investigation of generation of the surface of face teeth by two-parametrical motion of the generating line. J. Mashinovedenie **5**, 44–51 (1976)
159. Raikhman, G.N., Babichev, D.T.: Investigation of the shape of surfaces of teeth generated by two-parametrical motion of the line. J. Mashinovedenie **4**, 51–59 (1980)
160. Reshetov, L.N., Dogoda, M.I., Klin, M.V.: Several issues of geometry of spur and helical gears with the cycloidal tooth line. J. Izvestiya VUZov. Mashinostroenie **4**, 49–53 (1980)
161. Rodin, P.R.: Fundamentals of Surface Generation by Cutting. Kiev, Vyshcha shkola (1977)
162. Rodin, P.R.: Fundamentals of Design of Cutting Tools. Kiev, Vyshcha shkola (1999)
163. Sandler, A.I.: Profiling the grinding wheels for hob relieving. J. Met.-Cut. Meas. Tools **7**, 5–8 (1978). NIIMASH
164. Schultz, V.V.: Geometrical and energy theory of gearing. J. Vestnik Mashinostroeniya **8**, 26–27 (1990)
165. Schultz, V.V.: Shape of the Natural Wear of Machine Parts and Tools. Mashinostroenie, Leningrad (1990)
166. Schultz, V.V.: Tools for cutting wear resistant gearwheels. J. Gearing Transm. **1**, 40–41 (1991)
167. Schultz, V.V., Tikhomirov, V.V.: Geometrical optimization according to wear of cylindrical worm gears. J. Vestnik Mashinostroeniya **9**, 44–47 (1985)
168. Schultz, V.V., Tikhomirov, V.V.: Optimal contact lines of involute worm gears. J. Izvestiya VUZov. Mashinostroenie **8**, 18–21 (1985)
169. Segal, M.G.: Features of layouts of CNC machine-tools for machining the circular teeth of bevel and hypoid gears. In: Inter-University Scientific Contributed Volume "Investigation of Accuracy and Productivity of Gear-Cutting Machines and-Tools", pp. 19–23. Saratov, SPI (1985)
170. Segal, M.G.: Ways of applying the CNC in machine-tools for machining the circular teeth of bevel and hypoid gears. J. Izvestiya VUZov. Mashinostroenie **6**, 120–124 (1985)
171. Segal, M.G.: Evaluation of undulation of the surface machined by enveloping. J. Mach.-Tools Tool. **12**, 18–20 (1992)
172. Segal, M.G., Kovalev, V.G., Romalis, M.M.: Analysis of meshing quality and calculation of setting-up parameters for profiling the straight teeth of bevel gears. J. Izvestiya VUZov. Mashinostroenie **10**, 25–29 (1990)
173. Segal, M.G., Sheiko, L.I.: Classification of layout of multi-coordinate machine-tools for cutting bevel gearwheels with curvilinear teeth. J. Mach.-Tools Tool. **7**, 8–11 (1998)
174. Segal, M.G., Sheiko, L.I., Prikazchikov, S.Y.: Regularities of motion of actuators of multi-coordinate gear machining equipment for bevel gearwheels. In: Proceedings of International Conference on "Theory and Practice of Gears", pp. 263–268. Izhevsk, ISTU (1996)
175. Semenov, L.K.: Vector equation of the surface of the tooth of the straight bevel gears cut by circular broaching. J. Mashinovedenie **3**, 47–52 (1976)
176. Shalobaev, E.V.: Regulation of accuracy parameters of gearwheels and gears, choice of a new concept. In: Problems of Improving the Gears, pp. 149–157. Izhevsk-Moscow, ISTU (2000)
177. Sheveleva, G.I.: Solution of one task of the theory of enveloping. J. Mashinovedenie **6**, 48–53 (1976)
178. Sheveleva, G.I.: Criterion of accuracy for profiling of gear cutting hobs. J. Mach.-Tools Tool. **4**, 12–14 (1978)
179. Sheveleva, G.I.: Design of Gearing According to Local Conditions. Mashinostroenie, Moscow (1986)
180. Sheveleva, G.I.: Theory of Generation and Contact of Moving Bodies. Mosstankin, Moscow (1999)
181. Sheveleva, G.I., Bogolyubov, A.V.: Algorithms of numerical modeling of running-in for the gear part with the rack. J. Izvestiya VUZov. Mashinostroenie **10**, 44–47 (1984)

182. Sheveleva, G.I., Gundayev, S.A., Pogorelov, V.S.: Computer-aided modeling of the meshing process with circular teeth. J. Vestnik Mashinostroeniya **4**, 48–50 (1989)
183. Sheveleva, G.I., Novikova, T.A., Shukharev, E.A.: Technique for assessment of sensitivity of bevel gears to small shifts of gearwheels. J. Vestnik Mashinostroeniya **12**, 23–26 (1990)
184. Sheveleva, G.I., Shukharev, E.A.: Areas of allowable values of assembly errors of bevel gears with circular teeth. J. Vestnik Mashinostroeniya **8**, 13–15 (1995)
185. Sheveleva, G.I., Volkov, A.E.: Assessment of the influence of manufacturing errors on the quality of meshing of circular teeth of bevel gears. J. Vestnik Mashinostroeniya **7**, 8–12 (1995)
186. Shishov, V.P., Podroiko, V.I.: Synthesis of gears with extreme quality parameters of load carrying capacity. J. Vestnik Mashinostroeniya **8**, 33–35 (1985)
187. Shishov, V.P., Podroiko, V.I.: Worm gears with convex profile of the worm thread. J. Izvestiya VUZov. Mashinostroenie **5**, 32–35 (1985)
188. Sidorenko, A.K.: Features of Manufacturing of Coarse-Module Gears. Mashinostroenie, Moscow (1976)
189. Sidorenko, A.K.: Gear «70-NKMZ». Mashinostroenie, Moscow (1984)
190. Silich, A.A.: To geometrical analysis of cylindrical Novikov gears. In: Proceedings of Symposium on "Theory of Mechanisms, Strength of Machines and Equipment", pp. 13–26. Kurgan, KGU (1997)
191. Skvortsova, N.A., Panyukhin, V.V.: Self-locking gears with positive gear ratio. J. Izvestiya VUZov. Mashinostroenie **5**, 32–36 (1984)
192. Snesarev, G.A.: General issues of gearbox engineering. J. Gearing Transm. **1**, 5–7 (1991)
193. Soldatkin, E.P., Prokhorova, N.I.: Non-involute gearing with helical motion of the driven element. J. Izvestiya VUZov. Mashinostroenie **7**, 34–37 (1983)
194. Starzhinsky, V.E., Timofeev, B.P., Shalobaev, E.V., Kudinov, A.P.: Plastic Gearwheels in Mechanisms of Devices. Reference and scientific book. Gomel, IMMS NANB (1998)
195. Syzrantsev, V.N.: Issue of synthesis of real gearing of spur and helical gears with localized contact. In: Proceedings of 4th All-Union Symposium "Theory of Real Gearing. Part 1. Geometry and CAD of Real Gearing», pp. 17–23. Kurgan, KSU (1988)
196. Syzrantsev, V.N.: Synthesis of meshing of spur and helical gears with localized contact. DSc Thesis, Leningrad (1989)
197. Syzrantsev, V.N.: Methods of Experimental Assessment of Concentration of Cyclic Strains and Stresses on Surfaces of Machine Parts. Study guide, Kurgan, KMI (1993)
198. Syzrantsev, V.N.: Methods of synthesis of meshing of cylindrical gears with barrel-shaped, corset-shaped and arch-shaped teeth. J. Gearing Transm. **2**, 34–44 (1996)
199. Syzrantsev, V.N., Ratmanov, E.V., Rokhin, L.V.: Assessment of profile errors for shaping cutters ground by worm grinding wheels. In: Proceedings of International Conference on "Theory and Practice of Gears", pp. 193–198. Izhevsk, ISTU (1998)
200. Syzrantsev, V.N., Silich, A.A.: Technique for profiling the relieved hobs. In: Proceedings of International Conference on "Theory and Practice of Gears", pp. 389–392. Izhevsk, ISTU (1998)
201. Syzrantsev, V. N., Syzrantseva, K. V.: Analysis of Machine Parts by Methods of Boundary and Finite Elements. Kurgan (2000)
202. Timofeev, B.P.: Finding the gear ratio for a pair of gearwheels when solving the reverse task of the theory of gearing. In: Mechanics of Machines, vol. 61, pp. 30–36. Moscow, Nauka (1983)
203. Timofeev, B.P.: Characteristics of distribution of the error of gear ratio for a pair of gearwheels and a simple series of gearwheels. J. Izvestiya VUZov. Mashinostroenie **3**, 20–26 (1985)
204. Timofeev, B.P., Shalobaev, E.V.: State of art and prospects of regulation of the accuracy of gearwheels and gears. J. Vestnik Mashinostroeniya **12**, 34–36 (1990)
205. Trubachev, E.S.: To analysis of geometrical and kinematical parameters of meshing of spiroid gears at arbitrary arrangement of axes. In: Proceedings of International Conference on "Theory and Practice of Gears", pp. 381–385. Izhevsk, ISTU (1996)
206. Trubachev, E.S.: Invariant method for geometrical and kinematical investigation of worm type gears. In: Proceedings of 6th International Symposium on "Theory of Real Gearing. Informative Materials. Part 1", pp. 80–84. Kurgan, KSU (1997)

207. Tseitlin, N.I.: Synthesis of tooth meshing for harmonic gears. In: Harmonic Gears, pp. 103–114. Moscow, Mosstankin (1985)
208. Tsepkov, A.V.: Profiling of Relieved Tools. Mashinostroenie, Moscow (1979)
209. Varsimashvili, RSh: Meshing of bevel gears with variable gear ratio. J. Izvestiya VUZov. Mashinostroenie **10**, 70–75 (1976)
210. Varsimashvili, RSh: Planetary and Differential Gears with Non-Circular Gearwheels. Tbilisi, Metsniereba (1987)
211. Verkhovsky, A.V.: Worm gear with elliptical axial profile of the worm. J. Izvestiya VUZov. Mashinostroenie **2**, 45–48 (1978)
212. Verkhovsky, A.V.: New varieties of gears and worm gears. J. Vestnik Mashinostroeniya **8**, 24–28 (1985)
213. Verkhovsky, A.V.: Classification of general-type worm gears. In: Proceedings of 4th All-Union Symposium "Theory of Real Gearing. Part 1. Geometry and CAD of Real Gearing", pp. 34–37. Kurgan, KSU (1988)
214. Vinogradov, A.B., Pavlov, V.A.: Investigation of double-enveloping gear with the worm ground by the plane. J. Izvestiya VUZov. Mashinostroenie **11**, 30–34 (1984)
215. Volkov, A.E.: Mathematical modeling of flanks generating process for circular teeth of bevel gears and its features. J. Probl. Mach. Eng. Reliab. Mach. **4**, 74–83 (1999)
216. Volkov, A.E.: Technique of revealing the undercut of circular teeth of bevel and hypoid gearwheels. J. Probl. Mach. Eng. Reliab. Mach. **4**, 66–74 (2000)
217. Volkov, A.E., Gundayev, S.A., Sheveleva, G.I.: Triangulate algorithms of modeling the processes of generation and meshing of gears. J. Mashinovedenie, **6**, 60–65 (1986) (translated in J. Sov. Eng. Res. **5**(6), 51–56 (1986))
218. Volkov, D.P., Kraynev, A.F.: Harmonic Gears. Kiev, Technika (1976)
219. Volkov, GYu., Erikhov, M.L.: Criteria of serviceability and analysis of the worm gear with fixing balls. J. Vestnik Mashinostroeniya **10**, 20–23 (1986)
220. Vulgakov, E.B.: Aviation Gears and Gearboxes. Reference book, Moscow, Mashinostroenie (1981)
221. Vulgakov, E.B.: Theory of Involute Gears. Mashinostroenie, Moscow (1995)
222. Vulgakov, E.B., Vasina, L.M.: Involute Gears in Generalized Parameters. Reference book on geometrical analysis. Moscow, Mashinostroenie (1978)
223. Zablonsky, K.I.: Gears. Kiev, Technika (1976)
224. Zak, P.S.: Modification of double-enveloping gears with localized contact by method of machine-tool settings. Vestnik Mashinostroeniya **1**, 8–13 (1989)
225. Zalgaller, V.A., Litvin, F.L.: Sufficient condition of existence of the envelope to contact lines and the edge of regression on the surface of the envelope to the parametrical family of surfaces represented parametrical form. J. Izvestiya VUZov. Mathematics **3**(178), 20–23 (1977)
226. Zhuravlev, G.A.: To the influence of tooth shape on their contact strength. In: Proceedings of 3rd All-Union Symposium on "Theory and Geometry of Spatial Gearing", pp. 38–41. Kurgan (1979)
227. Zhuravlev, G.A., Iofis, R.B.: Hypoid Gears: Problems and development. Rostov-on-Don, RGU (1978)

Chapter 2
Bulgarian Experience in Applying and Improving Knowledge in the Field of Theory and Application of Modern Gears

Emilia Abadjieva, Valentin Abadjiev and Dimitar Karaivanov

Abstract Gear transmissions are mechanical systems with an extremely wide application in industry. The quality of the mechanical devices and machines, in which they are incorporated, depends on their characteristics. Among the largest traditional users of gear drives are the automotive industry, the production of cranes, hoists, winches, warehouse equipment, the shipbuilding industry, as well as the enterprises for the production of machine tools. Recently, the incorporation of the gear mechanisms in mechatronics and robotics are activated. The said above, requires their permanent improvement. The process of improvement includes not only upgrading of the traditionally existing gear set' types but also creating new innovative transmissions. The modern development of gear transmissions bears the mark of the introduction of computer technologies, which are characterized by the extremely high quality of the permanently changing hardware and software environment in which the production of mechanical transmissions are developed in the 21st century. The innovative development of the various sectors in the industry defines the increased requirements to the graduate students from engineering universities, whose professional qualification is oriented towards synthesis, design, and manufacture of gear drives. Contemporary experts (specialist) of gear mechanisms must have an optimal qualification that has to contribute to the effective application of: the emerging new mathematical models for their synthesis, new software products for the design and the new equipment for their elaboration. The lack of synchronization between the rates of innovative

E. Abadjieva (✉)
Graduate School of Engineering Science, Akita University, 1-1 Tegatagakuenmachi, Akita, Akita Prefecture 010-8502, Japan
e-mail: abadjieva@gipc.akita-u.ac.jp

E. Abadjieva · V. Abadjiev
Department of Mechatronics, Institute of Mechanics-BAS, Acad. G. Bonchev Str., Block 4, 1113 Sofia, Bulgaria
e-mail: abadjiev@imbm.bas.bg

D. Karaivanov
University of Chemical Technology and Metallurgy, Bul. "St. Kliment Ohridski", No 8, 1756 Sofia, Bulgaria
e-mail: dipecabg@yahoo.com

© Springer Nature Switzerland AG 2020   47
V. Goldfarb et al. (eds.), *New Approaches to Gear Design and Production*,
Mechanisms and Machine Science 81,
https://doi.org/10.1007/978-3-030-34945-5_2

development of the software and industrial technologies with the rates of adaptation of engineering education leads to a lag of engineers, experts in gear transmissions, in the formation of adequate professional competences. The present study can be treated formally as made up of two parts. The first part contains a brief review of the development of those branches in the industry in Bulgaria that are the major users of gear transmissions. For the first time it is realized an attempt to be defined approaches applicable to the training of mechanical engineers with knowledge in the field of synthesis, design, and manufacture of certain types of gear mechanisms. Special emphasis is put on the education of future engineers and Ph.D. students in the theory and practice of hyperboloid gear mechanisms. The second part is dedicated to the gearing mechanism education at Akita University for students. It represents a brief review of history of Gear Manufacture in Japan and the development of student education in the fields of Gearing at Akita University. It is presented an implementation of the Bulgarian experience and knowledge in student's training. At the end of each section, the material contains predictions for improving education on the theory and practice of different types of gear transmissions.

Keywords Gear transmission · Innovations · Engineering education · Technical universities

2.1 Gearing Education in Bulgarian Universities

2.1.1 Introduction

The 135.000 European Small and medium-sized enterprises (SMEs) [1] of the mechanical transmission sector producers of gears and gearing products rely on traditional technologies and are characterized by a general conservatism. However, over the years the end products in which the gears are used have become more complex and are pushing the state of the art in new technology. Thus the requirement for more sophisticated and reliable gears becomes extremely important. The gear industry has to implement major changes in gear design and gear fabrication techniques just to keep up with the changing needs of the end product.

The European production [2] decreased slightly in 2013 to €13.3 billion, after a peak of €13.5 billion in 2012. In 2014, the production recovered slightly, reaching €13.4 billion. Germany is the largest European producer of gears and gearing systems (43% share of European production), followed by Italy (27%) and Finland (5%) (see Fig. 2.1).

The total European export of gears and gearing systems increased by 1.7% per year between 2011 and 2015 to €7.8 billion (Fig. 2.2). European exports peaked in 2014 at €7.9 billion and declined slightly in 2015. The share of developing countries in exports is forecast to comprise around 20% in the coming years.

Germany is the largest European exporter of gears and gearing systems (€2.7 billion in 2015), followed by Italy (€1.4 billion). Together, they represented 53%

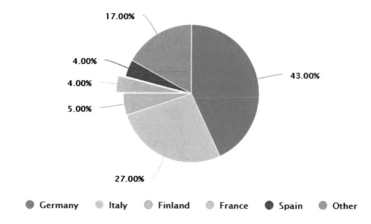

Fig. 2.1 Main European producers of gears and gearing systems (*Source* Trademap)

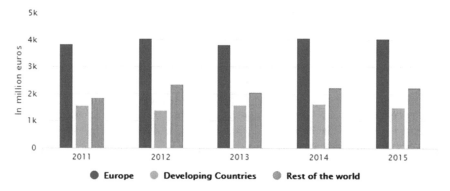

Fig. 2.2 European export of gears and gearing system to main destinations (*Source* Trademap)

of European exports. Other important exporters are Belgium (7% share), France (7%), Finland (6%) and Denmark (5%). German exports to developing countries are substantial, taking up almost 39% of the total European exports to developing countries. The European export of gears and gearing systems is expected to show a small growth of 0–2% in the next few years.

The process of Bulgaria's joining to the European Union is accompanied by a continuous and extreme increase in the importance of enhancing the competitiveness of the Bulgarian industry companies and their ability to withstand competitive pressure and economic factors. In this regard, the implementation of scientific achievements and new technologies as *well as the development of innovation potential are crucial for strengthening the Bulgarian* manufacture and hence for increasing employment and achieving economic growth [3]. In accordance with the objectives set in the Economic Reform Program accepted by the European Union (EU) in 2000, Lisbon, expanded in Gothenburg and improved in Stockholm and Barcelona, the actions of the countries, which are members of EU have to be focused on certain priority areas and the crucial importance between them has given to the *promotion of the inno-vations*. The Lisbon process requires instruments to be found in order to promote competitive manufactures with a potential for future development that could have a major impact on the restructuring of the economy. A key instrument for achieving high competitiveness of our economy is the elaboration and consistent application of a policy for the implementation of the Bulgarian innovations.

Without going into details on our national innovation policy, we will note that by Decision No 723/08.09.2004, "*The Innovative Strategy of the Republic of Bulgaria*" was accepted by the Council of Ministers of the Republic Bulgaria; and the constructed and further improved "*Innovative Strategy for Smart Specialization 2014–2020*" (Council of Ministers Decision No 384 of 13.07.2017) [4] again puts the emphasis on the development of the scientific-researches and innovation infrastructure of the Bulgarian industry, as well as on the technological modernization in the manufacturing sector. By Decision No 4/18.12.2018 of the meeting of the Council for Smart Growth is approved an update of the thematic sub-fields of the "*Innovative Strategy for Intelligent Specialization of the Republic of Bulgaria 2014–2020*".

One of the permanent *main goals* of the *Innovation Strategy* is to create conditions for stimulating researches, in order to create innovative technologies and products and their subsequent integration into companies. The commented strategy envisages a number of measures for its realization, among which the essential one is an *optimization of the science-technology-innovation relations*.

The objectives and measures, endorsed by the national innovation strategies, are pursued through the teachers and researchers (realized through the years) in the field of theory and practice gear transmissions.

2.1.2   Place of Gear Drives in the Bulgarian Industry

Gear transmissions are still widely used in almost all fields of the Bulgarian industry and, as a rule, they determine the technical and economic parameters of the corresponding units and machines in which they are incorporated. This requires, on one hand, that the existing types to be continuously improved and, on the other hand, a natural desire towards the creation of modern innovative transmissions to be formed. Expansion of computer hardware and software over the last 3–4 decades gave a certain boost to the development of the virtual design, simulation, and technology of the manufacture of existed and new types of gear mechanisms.

The said up to now explains the interest of the manufacturers and consumers to the various type of gears. The most significant markets for gear mechanisms in Bulgaria, both in the twentieth century and the twenty-first century, are characterized with the traditionally established production of: ICE forklifts, electro forklifts, electric platforms, tow tractors, diesel platforms (Balkancar Record—Plovdiv); cranes, hoists, winches, warehouse equipment (Balkancarpodem-Sofia, TILL INDUSTRIA—Gabrovo, Condra Bulgaria Ltd.—Dryanovo, Podemcran—Gabrovo (currently it occupies the fourth place in the world in the production of industrial cranes and components for them)), agricultural machinery and drive axels for trucks (Madara Group—Shumen) and others.

Gear transmissions for the above-mentioned products are manufactured in companies with long-standing traditions, such as: Modul PLC—Biala (produces cylindrical, worm and planetary gears and motor-gear box); ZMM-Nova Zagora manufactures cylindrical gears with straight involute and helical involute teeth, cylindrical and cylindrical-worm reductors; Balkancar ZP "Todor Petrov" Ltd.—Sofia (produces cylindrical and bevel gears for forklifts and agricultural machines); Bulmachinery Enterprises Ltd. (produces large gauging gears (up to 6 m in diameter) for the mining, cement and chemical industries). Over the past two decades, as successful manufacturers of various types of gears and transmissions, a number of small companies have been established, such as: Arco Ltd.—Rousse; Radina M Ltd.—Plovdiv; Karnas Ltd.—Rousse and others.

2.1.3 Higher Education Aspects of Gears Education in Technical Universities in the Republic of Bulgaria

Educational structures conducting gears training. The changes in the system for acquiring higher technical education in Bulgaria in the last 30 years are characterized by the introduction of bachelor's and master's qualification degrees for the graduate students in Mechanical Engineering and with the mass transformation of higher education institutions into universities. In Bulgaria, the following state universities are currently operating, and some of their departments conduct courses related to the theory and practice of gear mechanisms:

Technical University—Sofia (https://tu-sofia.bg/university/28). It also has two branches: The branch in Plovdiv includes: "Faculty of Mechanical Engineering and Instrument Building" and "Faculty of Electronics and Automation"; Branch in Sliven represents-the "Engineering and Pedagogical Faculty". In this university and in the bellow defined universities, gears, classified as plane and spatial (hyperboloid) transmissions, are studied in the disciplines (majors) "Theory of Mechanisms and Machines" and "Machine Elements".

From the moment of its establishment in 1974, the "Theory of Mechanisms and Machines" department deals with the theoretical questions related to the geometric and kinematic synthesis of plane gear mechanisms, of ordinary reductors and planetary reductors.

The main department, which educates students on the theory and practice of gear mechanisms, is called "Machine Elements and Non-Metallic Structures". Here, researches in the field of machine elements encompass scientific and applied studies dedicated to the loading capacity and gears performance.

The following researches are realized there:

– harmonic drives (Prof. G. Dimchev) and planetary mechanisms (Prof. K. Arnau-dov);
– load capacity and reliability of sintered gears, including model fatigue tests (Prof. P. Kolev) and gears with small modules (Prof. G. Dimchev and Prof. L. Dimitrov);
– theoretical and experimental studies on the fatigue and contact strength of tooth profiles (Prof. L. Dimitrov);
– development of test equipment, of methodologies for testing and conducting of tooth gear resource tests in order to evaluate new construction materials;
– geometrical synthesis of conjugated tooth profiles to obtain gears of a specific technological application.

The *Rousse University "Angel Kanchev"—Rousse* (https://uni-ruse.bg) has also two brunches: Branch in the city of Razgrad, where two departments—"Biotech-nology and Food Technologies" and "Chemistry and Chemical Technologies" are functioning; Branch in the Silistra city, including Department of "Philological and Natural Sciences"; Branch in the town of Vidin, which defines the following profes-sional fields, where the students are educated. They are: "Pedagogy of training in pair of subjects", "Informatics and Computer Sciences", "Machine Engineering", "Elec-trical Engineering, Electronics and Automation", "Communication and Computer Equipment", "Transport, Navigation and Aviation" and "General Engineering".

The "Machine Element" Department of this university was established in 1947. In 1975, to this department (now "Machine Engineering, Machine Elements, and Engineering Graphics") the "Industrial Scientific-Research Laboratory for Reductors Construction" is opened. As a scientific unit, the laboratory should be in service for the company "Modul"—Biala, which is specialized in the production of reductors and variators. At present, the subject "Design and Optimization of Gear Mechanisms and Reductors" [5] is studied in the above-mentioned department.

Technical University—Gabrovo (https://www.tugab.php/lang=bg) Sliven repre-sents: "Electronics and Electrotechnics"; "Machine building and instrumentation"; "Business"; Center for Postgraduate Qualification.

Technical University—Varna (https://www2.tu-varna.bg/index.php#). Here the students are educated in "Faculty of Electrical Engineering", "Faculty of Comput-ing and Automation", "Faculty of Mechanical and Technological Engineering" and "Shipbuilding Faculty".

Gear drives are studied in the majors of some Bulgarian technological universities (University of Chemical Technology and Metallurgy—Sofia, Mining University—Sofia, Forestry University—Sofia, etc.). Relatively detailed they are also considered in the Faculty of Fire Safety and Civil Protection at the Academy of Ministry of Interior of Bulgaria.

Textbooks and scientific content of the study material. In curriculums of the "Machine Elements" major, materials treating gear mechanisms have always occu-pied a significant place. After 1990, the volume of the studied material is reduced in accordance with the overall reduction in the number of hours on this subject and in relation to the division of education of the students into both degrees—bachelor and master. Along with the reduction of the educational material, treating gear mechanism, from the bachelor course, the master's programs do not, as a rule, include them in the curriculum. The Department of "Machine Science, Machine Elements and Engineering Graphics" at the University of Rousse "Angel Kanchev" is an exception in this respect, where the subject "Design and Optimization of Gears and Reductors" is taught to some of the master students [5].

Along with the obligatory existing chapters (sections) in the textbooks on *Machine Elements* and *Theory of Mechanisms and Machines*, such as classification of gears, gearing law, properties of the involute, geometric parameters of cylindrical gears having straight teeth and inclined teeth, calculation of load capacity of spur and helical gears (ISO 6336), bevel gears, worm gears, in some textbooks can be found other topics also. Usually this depends on the scientific interests of the authors or on the specifics of the university. For example, in the textbook of *Arnaudov, K., I. Dimitrov, P. Yordanov, and L. Lefterov. Machine elements. Sofia, Tehnika, 1980, 544 p.* (*in Bulgarian*) for both the internal gearing (nine pages) and the planetary and harmonic gear drives (four pages with a smaller font as additional material) occupy a certain place. In the textbook of *Duncheva G. Machine elements. Gabrovo: EKS-PRES, 2010, 272 p. ISBN 978-954-490-109-7* (*in Bulgarian*), the accuracy and measurement of gears and planetary gear drives are treated. The specificity of this textbook is imposed by the needs of the companies (in this city) that manufacture gear transmissions, planetary gears, and gear cutting tools.

However, in most university publishing houses in Bulgaria, there is a practice that the volume of textbooks is connected to the studied subject of the corresponding discipline. This prevents the elaboration and printing of scientific, guidebooks and educational books (such as [6–8]) whose content not only helps students to take the exam but also forms these materials as learning tools, applicable to the overall professional development of the mechanical engineer—expert on gear transmissions.

Gearing design projects. The design of gear drives has been preserved as a course project that combines the basic machine elements (gear set, shafts, bearings, spline joints/key joints, etc.). The depth of implementation of this project, as well as the application of modern calculation and design methods (specialized software, electronic catalogs, etc.), depends on the extent to which the teacher has scientific interests in the field of gear mechanisms and other machine elements. A good example is the Department of "Machine Science, Machine Elements and Engineering Graphics" at the University of Rousse, where at the course "Design and Optimization of Gears and Reductors", a software (developed from the lecturers from this department) is used [5].

2.1.4 The Usage of Software Products in Education of Gear Mechanisms

The wide variety of gear mechanisms used in industry and transport as reductors and multipliers, as well as the constant pursuit of researchers to create new and improved gearings on one hand and different and rapidly changing approaches to the mathematical modeling for the synthesis and design practically, make it difficult to create universal CAD systems. In connection with this, we have to note especially the extremely dynamic development of modern technical computing tools and software systems. This often requires a reassessment not only of the way in which computer programs are organized, but also leads to informal changes in applied mathematical models. In order to carry out research in the field of gearing theory and to provide adequate scientific support for this type production, computer design has evolved, forming three types of software:

First type. These programs, included here, are designed to research the impact of various kinematic, structural, technological and operational parameters on different quality characteristics of the studied gear sets. Essentially, this type of software is not subordinate to a particular strategy associated with the construction of CAD systems. The developed mathematical models, algorithms and computer programs are designed to determine the influence of some or other actual existing parameters on the quality characteristics of particular gear transmissions. However, the programs created in this case can be used as software modules, which are elements of a system of criteria for quality control of the synthesized gear mechanisms.

Second type. This group includes computer programs organized on the basis of algorithms contained in standardization documents, company methodologies or guides. Here, it should be included program products elaborated on algorithms for the geometric and robust calculation of traditional types of gear drives: cylindrical involute with external and internal mating, cylindrical worm gears, bevel gears with straight teeth, etc. It should be noted, in particular, that suitable algorithms in the majority of cases do not provide optimization in the synthesis and design of gears. For example, most of the published in journals algorithms for the synthesis of cylindrical involute, bevel and worm gears are able to provide only the technological and instrumental requirements for their manufacture in terms of standard defined modules (steps) and in terms of the occurrence of undercutting of their tooth surfaces under instrumental gearing or interference—in working mesh. Secondly, this category of software includes the ones that verify the strength characteristics of already geometrically and technologically synthesized gears. As such, these computer programs can be treated as tools for the analysis of gear systems.

Third type. Included in this category of computer program products are those based on mathematical models constructed on the basis of specially oriented to this scientific researches. For example, for Bulgaria, these are computer programs dealing with the synthesis and design of Spiroid[1] and Helicon gear sets, as well as conic and

---

[1] Spiroid and Helcion are a trademark registered by the Illinois Tool Works, Chicago, Ill.

hypoid gears—the Gleason type. For the current gear transmissions, including even the classical gears, treated in terms of current engineering requirements, it is required to be developed new mathematical approaches to their geometrical, technological and strength synthesis. The process of optimization synthesis, in this case, is realized by the application of the "directed searching method". This method ensures the possibility to reduce the number of calculated gears, constituting the synthesized toothed mechanism. The essence of the method is as follows:

- input parameters are defined, as well as those that will not be changed throughout the synthesis process;
- the variable parameters and the way of their change are defined respectively;
- the process of changing of the defined variable input parameters against their initial value continues until the defined optimization criteria are met;
- from the calculated variants of conjugated gear pairs, a final variant is chosen for which there is the best fulfillment of the defined conditions (restrictions/limitations) introduced in the mathematical model.

In other words, the process of optimization synthesis and design for the third type of software is based on adequate iterative procedures by which the desired solution is found along the path of variation of certain parameters.

The high requirements to the graduate students from Engineering Universities, who plan to devote their professional activity to the theoretical researches and manufacture of gears, (in particular the requirements for the qualification during the course of study, the acquired knowledge, abilities, and skills in the field of innovations), require knowledge not only of the licensed CAD products, but also of the defined above types of software products that they could use and create as students.

The first step in this direction is to familiarize the students with existing software for calculation and design of gears and reductors. This should be a mandatory part of the course project implementation. Some universities, such as the Technical University of Sofia, have licensed versions of KISS Soft® and Autodesk® (Invertor, Mechanical Desktop, AutoCAD Mechanical). As it has been already mentioned in the University of Rousse "Angel Kanchev", an authoring software for gear design is used. In the manual for its usage [5], seven self-contained educational-methodical materials are collected, accompanied by corresponding program units (see Fig. 2.3).

Fig. 2.3 Input screen of the main program units of [5] (Courtesy of Prof. Peter Nenov)

2.2 Gearing Education at Akita University for Students: A Brief Review of the Development of Student Education in the Fields of Gearing at the Akita University. Implementation of Bulgarian Experience and Knowledge in Student's Training

2.2.1 Introduction

It is difficult to say when the first gear mechanisms were invented. The initial information about the gears is found in old Chinese literature. They were also mentioned by Aristotle in some of his writings around 400 BCE. He defines gears as capable of reversing the direction of torque, and it can be found artifacts dating back almost this far with gears as central components, such as in wheels and rudimentary clocks. Gears are also shown in the innovative sketches by Leonardo da Vinci, around 1500. From that time till now the creation, development, and improvement of gear mechanisms progressed to such extended, so they are manufactured and used in almost all of the modern technological brunches: automobile, machine, military, and aviation industry. The application of the gears finds their adequate place in elaboration of robots for various needs and innovative devices, applicable in modern medicine.

According to the report [9], the world demands for gear production will rise up to 6% annually through 2019. This is a result of constantly increasing production of vehicles, long-life and durable devices, as well as production of energy-efficient units, such as 7- and 8-speed automatic transmissions, in the automotive market and gears used in a number of relatively small but fast-growing applications, such as wind and solar power, which are expected to increase significantly. Therefore, the gear manufacturing activity in North America, Western Europe Japan, and Russia is projected to increase going forward and generate additional demand for gears.

The existing contemporary requirements to industry to create powerful machines, noiseless, smoothly running, with high loading capacity and elaboration of new innovative mechanisms, to a great extent dictate the necessity of creating adequate teaching curricula of courses (in Japan) in which the gear mechanisms are studied.

One of the permanent *main goals* in front of the lecturers *in the Engineering Department in Graduate School of Engineering Science—Akita University* is to create conditions for stimulating students during their education, in order to develop them as successful engineers.

2.2.2 History of Gear Manufacture in Japan

Historically, Japan is the second larger gear manufacturer in the world [1]. The production and export of gearing in Asia is dominated by Japan. Japan's gear industry supports many of its internationally recognized industries, such as automobiles, shipbuilding, industrial machinery, and expanding aerospace industry. Japan's gear industry is developed after World War II, in conjunction with its automotive and machinery industries. According to [10] Japan's gear industry consists of about 350 firms and, in terms of production, is dominated by the captive gear operations of its automotive industry. In addition to the large captive producers of automotive gearing, there are four major independent producers of transmissions—Aisin-Warner Ltd., Japan Automatic Transmission Co., Ltd., Fuji Tekko Co., Ltd., and Aisin Seiki Co., Ltd.—and another 26 firms producing automotive transmission parts [11].

Captive production operations also produce gearing for large industrial corporations, such as Ishikawajima-Harima Heavy Industries (IHI), Mitsubishi, Komatsu, and Sumitomo. In Japan's growing aerospace industry, IHI, Mitsubishi, Kawasaki, and Fuji Heavy Industries produce most of the gears. Leading Japanese producers of gears and gear products are shown in Table 2.1, according to their applicable product sectors.

Japan is among the world leaders in gear research and development (R&D). The Ministry of International Trade and Industry (MITI) Agency of Industrial Science and Technology (AIST) has an active Mechanical Engineering Laboratory. However, MITI and the Japanese Government have not requested any gear research recently. The Japanese Gear Manufacturers Association (JGMA) does not fund or sponsor any gear research, except as might be required in developing product standards [USITC staff interview with JGMA officials, December 4, 1989, in page 91].

Table 2.1 Leading Japanese producer, by major sectors, 1989

Company	Industrial	Vehicle	Marine	Aerospace
Asano gear		×		
Aisin Seiki Co		×		
Fuji Heavy Industries	×	×		×
Fujikoshi ("Nachl")	×			
Fuji Tekko Co. Ltd		×		
Hasegawa	×			
Hitachi	×			
Honda		×		
IHI	×	×	×	×
Japan Automatic Transmission Co., Ltd		×		
Kawasaki	×	×	×	×
Maschinko	×			
Mitsubishi	×	×	×	×
Nissan		×		
Nissei Industrial	×			
Nippon Gear	×		×	
Osaka Selsa	×		×	
Sumitomo	×		×	
Toyota		×		
Yanmar Diesel	×	×	×	

Source Compiled by the staff of the U.S. International Trade Commission

The government funds research at several university gear research centers including the University of Tokyo, the Laboratory of Precision Machinery and Electronics of the Tokyo Institute of Technology, Kyoto University, and Kyushu University, which has a gear-making machine tool research laboratory. University research centers are usually very small, with teams of researchers dedicated to narrow research topics, such as noise or fatigue in gears. The universities typically do not perform direct research for companies, as they do not want to develop close corporate associations. Professors frequently conduct basic research, rather than application-specific research, and they are generally free to decide upon their own topics.

They can apply to the Ministry of Education for funding. The typical award is $39,000–$62,000 per year for three years, with a maximum of about $234,000 for a three-year project, although an additional $15,000–$16,000 per year may also be granted. Since such funding is relatively small, professors typically ask for donations of machinery from gear companies, such as test gears or testing equipment [USITC staff interviews with Prof. K. Umezawa, Research Laboratory of Precision Machinery and Electronics, Tokyo Institute of Technology, December 6, 1989, in page 92]. If companies want to adopt research results of professors, they usually ask permission and pay a nominal sum to the professor. The results of university research are generally published in the Journal of the Japan Society of Mechanical Engineers (JSME) and the Journal of the Japan Society of Precision Engineering, as well as being presented at international conferences. The SME sponsors gear research projects for which it solicits funds from companies. Typically, funding from companies for JSME projects totals $100,000–$125,000. The JSME also sponsors a conference every 4 years on mechanical power transmission machinery.

Proprietary research on gears is performed by larger companies, such as Toyota, Honda, Nissan, Mitsubishi, IHI, and Kawasaki. Mitsubishi and IHI are noted for marine gear research, but also conduct research on machinery and aerospace gearing. Industrial gearing research is minimal, since technology is frequently obtained through licensing arrangements with foreign companies. Industrial gear manufacturers are focusing their research on reducing noise, increasing gear reducer efficiency, and producing more compact and lighter gearboxes. For the automotive producers, the researches are focused on gears and production methods and productivity. Recent developments in Japanese automobile gearing include a reduced pressure angle on the gear, which allows for quieter gears that are also easier to manufacture, and ribbing on transmissions to reduce weight and decrease noise. Other research has focused on hard finishing methods, various cutting methods, and productivity improvements [USITC staff interviews with gear industry officials, Japan, December 4–11, 1989, in page 94].

According to [11] the total employment for the Japanese gear industry, including automotive and other vehicle gearing, is estimated at 39,000 persons. Many Japanese gear producers, as well as other heavy industry manufacturers, are having difficulty recruiting university graduates, especially engineers, and other types of skilled workers. In contrast, Japanese automotive producers report that they have not encountered problems recruiting production workers or engineers, possibly because they are large corporations that can offer lifetime employment. As in other industrialized countries, production workers in Japan's metalworking industries, including the gear industry, tend to be older, as recent graduates have shown a preference for the service sector over the manufacturing sector. In the gear industry, the average age of production workers is about 40–43 years and increasing. [Because of the current age of the workforce, there is some concern that the gear industry will eventually lose some of its skills. Worker training in the Japanese gear industry is left up to the individual companies, as the national apprenticeship program was eliminated in the 1970s. Large Japanese companies with captive gear operations are able to train employees, including cross-training in different aspects of the firm. Some companies rotate workers to different production assignments every 2 or 3 years. Automobile companies have extended the concept of cross-training to their gear design engineers, requiring them to learn to produce and test prototype gears, as well as design them [USITC staff interviews with automotive manufacturers, Japan, December 7–8, 1989, in page 100].

Hourly compensation costs for Japanese production workers in the industrial and commercial machinery manufacturing industries (SIC 35) rose 101.5% from $7.36 in 1984 to $14.83 in 1988. Measured in yen, they rose from ¥1,747 in 1984 to ¥1,902 in 1988, an increase of only 8.9%. Discussions with Japanese gear producers indicate that the typical factory worker with several years' experience earns about \$27,000–\$29,000 per year, including bonuses. Such bonuses are given to workers twice a year, and are generally based on economic conditions and individual capabilities. Bonuses may account for 5 months' salary or between 30 and 40% of worker's annual salary. In order to contain labor costs, Japanese gear producers use part-time workers in operations that require minimal training. Such areas include gear product assembly operations and selected office support services.

Even the smaller gear producers offer their employees many benefits, including health insurance, welfare pension insurance, labor insurance, and loan programs. Frequently, the company provides housing, either in company own dormitories usually for single male workers, or in company apartment buildings. Many firms sponsor club activities, including sports and English language groups, or informal groups outside of work that use company resources for product innovation[USITC staff interviews with gear industry officials, Japan, December 5–11, 1989, in page 103].

2.2.3 Educational Agenda of the Ministry of Education, Culture, Sports, Science, and Technology (MEXT)—Japan

As it is said [12], amid the rapidly changing circumstances in Japan and abroad surrounding universities, expectations and demands towards universities, such as the development of cultured human resources with deeply specialized knowledge, and contributions to the solution of various kinds of social issues, have become enlarged and diverse. Each university has made efforts to clarify its own originality and characteristics based on its educational principle, aiming to qualitatively maintain and improve its educational activities. However, on the other hand, under such circumstances as the increased percentage of students enrolling in universities, the diversified students' needs, the decrease in the population aged 18 and the progress of universities' cross-border education activities, it has become necessary to reexamine not only measures taken by each university but also how the entire higher education system should be. Under such circumstances, the Ministry of Education, Culture, Sports, Science, and Technology has made the following efforts to support universities' own educational activities.

- guaranteeing the quality of higher education through the establishment approval system and the quality assurance and accreditation system
- enhancing the quality of undergraduate and graduate school courses
- improving international competitiveness and etc.

According to [12] in order to be properly evaluated educational system in Engineering departments, the existing Accreditation System for Engineering Education operated by the JABEE (Japan Accreditation Board for Engineering Education) control whether the educational programs in engineering education are examined and accredited.

The main goals which are put in front of engineering education are: "(1) Improve engineering education, (2) Guarantee the international reference of engineering education and (3) Realize the mutual accreditation of engineering-related licenses with foreign organizations....." [12].

"Amid ongoing globalization, in order to develop an educational environment where Japanese people can acquire the necessary English skills and also international students can feel at ease to study in Japan, it is very important for Japanese universities to conduct lessons in English for a certain extent, or to develop courses where students can obtain academic degrees by taking lessons conducted entirely in English. In Japan, many universities have already established classes taught in English. Also, there are several undergraduate courses where students can graduate by taking only lessons conducted in English, and there are 50 or more graduate schools where students can graduate by taking only lessons conducted in English. Of course, such universities still also provide substantial Japanese-language education courses" [12].

In July 2008, in order to make Japan more open-minded and to maintain and develop the society, the Ministry of Education, Culture, Sports, Science and Technology and other ministries and agencies concerned, has created the Framework of the "300,000 International Students Plan". This plan sets the longer-term goal of accepting 300,000 international students by 2020.

One of the main points from "300,000 International Students Plan" is:

- To increase courses conducted only in English;
- To promote double degree programs, short-term overseas study programs, and others.

2.2.4 Theoretical and Practical Gearing Education at Akita University for Students from the Bachelor Program

The increased demands to industry to create more innovative devices and the recommendations from JABEE to Graduate School of Engineering Science, at Akita University some of the conducted classes, in the field of machine manufacturing and gear and machine design are taught in the English language.

In general, most of the courses which give the initial and substantial information in the field of gear design and manufacture are studied through the Bachelor Program in Graduate School of Engineering—Akita University. Those engineering courses are thought in Japanese (theoretical and practical ones) and in English Language (theoretical).

The subjects, which are conducted in English, are oriented to fulfill the needs of undergrad students from the 2nd and 3rd year from the two-division: *Mechanical Engineering Course and Creative Engineering Course* of the Department of Systems Design Engineering at Akita University.

Hence, there are elaborated two courses in the English language "*Design and Fabrication Methods for Manufacturing*" and "*Practical Machine Design*".

In both of the subjects, a big part of the material is dedicated to the gear design and its methods of manufacturing.

Practical Machine Design is perhaps the most important course for Mechanical Engineering students. It covers the needs of second-year undergrads from the Department of Systems Design Engineering. This course introduces the basics in practical design of machine parts and simple systems. It includes the integration of the main engineering disciplines, which are necessary for proper analysis, synthesis, and design of structures and various mechanisms. After finishing this course from the students from Department of Systems Design Engineering at Akita University is expected to:

- Understand and learn new English terminology, related to various basic theoretical knowledge, machine elements, systems, and machine design;
- Develop an ability to apply knowledge of mechanical science, and engineering;
- Create an ability to use the approaches, skills, and modern engineering tools necessary for engineering practice;
- Create an ability to identify, formulate, and solve engineering problems.

From the introduction part with definitions concerning machine design and in particular design of the gear mechanisms. There, the basic principles and stages of machine design are explained.

The part which is dedicated to the contemporary gearing education starts with basic introduction and definition of gears. It introduces classification of the gear mechanisms, on the basis of the location of their axes in the space, as well as on range of efficiency, type of load imposed on the support bearings, nominal range of reduction ratio, methods of manufacture, methods of refining and etc. Also the students become familiar with the basic terminology related to the gear science as: flanks, addendum, dedendum, tooth profile, pitch, gear ratio and etc.

Throughout this part of the course, students learn the specific terminology and characteristics of the gears, based on the classification of the location of their axes in the space. They receive basic knowledge about the different tooth profile design and the gear's advantages, depending on their application.

The types which are examined are as follows:

- Parallel Axis Gears: (External Spur Gears, Spur Pinion-Spur Rack Drives, External Helical Gears, External Herringbone or Double–Helical Gear) and Internal Gears: Internal Spur Gears, Internal Helical Gears;
- Intersecting Axis Gears: Face Gears, Beveloid Gearing (Beveloid is registered trade-mark of Vinco Corp., Detroit, Michigan);
- Nonintersecting-nonparallel Axis Gears (Hyperboloid Gears): *Crossed–Axis Helical Gears, Wormgears (single and double enveloping), Hypoid Gears,*
- Special Hyperboloid Gears (Planoid, Spiroid, Helicon, and Wildhaber worm gears (USA Patent US 3386305 A/1968); Planoid, Spiroid, and Helicon are registered trade-marks of Illinois Tool Works, Chicago, Ill).

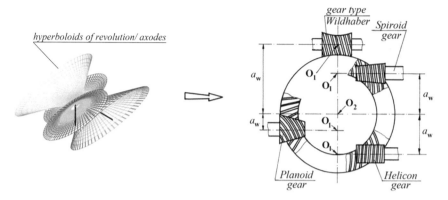

Fig. 2.4 Family of special hyperboloid gears

One of the accents in this course material is put on this special this special hyperboloid gears, since the biggest part of Dr. Abadjieva researches is dedicated to them.

The desire to transfer the Bulgarian knowledge of gear's design and Bulgarian experience in elaboration of those spatial motions transmissions and to motivate as well students to develop an interest in gear design is part of this course. On the basis of the realized classification shown in Fig. 2.4 [13–17], students obtain knowledge of a special group of hyperboloid gear mechanisms and the initial basic knowledge of the methods for their manufacture.

Here, special attention is put on Spiroid and Helicon gear mechanisms. Their geometric and technological features, which define them as special gear sets, are explained in detail.

Through the course, the student's accumulation of knowledge is control with regular reports and exams.

The process of manufacturing, assembling devices, delivering products and creating various machines has a great impact on the economic development of every country. The benefits of innovative gear systems, integrated into various machines are important to social development, as well. The first challenge to the students from every engineering discipline is to receive an understanding of all steps in the processes of manufacturing. The introduction part of the second course "Design and Fabrication Methods for Manufacturing" is oriented to 3rd years undergrad students, attending Engineering program of Department of Systems Design Engineering at Akita University. The objective goals of the students' education are:

- To make students understand the fabrication methods, which are included in manufacturing processes.
- To develop a technical intuition to define appropriate methods for manufacturing, which have to be included in production processes.

Hence, the content of this group of lectures is dedicated to the history of fabrication methods, gives something basic definitions of the processes related to gear's generation and the new requirements to the contemporary manufacturing processes.

As the course progress, the students obtained knowledge's in detail about various cutting and finishing processes related to the gear's production.

The evaluation of the obtained knowledge is strictly checked by weekly reports and exams.

2.3 Conclusion

2.3.1 Postgraduate and Ph.D. Education

The current trends of the industry expansion require the development of an educational concept that have to evolve and adapt to the changes in current specific manufacturing conditions. This reflects in the creation of up-to-date and timely innovations in the field of gear mechanisms. The effective progressive development of industrial manufacture demands a permanent interaction between employers and students. This will lead logically to the accumulation and application of new knowledge and improvement of the process of forming the professional competencies of the engineers—experts in the field of gear transmissions.

In order to obtain the qualification required for modern manufacture, it is not possible to rely solely on university education. The natural desire of teachers is to give as much as possible a better the theoretical basis of the specialized knowledge. Based on this, young engineer has to build on his/her future qualification, both independently and with the help of the employer, through highly eroded specialists from practice, academics and scientists. Some good directions can be made in this regard:

- Publishing specialized literature with an engineering focus on specific fields, which are not covered enough in the university education, such as—planetary gear mechanisms [18], an optimization of gears [5, 19]. The field of gear drives has always been up to date and attractive to scientific researches, in the form of scientific publications, dissertations, monographs. Much of them are not always adequate to be used by engineers from real practice. The editions quoted above are directed to them. For example, in [18] (with authors Prof. Dr. K. Arnaudov and Dr. D. Karaivanov) the reader will not be going to find complex differential equations, multi-mass dynamic models and multifactor optimization. However, it is possible to read clearly which factors affect the uneven load distribution between planets, the efficiency, which gear sets can be self-locking, how to measure the internal teeth with pin or balls, etc. The book of Prof. Linke [20], dedicated to gear mechanism, is a good example, that this extensive field cannot be covered only by university textbooks, and engineers have to increase their expertise in specialized literature.
- *Continuing Education* is usually understood as attending courses by engineers in order to raise their qualifications. The university lecturers are ready to conduct such

qualification courses, but companies are less interested in doing it. The problem, in this case, is that a small company can hardly be left by a specific specialist for a certain period of time. Practice shows that workplace training (educating) gives very good results. In this case, it is possible that the educating high-qualified specialists (scientists and university teachers) will take part in finding solution to the problems of manufacturing together with the engineers from the company. A good example for Bulgaria is the conducted test (a few years ago) in the field of gear mechanisms when the new engineers are recruited at "Podemcrane"—Gabrovo.

- The University teachers' efforts to raise interest and attract Ph.D. students provide a variable result. There is no more than one Ph.D. student per year in the field of gearing. It is worthwhile to mention, the experience and aspirations of the scientists from the Institute of Mechanics—the Bulgarian Academy of Sciences to spread ideas for the development of innovative engineering by creating doctoral thesis in the field of theory and practice of the spatial motions transformation. Objects of these studies are conic and hyperboloid gears of the Gleason type [21], Wildhaber, Spiroid and Helicon [15] (Spiroid and Helicon are registered trade-marks of Illinois Tool Works, Chicago, Ill) and spatial rack drives [22].

During the period 1994–2006, three types of Helicon reductors were created and implemented at Business Innovation Centre CIMEJS Co—Sofia city. The main technical data, characterizing each reductor-type are given in Table 2.2. The creation of these three types of Helicon reductors is an illustration of how through the theoretical and applied research, in the realization of Ph.D. dissertations, it is possible to achieve modern innovation spatial transmissions in the Bulgarian industry.

Table 2.2 Technical characteristics of motor-reductors type Helicon

Technical characteristics	RH 31	RH 45	RH 50
Offset, mm	31, 5	45	50
Gear ratio	13, 3…105	20…80	20…80
Maximum input power, kW	0, 370	1, 1	1, 5
Rated driving torque, Nm	50	100	150
Rated inlet min^{-1} revolutions,	1500	1500	1500
Theoretical efficiency, %	40…92	40…92	40…92
Weight without the motor, with:			
– cast iron housing parts, kg	12	34	38
– aluminum housing parts, kg	6	24, 5	26, 6

(a)   **(b)**

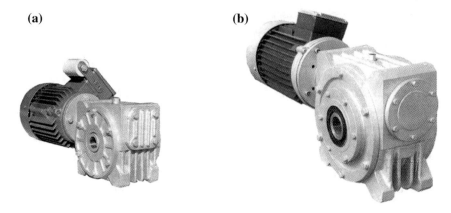

Fig. 2.5 Motor-reductor MRH 31 (**a**) and motor-reductorMPH 50 (**b**)

Each of the created type reductors is a result of the need to realize adequate rotations transformation for various mechanical systems. For example, the type RH 31 (Fig. 2.5) happened as a result of a request from the "BELIN" Ltd.—Sofia city to be driven children swings of type "Kiddie Rides" and Carousel type cradles. The type RH 45 was created for the manufacturing needs of the MELSIKON Ltd.—Sofia city (manufacturing grinding equipment), and the Helicon gear-sets of type RH 50 were designed to meet the manufacturing needs of SILOMA Ltd.—Silistra city for driving of the band-saw cutting machine of type OL220 DG. The motor-reductors MRH 31 and MRH 50 (Fig. 2.5) were experimentally implemented by the company SPESIMA Ltd.—Sofia city into the created and manufactured by its manipulator for proportioning and pouring of melts of aluminum and magnesium alloys in horizontal machines for casting under pressure FEEDMAT 1.

2.3.2 Predictions for Improving the Education on the Gear Transmissions

The permanent innovative development of the manufacturing objects of the current industrial branches (vehicles, robots, and robotic systems for realization of operations in engineering and medicine, aircraft and etc.) puts very high requirements on graduate students who plan to devote their professional activities on theoretical research, creation and practical realization of modern gear transmissions. Current trends and key aspects of the development of higher education in various countries are different in terms of the successful education of the gear experts in the universities within the bachelor's program [23]. Along with the study of various modern courses involving gear mechanism in engineering programs, a great deal of attention is paid in Japan to the English language competencies of the bachelor students, including future gear engineer-experts.

Gear transmissions' specialists have to improve continuously the engineering equipment, both by refining the design and technology of the creation of the existing gear transmissions and by elaborating innovative gear mechanisms.

According to Dr. Abadjieva, who educates undergrad students from Department of Systems Design Engineering—Graduate School of Engineering Science, Akita University—Japan, there are certain contradictions between the need to increase the efficiency of innovative manufacturing systems in industrial firms and the insufficient pace of reform in the education system. The development of professional competence of the expert-engineers requires defining of actions to develop innovative approaches in order to form professional competencies regarding the requirements to the education of the future expert engineers.

The analysis of the content of the conducted courses, related to the gear mechanisms within the education of the undergraduate engineers in the Bachelor's program shows the need to enrich the students' theoretical knowledge in the field of spatial transformation of regular motions, realized by means of adequate gear sets.

According to Dr. E. Abadjieva and Prof. Sc. D. V. Abadjiev—authors in this article, the first step in this direction is the introduction to the Japanese students, the usage of "3D Software Technology" for creation of hyperboloid gear transmission [24, 25]. In the cited researches, the theoretical approach to the synthesis of spatial small modules gear drives of type Spiroid and Helicon, elaborated for incorporation as driving into the fingers of a bio-robot hand (realized in Gifu University, Gifu, Japan) is shown [26, 27]. The illustrated possibility of their practical realization through 3D printing shows an example of the innovative realization of a special spatial transmission with high accuracy in a small production series (see Figs. 2.6, 2.7 and 2.8).

(a)   **(b)**

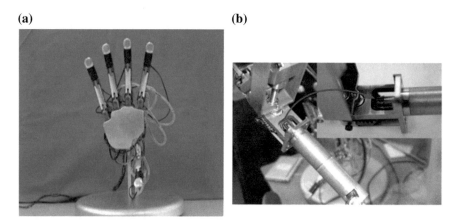

Fig. 2.6 Model of robot hand: **a** whole hand; **b** bevel gear with straight teeth with gear ratio $i_{12} = 4$; number of teeth—$z_1 = 10$, $z_2 = 40$; tooth module—$m = 0.5$ mm

(a) (b)

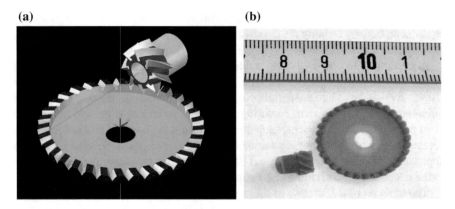

Fig. 2.7 Spiroid gear drive with offset 3,25 mm, gear ratio 32/8 (axial module 0.5 mm): **a** 3D CAD model; **b** 3D printed model (the shown scale is in mm)

(a) (b)

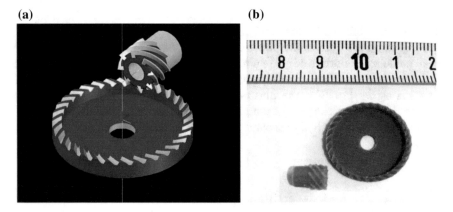

Fig. 2.8 Helicon gear drive with offset 3,25 mm, gear ratio 32/8 (axial module 0.5 mm): **a** 3D CAD model; **b** 3D printed model (the shown scale is in mm)

References

1. https://cordis.europa.eu/project/rcn/81679/factsheet/en
2. https://www.cbi.eu/market-information/motion-drives-control-automation/gears-gearing-systems
3. Annual report on the situation and development in the field of innovations. Ministry of Economy and Energy of the Republic of Bulgaria, Sofia, p. 36 (2005)
4. https://www.mi.government.bg/files/useruploads/files/innovations/ris3_26_10_2015_bg.pdf
5. Nenov, P., Anguelova, E., Varbanov, V., Ivanov, S.: Modernization of the educational process on the subject of machine elements with the help of wider usage of authors' software. In: Proceedings of Conference on Education and Information Systems, Technologies and Applications (EISTA 2004), vol. 2, pp. 48–53. Orlando, USA (2004)
6. Collins, J.A., Busby, H., Staab, G.: Mechanical Design of Machine Elements and Machines, 2nd edn. Wiley, Hoboken, NJ (2009)

7. Miltenovic, V.: Machine elements. Nis [Serbia]: Nis University Press (2009) (in Serbian)
8. Mott, R.L.: Machine Elements in Mechanical Design, 4th edn. Pearson, UppreSadle River, NJ (2004)
9. https://www.freedoniagroup.com/industry-study/world-gears-3320.htm
10. Competitive Position of The U.S. Gear Industry in U.S. and Global Markets: Report to the President on Investigation No. 332-275 Under Section 332 (g) of the Tariff Act of 1930 as amended, USITC Publication 2278, United States International Trade Commission Washington, DC 20436 (1990)
11. U.S. International Trade Commission, U.S. Global Competitiveness: The U.S. Automotive Parts Industry, USITC Publication 2037, Dec 1987, pp. 12–101 and Dodwell Marketing Consultants, The Structure of the Japanese Auto Parts Industry, 3rd ed., Oct 1986
12. Higher Education in Japan: Published by Higher Education Bureau, Ministry of Education, Culture, Sports, Science and Technology, (MEXT), http://www.mext.go.jp/en/policy/education/highered/title03/detail03/__icsFiles/afieldfile/2012/06/19/1302653_1.pdf
13. Abadjiev, V., Petrova, D., Abadjieva, E.: An optimization of Type Wildhaber gear sets based on loading capacity. Kinematic approach to criteria construction. In: Proceedings of 15th International Conference on Manufacturing Systems ICMaS, pp. 421–424. Editura Academia Romane, Bucharest (2006)
14. Abadjiev, V., Petrova, D., Abadjieva, E.: An optimization of Type Wildhaber gear sets based on loading capacity. A software estimate of the hydrodynamic loading capacity. In: Proceedings of 15th International Conference on Manufacturing Systems ICMaS, pp. 425–428. Editura Academia Romane, Bucharest (2006)
15. Abadjiev, V.: Gearing theory and technical applications of hyperboloid mechanisms, Sc. D. Thesis, Institute of Mechanics—BAS, Sofia (2007) (in Bulgarian)
16. Abadjieva, E., Abadjiev, V.: On the synthesis of hypoid gears with linear contact. Part I—geometric synthesis practices in the pitch contact point. In: Proceedings 10th Jubilee National Congress on Theoretical and Applied Mechanics, 1–6 Sept 2005 (2005)
17. Abadjieva, E., Abadjiev, V.: On the synthesis of hypoid gears with linear contact. Part II—geometric synthesis of the Planoid gearing. In: Proceedings 10th Jubilee National Congress on Theoretical and Applied Mechanics, 7–11 Sept 2005 (2005)
18. Arnaudov, K., Karaivanov, D.: Planetary gear trains, Sofia: Bulgarian Academy of Sciences Publ. "Prof. Marin Drinov", p. 368 (2017) (in Bulgarian, translated in English—https://www.routledge.com/Planetary-Gear-Trains-1st-Edition/Arnaudov-Karaivanov/p/book/9781138311855)
19. Nenov, P., Kaloyanov, B., Angelova., Varbanov, V.: Design of gear drives by using of GB contours. Ruse Univ. Publ., Ruse (2015) (in Bulgarian)
20. Linke, H., und andere: Stirnradverzahnung – Berechnung, Werkstoffe, Fertigung. 2. Auflage. München/Wien: Carl HanserVerlag (2010) (in German)
21. Minkov, K.: Mechanical and mathematical modeling of hyperboloid gears. Sc. D. Thesis, Sofia (1986) (in Bulgarian)
22. Abadjieva, E.: Mathematical models of the kinematic processes in spatial rack mechanisms and their application. Ph.D. Thesis, Bulgarian Academy of Sciences, Sofia (2009) (in Bulgarian)
23. Goldfarb, V., Krylov, E., Perminova, O., Barmina, N., Vasiliev, L.: Aspects of teaching "advanced gears" for future mechanical engineers within "Bachelor of Sciences" Programs at Technical Universities. Adv. Gear Eng., Mech. Mach. Sci. 51, 271–287 (2018) (Springer)
24. Abadjiev, V., Abadjieva, E., Petrova, D.: Synthesis of hyperboloid gear sets based on the pitch point approach. J. Mech. Mach. Theor. 55, 51–66 (2012) (Pergamon, USA)
25. Abadjieva, E., Abadjiev, V., Ignatova, D.: 3D Software technology, applicable in elaboration of the spatial face gear drives for incorporation into robot systems. World J. Eng. Technol. 4(3D), 91–99 (2016)

26. Abadjieva, E., Abadjiev, V.: 3D Software technology for practical realization of special hyperboloid gear mechanisms. In: Proceedings of EURO 2015 (27th European Conference on Operational Research) Conference in Glasgow, 12–15 July 2015 (online published http://euro2015. org/)
27. Abadjieva, E., Abadjiev, V., Kawazaki, H., Mouri, T.: On the synthesis of hyperboloid gears and technical application. In: Proceedings of the ASME 2013 International Power Transmission and Gearing Conference, Portland, Oregon, USA, 4–7 Aug 2013 (published on CD)

Chapter 3
Advanced Computer-Aided Gear Design, Analysis and Manufacturing

Claude Gosselin

Abstract Advances in numerical analysis coupled to increased computer power have opened the fields of gear design, analysis and manufacturing which were previously reserved to specialists with knowledge often gained over decades of hands-on work. Gear design, analysis and manufacturing involve defining the tools and machine adjustments for a specific application, confirming that the design is sound, and manufacturing the parts to given tolerances. Once a basic gear geometry has been defined, it must be tested to establish its capability under load and in the expected working environment, which includes relative displacements caused by manufacturing tolerances on the gearbox and/or deformations of the gearbox and support arbors when torque is applied. Modern gear design software incorporates the above functionalities where the gear set is tested "on-screen" in order to confirm that its behavior falls within the expected range. When the gear design engineer is satisfied with the behavior of the design, manufacturing takes over. Modern gear manufacturing machines are CnC based and therefore offer nearly unlimited control over tool and work piece motion as generation proceeds. In particular, the Closed Loop (i.e. Corrective Machine Settings) is essential in ensuring that the errors found in the machine and tooling are removed such that the expected quality is obtained. An offspring of the Closed Loop is Reverse Engineering (RE), where the machine settings of a measured gear set are established. RE is a powerful tool that can be used to analyze and debug existing gear sets; RE can also be used to define the machine settings of existing gears that must be replicated. This chapter presents the basic theory behind a general Tooth Flank Generator (TFG) that is used to model gears of any type: Cylindrical/Helical, Straight Bevel, Spiral Bevel, Face, Worm, Beveloid, even Hirth Couplings. Tooth Contact Analysis, without (TCA) and under load (LTCA) are then developed. Using the TFG, CnC manufacturing, Closed Loop and RE are introduced. Examples show the use of the TFG and the derived functions.

C. Gosselin (✉)
Involute Simulation Softwares Inc., Quebec, Canada
e-mail: HyGEARS@HyGEARS.com

© Springer Nature Switzerland AG 2020
V. Goldfarb et al. (eds.), *New Approaches to Gear Design and Production*,
Mechanisms and Machine Science 81,
https://doi.org/10.1007/978-3-030-34945-5_3

Keywords Gears · Spiral-Bevel · Straight-Bevel · Coniflex · Simulation · Tooth contact analysis · Loaded tooth contact analysis · Closed loop · Reverse engineering

3.1 Introduction

Gear design is a complex task that requires several qualities from the designer. In former days, say before the late 1970s, gear designs were essentially based on standards and tables found in gear handbooks and specialized literature.

Starting in the early 1980s, specialized software started appearing that would allow gear designers to interactively change parameters on a design and obtain some forms of numerical output from which the design would be "frozen". Even with the help of the early digital design software, long term experience was a fundamental requirement to produce good gear designs.

And then, there was no graphical output to, at the very least, allow "eyeballing" results.

As long as cylindrical gears were considered, the fundamentals being relatively "simple" in terms of geometry, normally the results would prove sound without too much modifications once in the testing phase.

However, when bevel gears were considered, especially spiral-bevel and hypoid gears, the design and development phases could be much longer. For example, besides the cutter definition, a typical Duplex Helical spiral-bevel pinion requires up to 10 machine adjustments, several of which are angular. Since older manufacturing machines were mechanical with several adjustments made with a Vernier angular precision of $0.5'$, i.e. 30 arc-secs, significant differences in the expected and actual contact patterns would appear.

This would lead to the "development" phase whereby an experienced designer would modify the actual machine settings one—or several—at a time until a satisfactory contact pattern would be obtained. The final machine settings would be recorded and would become the "developed" reference for all further production.

Of course, manufacturing the same pinion on a different machine could lead to different results depending on machine age, wear, and the operator making the machine adjustments.

Therefore, even a design expected to be "good" would not guarantee a "good" gear pair since manufacturing was involved and there was no guarantee that the design settings would replicate exactly in the actual machine.

The same would apply with contact and bending stresses: basic formulas, tables and graphs with curves would be used to calculate stresses, and it was expected that these would translate in operation, which in most cases did.

However, optimizing gear sets for a given task and increasing power density was an almost impossible task given the limited means.

The advent of fast personal computers with advanced graphics changed all this such that nowadays, even people with limited experience can produce good gear designs in a short time.

This chapter reviews some of the software fundamentals that are required to Design, Analyze, Manufacture and Assess a gear set, i.e. DAMA:

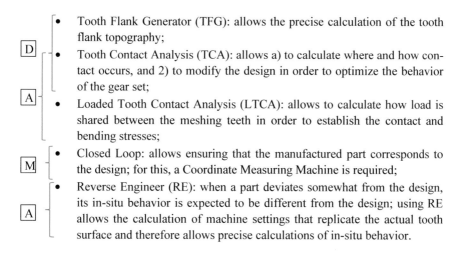

- Tooth Flank Generator (TFG): allows the precise calculation of the tooth flank topography;
- Tooth Contact Analysis (TCA): allows a) to calculate where and how contact occurs, and 2) to modify the design in order to optimize the behavior of the gear set;
- Loaded Tooth Contact Analysis (LTCA): allows to calculate how load is shared between the meshing teeth in order to establish the contact and bending stresses;
- Closed Loop: allows ensuring that the manufactured part corresponds to the design; for this, a Coordinate Measuring Machine is required;
- Reverse Engineer (RE): when a part deviates somewhat from the design, its in-situ behavior is expected to be different from the design; using RE allows the calculation of machine settings that replicate the actual tooth surface and therefore allows precise calculations of in-situ behavior.

In the following sections, the basics used in the HyGEARS [1] software are presented. HyGEARS has been in commercial use for 25 years now, and has consistently provided accurate results.

3.2   Tooth Modelling: The Tooth Flank Generator

A Tooth Flank Generator (TFG) requires the definition of the tool shape and its movements relative to the work piece; the resulting equations are solved within the boundaries of the blank. The vast majority of gears are generated, and therefore the generating process needs to be included in any TFG. However, non-generated gears can also be reproduced by preventing the generating motions in the following equations.

The generating process [2] is based on the concept of a cutter blade representing one tooth of a theoretical generating gear meshing with the work piece. The fundamental equation of meshing can be written as:

$$\overline{N} \cdot \overline{V}_r = 0 \qquad (3.1)$$

where the relative speed vector $\overline{V}_r$ of the contacting tool and work piece surface must be in a plane perpendicular to their common normal vector $\overline{N}$.

When applied using the reference frames of Fig. 3.1, Eq. (3.1) yields an unbounded generated surface in the work piece reference frame. The generated surface is a function of the machine settings and three surface variables, respectively cutter position α_c (angular or linear), work piece roll angle α_3 and position S of a point along the edge of the cutter blade:

$$S = f(\alpha_c, \alpha_3) \tag{3.2}$$

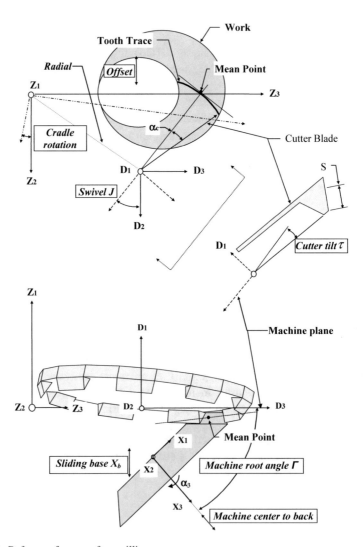

Fig. 3.1 Reference frames—face milling

Fig. 3.2 Face mill cutter
definition (actual cutter
photo courtesy Weiku.com)

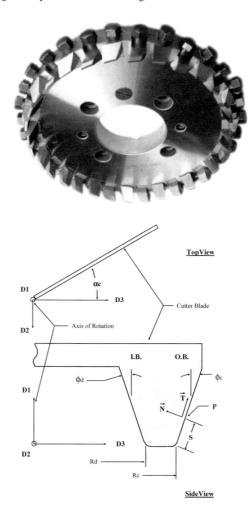

Fig. 3.3 Generated pinion

The solution of Eq. (3.2) is a series of contact points between cutter blade and
work piece describing a line along the path of the cutter blade. The envelope of a

series of such lines yields the generated pinion shown in Fig. 3.3. Equation (3.2) is solved in real time.

3.2.1 Spiral-Bevel Gears

The TFG includes work and tool adjustments and movements found in gear cutting machines. In CnC controlled machines, machine settings can be continuously altered during generation, thus allowing for significant improvements in the kinematics of gear sets.

Figure 3.1 represents the most general case in the simulation of cutting processes, and is therefore the basis for the Unified Model of the TFG. The implicit equation of the general tooth surface is:

$$\vec{X} = \vec{D}[\alpha_c]_1[\tau]_3[J]_1[R][L_{1m}]_1[\Gamma]_2[P][\alpha_3]_3[R_c]_3 \qquad (3.3)$$

$$\vec{D} = \begin{bmatrix} S\cos(\varphi) \\ 0 \\ (R \pm S\sin(\varphi)) \end{bmatrix} \qquad (3.4)$$

Vector $\vec{D}$ in Eq. (3.4) is the position of a point S along a Face Mill cutter blade—Fig. 3.2. In Eq. (3.3), vector $\vec{D}$ is rotated by cutter phase angle α_c, tilt angle τ and swivel angle J, translated to the origin of the machine by vector R, rotated by cradle angle L_{1m} and root angle Γ, translated to the origin of the work piece by vector P, rotated by roll angle α_3 and finally rotated by R_c, the Face Hobbing timing ratio (when applicable). Figure 3.2 shows the general definition of a Face Mill cutter blade.

Similarly, Eq. (3.5) defines vector $\vec{N}$, the unit vector normal to the cutter blade at point S, and Eq. (3.6) gives the transformations required between the reference frames of the cutter blade and the work piece, where translations R and P have been omitted.

$$\vec{N} = \begin{bmatrix} \sin(\varphi) \\ 0 \\ \mp\cos(\varphi) \end{bmatrix} \qquad (3.5)$$

$$\vec{N}_x = \vec{N}[\alpha_c]_1[\tau]_3[J]_1[L_{1m}]_1[\Gamma]_2[\alpha_3]_3[R_c]_3 \qquad (3.6)$$

In Eqs. (3.3) and (3.6), rotations and translations can be expanded in Taylor series to allow higher order manufacturing flexibility on CnC controlled machines.

3.2.2 Straight-Bevel Gears—2-Tool Generator

For straight-bevel gears cut in a 2-tool generator, the terms involving cutter tilt τ, cutter swivel J and Face Hobbing rotation R_c are dropped from Eqs. (3.3) and (3.6), and cutter phase angle α_c describes the translation of the tool along the face width. Figure 3.4 shows a straight-bevel gear set such as cut on a 2-tool generator. Equations (3.3) and (3.6) therefore become:

$$\vec{X} = \vec{D}[\alpha_c][R][L_1]_1[\Gamma]_2[P][\alpha_3]_3 \tag{3.3a}$$

$$\vec{N}_x = \vec{N}[L_1]_1[\Gamma]_2[\alpha_3]_3 \tag{3.6a}$$

While the cutter travel α_c is basically a straight line, it can also follow a curve such as to produce teeth with lengthwise crowning, a feature that guarantees localized contact on the tooth to avoid Toe and Heel edge contact.

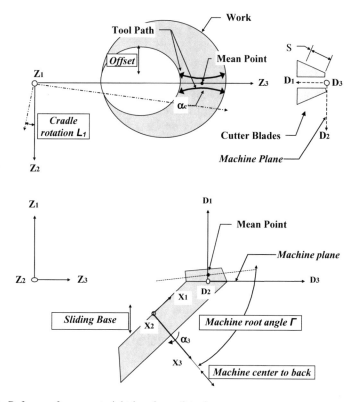

Fig. 3.4 Reference frames—straight-bevel gear 2-tool generator

Fig. 3.5 Straight-bevel gear
set—cut on a 2-tool
generator

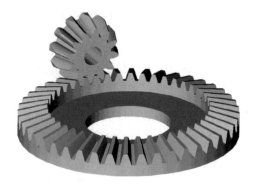

The 2-tool generators, which are mechanical machines, have widely been replaced by the Coniflex generators, also mechanical, and more recently by free form machines such as the Gleason Phoenix or 5Axis CnC machines (Fig. 3.5).

3.2.3 Coniflex™ Straight-Bevel Gears

Coniflex™ gears were originally cut on mechanical machines using 2 interlocking cutters, Fig. 3.6, each cutting one tooth flank of a gap. The cutters are located axially at the Cone Distance from the machine center; tilt is fixed, and the cutters can be offset; the work can be moved along the Sliding Base, but any Machine center to back changes can be accomplished only on CnC machines. The terms involving cutter swivel J and Face Hobbing rotation R_c are therefore dropped from Eqs. (3.3) and (3.6), which become:

$$\vec{X} = \vec{D}[\alpha_c]_1[\tau]_3[R][L_{1m}]_1[\varGamma]_2[P][\alpha_3]_3 \tag{3.3b}$$

$$\vec{N}_x = \vec{N}[\alpha_c]_1[\tau]_3[L_{1m}]_1[\varGamma]_2[\alpha_3]_3 \tag{3.6b}$$

While Coniflex machines are still widely found, they are being progressively replaced by free form machines such as the Gleason Phoenix or 5Axis CnC machines.

As can be assessed from Fig. 3.6, the cutter diameter is linked to the face width of the work because of the induced circular root line and therefore a practical limit in module is quickly reached when cutting Coniflex gears on 5Axis CnC machines because of the relatively low stiffness of the tool spindle (Fig. 3.7).

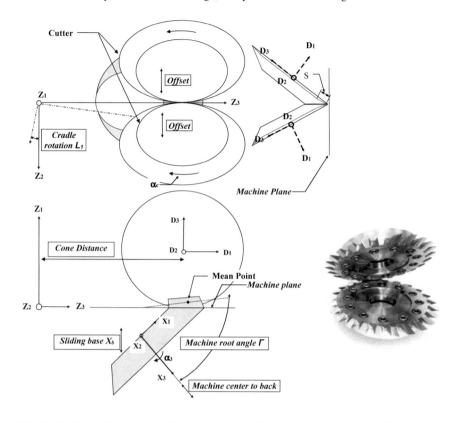

Fig. 3.6 Reference frames—coniflex bevel gear on coniflex generator (actual cutter picture courtesy of The Gleason Works, Rochester, NY, USA)

Fig. 3.7 Coniflex bevel gear set

3.2.4 Cylindrical Gears

Cylindrical gears are based on a rack moving across the face width while meshing with the generated surfaces, Fig. 3.8. Therefore, the terms involving cutter tilt τ, cutter swivel J and Face Hobbing rotation R_c are dropped from Eqs. (3.3) and (3.6), cutter phase angle α_c describes the translation of the rack along the face width, and rotation L_1 of the cradle becomes a translation of the rack as it meshes with the work while the Sliding Base becomes the Profile Shift.

A helical gear can be simulated by moving the rack along α_c pivoted at the desired helix angle ψ. Equations (3.3) and (3.6) now become:

$$\vec{X} = \vec{D}\,[\alpha_c][\psi]_1[R][L_1][P][\alpha_3]_3 \tag{3.3c}$$

$$\vec{N}_x = \vec{N}\,[\psi]_1[\alpha_3]_3 \tag{3.6c}$$

In the above paragraphs, examples were given of the use of the HyGEARS TFG for several common gear types, but the method is not limited to these, as Fig. 3.10

Fig. 3.8 Reference frames—spur/helical gear set

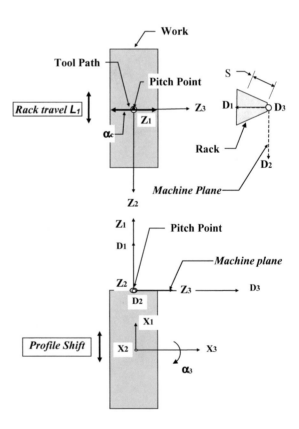

Fig. 3.9 Helical gear set with parallel axes

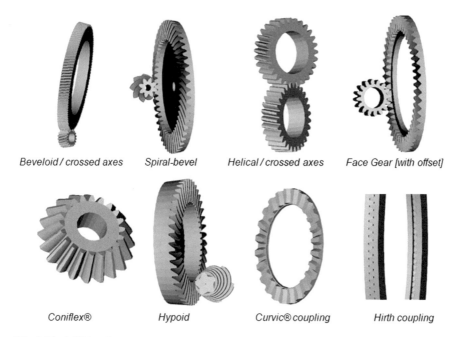

Fig. 3.10 Additional gear types

shows where several other gear types, modeled using the same method, are displayed. The method is therefore truly general and allows modelling any gear manufacturing process to obtain the exact tooth flank topography (Fig. 3.9).

Fig. 3.11 Comparison with
CAGE—8 × 37 face milled
spiral-bevel pinion

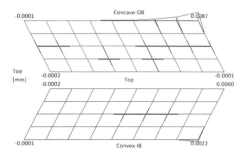

3.3   Calibration of the Tooth Flank Generator

The Tooth Flank Generator (TFG) aims to reproduce digitally the same tooth surfaces
as those cut on an actual machine.

In order to calibrate the TFG, its output is compared to data universally accepted
in the industry. For this purpose, two sources stand out, at least for spiral-bevel gears:
The Gleason Works and Klingelnberg GmbH.

3.3.1   Comparison with Gleason's CAGE

The output of the TFG described in the previous section is compared to the CMM
Nominal obtained from Gleason's CAGE software.

Using the Basic machine settings for an 8 × 39 Face Milled spiral-bevel pinion,
Fig. 3.11 below shows the differences between the CAGE simulation (red lines for
the OB, blue lines for the IB) and the simulation based on the presented TFG (black
lines). Except for two points at heel at the bottom of the OB and IB flanks of the
tooth (upper and lower right corners), the TFG output is identical to that of CAGE
within 0.0002 mm.

3.3.2   Comparison with Klingelnberg's KIMoS

Using the same Basic machine settings as for the above 8 × 39 Face Milled spiral-
bevel pinion, Fig. 3.12 below shows the differences between the KIMoS simulation
(red lines for the OB, blue lines for the IB) and the simulation based on the presented
TFG (black lines). The deviations noted with Gleason's CAGE near the fillet at
heel have disappeared, and the TFG output is identical to KIMoS' output within
0.0002 mm.

Fig. 3.12 Comparison with KIMoS—8 × 37 face milled spiral-bevel pinion

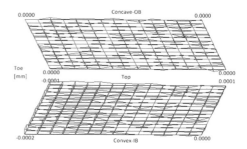

3.4 Tooth Contact Analysis—TCA

Tooth Contact Analysis, or TCA, is a technique where the simulated tooth surfaces are analyzed to find where they will be in contact as meshing proceeds, obtain the quality of rotation from the Transmission Error, and then establish the extent of the contact pattern when a marking compound of a given thickness is used.

3.4.1 General Approach to TCA

Contacting tooth surfaces can be classified in two basic categories: line and point contact. In both cases, for contact to occur, the same fundamental condition applies: in a common reference frame named Z, at any contact point, the coordinates on the pinion and gear teeth must be the same, and the normal unit vectors must be collinear and opposed in direction:

$$\vec{Z}_P = \vec{Z}_G \quad \vec{N}_P = \vec{N}_G \tag{3.7}$$

Given that the pinion and gear tooth surfaces depend on 2 independent variables each, and that their angular position must be determined to obtain a contact point, Eq. (3.7) results in a system of 6 variables in 5 independent equations. Thus, at least 1 variable is to be fixed to calculate a contact point.

One issue with this condition is that in line contact tooth surfaces, if, say, the pinion or gear angular position is fixed, there are an infinite number of solutions to Eq. (3.7) along the tooth face. Therefore, an alternate approach must be used—the Ease-Off Surface.

3.4.2 Ease-off Surface and Transmission Error

The Ease-Off Surface [2] is a general concept applicable to all gear types. It displays deviations in conjugacy between the meshing pinion and gear—Fig. 3.13.

Fig. 3.13 Ease-off surface
of a spiral-bevel gear set

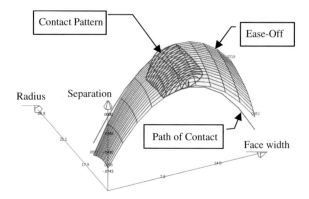

The Ease-Off is obtained by scanning the pinion and gear tooth surfaces for potential points of contact as per Eq. (3.7). At each contact point, the separation between the pinion and gear tooth surfaces is calculated and recorded.

A true conjugate point of contact exists where:

(i) tooth to tooth separation is minimum on the Ease-Off and
(ii) all components of the pinion and gear unit normal vectors are equal.

A series of true conjugate contact points yields the Path of Contact (PoC). Tooth surface measurements can be introduced in this approach to analyze existing gear sets.

The Transmission Error (TE) is the expression of the difference between the actual, or calculated, and theoretical angular positions of the gear member, as per Eq. (3.8):

$$\delta\varphi_3 = \varphi_3 - \theta_3 m_g \tag{3.8}$$

where $\delta\varphi_3$ is the TE, φ_3 is the calculated angular position of the gear, $\theta_3 m_g$ is the theoretical angular position of the gear, equal to the product of the pinion angular position by the gear ratio. A negative result to Eq. (3.8) means that the gear is late relative to the pinion, and is the desired result in order to prevent premature contact entry and late contact exit.

TE is typically displayed for 3 consecutive tooth pairs, as shown in Fig. 3.14 where each meshing tooth pair uses a different color. It is quite clear that:

• the center part of the TE curve is where contact occurs when there is no load;
• the right end of the pink curve (tooth pair 1) overlaps with the left end of the orange curve (tooth pair 3) and therefore contact ratio is expected to be between 2 and 3;
• the left and right ends of each curve are at the same level, indicating a centered contact pattern.

If the gear tooth flank is coated with a light marking compound and the gear pair is operated at slow speed under light torque, the succession of meshing contact points

Fig. 3.14 TE curves—3
consecutive tooth pairs

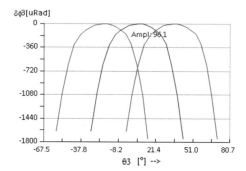

leaves a trace indicating which part of the tooth flank is expected to come in contact. This trace is called the contact pattern (CP).

The CP is calculated as follows:

- the Ease-Off surface is calculated; from this, the tooth profile separation is obtained for each point along the PoC;
- each contact point of the PoC is checked to test whether tooth profile separation is larger or smaller than the requested marking compound thickness. If tooth separation is smaller or equal, the tooth surfaces are scanned to find the extent of the contacting area.

From the above, one clearly can conclude that the Ease-Off, TE and CP are all linked.

This is shown in Fig. 3.15 where the PoC (red line), the CP (blue patch), and the TE curves for the gear Convex and Concave tooth flanks are lined up in 2 parallel columns. The green line at the bottom of the gear tooth flanks, top part of Fig. 3.15, is the fillet line, i.e. the boundary between active profile and fillet area.

3.5 Loaded Tooth Contact Analysis—LTCA

Loaded Tooth Contact Analysis, or LTCA, is the kinematic analysis of meshing tooth pairs under load [3, 4]. In LTCA, the tooth bending, shearing and contact deformations, tooth base rotation and gear body shearing can be used to calculate mesh stiffness along the PoC.

Equation (3.9), solved in the transverse plane, states that the rotation of each contacting tooth pair under load must be the same, while Eq. (3.10) states that the sum of the individual torque shares must equal the total applied torque. Equation (3.9) yields n-1 scalar equations, where n is the number of tooth pairs considered, and Eq. (3.10) yields the nth equation needed to iteratively solve the system.

Tooth bending and shearing may be obtained in various ways, such as a beam formula [4], the Finite Element Analysis, or the Finite Strips [5].

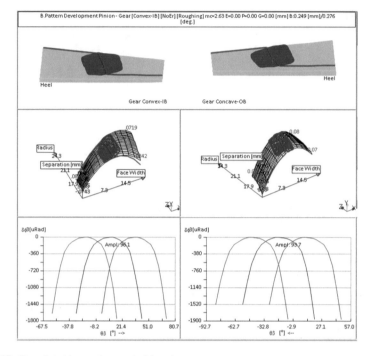

Fig. 3.15 Complete kinematics—spiral-bevel gear set

$$\left[\frac{F_j/K_j + \delta H_j + \delta W_j + \delta R_j + \delta I_j}{R_j}\right]_{j=-2,-1,+1,+2} = \left[\frac{F_i/K_i + \delta H_i + \delta W_i + \delta R_i + \delta I_i}{R_i}\right] \tag{3.9}$$

$$T_{tot} = \sum_{j=-2}^{j=+2} T_j \tag{3.10}$$

where i is the index of the main tooth pair, j is the index of a tooth pair on either side of the main tooth pair, F is the load applied on each meshing tooth pair, K is the tooth bending stiffness, δH is the Hertz contact deformation, δW is the displacement of a contact point caused by web shearing, δR is the displacement of a contact point caused by tooth base rotation, δI is the initial tooth separation, R_i is the radius of the point of contact and T_j is the torque share of each meshing tooth pair (Figs 3.16 and 3.17).

Contact deformations are obtained from the classical Hertz theory or from contact elements. While tooth bending and shearing, tooth base rotation and gear body shearing behave linearly with load, and thus can be computed only once, contact deformations must be recalculated at each iteration step because of their non-linearity.

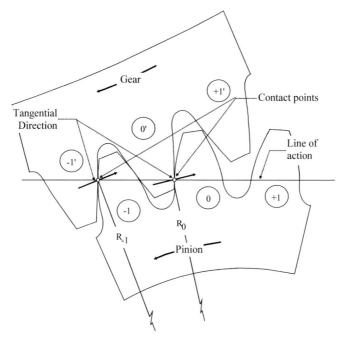

Fig. 3.16 LTCA tooth numbering

Fig. 3.17 Tooth load
sharing components

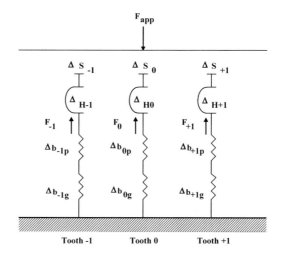

3.6   Closed Loop

The Closed Loop, an essential part of any gear manufacturing software, calculates changes in machine settings to remove surface errors caused by machine and tool, and match the manufactured tooth surface to the designed tooth surface [6–9].

The following averaged surface errors are generally considered adequate to describe the quality of a tooth flank and are used in the Closed Loop. Average surface errors are calculated as follows:

- pressure angle error : $\Phi = \dfrac{\sum_{col=1}^{j} \dfrac{\left[\sum_{row=1}^{i} \frac{\varepsilon_{i,j}-\varepsilon_{1,j}}{y_{i,j}-y_{1,j}}\right]}{i}}{j}$ (3.11)

- spiral angle error : $\Psi = \dfrac{\sum_{row=1}^{i} \dfrac{\left[\sum_{col=1}^{j} \frac{\varepsilon_{i,j}-\varepsilon_{i,1}}{x_{i,j}-x_{i,1}}\right]}{j}}{i}$ (3.12)

- crowning error : $\Xi = \dfrac{\sum_{row=1}^{i} \frac{(2\varepsilon_{i,mid}-(\varepsilon_{i,1}+\varepsilon_{i,j}))}{2}}{i}$ (3.13)

- bias error : $\zeta = \Phi_1 - \Phi_j$ (3.14)

- profile curvature error : $\xi = \dfrac{\sum_{col=1}^{j} \frac{(2\varepsilon_{mid,j}-(\varepsilon_{1,j}+\varepsilon_{i,j}))}{2}}{j}$ (3.15)

where:
- i is the index of row data along the face width,
- j is the index of column data depth wise,
- *mid* is the index of the mid-column or mid-row data,
- $\varepsilon_{i,j}$ is the error value at point ij of the measurement grid,
- $x_{i,j}$ is the distance between points along the face width,
- $y_{i,j}$ is the distance between measurement points depth wise.

Surface Matching, the algorithm used in Closed Loop, is based on the response of the error surface, i.e. the difference between the simulated and measured tooth surfaces, Fig. 3.18, to changes in selected machine settings. First order coefficients of sensitivity are obtained by changing machine settings, recalculating the error surface and solving Eqs. (3.11) to (3.15).

In the Surface Matching algorithm, a combination of machine settings is sought such that the theoretical surface matches the measured surface. To do so, the following objective functions are satisfied:

$$\Phi(m_i) - T_1 \leq L_1$$ (3.16a)

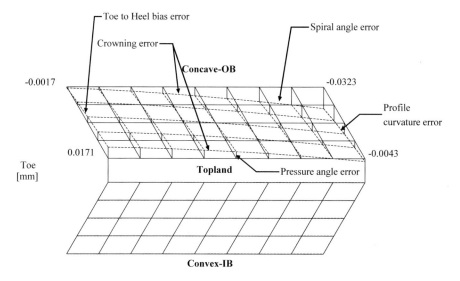

Fig. 3.18 Typical topography errors

$$\Psi(m_i) - T_2 \leq L_2 \tag{3.16b}$$

$$\Xi(m_i) - T_3 \leq L_3 \tag{3.16c}$$

$$\zeta(m_i) - T_4 \leq L_4 \tag{3.16d}$$

$$\xi(m_i) - T_5 \leq L_5 \tag{3.16e}$$

where:

- m_i are the considered machine settings,
- Φ and Ψ are the averaged pressure and spiral angle errors,
- Ξ and ζ are the lengthwise crowning and bias error values,
- ξ is the profile curvature error,
- T_i are target surface deviations,
- L_i are the tolerances within which the objective functions can be considered satisfied.

A Newton-Raphson based solution is used to solve the above functions.

3.7 TCA Optimization and Numerical Contact Pattern Development

Bevel gears operate in widely different environments. It is therefore fundamental for the designer to be able to tailor the TCA and CP over the expected operating range. HyGEARS offers an effective interface to allow the designer to develop the kinematics, i.e. the TE and CP.

In order to modify the kinematics, HyGEARS uses the Surface Matching algorithm outlined in Sect. 3.6 above, where the Ease-Off is modified in order to reflect the user's demands.

This means that 1st order coefficients of influence due to changes in machine settings are required to control the nth order solution. To illustrate how machine settings affect the Ease-Off, the following paragraphs show side by side the CP, TE, ease Off and Error Surface of a $Z = 19$ Duplex Helical pinion.

To control changes on the TCA, the following kinematic metrics are used:

- Bias: the angle made between the profile section of the PoC, red line below, and the perpendicular to the pitch cone;
- Horizontal Posn: axial location of the Mean Point;
- Vertical Posn: vertical location of the Mean Point;
- TE Amplitude: TE at the Transfer Point (Fig. 3.19).

Figure 3.20 below shows the reference kinematics and Error Surface of the pinion tooth compared to itself. The pinion's Concave tooth flank is considered here as it is the usual driving flank.

In Fig. 3.20, top right corner, the CP is seen to be centered on the tooth flank; TE is balanced between contact entry and exit; Bias, in the profile section of the PoC, is around $10°$; in the top left corner, the tooth flank is compared to itself, which means that the Error Surface (ES) is nil everywhere.

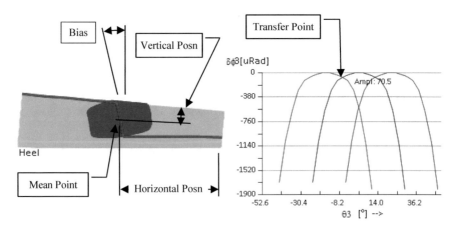

Fig. 3.19 Metrics for TCA development

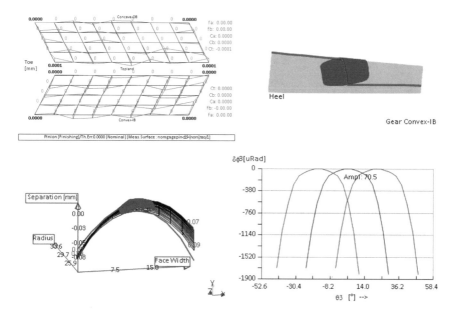

Fig. 3.20 Reference kinematics for TCA development

3.7.1 Change in Work Machine Root Angle

In Fig. 3.21, the pinion Machine Root Angle is increased by 0.5°; any change in Root Angle must be compensated by an appropriate change in Sliding Base to maintain tooth depth. In the top left corner, we can see that the Error Surface shows a combination of spiral and pressure angle errors, plus some warping; accordingly, top right corner, the CP moves towards Heel and fillet on the gear and the Bias of the PoC appears to have become negative; the TE at the Transfer Point, lower right corner, gets deeper and the shape is no more a perfect parabola but rather shows a sharp slope; finally, the Ease-Off, lower left corner, shows the CP and PoC moving towards Heel.

3.7.2 Change in Cutter Radial Distance

In Fig. 3.22, the Radial Distance is decreased by 0.2 mm. Radial Distance controls the spiral angle. In the top left corner, we can see that the Error Surface shows spiral angle error; accordingly, top right corner, the CP moves towards Heel on the gear; the TE at the Transfer Point barely changes; finally, the Ease-Off, lower left corner, shows the CP and PoC moving towards Heel.

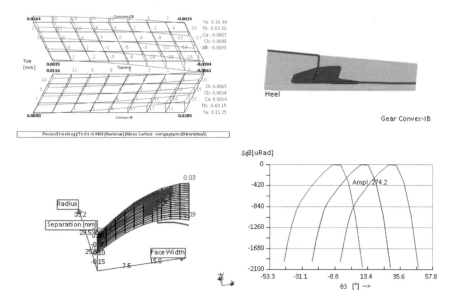

Fig. 3.21 Effects of 0.5° change in machine root angle

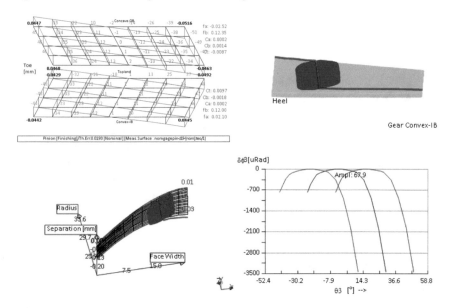

Fig. 3.22 Effects of −0.2 mm change in radial distance

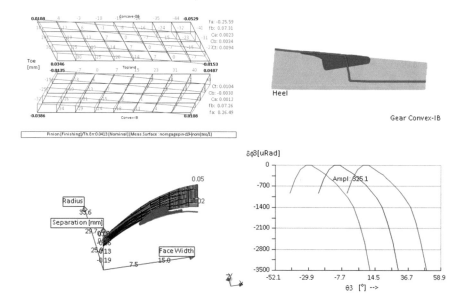

Fig. 3.23 Effects of 0.01 change in ratio of roll

3.7.3 Change in Ratio of Roll

In Fig. 3.23, the Ratio of Roll is increased from 2.083325 to 2.093325; ratio of Roll primarily controls the pressure angle, and also shows side effects on the spiral angle and ES warping, top left corner. Accordingly, top right corner, the CP sharply moves towards tip and Heel on the gear; as could be expected, the TE at the Transfer Point increases dramatically, which is reflected in the Ease-Off, lower left corner, where the CP and PoC have moved towards Heel and pinion fillet.

3.7.4 Change in Work Offset

In Fig. 3.24, work Offset is changed by -0.2 mm, from -0.3563 to -0.5563 mm; as can be seen, an Offset change induces pressure and spiral angle errors, some warping, and a slight measure of profile curvature, top left corner. Hence, top right corner, the CP moves towards tip and Toe on the gear; as could be expected, the TE at the Transfer Point increases dramatically, which is reflected in the Ease-Off, lower left corner, where the CP and PoC have moved towards Toe.

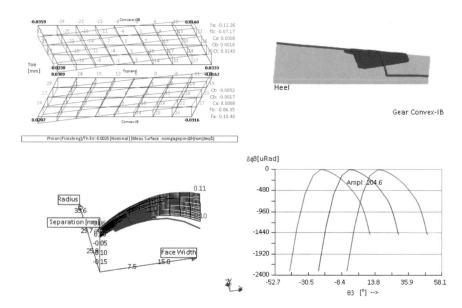

Fig. 3.24 Effects of 0.2 mm change in work offset

3.7.5 Change in Work Mounting Distance

In Fig. 3.25, work Mounting Distance (MD) is changed by −0.2 mm, from −0.2523

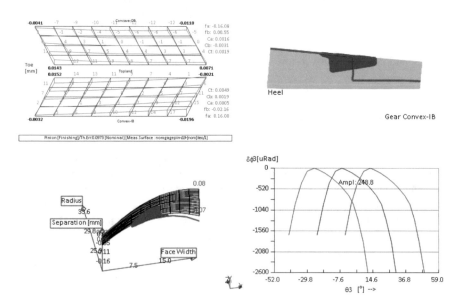

Fig. 3.25 Effects of 0.2 mm change in MCTB

to -0.4523 mm; clearly, a small change in MD, which must be compensated by an appropriate change in Sliding Base to maintain tooth depth, induces essentially pressure angle errors, some warping, and a slight measure of profile curvature, top left corner. Hence, top right corner, the CP moves towards tip but remains centered on the tooth flank; as could be expected, the TE at the Transfer Point increases dramatically, which is reflected in the Ease-Off, lower left corner.

3.7.6 Change in Cutter Tilt

In Fig. 3.26, cutter Tilt is increased by $0.2°$; in spiral bevel gears, cutter Tilt is used to control the pressure angle of the tooth flank, and the lengthwise crowning; cutter Tilt is supported by cutter Swivel which orients the tilt. Therefore, a change in cutter Tilt, which must be compensated by an appropriate change in Sliding Base to maintain tooth depth, induces principally pressure angle errors and some lengthwise crowning, but also some spiral angle and warping errors, top left corner. Hence, top right corner, the CP moves towards fillet but remains pretty much centered on the tooth flank; as could be expected, the TE at the Transfer Point increases dramatically, which is reflected in the Ease-Off, lower left corner.

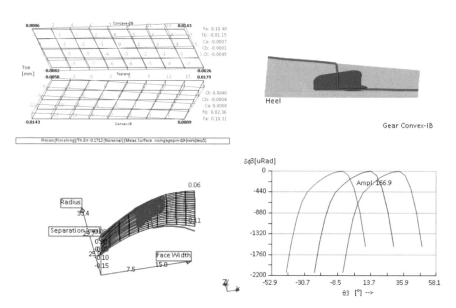

Fig. 3.26 Effects of $0.2°$ change in cutter tilt

3.7.7 Change in Helical Motion

In Fig. 3.27, Helical Motion is increased by 0.2 mm/rad; in Duplex Helical spiral-bevel gears, Helical Motion moves the work piece as generation proceeds, and is formulated as:

$$
\begin{aligned}
X_{bm} = X_b &+ B_1(C_r - \alpha_3 R_r) + B_2(C_r - \alpha_3 R_r)^2 \\
&+ B_3(C_r - \alpha_3 R_r)^3 + B_4(C_r - \alpha_3 R_r)^4 \\
&+ B_5(C_r - \alpha_3 R_r)^5 + B_6(C_r - \alpha_3 R_r)^6
\end{aligned}
\tag{3.17}
$$

Helical Motion is used as an additional DoF to control the shape of the CP and PoC; since it changes the slope followed by the cutter blades, Sliding Base and Machine Root Angle need be adjusted to control tooth depth when Helical Motion changes. Helical Motion induces primarily pressure and spiral angle errors and some warping errors, top left corner. Hence, top right corner, the CP moves towards tip but remains centered on the tooth flank; as could be expected, the TE at the Transfer Point increases somewhat, which is reflected in the Ease-Off, lower left corner.

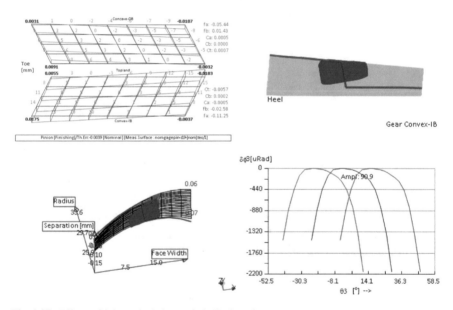

Fig. 3.27 Effects of 0.2 mm/rad change in helical motion

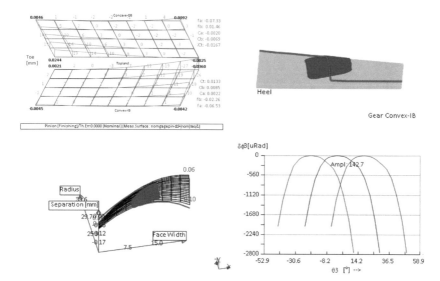

Fig. 3.28 Effects of 0.02 change in modified roll 2C

3.7.8 Change in Modified Roll 2C

In Fig. 3.28, 2nd order Modified Roll (MR) coefficient 2C is changed from 0 to 0.02; MR is used to change the work to cradle Ratio of Roll during generation; it is formulated as:

$$L_{1m} = \alpha_3 R_r + 2C(C_r - \alpha_3 R_r)^2 - 6D(C_r - \alpha_3 R_r)^3$$
$$+ 24D(C_r - \alpha_3 R_r)^4 - 120E(C_r - \alpha_3 R_r)^5$$
$$+ 720F(C_r - \alpha_3 R_r)^6 \tag{3.18}$$

MR therefore offers some measure of lengthwise crowning, pressure angle error, and a warping side effect, as can be seen, top left corner. Hence, top right corner, the CP moves towards tip but remains centered on the tooth flank; as could be expected, the TE at the Transfer Point changes somewhat, which is reflected in the Ease-Off, lower left corner, where larger crowning is visible.

3.7.9 Change in Modified Roll 6D

In Fig. 3.29, 3rd order Modified Roll (MR) coefficient 6D is changed from 0 to 0.2; again, it offers some measure of lengthwise crowning, and a warping side effect, as can be seen, top left corner. Interestingly, while the effect remains in the same direction on the Concave flank, upper half—top left corner, it is inverted from 2C for

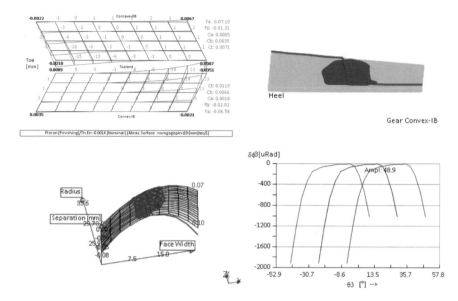

Fig. 3.29 Effects of 0.2 change in modified roll 6D

the Convex flank, lower half—top left corner. The CP slightly moves towards fillet but remains centered on the tooth flank; as could be expected, the TE at the Transfer Point changes both in amplitude and shape, which is reflected in the Ease-Off, lower left corner.

Modified Roll is currently defined up to the 6th order in HyGEARS; given the limited space in this chapter, we will stop at the 3rd order, i.e. coefficient 6D, but

3.7.10 Summary

The above paragraphs aimed at illustrating how machine settings can affect the kinematics of spiral-bevel gears, both what is immediately visible such as the CP, and also what is obtained by computation such as the TE and Ease-Off.

Therefore, it is a straightforward affair to adapt the Surface Matching algorithm to control the Ease-Off when developing the kinematics of a gear set.

While this section focuses on spiral-bevel gears, the same applies to any type of gear.

3.7.11 Sample Contact Pattern Development

Table 3.1 gives the machine settings used for the above example of a 19 × 35 spiral-bevel gear set, while Fig. 3.30 shows the associated kinematics.

Table 3.1 Initial machine settings

Pinion [finishing] cutter specifications	(O.B.)		(I.B.)
Average diameter		4.5000	
Blade angle	14.5000		25.5000
Blade edge radius		0.0250	
Point width		0.0703	
TopRem letter	No		
TopRem length	0.0000		0.0000
TopRem angle	0.00.00		0.00.00
Cutter gaging	0.0000		0.0000
Pinion [finishing]: duplex helical			
Machine settings—phoenix			
Mean radius		28.2624	
Radial distance		53.3900	
Cutter tilt		1.1545	
Swivel angle		25.6907	
Blank offset		−0.3563	
Machine root angle		23.4349	
Machine center to back		−0.2523	
Sliding base		2.0592	
Rate of roll		2.08333	
Cradle angle		62.6394	
Helical motion [mm]/Rad			
…1st	4.29888		
Gear [finishing] cutter specifications	(I.B.)		(O.B.)
Average diameter		4.5000	
Blade angle	25.5000		14.5000
Blade edge radius		0.0250	
Point width		0.0628	
TopRem depth	0.0000		0.0000
TopRem radius	0.0000		0.0000
Cutter gaging	0.0000		0.0000
Gear [finishing]: spread blade Machine settings—phoenix			
Mean radius		52.0625	
Radial distance		53.6173	
Cutter tilt		3.7879	
Swivel angle		198.3400	

(continued)

Table 3.1 (continued)

Pinion [finishing] cutter specifications	(O.B.)		(I.B.)
Blank offset		0.0000	
Machine root angle		57.8101	
Machine center to back		0.0632	
Sliding base		3.3288	
Rate of roll		1.13581	
Cradle angle		60.4015	

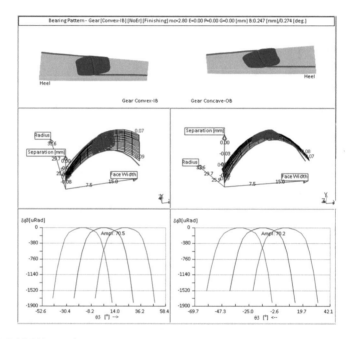

Fig. 3.30 Initial kinematics

3.7.11.1   Move Contact Pattern to 40% of Face Width

Under load, it is usual for the CP to move towards Heel; hence, designers will often locate the CP somewhat more towards Toe to guarantee that under load, when it moves, the CP remains centered. Table 3.2 lists the modified pinion machine settings with the CP moved to 40% of face width on both flanks. Only the pinion machine settings were modified. Figure 3.31 shows the resultant kinematics.

Table 3.2 Modified pinion machine settings

Pinion [finishing] cutter specifications	(O.B.)		(I.B.)
Average diameter		4.5000	
Blade angle	14.5000		25.5000
Blade edge radius		0.0250	
Point width		0.0703	
TopRem letter	No		
TopRem length	0.0000		0.0000
TopRem angle	0.00.00		0.00.00
Cutter gaging	0.0000		0.0000
Pinion [finishing]: duplex helical			
Machine settings—phoenix			
Mean radius		28.2624	
Radial distance		53.3997	
Cutter tilt		1.3518	
Swivel angle		25.7010	
Blank offset		−0.3563	
Machine root angle		23.0883	
Machine center to back		−0.2419	
Sliding base		2.1967	
Rate of roll		2.08287	
Cradle angle		62.6497	
Helical motion [mm]/Rad			
…1st		4.29888	

3.7.11.2   Increase PoC Bias to 50°

Increasing the Bias of the PoC extends the duration of the profile contact, ensuring smoother motion transfer between tooth pairs and better load sharing. In this case, the initial Bias of ~10° is increased to 50°. Table 3.3 lists the modified pinion machine settings. Figure 3.32 shows the resultant kinematics.

3.7.11.3   Increase TE to 150 μRad

When the gear set of Sect. 3.7.11.2 is analyzed under a 200 Nm pinion torque, Fig. 3.33, we see that the TE under load is around 140 μRad, red curve, which translates as 7.3 μm; contact stress is 1.7 GPa.

It is desired to reduce the TE under load in order to reduce the dynamic excitation. The un-loaded TE is therefore increased from the initial 77 μRad to 150 μRad—which brings down the Transfer Point—with the net effect that TE under load

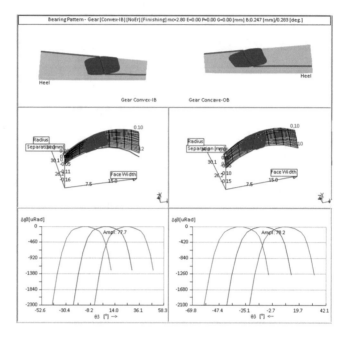

Fig. 3.31 Kinematics with CP at 40% of Face Width

becomes 89 μRad, i.e. 4.6 μm, rather than the initial 7.3 μm, and a quieter gear set is expected under load.

The resulting un-loaded kinematics appear in Fig. 3.34 and the LTCA results appear in Fig. 3.35. We note that the contact stresses have barely changed in the process.

Table 3.4 lists the resulting pinion machine settings; MR coefficients up to 5th order are clearly visible.

3.8 Reverse Engineering—RE

Reverse Engineering (RE) of gears is a technique where the parametric definition of a known or unknown part is obtained by

(a) creating the part model from a file defining the geometry, data on drawings, a table of parameters, or in the worst case, taking basic dimensions such as Z, Face Angle, approximate spiral-angle, OD, etc. from an actual part;
(b) creating a CMM inspection file;
(c) measuring the part using the CMM inspection file;
(d) applying the Surface Matching algorithm of the above section "3.6 Closed Loop".

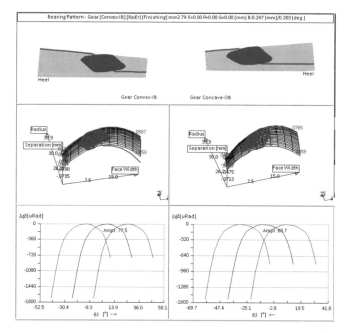

Fig. 3.32 Kinematics with 50° bias of the PoC

Fig. 3.33 LTCA—200 Nm
pinion torque—50° bias

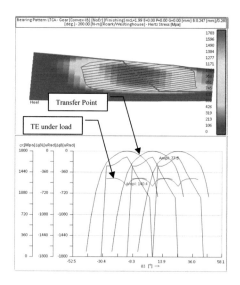

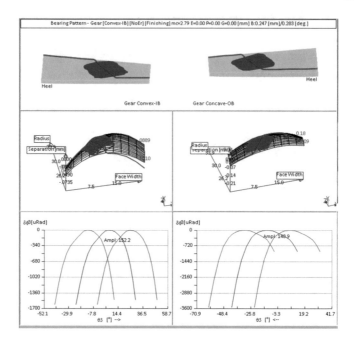

Fig. 3.34 Kinematics with 150 μRad TE

Fig. 3.35 LTCA—200 Nm
pinion torque—150 μRad
TE

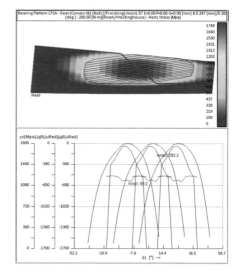

Table 3.3 Modified pinion machine settings

Pinion [finishing] cutter specifications	(O.B.)		(I.B.)
Average diameter		4.5000	
Blade angle	14.5000		25.5000
Blade edge radius		0.0250	
Point width		0.0703	
TopRem letter	No		
TopRem length	0.0000		0.0000
TopRem angle	0.00.00		0.00.00
Cutter gaging	0.0000		0.0000
Pinion [finishing]: duplex helical			
Machine settings—phoenix			
Mean radius		28.2624	
Radial distance		53.3909	
Cutter tilt		0.3491	
Swivel angle		205.6642	
Blank offset		−0.3563	
Machine root angle		24.1312	
Machine center to back		−0.1129	
Sliding base		2.7702	
Rate of roll		2.08223	
Cradle angle		62.6129	
Helical motion [mm]/Rad			
…1st		2.90728	

When using the Surface Matching algorithm for RE, the changes in machine settings are added rather than subtracted as in Closed Loop. Once RE has been applied, it is possible to use the parametric model for the usual applications: TE, CP development, LTCA, manufacturing and Closed Loop, etc.

Typical situations where RE is required:

1. a spiral-bevel gear set needs to be replicated on standard Gleason machines; unfortunately, the Summary defining the blank, cutter and machine settings is lost;
2. a large number of Powder Metallurgy (PM) bevel gears has been outsourced; upon reception, it is found that parts deviate so much from target that when meshed with the pinion cut as designed, bad contact patterns are observed and noise is emitted; obtaining a new series of PM gears would take at least 2 months and therefore an alternate solution is required;
3. a differential straight-bevel gear set is forged by a supplier; it is desired to analyze the gear set in order to evaluate in situ bending and contact stresses.

Table 3.4 Modified pinion machine settings

Pinion [finishing] cutter specifications	(O.B.)		(I.B.)
Average diameter		4.5000	
Blade angle	14.5000		25.5000
Blade edge radius		0.0250	
Point width		0.0703	
TopRem letter	No		
TopRem length	0.0000		0.0000
TopRem angle	0.00.00		0.00.00
Cutter gaging	0.0000		0.0000
Pinion [finishing]: duplex helical			
Machine settings—phoenix			
Mean radius		28.2624	
Radial distance		53.3819	
Cutter tilt		0.3734	
Swivel angle		205.6682	
Blank offset		−0.3563	
Machine root angle		24.1504	
Machine center to back		−0.1128	
Sliding base		2.7688	
Rate of roll		2.07935	
Cradle angle		62.6168	
Helical motion [mm]/Rad			
…1st		2.90728	
MRoll 2C		−0.00200	
MRoll 6D		−0.31050	
MRoll 24E		−0.59400	
MRoll 120F		1.35000	

3.8.1 Replicate an Unknown Existing Part

An existing spiral-bevel pinion is to be replicated. The original machine settings are unknown, but the pinion and gear drawings are available, and the Duplex Helical process is specified for the pinion. Therefore, an initial design is created with HyGEARS from data given on the drawings. The CMM Nominal inspection files are generated and used to measure the pinion. The pinion CMM result based on the initial design is shown in Fig. 3.36, and it appears that the 6" original cutter diameter assumption was wrong.

Fig. 3.36 1st pinion
CMM—based on initial
design

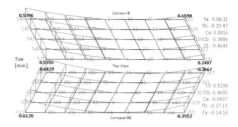

Fig. 3.37 2nd pinion
CMM—based on 2nd design

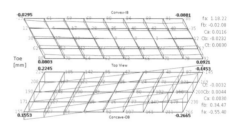

Using the RE algorithm, it is found that the cutter diameter is rather 7.5". A new spiral-bevel gear set is therefore created using 7.5" cutters, the pinion CMM inspection file is re-created, and the new measurement appears in Fig. 3.37; cutter diameter is now correct, but pressure angle, spiral angle and thickness errors are visible.

Applying RE to the result of Fig. 3.37 reveals that the OB-IB blade angles should be closer to 16.5−23.5° rather than the initially assumed 18−24.5°. A third design is therefore made using OB-IB blade angles of 16.5−23.5°; the pinion CMM inspection file is re-created, and measurement appears in Fig. 3.38. It is now obvious that the correct definition of the pinion has been obtained. The final machine settings and cutter definition can then be used to cut the part on any Duplex Helical capable spiral-bevel generator, either mechanical, free form or 5Axis CnC.

Fig. 3.38 3rd pinion
CMM—based on 3rd design

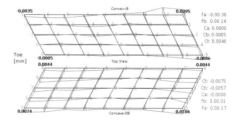

3.8.2 Pinion Re-Design Based on Existing Gear

A large amount of spiral bevel gear wheels had been manufactured by Powder Metallurgy (PM). The Summaries of both the pinion and gear are available. Typically,
PM gears deviate from the design since the mold is obtained by EDM of a copper
electrode and a solid block of hard steel. In this case, when the pinion was cut to
design and meshed with a PM gear, the contact patterns shown in Fig. 3.39 were
obtained which are low on the gear tooth and therefore high on the pinion tooth.

When such contact patterns are obtained, the TE deviates from a parabolic shape
and in this case resulted in significant noise issues. Both the pinion and gear were
measured and RE to obtain the contact patterns displayed in Fig. 3.40; as is evident, the actual contact patterns of Fig. 3.39 and the simulated contact patterns of
Fig. 3.40—after RE—correlate very well.

The Duplex Helical pinion machine settings were then modified to produce the
good contact patterns and associated TE shown in Fig. 3.41. The initial loud whining
noise also came down to more usual levels. This saved the production and eliminated
lost time and money waiting for a new series of gear members to arrive—i.e. months
in this case

(a)   **(b)**

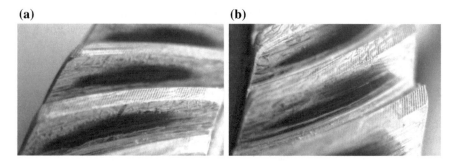

Fig. 3.39 Measured contact patterns

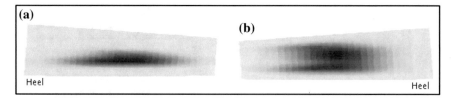

Fig. 3.40 Contact patterns calculated after RE

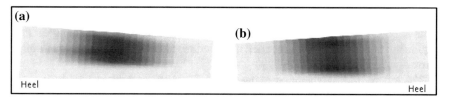

Fig. 3.41 Re-developed contact patterns

3.8.3 *In Situ Behavior—Coniflex™ Straight-Bevel Gear Set*

A Coniflex™ straight-bevel gear set is used in a high power-density gearbox. Machine settings and blank dimensions are known. Given the pinion and gear rotate little since they are differential gears, their behavior is expected to be quasi-static. However, the actual contact pattern differs from the design and it is desired to (1) understand why and (2) analyze the resulting kinematics in situ.

The machine settings and blank dimensions provided in the Summary are entered into HyGEARS; the original kinematics, based on the Summary data, are displayed in Fig. 3.42. The contact pattern is a bit Heel-heavy and high on the gear tooth flank; this is consistent with the shape of the TE, on both tooth flanks.

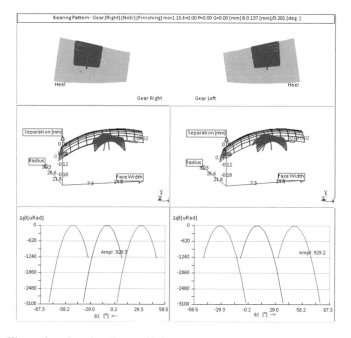

Fig. 3.42 Kinematics—based on the provided summary

CMM inspection files are generated from the provided machine settings, and the parts are measured as shown in Figs. 3.43 and 3.44. Both the pinion and gear show significant helix angle errors, inverted in direction, combined to some pressure angle error.

The RE algorithm is applied to the CMM data of Figs. 3.43 and 3.44, the result of which appears in Figs. 3.45 and 3.46. Profile waviness can be seen on the left flank (blue lines) of the pinion, Fig. 3.45; although also visible on the right flank (red lines), it is of a lesser amplitude.

Profile and lengthwise crowning can be seen on both flanks of the RE gear, Fig. 3.46.

Figure 3.47 shows the kinematics of the gear set after RE, where we can see that the contact patterns have shifted towards Toe, and that they are now rather low on

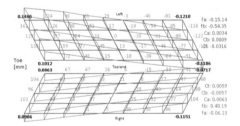

Fig. 3.43 Pinion CMM results

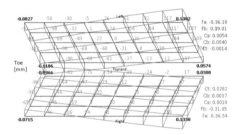

Fig. 3.44 Gear CMM results

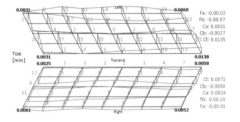

Fig. 3.45 Pinion CMM after RE

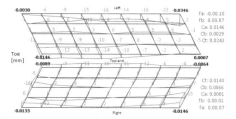

Fig. 3.46 Gear CMM after RE

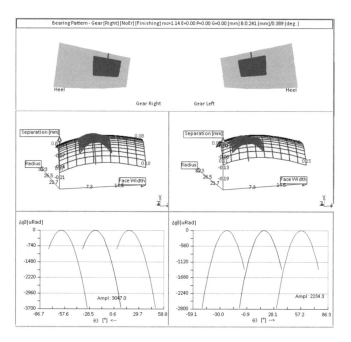

Fig. 3.47 Kinematics—after RE

the tooth flank of the gear, which again is consistent with the TE at the bottom of Fig. 3.47.

This also means that the contact patterns will be high on the pinion tooth flank, as is shown in Fig. 3.48 where the simulated and actual contact patterns of the left pinion tooth flank are compared and correlate well.

From the above, it is now clear that the simulation after RE can be used to calculate the contact and bending stresses that will actually take place in the gearbox.

While contact stresses may not change dramatically, it is expected that the bending stresses will, given the Toe location of the contact pattern.

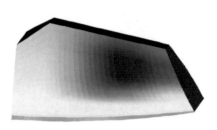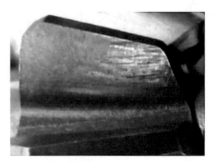

Fig. 3.48 Simulated contact pattern after RE and actual contact pattern

3.9 Conclusions

The methods presented in the previous sections describe a system to simulate gear manufacturing processes, analyze the kinematics without and under load, modify the kinematics to user desire, control manufacturing through Closed Loop (i.e. Corrective Machine Settings) and Reverse Engineer actual tooth surfaces to allow in situ analysis.

The presented tools allow gear designers to define tooling and machine adjustments that will produce the best possible designs for a given application. The obtained machine settings for spiral and straight bevel gears apply to most machines found in the industry, whether mechanical, free form or multi-axis CnC.

While the chosen examples are spiral and straight bevel gears, the method is also currently available for many other gear types.

In practice, the presented Tooth Flank Generator, TCA and LTCA, Closed Loop and RE algorithms have been in industrial use for more than two decades, and have proven effective and reliable.

Acknowledgements Fixed Setting, Modified Roll, Duplex Helical, Spread Blade, Formate, Coniflex, TopRem all Trade Marks of The Gleason Works, Rochester, NY, USA.

References

1. www.HyGEARS.com
2. Gosselin, C., Thomas, J.: A unified approach to the simulation of gear manufacturing and operation. In: International Conference on Gears, 7–9. T.U.M., Munich (2013)
3. Gosselin, C., Nguyen, D., Cloutier, L.: A general formulation for the calculation of the load sharing and transmission error under load of spiral bevel and hypoid gears, pp. 433–450. IFTOMM (1995)
4. Krenzer, J.T.: Tooth contact analysis of spiral bevel gears under load, Gleason Publication SD3458, April 1981
5. Gosselin, C., Gagnon, P., Cloutier, L.: Accurate tooth stiffness of spiral bevel gear teeth by the finite strip method. ASME J. Mech. Des. **120** (1998)

6. Gosselin, C., Thomas, J.: Integrated closed loop in 5Axis CnC gear manufacturing. In: International Conference on Gear Production 2015, 5–6. T.U.M., Munich (2015)
7. Gosselin, C., Shiono, Y., Kagimoto, H., Aoyama, N.: Corrective machine settings of spiral bevel and hypoid gears with profile deviations, 16–18. World Congress on Gearing, Paris (1999)
8. Gosselin, C., Nonaka, T., Shiono, Y., Kubo, A., Tatsuno, T.: Identification of the machine settings of real hypoid gear tooth surfaces. ASME J. Mech. Des. **120** (1998)
9. Krenzer, T.J.: Computer aided corrective machine settings for manufacturing bevel and hypoid gear sets, AGMA Paper 84-FTM-4 (1984)

Chapter 4
Analytical Simulation of the Tooth Contact of Spur Gears

José I. Pedrero, Miguel Pleguezuelos and Miryam B. Sánchez

Abstract The geometry of the gear teeth, as well as that of the tooth contact, is complex and not easy to simulate. Thus, computer aided techniques, as finite element analyses or complex tooth contact analyses, should be applied for strength calculations or simulations of contact conditions. These techniques provide reliable results, but usually requires long time preparation and high computational cost. Simple models of the linear theory of elasticity or theoretical, conjugate contact conditions may be not sufficiently accurate for modern gears whose power-capacity to size ratio is continuously increasing. However, these simple models, even simpler approximate equations describing them, could provide preliminary information on the influence of the geometrical parameters on the design factors, and therefore optimization criteria from different points of view. In this work, an approximate equation of the meshing stiffness of spur pairs is applied to the analysis of the load sharing ratio, load capacity, and quasi static transmission error of spur gears. The study extends to both standard and high contact ratio spur gears and considers gear teeth with profile modifications.

Keywords Spur gears · Meshing stiffness · Load sharing ratio · Load carrying capacity · Transmission error

4.1 Introduction

Presently, the growing competitiveness of the markets, environmental demands, and required quality levels, force the development of increasingly accurate and reliable gear calculation methods. Modern computer tools, as analysis packages by the Finite

J. I. Pedrero (✉) · M. Pleguezuelos · M. B. Sánchez
Departamento de Mecánica, UNED, C./Juan Del Rosal 12, 28040 Madrid, Spain
e-mail: jpedrero@ind.uned.es

M. Pleguezuelos
e-mail: mpleguezuelos@ind.uned.es

M. B. Sánchez
e-mail: msanchez@ind.uned.es

© Springer Nature Switzerland AG 2020
V. Goldfarb et al. (eds.), *New Approaches to Gear Design and Production*,
Mechanisms and Machine Science 81,
https://doi.org/10.1007/978-3-030-34945-5_4

Element Method (FEM), provides precise results on stresses levels, deflections, stiffness, etc.; however, the FEM model preparation requires long time and considerable effort, while FEM calculation requires high computational cost [4, 9].

Though the final validation by the FEM is unavoidable in modern gear design, for previous steps and preliminary decisions, easier analytic calculations based on simple elastic models may be enough. Gear Standards ISO [6] and AGMA [1] provides calculation methods for the load carrying capacity based on the Navier equation and the Hertz contact model. Nevertheless, even these simple elastic models are difficult to apply due to the relatively complex geometry of the teeth, particularly at the root trochoid [1, 6, 12].

Despite this, many studies have been carried out to find analytical methods for gear design, and specifically for the determination of the meshing stiffness. Chen and Shao [2] proposed a calculation method of the meshing stiffness of internal spur gear pairs, based on the potential energy (PE) method. Chen et al. [3] extended the study to teeth with profile shift. The same PE method was used by Wang et al. [19] to develop a model of the time-varying mesh stiffness for helical gears considering the axial mesh force. Lei et al. [8] evaluated the time-varying mesh stiffness from a probability distribution model of the tooth pits. Ma et al. [10] developed an analytical method which considers tip relieved teeth. Saxena et al. [18] studied the influence of shaft misalignments and friction forces on the mesh stiffness.

Many researches have also been developed by FEM techniques. Wilcox and Coleman [21] applied FE techniques to evaluate the gear tooth stresses. González-Pérez et al. [5] developed a FE model for the stress analysis of lightweight spur gears. Roda-Casanova et al. [13] evaluated the ISO face load factor by the FEM. Kiekbusch et al. [7] developed two- and three-dimensional FE models to calculate the torsional mesh stiffness. Fernández del Rincón et al. [4] proposed a model for the meshing stiffness of spur gears which combines a global term, calculated by using FEM simulations, and a local term, obtained from the Hertzian contact theory.

Nowadays, the power-capacity to size ratio is in continuous growth. This means that very accurate calculations are essential, and many rating factors should be considered to account many influences on the actual operating conditions. The hypothesis of teeth behavior as rigid bodies may be acceptable for stress calculations from simple elastic models, as ones considered by ISO [6] and AGMA [1]. However, the teeth deflections induce an earlier start of contact and a fluctuation of the output shaft velocity [12]. The earlier start of contact results in a shock between the root of the driving gear tooth and the tip of the driven gear tooth. The velocity fluctuation produces accelerations and decelerations of the gear, resulting in vibrations and dynamic load [17]. Both the effects could be dangerous, but the first one may be completely unacceptable in many cases.

To avoid the shock, and to control the fluctuation of the output velocity—which can not be avoided but can be moderated-, the profile modification is often used [12, 17]. In fact, a suitable relief at the driven tooth tip delays the actual start of contact to the proper theoretical inner point of contact. Suitable shape and length of profile modification influence the teeth deflections and therefore the output velocity.

Profile modifications must be considered for gear calculations. Their influence on the load sharing, the dynamic load, the resonance effects, etc., will have decisive importance on the load carrying capacity evaluation. Obviously, FE models can consider—or even better, must consider—the modified geometry of the teeth. Nevertheless, analytical calculations methods of gear standards [1, 6] have not conveniently addressed the problem yet. In this paper, some analytical equations for the load sharing, the load carrying capacity, and the quasi static transmission error (QSTE), which account the influence of profile modifications, are provided. All of them are based on a simple, approximate equation of the meshing stiffness of the tooth pair [16], which provides very accurate results. With these equations, analytical calculation methods existing in literature can be improved to consider profile modifications, what may be useful for pre-design studies or standardization purposes.

4.2 Meshing Stiffness, Load Sharing, and Tooth Deflection

The meshing stiffness K_M can be computed from the following equation:

$$K_M = \left(\frac{1}{k_{x1}} + \frac{1}{k_{s1}} + \frac{1}{k_{n1}} + \frac{1}{k_{x2}} + \frac{1}{k_{s2}} + \frac{1}{k_{n2}} + \frac{1}{k_H} \right)^{-1} \tag{4.1}$$

where k_x is the bending stiffness, k_s is the shear stiffness, k_n is the compressive stiffness, k_H is the contact stiffness, and subscripts 1 and 2 denote the driving gear and the driven gear, respectively. Some authors [11, 20] consider an additional term owing to the gear body stiffness, which does not have a big influence. There is a wide consensus on calculating the bending, the shear, and the compressive stiffness from the simple equations of the linear theory of elasticity, as described in [16]; nevertheless, consensus is not so wide for the contact stiffness. Some authors—the authors of this paper among them—use the Hertzian approach; other authors use the Weber-Banaschek approach; other ones consider the stiffness corresponding to a fraction of the deformation according to Weber-Banaschek, typically a 50% as recommended by KissSoft [11].

For the aim of this work, the discussion on the contact stiffness is not critical. Figure 4.1 shows the evolution of the meshing stiffness along the path of contact, considering three contact stiffness approaches above. A fourth curve neglecting the contact stiffness has been also represented. All the curves have the same shape, and therefore similar results would be obtained from any of them, although the numerical values calculated form each approach may present discrepancies. In this work, the Hertzian contact stiffness will be considered. It is remarkable the similarity between the curves corresponding to Hertzian approach and 50% of Weber-Banaschek deformation approach.

The meshing stiffness computed from Eq. (4.1) and the Hertzian contact stiffness, can by approximated by the following expression [16]:

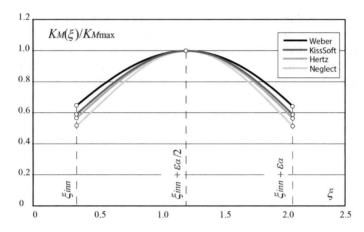

Fig. 4.1 Curves of meshing stiffness with different contact stiffness approaches

$$K_M(\xi) = K_{Mmax} \cos(b_0(\xi - \xi_m)) \tag{4.2}$$

where:

$$b_0 = \left[\tfrac{1}{2}\left(1.11 + \tfrac{\varepsilon_\alpha}{2}\right)^2 - 1.17\right]^{-1/2}$$
$$\xi_m = \xi_{inn} + \tfrac{\varepsilon_\alpha}{2} \tag{4.3}$$

being ε_α the theoretical contact ratio. ξ is the contact point parameter and describes the contact position in the path of contact. It is expressed as follows:

$$\xi = \frac{z_1}{2\pi}\sqrt{\frac{r_{c1}^2}{r_{b1}^2} - 1} \tag{4.4}$$

in which z_1 is the number of teeth on the driving gear, r_{b1} the base radius, and r_{c1} the radius of the point C, which is defined by the intersection of the driving tooth involute profile (or its prolongation) and the pressure line. Inside the contact interval, point C is the contact point. ξ_{inn} corresponds to the inner point of contact and ξ_m to the midpoint of the interval of contact. According to Eq. (4.4), the contact point parameter corresponding to the outer point of contact ξ_o is given by:

$$\xi_o = \xi_{inn} + \varepsilon_\alpha \tag{4.5}$$

Similarly, the difference between the contact point parameter of two consecutive teeth is equal to 1, and consequently:

$$\xi_{(i+j)} = \xi_{(i)} + j \tag{4.6}$$

The load sharing ratio $R(\xi)$ can be expressed as follows:

$$R(\xi) = \frac{F(\xi)}{F_T} = \frac{K_M(\xi)}{\sum_j K_M(\xi + j)} \tag{4.7}$$

where $F(\xi)$ is the load at tooth pair in contact at point ξ, F_T is the total load, and the sum is extended to all the teeth in simultaneous contact, i.e., all the entire values of j between 0 and z_1-1 verifying:

$$\xi_{inn} \le \xi + j \le \xi_{inn} + \varepsilon_\alpha \tag{4.8}$$

The tooth deflection is given by the expression:

$$\delta(\xi) = \frac{F(\xi)}{K_M(\xi)} \tag{4.9}$$

which, according to Eq. (4.7), can be expressed as follows:

$$\delta(\xi) = \frac{F_T}{\sum_j K_M(\xi + j)} \tag{4.10}$$

From Eq. (4.10), it can be concluded that $\delta(\xi)$ does not depend on j, which means that the tooth deflection is equal for all the teeth in simultaneous contact, at every contact position. Figure 4.2 presents the curves of meshing stiffness, load sharing ratio, and tooth pair deflection for standard and high contact ratio (HCR) spur gears.

4.3   Extended Contact Interval

As discussed above, the teeth deflections induce a delay angle on the driven gear respect to the driving one. Owing to this offset angle, the root of the driving tooth hits the tip of the driven one, resulting in a shock. This actual start of contact occurs before the theoretical inner point of contact and outside the line of action. Figure 4.3 presents the actual start of contact at point I.

The length of the additional contact interval is described by the distance between points b and e in Fig. 4.3, which is named $(\Delta-\delta)$. In previous works [12, 17], the authors studied the relation between the additional length of contact $(\Delta-\delta)$ and the tooth pair deflection δ, which is described in Fig. 4.3 by the distance between points a and b,—and between points c and d as well. From this relation, it is very simple to calculate δ from the value of $(\Delta-\delta)$ but calculating $(\Delta-\delta)$ from the value of δ requires solving a highly non-linear equation. In the mentioned works [12, 17], an approximate parabolic equation was found, which provides very accurate results. It is expressed as follows:

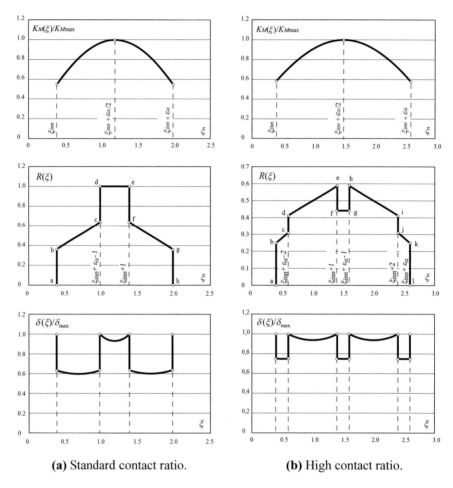

(a) Standard contact ratio.   **(b)** High contact ratio.

Fig. 4.2 Curves of meshing stiffness, load sharing ratio and tooth-pair deflection

$$\left(\frac{\Delta - \delta}{r_{b1}}\right) = \sqrt{\frac{1}{C_p}\left(\frac{\delta}{r_{b1}}\right)} \tag{4.11}$$

Obviously, a similar additional contact interval arises at the end of contact, whose length can be approximated by the following equation [17]:

$$\left(\frac{\Delta'}{r_{b1}}\right) = \sqrt{\frac{1}{C'_p}\left(\frac{\delta'}{r_{b1}}\right)} \tag{4.12}$$

The calculation of the parabola coefficients C_p and C'_p can be easily performed, as described in [17]. Figure 4.4 presents the good agreement between the approximate Eqs. (4.11) and (4.12) and their respective geometrical relations.

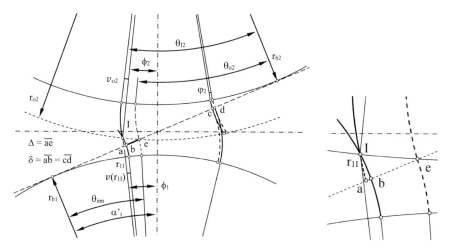

Fig. 4.3 Actual start of contact of loaded tooth

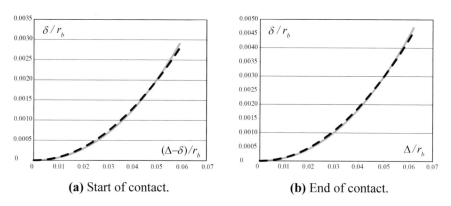

(a) Start of contact.   **(b)** End of contact.

Fig. 4.4 Relation between tooth-pair deflection and contact interval extension. (Approximated parabolas in dashed lines)

The meshing stiffness of the tooth pair along both the additional contact intervals can also be computed using Eq. (4.1). The result is shown in Fig. 4.5. The meshing stiffness is almost uniform along each additional interval. Thus, the meshing stiffness along the extended contact interval can be approximated by the following equation:

$$
\begin{aligned}
K_M(\xi) &= K_{M\max} \cos\!\left(b_0 \tfrac{\varepsilon_\alpha}{2}\right) & \xi_{\min} \le \xi \le \xi_{inn} \\
K_M(\xi) &= K_{M\max} \cos\!\left(b_0(\xi - \xi_m)\right) & \xi_{inn} \le \xi \le \xi_{inn} + \varepsilon_\alpha \\
K_M(\xi) &= K_{M\max} \cos\!\left(b_0 \tfrac{\varepsilon_\alpha}{2}\right) & \xi_{inn} + \varepsilon_\alpha \le \xi \le \xi_{\max}
\end{aligned}
\qquad (4.13)
$$

Fig. 4.5 Extended meshing stiffness

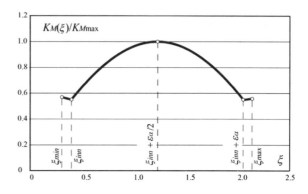

4.4   Actual Contact Conditions

For loaded teeth, the meshing stiffness is described by Eq. (4.13) and Fig. 4.5. The contact ratio ε_α and the theoretical inner point of contact ξ_{inn} are known, as they depend on the geometry of the considered gear pair. Coefficients b_0 and ξ_m can be computed with Eq. (4.3). The limits of the extended contact interval, ξ_{min} and ξ_{max}, can be computed with Eqs. (4.11) and (4.12). Consequently, the meshing stiffness is perfectly known at any point of the complete contact interval.

The actual load sharing ratio and teeth deflection can be obtained by accounting that the delay of the driven teeth is the same for all the teeth in contact. Accordingly, the load at each tooth pair when one of them is in contact inside the inner additional interval, can be expressed as follows:

$$F_0(\xi) = K_M(\xi)(\delta(\xi) - \delta_G(\xi))$$
$$F_{j>0}(\xi) = K_M(\xi + j)\delta(\xi)$$

(4.14)

$\delta_G(\xi)$ is the distance the driving tooth should approach to the driven one to contact it at contact point ξ. Its value can be computed with Eq. (4.11)—after some calculations, as $\delta_G(\xi)$ corresponds to the distance $(\Delta-\delta)$ for $\delta = \delta(\xi)$.

The same may occur at the outer additional interval, in which $\delta_G(\xi)$ is described by Eq. (4.12). It is even possible, in specific cases, having a tooth pair in contact inside the inner additional interval and other pair in contact inside the outer additional interval, simultaneously. In general, the load at any tooth pair can be described by the following equation:

$$F_j(\xi) = K_M(\xi + j)(\delta(\xi) - \delta_G(\xi + j))$$

(4.15)

with j from 0 to the integer part of the contact ratio E_α—which describes the last tooth in contact and:

$$\delta_G(\xi) = r_{b1}\sqrt{\frac{1}{C_p}\left(\frac{\delta(\xi)}{r_{b1}}\right)} \quad \xi_{min} \leq \xi \leq \xi_{inn}$$

$$\delta_G(\xi) = 0 \qquad\qquad\qquad \xi_{inn} \leq \xi \leq \xi_{inn} + \varepsilon_\alpha \qquad (4.16)$$

$$\delta_G(\xi) = r_{b1}\sqrt{\frac{1}{C'_p}\left(\frac{\delta(\xi)}{r_{b1}}\right)} \quad \xi_{inn} + \varepsilon_\alpha \leq \xi \leq \xi_{max}$$

Since the total load F_T is known, the teeth deflection $\delta(\xi)$ can be obtained by adding all the Eqs. (4.15)—for all the values of j—and equalizing to the total load, which results in:

$$\delta(\xi) = \frac{F_T + \sum_{j \geq 0} K_M(\xi + j)\delta_G(\xi + j)}{\sum_{j \geq 0} K_M(\xi + j)} \qquad (4.17)$$

The actual load sharing ratio is obtained from the combination of Eqs. (4.15) and (4.17):

$$R(\xi) = \frac{K_M(\xi)}{\sum_{j \geq 0} K_M(\xi + j)}\left(1 + \frac{\sum_{j \geq 0} K_M(\xi + j)\delta_G(\xi + j)}{F_T}\right) \qquad (4.18)$$

The result is presented in Fig. 4.6.

Equation (4.17) also describes the QSTE. In fact, the offset angle φ_2 between both the gears corresponds to the distance δ in the line of action, as shown in Fig. 4.3. But this offset angle is the difference between the theoretical—conjugate—angular position of the driven gear and the actual one, which is, by definition, the QSTE. Accordingly, the QSTE is given by the equation:

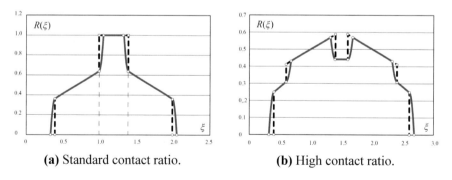

(a) Standard contact ratio.   (b) High contact ratio.

Fig. 4.6 Theoretical (dashed) and actual (solid) load sharing ratio

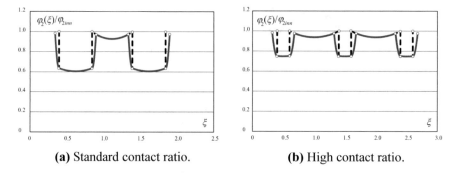

(a) Standard contact ratio.   **(b)** High contact ratio.

Fig. 4.7 Theoretical (dashed) and actual (solid) quasi-static transmission error

$$\varphi_2 = \frac{\delta(\xi)}{r_{b2}} = \frac{F_T + \sum_{j\geq0} K_M(\xi+j)\delta_G(\xi+j)}{r_{b2}\sum_{j\geq0} K_M(\xi+j)} \tag{4.19}$$

The result is presented in Fig. 4.7.

4.5 Profile Modifications

A tip relief at the driven gear teeth can be used to avoid the shock at the beginning of the contact. From Fig. 4.3, if the driven tooth tip were relieved, contact would not occur at point I, but later. In addition, if the amount of relief is suitable, the start of contact would be delayed to the theoretical inner point of contact, point e.

Figure 4.8 presents the geometry of a tooth with tip relief, which is the most usual profile modification. The shape of modification describes the modified profile, and

Fig. 4.8 Tip relief

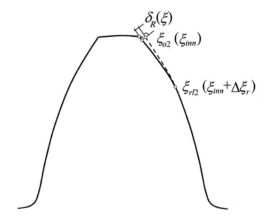

is expressed as a function of the contact point parameter, $\delta_R = \delta_R(\xi)$. The amount of modification is the modification corresponding to the tooth tip, and therefore it will be given by $\delta_R(\xi_{inn})$—for the driven tooth relief—or by $\delta_R(\xi_o)$—for the driving tooth relief. The length of modification $\Delta\xi_r$ is the length of the interval of modification, expressed in terms of ξ. Eventually, the profile of both the mating teeth may be modified. Accordingly, the profile modification may be expressed as follows:

$$
\begin{aligned}
\delta_R(\xi) &= \delta_{R-inn}(\xi) \quad \xi_{inn} \leq \xi \leq \xi_{inn} + \Delta\xi_{r-inn} \\
\delta_G(\xi) &= 0 \qquad\qquad\quad \xi_{inn} + \Delta\xi_{r-inn} \leq \xi \leq \xi_{inn} + \varepsilon_\alpha - \Delta\xi_{r-o} \\
\delta_G(\xi) &= \delta_{R-o}(\xi) \quad \xi_{inn} + \varepsilon_\alpha - \Delta\xi_{r-o} \leq \xi \leq \xi_{inn} + \varepsilon_\alpha
\end{aligned}
\tag{4.20}
$$

To avoid the shock at the beginning of the contact, the amount of modification at the driven tooth, $\delta_R(\xi_{inn})$, should be equal to the tooth deflection just before the start of contact of the new pair at the theoretical inner point of contact, $\delta(\xi_{inn}^-)$. Additionally, at any moment of the meshing cycle there should be at least one contact point at the involute, non-modified interval of both the profiles. In consequence, it should be verified the following:

$$
\Delta\xi_{r-inn} + \Delta\xi_{r-o} < \varepsilon_\alpha - 1
\tag{4.21}
$$

The load sharing ratio and the QSTE for modified profiles can be determined by following the same procedure as in the previous cases. The load at any tooth pair can be described by the following equation:

$$
F_j(\xi) = K_M(\xi + j)(\delta(\xi) - \delta_G(\xi + j) - \delta_R(\xi + j))
\tag{4.22}
$$

Since the sum of all these loads should be equal to the total load, the tooth pair deflection will be given by the equation:

$$
\delta(\xi) = \frac{F_T + \sum\limits_{j \geq 0} K_M(\xi + j)(\delta_G(\xi + j) + \delta_R(\xi + j))}{\sum\limits_{j \geq 0} K_M(\xi + j)}
\tag{4.23}
$$

Accordingly, the load sharing ratio and the QSTE are given by the following expressions:

$$
R(\xi) = \frac{K_M(\xi)}{\sum\limits_{j \geq 0} K_M(\xi + j)} \left(1 + \frac{\sum\limits_{j \geq 0} K_M(\xi + j)(\delta_G(\xi + j) + \delta_R(\xi + j))}{F_T} \right)
\tag{4.24}
$$

$$
\varphi_2 = \frac{\delta(\xi)}{r_{b2}} = \frac{F_T + \sum\limits_{j \geq 0} K_M(\xi + j)(\delta_G(\xi + j) + \delta_R(\xi + j))}{r_{b2} \sum\limits_{j \geq 0} K_M(\xi + j)}
\tag{4.25}
$$

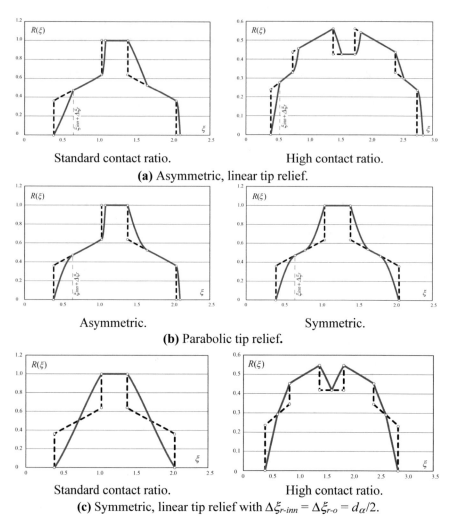

Fig. 4.9 Load sharing ratio for different profile modifications

Figures 4.9 and 4.10 present the load sharing ratio and the QSTE of several examples combining standard and high contact ratio, symmetric and asymmetric reliefs, and linear and parabolic reliefs.

4.6 Specific Applications

The equations above allow to obtain the load sharing ratio and the QSTE from a given—or being checked—profile modification. However, all these equations are

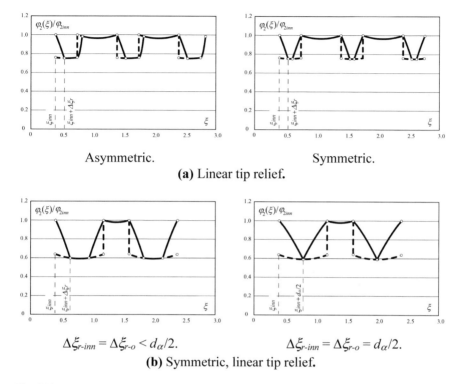

Asymmetric.   Symmetric.

(a) Linear tip relief.

$\Delta\xi_{r\text{-}inn} = \Delta\xi_{r\text{-}o} < d_\alpha/2.$ $\Delta\xi_{r\text{-}inn} = \Delta\xi_{r\text{-}o} = d_\alpha/2.$

(b) Symmetric, linear tip relief.

Fig. 4.10 Quasi-static transmission error for different profile modifications

sufficiently simple to allow the opposite approach as well: obtaining the profile modification required to ensure a given load sharing ratio or QSTE. Obviously, there are a lot of restrictions in the preestablished functions of load sharing ratio or QSTE, but some specific requirements could be easily regarded with the simplified model presented in this work. A couple of examples are presented below.

4.6.1   Smooth Fluctuation of the QSTE

From Fig. 4.10, the fluctuation of the QSTE can not be reduced with a length of relief smaller than the fractional part of the contact ratio, denoted by d_α. Consequently, and according to Eq. (4.21), tip relief cannot be used to reduce the QSTE fluctuation in standard contact ratio spur gears. However, for HCR gears, the Eq. (4.21) could be regarded even for a length of modification greater than d_α.

In the case of asymmetric relief, as the length of modification increases (above d_α), the minimum QSTE increases, but the maximum one increases too. The fluctuation of the QSTE can be reduced, but not significantly. In addition, the maximum slope of the curve of QSTE is not also reduced, and consequently the maximum variation

Fig. 4.11 Quasi-static transmission error of minimum dynamic load

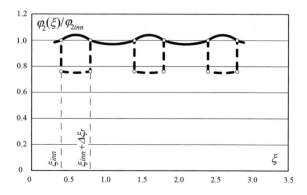

of the velocity—i.e., the maximum acceleration and thus the maximum dynamic load– keeps its value.

For symmetric relief (at the driven and the driving teeth, each one longer than $d_\alpha/2$), the minimum QSTE increases, but a local maximum arises at the midpoint of the intervals of tow pair tooth contact, which becomes the absolute maximum. The slope of the QSTE curve is progressively reduced and reaches the minimum value for a length of relief equal to d_α at both the teeth. Figure 4.11 presents the QSTE of a gear with contact ratio 2.4 and symmetric linear tip relief with $\Delta \xi_r = 0.4$.

4.6.2 Balancing Tooth-Root and Pitting Load Carrying Capacities

Figure 4.12 presents the load sharing ratio for a HCR spur gear, with linear tip relief at the driving tooth tip and length of relief greater than d_α. The curve has been obtained from Eq. (4.24). Along the interval of modification, $\xi_{inn} \leq \xi \leq \xi_{inn} + \Delta \xi_r$, the load is smaller than that without profile modification. Since the interval of modification

Fig. 4.12 Load sharing ratio for HCR spur gear with long tip relief

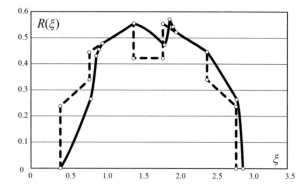

reaches the inner interval of two pair tooth contact, the load at the corresponding sub-interval of the upper two pair tooth contact interval is greater than that for unmodified teeth. Therefore, a peak of load arises at the upper interval of two pair tooth contact.

In many cases, the critical contact stress of HCR spur gears with unmodified profiles is located at the inner limit of the inner interval of two pair tooth contact [14]. According to Fig. 4.11, for a tip relief longer than d_α (named long tip relief), the critical contact stress will decrease, thus the pitting load carrying capacity will increase.

The critical tooth-root stress is always located inside the upper interval of two pair tooth contact, throughout which the stress keeps a quite uniform value [15]. Accordingly, the peak of load owing to the long tip relief will increase the critical tooth-root stress and will decrease the tooth-root load carrying capacity.

In conclusion, long tip reliefs at the driven tooth tip can be used to balance the pitting and tooth-root load carrying capacities, and consequently to increase the final load capacity.

4.7 Conclusions

A simple, analytic model for the load sharing ratio and quasi-static transmission error for spur gears with profile modification has been presented. It is based on the equal tooth-pair deflection for all the teeth in simultaneous contact, at any position of the meshing interval. Both standard and high contact ratio spur gears have been considered. For unmodified profiles or gear pairs with profile modification only at the beginning of the contact, the enlargement of the actual contact interval and the load transfer throughout the additional intervals have been also studied.

To illustrate specific applications of this model, the symmetric tip relief required to ensure smooth quasi-static transmission error—and accordingly minimum peak of instantaneous dynamic load—and the length of long profile modification to improve the load carrying capacity, have been provided.

Acknowledgements The authors express their gratitude to the Spanish Council for Scientific and Technological Research for the support of the project DPI2015-69201-C2-1-R, "Load Distribution and Strength Calculation of Gears with Modified Geometry", as well as the School of Engineering of UNED for the support of the action 2019-MEC24, "Simulation of the transmission error of spur gears with profile modification".

References

1. AGMA (American Gear Manufacturers Association): Fundamental Rating Factors and Calculation Methods for Involute Spur and Helical Gear Teeth, Standard 2001–D04. AGMA, Alexandria, VA (2004)
2. Chen, Z., Shao, Y.: Mesh stiffness of an internal spur gear pair with ring gear rim deformation. Mech. Mach. Theory **69**, 1–12 (2013)
3. Chen, Z., Zhai, W., Shao, Y., Wang, K.: Mesh stiffness evaluation of an internal spur gear pair with tooth profile shift. Sci. China Technol. Sci. **59**(9), 1328–1339 (2016)
4. Fernández del Rincón, A., Viadero, F., Iglesias, M., García, P., de Juan, A., Sancibrián, R.: A model for the study of meshing stiffness in spur gear transmissions. Mech. Mach. Theory **61**, 30–58 (2013)
5. Gonzalez-Perez, I., Fuentes, A., Roda-Casanova, V., Sanchez-Marin, F.T., Iserte, J.L.: A finite element model for stress analysis of lightweight spur gear drives based on thin-webbed and thin-rimmed gears. In: Proceedings of International Conference on Gears, VDI-Society for Product and Process Design, pp. 75–86. Garching, Germany (2013)
6. ISO (International Organization for Standardization): Calculation of Load Capacity of Spur and Helical Gears, Standard 6336, Parts 1, 2, and 3. ISO, Geneva, Switzerland (2006)
7. Kiekbusch, T., Sappok, D., Sauer, B., Howard, I.: Calculation of the combined torsional mesh stiffness of spur gears with two-and three-dimensional parametrical FE models. J. Mech. Eng. **57**(11), 810–818 (2011)
8. Lei, Y., Liu, Z., Wang, D., Yang, X., Liu, H., Lin, J.: A probability distribution model of tooth pits for evaluating time-varying mesh stiffness of pitting gears. Mech. Syst. Signal Process. **106**, 355–366 (2018)
9. Li, S.: Finite element analyses for contact strength and bending strength of a pair of spur gears with machining errors, assembly errors and tooth modifications. Mech. Mach. Theory **42**, 88–114 (2007)
10. Ma, H., Zeng, J., Feng, R., Pang, X., Wen, B.: An improved analytical method for mesh stiffness calculation of spur gears with tip relief. Mech. Mach. Theory **98**, 64–80 (2016)
11. Mahr, B., Kissling, U.: Comparison between different commercial gear tooth contact analysis software packages. https://www.kisssoft.ch/english/downloads/documentation_kisssoft.php
12. Pedrero, J.I., Pleguezuelos, M., Sánchez, M.B.: Control del error de transmisión cuasi-estático mediante rebaje de punta en engranajes rectos de perfil de evolvente. Revista Iberoamericana de Ingeniería Mecánica **22**, 71–90 (2018)
13. Roda-Casanova, V., Sanchez-Marin, F.T., Gonzalez-Perez, I., Iserte, J.L., Fuentes, A.: Determination of the ISO face load factor in spur gear drives by the finite element modeling of gears and shafts. Mech. Mach. Theory **65**, 1–13 (2013)
14. Sánchez, M.B., Pedrero, J.I., Pleguezuelos, M.: Contact stress calculation of high transverse contact ratio spur and helical gear teeth. Mech. Mach. Theory **64**, 93–110 (2013)
15. Sánchez, M.B., Pleguezuelos, M., Pedrero, J.I.: Tooth-root stress calculation of high transverse contact ratio spur and helical gears. Meccanica **49**, 347–364 (2014)
16. Sánchez, M.B., Pleguezuelos, M., Pedrero, J.I.: Approximate equations for the meshing stiffness and the load sharing ratio of spur gears including hertzian effects. Mech. Mach. Theory **109**, 231–249 (2017)
17. Sánchez, M.B., Pleguezuelos, M., Pedrero, J.I.: Influence of profile modifications on meshing stiffness, load sharing, and transmission error of involute spur gears. Mech. Mach. Theory **139**, 506–525 (2019)
18. Saxena, A., Parey, A., Chouksey, M.: Effect of shaft misalignment and friction force on time varying mesh stiffness of spur gear pair. Eng. Fail. Anal. **49**, 79–91 (2015)
19. Wang, Q., Zhao, B., Fu, Y., Kong, X., Ma, H.: An improved time-varying mesh stiffness model for helical gear pairs considering axial mesh force component. Mech. Syst. Signal Process. **106**, 413–429 (2018)

20. Weber, C., Banaschek, K.: Formänderung und profilrücknahme bei gerad und schrägverzahnten rädern. Schriftenreihe Antriebstechnik. Vieweg Verlag, Braunschweig, Germany (1955)
21. Wilcox, L., Coleman, W.: Application of finite elements to the analysis of gear tooth stresses. ASME J. Eng. Ind. **95**(4), 1139–1148 (1973)

Chapter 5
Automation of Technological Preproduction of Straight Bevel Gears

V. Medvedev, A. Volkov and S. Biryukov

Abstract The paper describes an approach to the automated technological preparation for production of straight bevel gears with the localized contact. The flank of the driven gear is a conical involute surface. The contact localization in the gear is achieved by using the modified conical involute surface for the driving gear flank. The contact localization is carried out by the synthesis parameters. The paper proposes the technique to make a meaningful choice of synthesis parameters. The strength checking calculation is carried out using the Hertz solutions of the problem of contact between two bodies bounded by surfaces of the second order. The examples of calculation for two straight bevel gears are presented.

Keywords Straight bevel gears · Modified conical involute surface · Synthesis parameters · Contact problem · Localized contact

5.1 Introduction

Technological preparation for production of straight bevel gears has evolved considerably over the recent decades. In the middle of the last century calculation of gear geometry was carried out first with the determination of contour dimensions and measurement parameters of the gears. During the geometry calculation, the parameters of gear pair were determined, as well as the number of teeth of the equivalent spur gear. They were then used in calculations of contact and bending strength and

The Author V. Medvedev was deceased.

V. Medvedev · A. Volkov (✉) · S. Biryukov
MSTU «STANKIN», Moscow, Russia
e-mail: volkov411@gmail.com

V. Medvedev
e-mail: vladimir.ivanovich.medvedev@gmail.com

S. Biryukov
e-mail: bserg1234@mail.ru

© Springer Nature Switzerland AG 2020 133
V. Goldfarb et al. (eds.), *New Approaches to Gear Design and Production*,
Mechanisms and Machine Science 81,
https://doi.org/10.1007/978-3-030-34945-5_5

durability of gears using approximate dependencies taken from the methodology of strength calculation of involute spur gears.

The task of the technologist was to realize the gear that satisfies the results of the design calculation. It was also necessary to take into account that the gear should work despite the unavoidable manufacturing and operational errors. This is achieved by avoiding the strict conjugacy of surfaces. For this purpose, the contact localization in the profile and longitudinal directions is introduced [2, 5, 9, 10]. The problem was with the implementation of contact localization on the existing machine equipment.

The development of computer technology has led to the creation of computer software that can simulate the processes of tooth flanks formation and gears operation. It became possible in the gear design to take into the account the shape of the flank that can be obtained on the gear processing machine. In this case, the strength calculation is performed using such mathematical methods as the boundary element method or the finite element method.

In parallel with the development of computer technology, multi-axis computer numerical control machines were created. Such machines can provide theoretically accurate, i.e. in the absence of errors, reproduction of the calculated flank. For example, end mill processing can be used for that purpose.

The modern technologist can recreate almost any given surface on the CNC machine. Therefore, you can use the following approach. The shape of the flank of the driven gear is set. The pinion tooth flank is defined as an envelope to the family of the gear tooth flanks, obtained during its movement relative to the fixed pinion [1, 3, 6, 7, 14]. Then the pinion flank crowning is applied to provide profile and longitudinal localization of contact. And finally, the verification strength calculation is carried out using the mathematical modeling [12, 13, 16]. According to its results, the degree of the contact localization can be adjusted. And only then we determine the shape of the surfaces that we need to be made on the CNC machine.

This article describes an approach to the automation of technological preproduction of straight bevel gears with localized contact. The flank of the driven gear is a conical involute surface. The contact localization in the gear is achieved by modifying the driving gear flank. The strength checking calculation is carried out using the Hertz solutions of the problem of contact between two bodies bounded by surfaces of the second order.

The article does not cover the problem of obtaining the tooth fillet. It is assumed that there will be a smooth transition at the boundary between the active surface and the fillet.

5.2   Synthesis of Straight Bevel Gear Flanks

5.2.1   Conical Involute Surface

The meshing surfaces of teeth will be built upon conical involute surfaces. Such surfaces provide ideal meshing and linear contact [17].

From here on out, values relating to the driven gear will be marked by the subscript 1, and the values relating to the driving gear—by the subscript 2.

Consider the fixed driven coordinate system $Ox_1y_1z_1$ (Fig. 5.1a). The origin O is located at the pitch cone apex. The coordinate axis z_1 is the axis of rotation of the driven gear. The fixed coordinate system $Ox_2y_2z_2$ is connected to the driving gear (Fig. 5.1b). Both systems have a common origin O. The axis z_2 is the axis of rotation of the driving gear. The axis x_2 coincides with the axis y_1. The axis y_2 coincides with the axis z_1. The axis z_2 coincides with the axis x_1. The coordinate axes x_1 and x_2 are chosen so as the edge of regression of each conical involute surface is located in the corresponding plane Oxz.

The flank of the tooth of the driven gear is defined in the coordinate system $Ox_1y_1z_1$. The flank of the tooth of the driving gear is defined in the coordinate system $Ox_2y_2z_2$. Both conical involute surfaces can be written uniformly using the relations ($i = 1, 2$) [18]

$$x_i = L(\sin \varphi_i \sin \psi_i + \cos \varphi_i \cos \psi_i \sin \delta_{bi})$$
$$y_i = L(-\cos \varphi_i \sin \psi_i + \sin \varphi_i \cos \psi_i \sin \delta_{bi}) \qquad (5.1)$$
$$z_i = L \cos \psi_i \cos \delta_{bi}$$

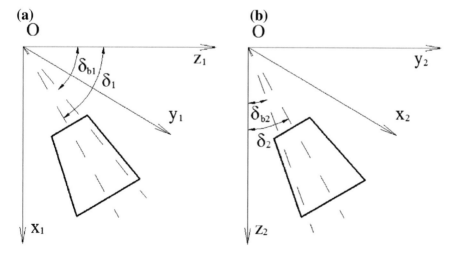

Fig. 5.1 Fixed coordinate system **a** driven; **b** driving

Here, the surface coordinates of the point of the conical involute surface are the cone distance L and the parameter φ_i of the spherical involute of the i-th gear on the sphere of radius L. In (5.1) the following relation between the parameters takes place:

$$\psi_i = \varphi_i \cos \delta_{bi}, \tag{5.2}$$

where δ_{bi}—the base cone angle, associated with pressure angle α and the pitch cone angle δ_i with the relation $\sin \delta_{bi} = \cos \alpha \cdot \sin \delta_i$.

5.2.2 Synthesis Parameters

The surface, defined by relations (5.1), provide constant transmission ratio in meshing. Disadvantages of such meshing were reviewed in [2, 4, 9, 15]. In addition, surfaces (5.1) do not provide localized contact. Without contact localization the bearing contact can shift to the edges of the teeth in the presence of inevitable errors. This can lead to a sharply increased concentration of contact pressure on the edges and contribute to rapid wear and breakage of the teeth.

To localize the contact we modify the surface of the driven gear in two directions: along the generatrix of the pitch cone and in the perpendicular direction. This will provide point contact of surfaces under slight load and localized contact under load.

The notional center P of the future bearing contact will be set on the flank of the driven gear by two parameters (Fig. 5.2):

– the radius L_c of the sphere centered at the pitch cone apex;

Fig. 5.2 Parameters of modification of the driven gear flank

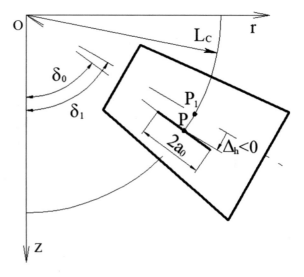

– the shift Δh in the direction perpendicular to the generatrix of pitch cone.

The shift Δh determines the position of the contact point P relative to the tooth surface point P_1 of the driven gear. The point P_1 is located at the intersection of the pitch cone and a sphere with a radius of L_c. The parameters L_c and Δh thus determine the position of the single point P where the involute and the modified surfaces coincide.

Two types of modification are used for contact localization: profile and longitudinal modification. They are defined by the following parameters (Fig. 5.2):

– the desired semi-axis a_0 of the instantaneous contact ellipse at the specified thickness ξ of the paint layer;
– the coefficient C of profile modification.

The above four parameters—L_c, Δh, a_0, C—will be referred to as synthesis parameters.

5.2.3 Modified Conical Involute Surface

In the proposed work only the conical involute surface of the driven gear will be subjected to modification. Profile modification is determined by shifting all points of the tooth surface of the driven gear at an angle

$$\delta\theta_1 = -C(\varphi_1 - \varphi_0)^2.$$

The shift is carried out along the arc of a circle, lying in a plane perpendicular to the axis of rotation and centered on this axis.

With the profile modification, the linear contact is maintained. The transmission ratio is equal to its nominal value only if the value of the parameter φ_1 of a contact point is equal to the specified value φ_0. The value φ_0 is calculated given value Δh as follows.

Firstly, the angle δ_0 at the cone apex of zero profile modification is determined (Fig. 5.2). This cone contains the point P of zero modification and has a common origin and axis with the pitch cone. The angle is therefore calculated

$$\delta_0 = \delta_1 + \Delta h / L_c$$

Then the angle ψ_0 is determined from the equation

$$\sin^2 \psi_0 = \left(\sin^2 \delta_0 - \sin^2 \delta_b\right)$$

Finally, knowing ψ_0 with the help of formula (5.2) we calculate φ_0.

By changing the value Δh, we can shift the bearing contact along the tooth depth. A positive value causes a shift of the bearing contact on the tooth surface of the driven gear upwards the addendum, a negative value—downwards the dedendum. Note that the shift of the bearing contact on the tooth surface of the driving gear is in the opposite direction.

The value of the profile modification is proportional to the coefficient C. An increase in this coefficient leads to an increase in the irregularity of the transmitting rotation and a size decrease of the bearing contact along the tooth depth.

The longitudinal modification is intended to adjust the size of the instantaneous contact area using the value a_0 (Fig. 5.2). This modification does not affect the irregularity of the transmitting rotation. The longitudinal modification is made by shifting all points of the pinion surface by an angle

$$\delta\theta_L = -\xi(L - L_c)^2 / (r_1 a_0^2),$$

where r_1—the distance from the current point to the axis of rotation of the driven gear, L—the distance from the current point to the pitch cone apex. In case of longitudinal modification, the contact of surfaces is possible only on the intersection of the flank with the sphere of radius L_c. In case of longitudinal modification, each point shifts by an angle $\delta\theta_L$ along the arc of a circle, lying in the plane perpendicular to the axis of rotation and centered on this axis.

Thus, the modified surface of the driven gear is determined by the equations

$$x_{1m} = x_1 \cos\delta\theta_{1m} - y_1 \sin\delta\theta_{1m};$$
$$y_{1m} = x_1 \sin\delta\theta_{1m} + y_1 \cos\delta\theta_{1m} \tag{5.3}$$

$$z_{1m} = z_1.$$

In Eq. (5.3) if there are both types of modification

$$\delta\theta_{1m} = \delta\theta_1 + \delta\theta_L.$$

The coordinates x_1, y_1, z_1 are determined using the Eq. (5.1).

Equation (5.3) determine the surface of the teeth of the driven gear, and Eq. (5.1) while $i = 2$ determine the surface of the teeth of the driving gear. Gears with such flanks are to be manufactured.

5.3 The Mathematical Model of Tooth Meshing

Consider the conditions of meshing of the tooth flank of the driven gear with the tooth flank of the driving gear. The motion of the gears in meshing will be examined in a stationary coordinate system $Ox_L y_L z_L$ of the assembly, coinciding with the driven coordinate system (Fig. 5.1a).

The position of the teeth of the driven gear, determined by the relations (5.3), and the position of the teeth of the driving gear, determined by the relations (5.1) where $i = 2$, is called the initial position.

5.3.1   Calculation of the Contact Point of the Teeth

Suppose both the driving gear and the driven gear have only one tooth per gear. If the teeth are in the initial position, they do not touch each other. In order for the surfaces of the teeth gear to touch, the gears need to be rotated around their axes of rotation. Such turn is divided into two parts.

Let P_i be the point of the unmodified tooth surface of the gear under the number i, lying at the intersection of the sphere with the radius L and the pitch cone (Fig. 5.2).

Turn the driving gear and the driven gear around their axes to overlap the points P_1 and P_2. The driven gear will rotate around the axis z_1 by an angle λ_1, and the driving gear—by an angle λ_2 around the axis z_2, coinciding with the axis x_1. Because the points P_1 and P_2 converge in the plane $Ox_L y_L$, to determine the angles λ_1 and λ_2 two equations can be made

$$y_{p1}(\lambda_1) = x_{p1} \sin \lambda_1 + y_{p1} \cos \lambda_1 = 0,$$

$$x_{p2}(\lambda_2) = x_{p2} \cos \lambda_2 - y_{p2} \sin \lambda_2 = 0.$$

Here x_{p1}, y_{p1}, x_{p2}, y_{p2}—the coordinates of the points P_1 and P_2 in the initial position, and $y_{p1}(\lambda_1)$, $x_{p2}(\lambda_2)$—the coordinates of the same points after rotating the gears at angles λ_1 and λ_2 accordingly.

The position of the gears after the rotation by angles λ_1 and λ_2 from the initial position will be called the starting position. The rotation angles θ_1 and θ_2 of the driven gear and the driving gear, respectively, will be calculated from the starting position of the gears.

Let's also take into account the additional rotation angle $\delta\theta_{1m}$. As a result of all the rotations, we obtain the surfaces defined by equations

$$\theta_i^* = \lambda_i + \theta_i + \delta\theta_{i\,m}; \tag{5.4}$$

$$x_i^* = x_i \cos \theta_i^* - y_i \sin \theta_i^*; \quad y_i^* = x_i \sin \theta_i^* + y_i \sin \theta_i^*; \quad z_i^* = z_i. \tag{5.5}$$

Note that $\delta\theta_{2m} = 0$, and the values x_i, y_i, z_i are determined by the relations (5.1).

We might consider that the coordinates of the contact point depend on only one variable

$$\begin{aligned} x_1^* = x_1^*(\varphi_1); \quad y_1^* = y_1^*(\varphi_1); \quad z_1^* = z_1^*(\varphi_1); \\ x_2^* = x_2^*(\varphi_2); \quad y_2^* = y_2^*(\varphi_2); \quad z_2^* = z_2^*(\varphi_2). \end{aligned} \tag{5.6}$$

In the following paragraphs, the methods of determining such meshing characteristics as the transmission error curve, the contact path, the instantaneous contact areas
and the bearing contact are described. Two conditions are met for this. First—the
coordinates of the contact points of the surfaces of the gears are the same. Second—
the intersection lines of the contacting surfaces with the sphere of radius L have a
common tangent.

5.3.2 The Calculation of the Transmission Error Curve and the Contact Path

The transmission error curve (Fig. 5.3) is a graph of the function $\Delta\theta_2$ of θ_1 [2, 4, 11]:

$$\Delta\theta_2 = -\theta_2 - \theta_1 N_1 / N_2. \tag{5.7}$$

Here N_1, N_2 are the number of teeth of the driven gear and the driving gear
respectively. The angle θ_2 is equal to the rotation angle of the driving gear to the
touch of the flanks. The angle $\theta_1 N_1 / N_2$ is equal to the angle of rotation of the
driving gear with ideal meshing. The value $\Delta\theta_2$ is called the transmission error. A
minus sign before θ_2 in the ratio (5.7) is related the fact that gears rotate in opposite
directions in meshing.

The rotation angle of the driving gear in radians is shown on the x-axis in Fig. 5.3.
And on the y-axis—the transmission error of the driven gear in radians. The maximum
transmission error at slight load is $A = 3.59 \cdot 10^{-4}$ rad (Fig. 5.3).

Fig. 5.3 The transmission
error curve

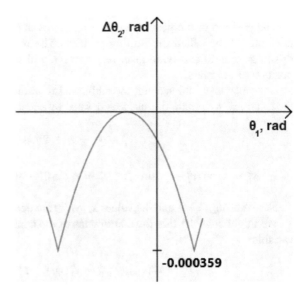

To calculate the contact path and the transmission error curve, the conditions of touch of the surfaces of the teeth are used. They can be written as three equations

$$\frac{\partial y_2^*}{\partial \varphi_2}\frac{\partial y_1^*}{\partial \varphi_1} - \frac{\partial z_1^*}{\partial \varphi_1}\frac{\partial x_2^*}{\partial \varphi_2} = 0;$$
$$y_1^* - x_2^* = 0;$$
$$z_1^* - y_2^* = 0.$$
(5.8)

The first of them expresses the condition of touch of surfaces, and the second and third express the condition of equality of two coordinates of surfaces at the contact point. The equality of the third coordinate is ensured by the fact that the contact points lie on the same sphere.

To find the dependencies (5.6) and (5.7) we assume that $L = L_c$. Equation (5.8) with a given value θ_1 become three equations with the three unknowns φ_1, φ_2, θ_2.

$$F_j(\varphi_1, \varphi_2, \theta_2) = 0, \quad (j = 1, 2, 3).$$
(5.9)

Note that in Eq. (5.9) only the profile crowning might be taken into account. That is due to the fact that the longitudinal crowning has no effect on the equation of the contact path and on shape the transmission error curve.

Newton's method is used to solve the system of non-linear Eq. (5.9).

5.3.3 The Plotting of the Instantaneous Contact Areas and the Bearing Contact

When determining the boundaries of the instantaneous contact patches, we assume that the surfaces of the teeth are covered with a layer of paint. The thickness of the paint is constant along the axis y_i and equal to ξ. So the equations of surfaces with the paint will be described by $(i = 1, 2)$

$$x_i = L(\sin \varphi_i \sin \psi_i + \cos \varphi_i \cos \psi_i \sin \delta_{bi})$$
$$y_i = L(- \cos \varphi_i \sin \psi_i + \sin \varphi_i \cos \psi_i \sin \delta_{bi}) + \xi .$$
$$z_i = L \cos \psi_i \cos \delta_{bi}$$
(5.10)

If we pick the value of the variable θ_1, then the Eq. (5.8) together with the relations (5.10), (5.4) and (5.5) can be viewed as a system of equations

$$f_j(\varphi_1, \varphi_2, L) = 0, \quad (j = 1, 2, 3)$$
(5.11)

with the respect to three unknown variables: φ_1, φ_2, L. From this system it is possible to determine the internal $L_{in}(\theta_1)$ and external $L_{ex}(\theta_1)$ boundaries of the instantaneous contact areas at given values ξ, θ_1, $\theta_2(\theta_1)$ and L_c.

Fig. 5.4 The bearing contact
under slight load

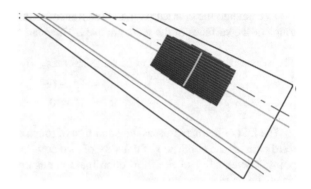

We take into account both profile and longitudinal crowning in system (5.11). The solution of the system (5.11) is also obtained by Newton's method.

The combination of all instantaneous contact areas is a bearing contact under slight load at a given ξ thickness of the paint layer (Fig. 5.4). Baxter [2] named it the "tooth bearing". On the Fig. 5.4 the tooth contour is outlined by a quadrangle. Multicoloured lines in Fig. 5.4 represent the instantaneous contact areas at different phases of meshing. Note that the length of the instantaneous areas is constant in the interval of rotation angles matching the teeth contact. The light line in the middle of the bearing contact is the contact path.

The methods described above were implemented as a software module designed to analyze the tooth meshing of a straight orthogonal bevel gears under slight load. Its produced result is the transmission error curve and bearing contact on the surface of the meshing teeth when tested on a bevel-gear tester.

The software module allows by varying the synthesis parameters to obtain the gearing with the required size and position of the bearing contact and required irregularity of the transmitting rotation.

5.4 Methods of Selection of Synthesis Parameters

The search method of the optimal values of the synthesis parameters consists of three stages. At the first stage, the initial values of the synthesis parameters are determined.

The recommended value for parameter L_c is

$$L_c = R_e - b/2, \tag{5.12}$$

where R_e—the outer cone distance; b—the face width.

The initial approximation of the shift Δh is necessary for the point P to be in the middle of depth of the flank. The maximum modulo Δh value is calculated for this, which can be estimated as follows

$$\Delta h_{bound} = L_c\left(\frac{\delta_{a1} + \delta_{f1}}{2} - \delta_1\right),$$ (5.13)

where δ_{a1}—the face angle; δ_{f1}—the root angle of the driven gear.

Note that the right part of Eq. (5.13) can be either positive or negative. Given that the contact localization is provided by modifying the flanks of the larger gear, $\Delta h_{bound} \leq 0$. The recommended starting shift value Δh is equal to

$$\Delta h = \Delta h_{bound}/2.$$ (5.14)

For the semi-axis a_0 of the instantaneous contact ellipse, we recommend the following as an initial approximation

$$a_0 = b/5.$$ (5.15)

The initial value of the coefficient C is equal to

$$C = 0.01.$$ (5.16)

For the selected initial values of the synthesis parameters, by methods described above we determine the bearing contact and the transmission error curve.

At the second stage, the obtained meshing characteristics are analyzed and, if necessary, new values of the synthesis parameters are determined. The main purpose of adjusting the values of the synthesis parameters is to avoid the edge contact and to get the maximum size of the bearing contact at the same time.

The main idea of the varying of the values of the parameters of the synthesis is as follows. If the bearing contact is too large and close to the both toe and heel, the parameter a_0 should be reduced. If the bearing contact touches the tip edge of both gears at the same time, it is necessary to increase the coefficient C.

If the bearing contact reaches the toe, L_c needs to be reduced, and if it reaches the heel we increase L_c. If the bearing contact crosses the tip edge of the driving gear tooth, then the value Δh needs to be increased, and if it crosses the tip edge of the driven gear—Δh needs to be reduced.

Increase in size in the longitudinal direction of the bearing contact is possible by increasing the value of a_0. Increase in size of the bearing contact in the profile direction can be achieved by decreasing the value of C.

The selection of values of synthesis parameters is done interactively to obtain satisfactory quality of teeth in meshing.

The recommended paint layer thickness when tested on a bevel-gear tester is $\xi = 0.005$ mm. This value is recommended for obtaining preliminary bearing contact.

At the third stage the contact problem of the elasticity theory is solved. If the contact characteristics are unsatisfactory, then new values of the synthesis parameters are determined. To achieve this the same technique as at the second stage can be used.

5.5   The Contact Problem Statement

As a result the final judgment concerning the quality of the gear is made by solving the contact problem of the elasticity theory under the specified load.

The initial data is as follows:

– tooth flank of gears;
– modulus of elasticity E_1 and E_2 of the material of the gear teeth;
– Poisson's ratio v_1 and v_2;
– torque T_1 on the driving gear shaft.

The results of the solution of the contact problem are the bearing contact and the maximum contact pressure P_{max} during operation of the gear pair.

Several meshing phases are examined in the meshing process of one pair of teeth:

$$\theta_{1min} \leq \theta_{1i} \leq \theta_{1max},$$

where i—sequence number of the meshing phase.

At each meshing phase θ_{1i}, the instantaneous contact area is determined by solving the Hertz problem and the maximum contact pressure is calculated. The bearing contact is a combination of the instantaneous contact areas calculated for all meshing phases. The maximum contact pressure P_{max} during the operation of the gear pair is calculated as:

$$P_{max} = \max_i \sigma_i.$$

5.6   Calculation of Tooth Contact Characteristics in an Arbitrary Meshing Phase

Consider the relative position of the gear pair teeth at some meshing phase. The meshing phase is determined by the value of rotation angle θ_1 of the driving gear and the value of the rotation angle θ_2 of the driven gear. The rotation angles are linked by the formula (5.7). The flanks in this position touch at the point C (Fig. 5.5a).

The instant contact area under load and the maximum contact pressure at the referred phase will be calculated using the Hertz solution of the problem of contact between two bodies bounded by surfaces of the second order [8].

Consider the contact of the teeth in coordinate system $Cuvt$, where C is the contact point. Axis t coincides with the common normal to the surfaces at the point C. The axes u, v are located in plane tangential to the contacting surfaces. The backlash δt_i between the teeth surfaces is measured along the common normal. The backlash at the moment of touch is represented by a quadratic form of coordinates u, v

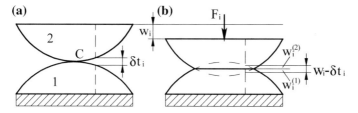

Fig. 5.5 The position of the tooth flanks **a** at the moment of touch; **b** after applying force F_i

$$\delta t_i = k_{uui} u^2 + k_{vvi} v^2. \qquad (5.17)$$

Values k_{uui}, k_{vvi} will be called the principal reduced curvatures, i.e. principal curvatures of the backlash surface.

The condition of tooth contact can be written as (Fig. 5.5)

$$w_i - \delta t_i = w_i^{(1)} + w_i^{(2)}, \qquad (5.18)$$

where w_i—the elastic displacement of the bodies along the common normal under the effect of normal force F_i.

Because of the deformation, the points of the surfaces 1 and 2 have gotten displacements $w_i^{(1)}$ and $w_i^{(2)}$ (Fig. 5.5b).

Consider the elastic displacement $w_{\tau i}$ in the direction of the tangent to the circle along which the point C moves. It is associated with an elastic displacement w_i by the relation

$$w_{\tau i} = w_i \cos \alpha, \qquad (5.19)$$

where α is the profile angle at the contact point.

The relationship between the elastic displacement $w_{\tau i}$, the normal force F_i and the semi-axes a_i, b_i of the instantaneous contact ellipse is as follows [8]:

$$F_i = \gamma_i w_{\tau i}^{3/2}; \quad a_i = \chi_i F_i^{1/3}; \quad b_i = \mu_i a_i. \qquad (5.20)$$

Here γ_i, χ_i, μ_i are the coefficients whose values depend on the angle θ_1.

The ratio $\mu_i = b_i / a_i$ of the semi-axes of the instantaneous contact ellipse is determined from the equation [6, 7]

$$J_2(\mu_i)k_{uui} - J_1(\mu_i)k_{vvi} = 0. \qquad (5.21)$$

In Eq. (5.21) the following notations are used

$$J_1(\mu_i) = \int_0^{\pi/2} \frac{\cos^2 \lambda \cdot d\lambda}{(\sin^2 \lambda + \mu_i \cos^2 \lambda)^{1/2}}; \qquad (5.22)$$

$$J_2(\mu_i) = \int\limits_0^{\pi/2} \frac{\cos^2 \lambda \cdot d\lambda}{(\sin^2 \lambda + \mu_i \cos^2 \lambda)^{3/2}}. \tag{5.23}$$

The coefficients of the relationships (5.20) are defined as follows

$$\chi_i = \sqrt[3]{\frac{3 J_1(\mu_i)}{\pi K k_{uui}}}; \tag{5.24}$$

$$\gamma_i = \frac{\pi K}{3} \sqrt{\frac{J_1(\mu_i)}{k_{uui} J_0^3(\mu_i) \cos^3 \alpha}}. \tag{5.25}$$

Here

$$K = 2 \left[\frac{1 - v_1^2}{E_1} + \frac{1 - v_2^2}{E_2} \right]^{-1};$$

$$J_0(\mu_i) = \int\limits_0^{\pi/2} \frac{d\lambda}{(\sin^2 \lambda + \mu_i \cos^2 \lambda)^{1/2}}.$$

The maximum contact pressure is calculated according to the formula

$$\sigma_i = \frac{3 F_i}{2 \pi a_i b_i}. \tag{5.26}$$

The normal force F_i depends on the torque M_i on the driving gear shaft at the i-th meshing phase

$$F_i = \frac{M_i}{r \cos \alpha}, \tag{5.27}$$

where r is the distance from the contact point C to the axis of rotation of the driving gear.

The torque M_i is determined by a given torque T_1 at each phase. The number of contacting tooth pairs is taken into account. This technique is described in detail in [6, 7, 16].

Thus, the relations (5.17)–(5.27) allow to determine the maximum contact pressure and instantaneous contact area for the torque M_i on the shaft of the driving gear.

The above solution of the contact problem of elasticity theory for straight bevel gears is implemented as a software module. Its produced result is the transmission error curve, graph showing maximum contact pressure over meshing phase, bearing contact under specified load.

The software module is designed to test the functionality of straight bevel gears at given torque on the shaft of the driving gear.

5.7   Numerical Examples

We present the results of calculations obtained using the methods described above for two orthogonal straight bevel gears.

5.7.1   Example 1

Geometric parameters and data for strength calculation for the gear 1 are shown in Table 5.1.

The values synthesis parameters derived from the formulas (5.12)–(5.16) are presented in Table 5.2 for step 1. These values are selected at the first stage of the synthesis parameters selection. The calculation results under slight load for the gear 1 are given in Figs. 5.3 and 5.4.

Table 5.1   Geometric parameters and data for gear 1

Gear parameters	Driving gear	Driven gear
Number of teeth	12	22
Outer transverse module (mm)	5	
Pressure angle (°)	20	
Face width (mm)	20	20
Outer cone distance (mm)	62.65	62.65
Nominal torque T_1 (Nm)	120	
Modulus of elasticity (MPa)	214,000	214,000
Poisson's ratio	0.28	0.28

Table 5.2   Synthesis parameters depending on step

Step	L_c	a_0	Δh	C	A_0	A_{120}	P_{max}
1	52.6	4	−0.83	0.01	$3.59 \cdot 10^{-4}$	$1.86 \cdot 10^{-4}$	1767
2	52.6	5	−0.83	0.01	$3.59 \cdot 10^{-4}$	$2.02 \cdot 10^{-4}$	1646
3	53.6	5	−0.83	0.01	$3.59 \cdot 10^{-4}$	$2.06 \cdot 10^{-4}$	1624
4	53.6	5.5	−0.83	0.01	$3.59 \cdot 10^{-4}$	$2.13 \cdot 10^{-4}$	1572
5	53.3	5.5	−0.83	0.01	$3.59 \cdot 10^{-4}$	$2.11 \cdot 10^{-4}$	1575
6	53.3	5.9	−0.83	0.01	$3.59 \cdot 10^{-4}$	$2.16 \cdot 10^{-4}$	1542
7	53.3	5.9	−0.95	0.01	$3.58 \cdot 10^{-4}$	$2.17 \cdot 10^{-4}$	1520

The results of the calculation under load for gear 1 are given in Figs. 5.6, 5.7, 5.8 and 5.9.

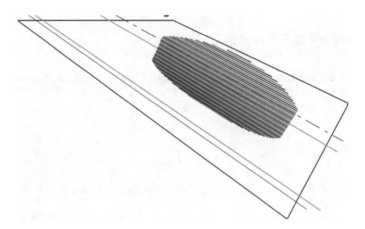

Fig. 5.6 The bearing contact on the driven gear flank under load in step 1

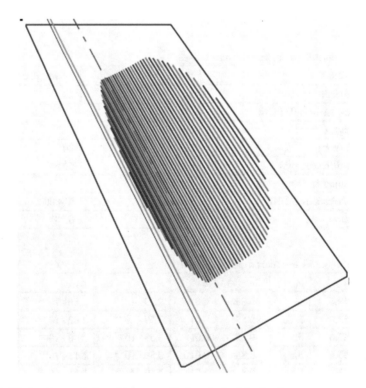

Fig. 5.7 The bearing contact on driving gear flank under load in step 1

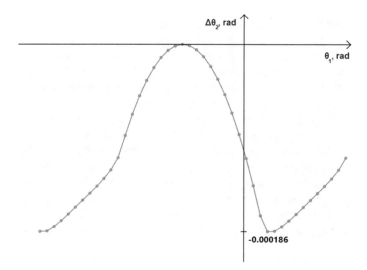

Fig. 5.8   The transmission error curve under load in step 1

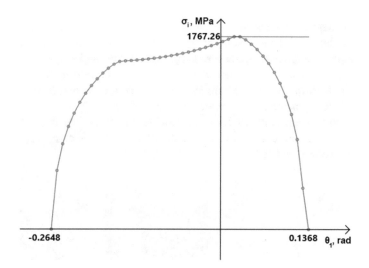

Fig. 5.9   The graph showing maximum contact pressure of the meshing phase in step 1

Figures 5.6 and 5.7 show the bearing contacts obtained using the initial values of the synthesis parameters (Table 5.2, step 1) and at torque $T_1 = 120$ Nm. The size of the bearing contacts is larger than similar bearing contacts under slight load (Fig. 5.4). It should be noted that the bearing contacts never cross the tooth edges. Thus, there is no edge contact under specified load. The second stage of the synthesis parameters selection is considered to be completed.

The maximum transmission error under load $A_{120} = 1.86 \cdot 10^{-4}$ rad (Fig. 5.8) decreased compared to similar error under slight load (Fig. 5.3) by almost two times.

In Fig. 5.9 the rotation angle of the driving gear is shown on the x-axis in radians and maximum contact pressures are shown in MPa on y-axis. The graph showing maximum contact pressure of the meshing phase can be divided into three parts. Two local maximums are clearly visible on the graph. They represent the boundaries between the single pair and two pair contact areas. In the two pair contact area at the start of tooth meshing (on the left in Fig. 5.9) the contact pressure increases from zero to 1560 MPa. In the two pair contact area at the end of tooth meshing (on the right in Fig. 5.9) the contact pressure decreases from 1767 MPa to zero. Between two local maximums there is one-pair contact area. In this area, as seen from Figs. 5.6 and 5.7, the instantaneous contact areas have almost the same length. In two pair contact areas their length changes significantly.

The maximum contact pressure in the gear, obtained in step 1 is $P_{max} = 1767$ MPa (Fig. 5.9).

As seen in Figs. 5.6 and 5.7, it is possible to increase the size of the bearing contact and reduce the maximum contact pressure. We will carry out the third stage of the synthesis parameters selection. To that end we increase the value a_0. To avoid the shift of the bearing contact to the tooth toe of the driven wheel, we increase the value L_c.

Table 5.2 shows the step-by-step change of synthesis parameters at the third stage. The calculation results for each step are also presented in the table: A_0—the maximum transmission error under slight load; A_{120}—the maximum transmission error under load at nominal torque $T_1 = 120$ Nm; P_{max}—the maximum contact pressure.

In step 7, the contact characteristics shown in Figs. 5.10, 5.11 and 5.12 were obtained. The bearing contacts takes up most of the flank, but there is no edge contact. The maximum contact pressure was reduced by 14% compared to step 1. The overlap factor stood at 1.31.

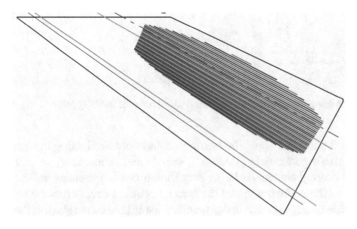

Fig. 5.10 The bearing contact on the driven gear flank under load in step 7

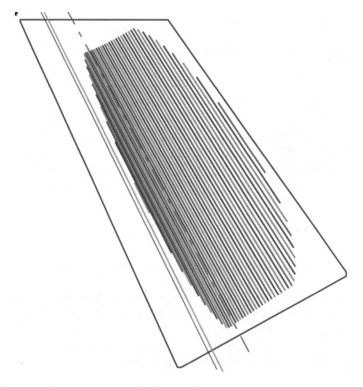

Fig. 5.11 The bearing contact on the driving gear flank under load in step 7

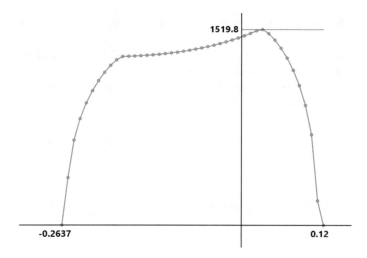

Fig. 5.12 The graph showing maximum contact pressure of the meshing phase in step 7

Table 5.3 Geometric parameters and data for gear 2

Gear parameters	Driving gear	Driven gear
Number of teeth	10	41
Outer transverse module (mm)	8.568	
Pressure angle (°)	24	
Face width (mm)	43	43
Outer cone distance (mm)	180.79	180.79
Nominal torque T_1 (Nm)	650	
Modulus of elasticity (MPa)	214,000	214,000
Poisson's ratio	0.28	0.28

5.7.2 Example 2

Geometric parameters and data for strength calculation for the gear 2 are shown in Table 5.3.

The values of the synthesis parameters obtained by formulas (5.12)–(5.16) are presented in the Table 5.4 for step 1. These values are selected at the first stage of the synthesis parameters selection. Figure 5.13 shows the bearing contacts on the driving gear tooth obtained using the synthesis parameters in step 1 under slight load (Fig. 5.13a) and at torque $T_1 = 650$ Nm (Fig. 5.13b).

The bearing contact under slight load is well-localized. But the bearing contact under load crosses to the tip edge both on the driving gear tooth and on the driven gear tooth. Thus, there is an edge contact. In this case, the Hertz solution does not apply. It is for that reason that there are dashes in last two columns in Table 5.4 for step 1.

To exclude the edge contact, we first increase the value of C. The new value of C = 0.02 in step 2. The overlap factor is 1.25.

Table 5.4 shows step-by-step changes of synthesis parameters and gear performance characteristics.

Table 5.4 Synthesis parameters depending on step

Step	L_c	a_0	Δh	C	A_0	A_{650}	P_{max}
1	159.3	8.6	−1.5	0.01	$2.38 \cdot 10^{-4}$	–	–
2	159.3	8.6	−1.5	0.02	$4.69 \cdot 10^{-4}$	$3.21 \cdot 10^{-4}$	1587
3	159.3	10	−1.5	0.02	$4.69 \cdot 10^{-4}$	$3.32 \cdot 10^{-4}$	1510
4	159.3	10	−1.8	0.02	$4.68 \cdot 10^{-4}$	$3.35 \cdot 10^{-4}$	1470
5	159.3	10.5	−1.8	0.02	$4.68 \cdot 10^{-4}$	$3.38 \cdot 10^{-4}$	1443

(a)   **(b)**

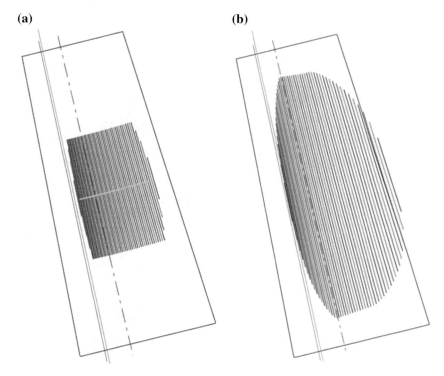

Fig. 5.13 The edge bearing contact on the driven gear flank **a** under slight load; **b** under load in step 1

The bearing contact on the driving gear flank for step 5 is shown in Fig. 5.14. The bearing contact is localized, there is no edge contact. The maximum contact pressure is 1443 MPa.

5.8   Conclusion

The technique described in the article enables to make a reasonable choice of synthesis parameters and the shape of tooth flanks to provide localized contact in straight bevel gears.

The described technique can be used to obtain localized contact in other gears.

Fig. 5.14 The bearing contact on driving gear flank under load in step 5

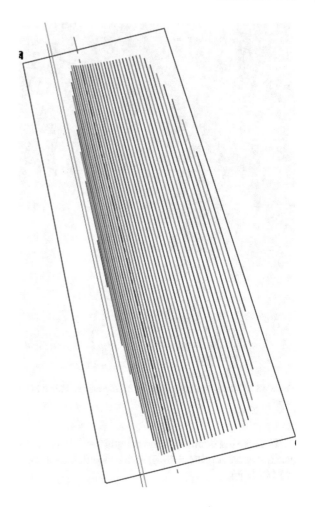

References

1. Akimov, V., Lagutin, S., Volkov, A.: New approach to the local synthesis of spiral bevel gears. In: Proceedings of the ASME International Design Engineering Technical Conferences and Computers and Information in Engineering Conference, IDETC/CIE2007. Mechanisms and Robotics Conference, Las Vegas, NV, pp. 13–17 (2008)
2. Baxter, M.: Basic Geometry and Tooth Contact of Hypoid Gears, Industrial Mathematics, vol. 11, Part 2, pp. 19–42 (1961)
3. Goldfarb, V., Trubachov, E.: Manufacturing synthesis of spiroid gearing. In: Proceedings of the Eleventh World Congress in Mechanisms and Machine Science, vol. 2, pp. 901–905. China Machine Press, Tianjin (2004)
4. Litvin, F., Fuentes, A.: Gear Geometry and Applied Theory, 2nd edn, 800 p. University Press, Cambridge (2004)
5. Lopato, G., Kabatov, N., Segal, M.: Spiral Bevel and Hypoid Gears, 423 p. Mashinostroenie, Moscow (1977) (in Russian)

6. Medvedev, V., Matveenkov, D., Volkov, A.: Synthesis of contact-optimal spiral bevel gears. Russ. Eng. Res. **35**(1), 51–56 (2015)
7. Medvedev, V., Volkov, A., Volosova, M., Zubelevich, O.: Mathematical model and algorithm for contact stress analysis of gears with multi-pair contact. Mech. Mach. Theory **86**, 156–171 (2015)
8. Rabotnov, Y.: Mechanics of Deformed Solids, 744 p. Nauka, Moscow (1979) (in Russian)
9. Sandler, A., Lagutin, S.: Technique of profile localization of bearing contact in worm gears. In: Proceedings of the International Conference on Gears, Munich, Germany, VDI-Berichte 2108.2, pp. 1233–1244 (2010)
10. Sandler, A., Lagutin, S., Gudov, E.: Design and Technological Approach to Creating of Real Worm Gears. In: International Symposium "Theory and practice of gearing—2014", pp. 381–386. Izhevsk, Russia (2014) (in Russian)
11. Sheveleva, G.: Theory of Generation and Contact of Meshing of Moving Bodies, 494 p. Stankin, Moscow (1999) (in Russian)
12. Sheveleva, G., Medvedev, V., Volkov, A.: Mathematical simulation of spiral bevel gears production and processes with contact and bending stressing. In: Proceedings of the Ninth World Congress on the Theory of Machines and Mechanisms, vol. 1, pp. 509–513. Politecnico di Milano, Italy (1995)
13. Simon, V.: Load distribution in spiral bevel gears. ASME J. Mech. Des. **129**, 201–209 (2007)
14. Syzrantsev, V., Golofast, S., Fedchenko, E.: Geometrical synthesis of high-speed transmissions with a closed system of rolling bodies. Int. J. Mech. Eng. Comput. Appl. (IJMCA) **2**(2), 19–27 (2014)
15. Volkov, A., Sheveleva, G.: Computer calculation of tooth-broaching heads for machining of straight-tooth bevel gears. Sov. Eng. Res. **10**(11), 97–101 (1990)
16. Volkov, A.: Analysis of heavily-loaded gearing with account for simultaneous operation of three pairs of teeth. J. Mach. Manuf. Reliab. **6**, 84–91 (2000)
17. Volkov, A., Medvedev, V.: Designing and technological calculations of spiral bevel gears. Tutorial, 151 p. Publishing House of the MSTU "STANKIN", Moscow (2007) (in Russian)
18. Volkov, A., Medvedev, V., Biryukov, S.: Algorithms for the synthesis and analysis of the meshing of involute straight bevel gears with localized contact, Vestnik MSTU "Stankin", vol. 1, pp. 98–105 (2019) (in Russian)

Chapter 6
Computer-Aided Design of Gears and Machine-Tool Meshing with Application of New Concepts, Images and Indices

Dmitry T. Babichev and Natalya A. Barmina

Abstract The manuscript presents the short review of works on the theory of gearing—from the formation of the theory of gearing (TG) in the 19th century to computer-aided design of gears nowadays. Difficulties of analyzing the processes of kinematic generation are specified; they are caused by incomplete adequacy of basic concepts of the classic three dimensional geometry and theoretical mechanics to the concepts that are required at determination of tooth flanks. The ways of development of the classic TG are analyzed. New and updated basic concepts are proposed: geometrical (three types of the set of normal lines—sector, prism and pyramid) and kinematic (parameters of feeding-in of points of the generating surface into the generated one—velocity, acceleration, etc.) These concepts allowed for developing a new *kinematic method* for computer-aided simulation of processes of *kinematic generation*. Fundamentals of this kinematic method are stated; it allows for determining each and all elements generated by a tool on the product item: fragments of surfaces, segments of lines, and various jogs of surfaces and lines. The method is based on the "triad": jogs + multi-parametric envelopes + new presentations on curvilinear coordinates. The manuscript gives the review of works and problems solved on the basis of the new concepts described here. 50 years ago the classic TG was formed with kinematic methods at its basis; 20 years ago the Integrated Gear Design was formed with non-differential methods at its basis. Nowadays IGD is the core of the actively progressing computer-aided design (CAD). We suppose that the next stage of the development of TG and CAD can be the stated in this manuscript method of studying the kinematic generation of tooth flanks that: has the reliability of non-differential methods, requires less intensive computing; allows for determining curvatures in meshing, and gives more illustrative results. An important issue of the manuscript is an attempt to give the scientifically substantiated, comprehensible

D. T. Babichev (✉)
Tyumen Industrial University, Volodarsky str. 38, P.O. Box 62500, Tyumen, Russia
e-mail: babichevdt@rambler.ru

N. A. Barmina
Institute of Mechanics, Kalashnikov Izhevsk State Technical University, Studencheskaya str. 7, P.O. Box 426069, Izhevsk, Russia
e-mail: barmina-nat@mail.ru

© Springer Nature Switzerland AG 2020
V. Goldfarb et al. (eds.), *New Approaches to Gear Design and Production*,
Mechanisms and Machine Science 81,
https://doi.org/10.1007/978-3-030-34945-5_6

and maximum identical Russian and English terms on the proposed new and several related concepts.

Keywords Computer-aided design · Gear · Machine-tool meshing · Kinematic generation · Kinematic method · Jogs of surfaces and profiles

Definitions, Designations and Abbreviations
The following names for elements and machine-tool meshing, their designations and abbreviations are used in the manuscript:

1. CAD—computer-aided design of gears, machine-tool meshing and manufacturing processes of gear machining.
2. MESHING—a three-element mechanism with the kinematic pair of higher degree. One should distinguish the operating meshing of the pinion and gearwheel and the machine-tool meshing of the initial tool surface (ITS) with the part machined by the enveloping method. All types of meshing consist of three basic elements: Σ_1, Σ_2 and Σ that are surfaces or lines (Fig. 6.1).
3. NORMAL LINE **N** and UNIT NORMAL VECTOR **n**—vectors perpendicular to the surface (profile) of the tooth at the considered point; they are always *directed outwards the tooth body* (Fig. 6.1).
4. SURFACES (PROFILES) OF TEETH Σ_1 and Σ_2—two-dimensional (one-dimensional) related set of points separating the tooth body from the environment (Fig. 6.1). *The tooth surface* has two curvilinear coordinates (u and v), and *the tooth profile*—only one (u or v).
5. SURFACE (LINE) OF MESHING Σ—the trajectory of the line (point) of tooth contact relative to the base member—a fixed element in the three-element mechanism (Fig. 6.1).

Fig. 6.1 Elements of meshing

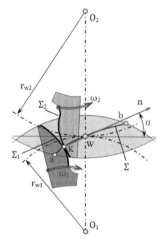

6. PITCH POINT W—the point in plane meshing where linear velocities of points
    on the pinion and gearwheel *are equal by the value and direction.* W is the instant
    center of relative velocities.
7. COORDINATE SYSTEMS (BASES Q_1, Q_2, Q)—Cartesian coordinate systems
    rigidly fixed to basic elements of meshing. In planar meshing (Fig. 6.1) bases
    are fixed to: Q_1 to Σ_1 at the point O_1 on the axis of rotation, Q_2 to Σ_2 at the
    point O_2 on the second axis of rotation, Q to Σ at the pitch point W. In spatial
    meshing there can be additional operating bases on some elements, for instance,
    at the center of the tooth profile curvature or on the axis of rotation.
8. CONJUGATED POINTS—points K_1 on Σ_1 and K_2 on Σ_2 which are contact
    ones at the point K on Σ during gear operation (Fig. 6.1).
9. CONJUGATED SURFACES OF ELEMENTS—surfaces of interacting teeth
    that provide the assigned gear ratio at all phases of meshing $i_{12} = \omega_1/\omega_2 =
    f(\varphi_1)$; where ω_1, ω_2 are angular velocities of elements, φ_1 is the angle of the
    input element rotation, $f(\varphi_1)$ is the law of motion of the mechanism (in gears
    one should usually obtain $f(\varphi_1) = $ const).
10. TG—Theory of Gearing. Two TGs are distinguished: the classic TG based
    on classic geometrical and kinematic concepts of mathematics and mechanics;
    and the alternative TG that additionally applies new geometric and kinematic
    concepts, such as: the jog of the body surface; sectors, prisms and pyramids of
    normal and tangent lines; velocity and acceleration of feeding-in and other.
11. TKG (Theory of Kinematic Generation)—an important part of the TG singled
    out into a separate section of the science on surfaces. TKG comprises the issues
    of determining and analyzing the properties of surfaces generated by mov-
    ing bodies. The generating elements here can be: initial tool surfaces (ITS) in
    the machine-tool meshing or tooth surfaces in gears, that is, one of surfaces
    Σ_1 or Σ_2.

6.1   Statement of the Problem

6.1.1   Overview of the Development of Computer-Aided Design of Gears

Formation of CAD can be conditionally divided into the following stages:

Stage 1 (till the end of the 19th century)—Development of the basis of TG as
the set of means of analysis and synthesis of gears and machine-tool meshing. Great
contribution to the development of the basis of TG was made by: L. Euler, L. Camus,
T. Olivier, R. Willis, F. Reuleaux, Ch. I. Gochman and other. For more details see
Volkov and Babichev [32], Babichev et al. [12].

Stage 2 (till the mid-1970s)—Development of the classic TG. Great contribution
to its development was made by Soviet scientists, especially F. L. Litvin. For more
details see Babichev et al. [12].

Stage 3 (to the beginning of the 21st century)—Development of the Integrated Gear Design (IGD)—in Russia, Belarus and Ukraine this stage is called "Theory of real gearing". Note, that at this stage "world-wide" and "Russian" directions of works were distinguished here. The Russian direction was supported, as a rule, by university teachers who were highly-skilled experts in TG; however, they lost the cooperation with manufacturers in 1980–2000. The researches were characterized by a high *theoretical* level. This stage is described in more details in Babichev et al. [13], Volkov and Babichev [32].

Stage 4 (till nowadays)—Development and implementation of computer-aided design in general and gears in particular; it has been especially intensively carried out in the last two decades. The leader in the development of CAD in Russia became the Institute of Mechanics—scientific department of Kalashnikov Izhevsk State Technical University (V. I. Goldfarb, E. S. Trubachev, and other). They have developed, are supporting and advancing the CAD system "SPDIAL+" for the engineering design of worm-type gears. The list of problems to be solved within the CAD is being constantly enlarged. We do not enumerate and analyze them in this manuscript. Let us note only a number of features of the *optimization design*:

- At *optimization* synthesis it is necessary to solve thousand-fold the main problems of the classic TG: to determine the conjugated surfaces (profiles) of teeth Σ_1 and Σ_2, and to compute the qualitative parameters of their contact.
- Surfaces Σ_1 and Σ_2 are determined in most cases by non-differential methods, though the basis of the classic TG is the kinematic method related to the class of differential ones.

6.1.2 Problems Arising at Solving the Main Issues of TG

The reason of application of non-differential methods within CAD is the inadequacy of numerical duplexes formed by kinematic methods of the classic TG to the real process of generation. Note, that non-differential methods are operating slower that kinematic ones by two orders; and they do not allow for determining the curvature of generated tooth surfaces.

In 1980–2000s the complex theoretical research on improving the kinematic method of analysis of kinematic generation processes was carried out in Tyumen in close cooperation with Izhevsk (both Russian cities). Approaches, methods, techniques, mathematical models and algorithms have been developed that allow for making the universal programs for computer-aided simulation of generation processes and of interaction of elements of kinematic pairs of higher degree:

- adequate to real processes of generation and contact of bodies;
- possessing the reliability of modern non-differential methods;
- covering the problems that are solved now only by differential methods.

The new method of analysis of generation processes allows for determining every and all elements formed by the tool on the product item: both fragments of surfaces and their boundaries—jogs of different origin. This method is based on the "triad": jogs + multi-parametric enveloping lines + new representations of curvilinear coordinates. Withdrawal of any element from this "triad" makes simulation of the generation process incomplete:

(1) when there is no set of normal lines in the jog of the profile or surface, a computer will lose surfaces generated by jogs;
(2) when there are no multi-parametric enveloping lines, it will be impossible to reveal cutoffs during tool approach/withdrawal;
(3) when there is no proposed system of operating with curvilinear coordinates, the computer will not form jogs on the machined surface and there will be no complete analysis of both consequent envelopes and contact of surfaces with jogs.

Works on development of the new method of analysis of kinematic generation revealed the main reason of inadequacy of methods of classic TG and TKG to the reality of generation processes. It is the absence of all components of the mentioned "triad" in the classic TG. Therefore, an alternative TKG and a new kinematic direction in TG have been in fact developed. Results of these investigations are described in DSc thesis [7] and in Babichev et al. [1, 2, 4, 5, 8, 9, 11–13, 15, 16], "Concept", "Calculation").

6.1.3 Linguistic Problems

Discussions with English-speaking editors, reviewers and readers revealed that fundamentals of the developed method (alternative theories TG and TKG) are uneasily understood and are not of great interest to developers of the software for gearing CAD. One of the reasons is, to our opinion, the absence of the unified and fixed terminology in English and Russian languages for new primary geometry and kinematic concepts. So, the reverse *computer-aided* translation of our manuscripts on this theme from English into Russian gives mishmash which is absolutely incomprehensible to Russian-speaking experts. And nothing helps here: neither the four-language dictionary on gears [30], nor reading publications on the same theme.

6.1.4 Problems Studied in This Manuscript

Basing on the described above, two problems are stated to be considered in the present manuscript:

Problem 1 To state simply and accessibly the information on the most important images, concepts and statements that are the basis of alternative TG and TKG. The information should be stated to be comprehensible also for those who are not familiar with details of gear geometry and gear machining tools. The main attention is paid not to *applying these concepts* within CAD, but to the *number of problems (previously unsolvable) that can be solved* by applying these or those new concepts and images. And instead of clarifying how these concepts work when dealing with specific problems of CAD, we should give references to publications (basically in English) where these problems are considered. The reader will be able to find confirmation of the adequacy of the obtained results to realities of kinematic generation processes and to see the reason of increasing the reliability and decreasing the labor intensity of the solution.

Problem 2 To try *one of the techniques* of developing the technical names for new concepts appearing in the science on geometry and kinematics of gearing. The feature of the technique is that when developing new names in the pair of languages (in this manuscript—Russian and English) the basis of the new name is one and same concepts-prototypes from two languages. Selection of these concepts is carried out by both native speakers of these languages (experts working in this sphere of science and professional translators), and scientists who have proposed and developed these concepts or apply them in their work.

6.2 Development of the Classic Theory of Gearing

6.2.1 Main Problems of Classic TG and TKG

The monograph by Litvin [22] is the hand book for all gear experts. It symbolizes the formation of the classic TG as an independent part of the mechanism science. It consist of two parts: "Theory of gearing" and "Gear geometry". The ratio of their volume is 1:2. In part 1 two main problems of TG are stated:

Problem 1 The scheme of the three-element mechanism is assigned and the law of motion of its elements is known. Knowing the surface Σ_1 of teeth of the 1st element, one should determine the surface Σ_2 of teeth of the 2nd element.

Problem 2 The scheme of the mechanism is assigned, surfaces Σ_1 and Σ_2 of teeth of both elements are known. One should determine the law of motion of these elements, that is, the function that correlates their displacements.

These two problems should be supplemented with two more:

Problem 3 To determine qualitative parameters of operation of the meshing. This problem was not enumerated by F. L. Litvin, but it was considered and solved in [22], in particular, radii of curvature of Σ_1 and Σ_2 were determined.

Table 6.1 Solution of the main problem of TG—to determine Σ_2 at one-parametric enveloping

Classic form of the theory of gearing	Operator form of representation
$\mathbf{r}_1 = \mathbf{r}_1(u, v)$—generating Σ_1 (1к) $\mathbf{V}_{12} \cdot \mathbf{n} = 0$—equation of meshing (2к) $\mathbf{r}_2 = \mathbf{M}_{21}(\varphi) \cdot \mathbf{r}_1(u, v)$—enveloping Σ_2 (3к) Here $\mathbf{M}_{21}$ is the transition matrix from Σ_1 to Σ_2	$\mathbf{r}_1 = \mathbf{r}_1(u, v) - \sum$ (1) $V_N = F(u, v, j) = \mathbf{V}_{12} \cdot \mathbf{n} = 0 - \text{EM}$ (2) $\mathbf{r}_2 = \mathbf{r}_2(u_2, v_2) =$ $\mathbf{M}_{21}(\varphi) \cdot \mathbf{r}_1(u, v) - \sum_2$ (3) $\mathbf{n}_1 = \mathbf{n}_1(u, v) = \mathrm{P_N}(\mathbf{r}_1(u, v)) - \mathbf{n}$ (4) $\mathbf{V}_{12} = \mathbf{V}_{12}(u, v, j) = \mathbf{V}_{12}(\mathbf{r}_1, j)$ $\qquad = \mathrm{P_V}(\mathbf{r}_1(u, v), \mathbf{M}_{21}(j)) - \mathbf{V}_{12}$ (5)

Comments to the Table 6.1: (2к, 2)—Equations of meshing: the vector of the relative velocity $\mathbf{V}_{12}$ is perpendicular to the normal line $\mathbf{n}$ to Σ_1; in (2) φ is the enveloping parameter. (4)—it is the comment, that the vector of the normal line $\mathbf{n}_1$ to Σ_1 is determined as the action of some operator $\mathrm{P_N}$ (by the way, based on methods of differential geometry) on the Eq. (6.1). (5)—it is the comment that the vector $\mathbf{V}_{12}$ is determined as the action on the vector $\mathbf{r}_1$ and matrix $\mathbf{M}_{21}$ of the operator $\mathrm{P_V}$ (by the way, based on vector or matrix operations including differentiation

Problem 4 Knowing the surface (line) of meshing Σ, one should determine the surfaces Σ_1 and Σ_2 of teeth of both elements. When Σ_1 or Σ_2 is known, the surface Σ is concurrently determined when solving the Problem 1. But within optimization synthesis sometimes the optimal Σ is synthesized not knowing Σ_1 or Σ_2 [27, 29]; then the solution of the Problem 4 is much required.

Table 6.1 presents two forms of the main design equations applied when determining the envelopes of surfaces at one-parametric enveloping: the traditional classic form and operator form (more advanced). We suppose, that the operator form is preferable for non-experts in TG. The operator form directs them that in order to understand the fundamentals of TG and TKG it is necessary to consider and master the action of two operators $\mathrm{P_N}$ and $\mathrm{P_V}$.

6.2.2   Development of Methods of Solving the Main Problems of TG and TKG

The classic TG was formed by the mid-1970s of the 20th century and further it began developing as the theory of real gearing (IGD). Among the works of the classic TG the most valuable for the theory of real gearing are, to our opinion, publications of: V. A. Shishkov who has already proposed the concept "acceleration of feeding-in" by 1950 [28]; F. L. Litvin who developed the kinematic and matrix methods of analysis of gearing [22, 23]; L. V. Korostelev, who first began to consider the kinematic parameters of load capacity of spatial gearing [19]; V. A. Zalgaller who made the analysis of the classic theory of envelopes [34]; G. I. Sheveleva who proposed the concept "covering" and who advanced non-differential methods and developed new techniques of contact stress analysis [27]; M. G. Segal who developed an efficient method of tooth contact localization [26]; S. A. Lagutin who introduced the concept "meshing space" at synthesis of gears [20]; V. I. Goldfarb, E. S. Trubachev et al. who made various researches of worm type gears and proposed new types of tools for tooth cutting on gearwheels and worms [17, 18, 31]. More detailed description of achievements of Soviet and Russian experts in development of the theory of real gearing is presented in Babichev [12, 13].

6.2.3   Basic Concepts Used in Mechanics, Geometry, TG, TKG

In [7] the analysis of kinematic methods is carried out in three points:

- purpose of methods and functions performed by them (types of problems to be solved);
- used data (systematization of basic concepts of these methods);
- contents of methods (systematization of techniques and algorithms).

There are 8 known methods of analysis of geometry of meshing [12]. However only three of them (screw, kinematic and non-differential) can be considered as the methods of TG. The rest methods are, in their essence, parts of mathematics applied in studies of meshing. In particular, the vector and matrix calculus are also applied in TG [22]. But the main methods of the *classic* TG are kinematic methods. They are commonly used in solving a great number of problems: from determining points on tooth surfaces to calculation of the curvature. However, kinematic methods are characterized by crucial drawbacks pointed above, that is why, they have been replaced by non-differential methods as more reliable though slower ones.

Let us consider several reasons of low reliability of kinematic methods of TG.

6.2.3.1   Reason 1—Incomplete Adequacy of Basic Concepts of the Classic Geometry to the Requirements of TKG and TG

TG is the science on interaction of surfaces. It is based on parts of mathematics that operate with surfaces, lines, vectors, matrices etc. Along with the mathematical apparatus, the basic geometrical concepts moved from mathematics to TG. They are primitives that are used in geometry: surface, line, tangent line, normal line, curvature, curvilinear coordinate etc. Nevertheless, real gears and machine-tool meshing have geometrical properties that are absent in the list of geometrical primitives. Correspondingly, there is no mathematical apparatus that considers and analyzes these properties.

In order to prove it, Fig. 6.2 shows the example of generation of the surface 2 by the moving body 1 that has edges, for example, by the cutter. Here points A, B and C are the generating ones that remain them when changing the direction of motion. The generating point B can be determined by the kinematic method of the classic TG

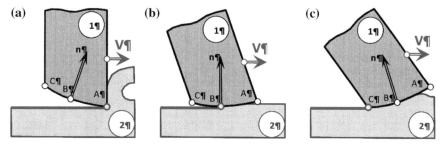

Fig. 6.2 Generation of the surface on the body 2 by the surface of the body 1 that has jogs. Generating points: **a** point of jog A; **b** point B on the line AC; **c** point of jog C

in accordance with the condition that at the contact point the vector of the relative velocity V_{12} is perpendicular to the normal vector $\mathbf{n}$ (or unit vector) to the generating element 1. For this purpose the equation of meshing is written and solved:

$$\mathbf{V} \cdot \mathbf{n} = F(u, t) = 0 \tag{6.1}$$

Here u is the curvilinear coordinate, and t is the enveloping parameter.

Note, that there is no *mathematical* answer in the classic TG to the question: "What are the conditions when the points of jog A and C become generating". The reason for this is that in accordance with the laws of geometry there are no derivatives at special points of surfaces. Therefore, there are no tangent lines and normal line to the surface at special points; and it is impossible to write the equation of meshing (6.1) for these points. That is why, many experts on kinematic generation suppose that *there is no enveloping line* for special points on generating surfaces, since there is no vector of the normal line to the generating surface at these points [27, page 464 and Fig. 14.21d].

What should be done in cases when it is necessary to investigate surfaces generated by special points A and C? There are two ways:

- Non-differential methods are applied that imply the refusal from application of differential methods (with the loss of the ability to calculate the curvature, etc.). Here the amount of calculations is rapidly increased (by two or more orders); points on the "covering" surfaces are determined rather than on the enveloping one [27]; there are problems with curvilinear coordinates on Σ_2 and other.
- Trajectories of motion are tracked (analytically or numerically) for special points on the generating element Σ_1 with respect to the basis Q_2 of the generated Σ_2 on the body 2. It is not easy and it is poorly implemented in the universal computer software.

Besides, all generating points in Fig. 6.2 can be determined basing on Fig. 6.3. It presents the same generating body 1 with points A, B, C in one of positions. All the boundaries of the body 1 shows normal lines n directed from the body 1 and vectors of relative velocities V (they are transferred from Fig. 6.2). In Fig. 6.3 the body 1

Fig. 6.3 Implementation of sectors of normal lines at points of jogs allows for determining all generating points by solving the equation of meshing $\mathbf{V} \cdot \mathbf{n} = 0$

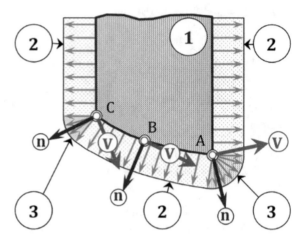

is the body of the generating element at cutting, grinding, shaving; 2 are brushes of normal lines to lines of the generating element; 3 are sectors of normal lines at jogs of the generating element; **V** are vectors of relative velocities; **n** are normal vectors or unit vectors to the body (always directed outwards the body solid).

Normal lines **n** *on lines* of the body contour form *brushes of normal lines*; and *in jogs (points A and C) they* form plane *sectors of normal lines*. And the most essential conclusion from Fig. 6.3: if *there is the normal line* **n** *in the sector of normal lines which is perpendicular to* **V**, *this jog is the generating one and the point of jog lies on the generated surface* Σ_2 of the body 2. The techniques of determining the contact normal lines **n** in jogs and the methodology of solving the equation of meshing are not considered in this manuscript. For plane jogs these issues are considered in Babichev [7, 8].

Therefore, introduction of the set of normal lines in jogs (sectors of normal line in jogs of profiles in Fig. 6.3) crucially solves two problems of the TKG:

- First, at kinematic generation, the special points on profiles and surfaces Σ_1 form the profile or surface Σ_2 that are the enveloping lines of the family of these special points.
- Second, it allows for determining the generated surface or profile Σ_2 by solving the classic equation of meshing of TG: $\mathbf{V} \cdot \mathbf{n} = 0$.

Basing on the concept "jogs of the body surfaces" the following procedures have been done: systematization of types of jogs, analysis of parameters of jogs, development of the system of curvilinear coordinates in jogs, their relation to sets of normal lines for all types of jogs on Σ_1, development of the technique of determining the points on Σ_2 and unit normal vectors to Σ_2. All possible types of sets of normal lines in jogs have been considered here: "sector of normal lines" for the point of the jog on the profile; "prism of normal lines" for the jog in the form of a rib; "pyramid of normal lines" for the apex (see in more detail in part 3.2, Fig. 6.4). Therefore, the foundation of a new TKG has been developed: (a) basing on differential approaches;

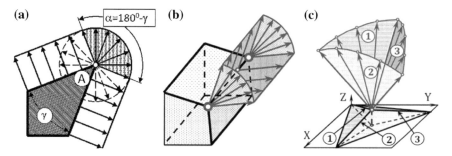

Fig. 6.4 Types of jogs of the body surface and new geometrical images: **a** *point of the profile jog* and the plane sector perpendicular to the line of jog at the point A; **b** *edge of the jog* and prism of normal lines; **c** *vertex of the jog* and pyramid of normal lines

(b) unified for all types of generating elements (surfaces, lines, jogs); (c) focused on application within universal software programs.

Note, that the concept of jogs as of specific features of generating surfaces comes from, evidently, P. R. Rodin. He wrote: "…the surface of a part that consists of a number of adjacent segments can be considered as the unified surface. The point of the jog of the surface profile located on the boundary of adjacent segments can be considered as the segment of the arc of a circle with the radius tending to zero" [24, page 52]. But, as far as we know, this idea was not further developed and no complex mathematical models were made on its basis, though the monograph [25] was published.

6.2.3.2   Reason 2—Incomplete Adequacy of Basic Concepts of Kinematics to Requirements of TKG

When analyzing the reason 2, let us consider the basic kinematic concepts of the theoretical mechanics and parameters of motions applied there—see Table 6.2. This Table is based on the idea that kinematics deals with *three basic concepts (objects): point, solid body and basis* (Cartesian coordinate system).

It follows from Table 6.2, that two types of points are used in mechanics; an individual point and a point of the solid body, that is, in the basis of this body. Three vectors are the parameters for each of these points: geometrical parameter $\mathbf{r}$; kinematic parameters $\mathbf{V}$, $\mathbf{a}$.

Table 6.2 does not present points *lying on the surface of the solid body*. It is probably accepted for the theoretical mechanics. But at kinematic generation, the surface Σ_2 on the final product is formed exactly by this type of points on Σ_1. Points on Σ_1 have additional *geometrical parameters* such as: the unit normal vector $\mathbf{n}$ to Σ_1; radii R_u and R_v of curvature Σ_1 in two directions; unit vectors $\mathbf{c_u}$ and $\mathbf{c_v}$ of tangent lines to Σ_1 that assign these directions; the radius ρ_u of geodesic torsion in the direction of $\mathbf{c_u}$; curvilinear coordinates of this point on Σ_1; and other. And there

Table 6.2 Basic kinematic concepts of mechanics and parameters of motion of objects

Object	Parameters of motion	Comments
1. Point	Radius-vector ($\mathbf{r} = \mathbf{r}(t)$), linear velocity $\mathbf{V}$ and linear acceleration $\mathbf{a}$	*There are*: $\mathbf{V}$ and $\mathbf{a}$, $\boldsymbol{\omega}$ and $\boldsymbol{\varepsilon}$—relative, absolute; $\mathbf{a}$—normal, tangential, Coriolis
2. Solid body	Parameters of the basic *point*: $\mathbf{r} = \mathbf{r}(t)$, $\mathbf{V}$, $\mathbf{a}$; three angular coordinates $\boldsymbol{\varphi} = \boldsymbol{\varphi}(t)$; angular velocity $\boldsymbol{\omega}$ and angular acceleration $\boldsymbol{\varepsilon}$	*There are motions of the point*: in space, along the surface, along the line
3. Basis (coordinate system)	Screw of the relative motion: $\mathbf{r}$, $\mathbf{V}$, $\boldsymbol{\omega}$ and two angular coordinates $\boldsymbol{\alpha}$; screw of the generating wheel ($\mathbf{r}$, $\boldsymbol{\alpha}$, $\mathbf{V}$, $\boldsymbol{\omega}$); axodes	Two bases are always considered: "fixed" and "movable", related to two "bodies"

are no parameters among kinematic basic concepts in the theoretical mechanics that depend on the enumerated geometrical parameters of the surface.

Conclusion. The set of parameters of motion of points in mechanics is limited and it *does not allow for complete describing the motion of points of the generating surface* Σ_1 in the basis Q_2 of the generated Σ_2. V. A. Shishkov [28] was the first to speak on the usefulness and necessity of such a parameter—the *velocity of feeding-in* V_N of the surface Σ_1 to the basis Q_2 of the generated Σ_2. Shishkov called the velocity V_N as the velocity of mutual approach (removal) of the generating surface Σ_1 in the basis of the generated surface Σ_2. The feeding-in velocity V_N is determined as the projection of the velocity of the tool point motion relative to the machined part on the direction of the normal line to the initial tool surface (ITS) at this point, that is, by the scalar product:

$$V_N = \mathbf{V_{12}} \cdot \mathbf{n} = V_{12x} \cdot n_x + V_{12x} \cdot n_x + V_{12x} \cdot n_x \qquad (6.2)$$

The velocity V_N can be calculated by (6.2) at any point on Σ_1 and at any position of this element 1. In general case, the velocity V_N can be: (1) *positive*—the tool is fed into the workpiece solid, that is, the metal removal takes place if it is available at this area of the workpiece; (2) *negative*—the tool is removed from the workpiece and machining at this area is impossible; (3) *zero*—at this time instant the point of the tool is contacting with the enveloping line of the family of ITS.

In 1970s a new useful concept was proposed—the acceleration of feeding-in. This acceleration a_n shows the pace of time variation of the feeding-in velocity V_N at the point of that coordinate system which the generated surface (in spatial meshing) or the generated line (in plane meshing) belongs to. The acceleration of feeding-in a_N is the time derivative t on the velocity of feeding-in V_N, that is:

$$a_N = \frac{dV_N}{dt} = \frac{d}{dt}(\mathbf{V_{12}} \cdot \mathbf{n}) = \mathbf{a_{12}} \cdot \mathbf{n} + \mathbf{V_{12}} \cdot \mathbf{n} \qquad (6.3)$$

where **n** is the unit normal vector to the generating element; n' is its derivative; $\mathbf{a}_{12}$ is the acceleration of the point of the generating element relative to the *coordinate system of the generated element* (formulas for calculation n' and $\mathbf{a}_{12}$ are given in Babichev [2, 7], Babichev and Storchak [16].

Application of a_n [21] allowed for developing the technique and software program for determining the thickness of cut layers at tooth machining by any tools (from grinding wheels to module hobs and cutting heads) operating with one or two parameters of running-in. It also allowed for establishing the laws for the height of scallops with a tape shape and the height of pyramids for the scaled shape of surface roughness at one and two parameters of running-in.

Later on, the acceleration a_n allowed for obtaining a simple basic formula [2, 7] for determining the radius of the reduced curvature R_Σ at any normal sections at the point of contact of two moving bodies (at linear contact):

$$\frac{1}{R_\Sigma} = \frac{-\omega_t^2}{a_N} \tag{6.4}$$

where ω_t is the angular velocity of rolling of Σ_1 over Σ_2 in the plane of section where the curvature is calculated; a_N is the acceleration of feeding-in.

The fundamental character of the obtained basic formula (6.4) implies that:

- In its generality it is comparable with kinematic formulas of Rodrigues and Frenet, that are common in the differential geometry and applied for determining the radii of curvature of surfaces *assigned by equations* [22]. And the formula (6.4) is preferable for surfaces *generated by enveloping methods.*
- The curvature at any normal section of all surfaces generated by enveloping methods depends only on two parameters, and *one of them (the acceleration of feeding-in a_N) does not depend on the direction of the section.*
- When $a_N > 0$, the enveloping Σ_2 is generated at the assigned point inside the enveloped Σ_1 which is physically impossible to implement.

It is known that the first derivatives are used in geometry for determining the tangent and normal lines to lines and surfaces. The second derivatives are applied for determining the radii of curvature of lines and surfaces. The absence of the feeding-in acceleration a_N in the classic TG is nonsense: without this important parameter of feeding a solid into space, it is impossible to carry out the analysis of radii of curvature at generation of many types of gearing—there are no calculation formulas in manuscripts and monographs. For instance, one should try to calculate the radii of curvature in the spatial cam mechanism in accordance with formulas [22, Sect. 34], if it is accepted in Sect. 33, page 141 in derivation of formulas 33.3–33.6 that: "the position of axes of rotation of elements in the fixed space does not vary and the distance between axes is constant". That is, the calculation formulas are available only for a small portion of all the variety of meshing, for those with the simplest motions. And formulas are non-versatile there—derivation of other formulas will be necessary for other motions and coordinate systems.

6.2.3.3   Works by Shishkov Are the Important Part of Development of Classic TG and TKG

Shishkov was the first gear-cutting expert who proposed a new parameter "*velocity of feeding-in* V_N" of motion of *the point of the surface* of one element relative to the basis of the other element. He used V_N "to determine the thickness of chips by coordinates of the point of the cutting surface and direction of its normal line ... without knowing of these surfaces of cutting" [28]. It appears that V. A. Shishkov was also the first to state the kinematic principle of determining the contact points in meshing (approximately in this way): "At the point of contact of Σ_1 with Σ_2 the velocity of feeding-in V_N is equal to zero". Later, it turned out [1, 7–9] that the velocity of feeding-in V_N *as the geometrical kinematic parameter* is applicable for all types of generating elements: surfaces, lines, points in jogs of bodies, and multi-parametric meshing. When solving several problems of kinematic generation, the second and higher derivatives of the feeding-in velocity V_N became useful.

6.2.4   Update of Basic Concepts—The Next Stage of Development of TG and TKG

The proposed new geometrical concepts—jogs and images generated by them (sectors of normal lines and other)—allow for mathematical representing [1, 5] of each body of its generating elements (including those for all types of common blade tools [5, 7]) as a continuous smooth surface differentiated at all its points. This way of representation is focused on computer-aided implementation in universal software programs of analysis of generating processes. Smoothness and derivability of such a surface (at its any point there are the normal vector and curvature) guarantee the applicability of differential methods and, first of all, kinematic ones. New geometrical concepts like "sector of normal lines" allow for developing the mathematical models of geometry of bodies and generation processes that are more adequate to real properties of bodies and generation processes than the existing models developed by means of concepts of the geometry.

   Updated and proposed geometrical kinematical concepts—the feeding-in velocity V_N, the feeding-in acceleration a_n and other—allow for determining all points on the generated surface by the kinematic method [1, 5]. Here one can either use the classic equation of meshing (6.1) and algorithms from Table 6.1, or calculate the feeding-in velocity V_N by the formula (6.2) and determine points with $V_N = 0$ on all segments of the generating surface and for all types of jogs. Application of the feeding-in acceleration a_n allows for determining reliably the radii of curvature in meshing for all types of generating elements and in accordance with universal algorithms—see basic (6.4) and design (6.2)–(6.3) formulas.

   New and updated concepts and images—geometrical (jogs, sectors of normal lines, etc.) and kinematic (velocity, acceleration, etc.) are the basis for development of a *reliable kinematic method for TG*. Note, that when determining the envelope of the family of generating surfaces and lines, the differential methods of the classic mathematics include the envelopes and trajectories of lines and points of self-intersection of generating surfaces and lines [12, 34]. It creates serious difficulties, for instance, when analyzing the consequent enveloping processes.

6.3 Fundamentals of the Alternative Theory of Gearing

6.3.1 Axioms of Alternative TG and TKG

We call axioms the main statements of generation that are the basis of the alternative TG and TKG.

Axiom 1—on geometry of bodies. The surface of any tool or gearwheel always consists of the number of segments of different surfaces, ribs and apexes. *This number can always be represented as an integrated continuous surface with two continuous curvilinear coordinates on it.* From mathematical point of view it is the smooth surface differentiable at all its points.

Axiom 2—the main law of generation. For all methods of surface mechanical generation by solid tools—by cutting, pressing, grinding—and also for all relative motions of the workpiece and tool, *the surface of the product item is finally generated only by those points of the initial tool surface (ITS) and only at that time instant, when the velocity of feeding-in of those points of ITS into the workpiece solid is equal to zero.*

These statements are not strictly proved here, that is, they are taken proofless. Though in Babichev [1, 5] there were attempts to represent these statements as two provable theorems applying two dozens of basic axioms (on the geometry of body surfaces and on the relative motion of surfaces of two bodies). In recent years we did not manage to choose a meshing for which both (or at least one) stated axioms are not valid.

6.3.2 Types of Jogs on Teeth and Images Generated by Jogs

We suppose that there are only three types of jogs on lines and surfaces of teeth. They are shown in Fig. 6.4: (a) *jog of the profile* (Fig. 6.4a)—the point of intersection of two lines; (b) *edge of the surface jog* (Fig. 6.4b)—the line of intersection of two surfaces, (c) *vertex of the surface jog* (Fig. 6.4c)—the point of intersection of three surfaces. This Figure also shows geometrical images generated by these types of jogs: (a) *plane sector of normal lines* in the profile jog; (b) *prism of normal line* to the jog edge; (c) *pyramid of normal lines* to the jog vertex. Here, vertexes of polygonal pyramids of normal lines are considered as the set of three-edged pyramids with the common vertex or the set of three-edged pyramids connected by edges of jogs. This important statement is not proved in this manuscript; note only, that it allows for applying only three types of jogs.

Figure 6.5a shows all main parameters of the jog of the tooth profile. Figure 6.5b presents two adjacent surfaces with: 1—continuous coordinate u-line with the jog on the edge; 2—brushes of normal lines to surfaces on the u-line; 3—fan of normal lines on the edge (along v-line); $\mathbf{n_A}$, $\mathbf{n_B}$—unit normal vectors at the beginning and end of the jog on u-line. Figure 6.5c shows *the prism of normal lines* on the edge and two curvilinear coordinates u and v on it. Figure 6.5d presents *the pyramid of normal lines* on the vertex and its curvilinear coordinates v_1 and v_2.

Comments on the dimensionality of all three types of sets of normal lines. Note first of all, that for any of these three sets the lengths of all normal lines are the same and equal to one for *unit* normal vectors.

Let us start with the *plane sector of normal lines* in the profile jog—see Figs. 6.4a and 6.5 (in Fig. 6.3 sectors of normal lines are present in all four jogs). In order to

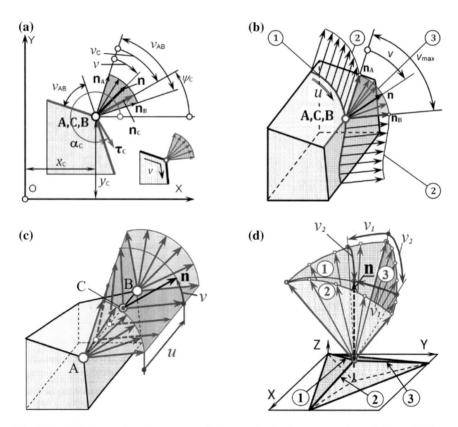

Fig. 6.5 Profile jog and curvilinear coordinates on the tooth: **a** parameters *of the profile jog*; **b** continuous line on the tooth and continuous curvilinear coordinate u on the tooth; **c** *prism of normal lines* on the edge; **d** *pyramid of normal lines* on the vertex

specify the normal line in the sector, only *one* number should be assigned—the angle *v*. That is why, *the sector of normal lines is one-dimensional.* Though the sector of the circumference where the sector of normal lines is arranged is two-dimensional.

As for the dimensionality of surface jogs shown in Figs. 6.4b, c and 6.5c, d (that is, of prisms and pyramids of normal lines), the comments are needed with application of machine-tool gearing terminology. Figure 6.5c shows *the prism of normal lines* on the edge AB, that is, on the line of intersection of the face and the end edges of the blade tool. Here, the edge AB can be the spatial curve and the edges are not planes, thus resulting in sectors of normal lines with smoothly varying angle of the sector along AB. In order to specify the normal line in the prism, *two* numbers should be assigned—the distance *u* along AB and the angle *v* in the sector at the point C (Fig. 6.5c).

That is why, *the prism of normal lines is the two-dimensional set.* Though within the usual three-dimensional geometry "prism" is the three-dimensional object. *The pyramid of normal lines is also the two-dimensional set*, since in order to specify a certain normal line in the pyramid of normal lines, one should assign *two* numbers— the angles v_1 and v_2 (curvilinear coordinates) in two sectors (Fig. 6.5d). Note, that it is better to assign these angles in sectors with the maximum and minimum angles at the apex of the sector, that is, on those edges of the tool, where the angles of the jog are maximum and minimum.

Table 6.3 presents the description of all these types of jogs, the names of elements of jogs and the main properties specific to each type of jogs.

Table 6.4 describes geometrical images generated by all types of jogs. It contains comments to three primary images: SECTOR, PRISM AND PYRAMID of normal lines and tangent lines and also to images-sets: two one-dimensional (BRUSHES and SECTORS of normal lines) and three two-dimensional (FIELD, PRISM and PYRAMID of normal lines).

Table 6.3 Types of jogs on surfaces and profiles of teeth of gearwheels and tools

1. JOG OF THE TOOTH PROFILE is the *point* of intersection of two neighboring segments of the profile (Figs. 6.4a and 6.5). *The point* A in Fig. 6.4a is considered to be plane *line*—the arc of the circumference of zero radius connecting the neighboring segments of profiles. There is *one curvilinear coordinate u* or *v* and one-dimensional sets of unit normal vectors **n** and tangent lines **τ**

2. JOG OF THE BODY SURFACE is the *line* of intersection of two surfaces (Figs. 6.4b and 6.5c) or the *point* of intersection of three surfaces of the body (Figs. 6.4c and 6.5d). Here, the *line* of intersection AB in Fig. 6.5c is called the *edge;* and it is considered to be the *surface*—a part of "tube" of zero radius covering the jog line from outside. *The point* of intersection of three planes on the tooth (Fig. 6.5d) is called the *vertex* (it is the three-edged angle on the tooth); and it is also considered to be the *surface. The vertex* is the point in which three lines of jog are met (three edges on the tooth surface). It is important to note, that *there are two-dimensional sets of unit normal vectors* **n** *and tangent lines* **τ** in these two types of jogs of *surfaces of bodies.* See Table 6.4 for more details on normal vectors $\mathbf{n_A}$, $\mathbf{n_B}$ at extreme points A and B on the jog edge and on the normal line **n** in the current point C on the jog edge (Fig. 6.5a/b/c)

Table 6.4 New geometrical images generated by jogs

1. SECTORS OF NORMAL LINES AND TANGENT LINES—*one-dimensional sets* of unit normal vectors (Figs. 6.4a and 6.5a) and unit vectors of tangent lines in the JOG OF THE TOOTH PROFILE. Three points are distinguished in each of these jogs (Fig. 6.5a, b, c): A—the origin of the jog (n_A is the unit normal vector at the point A to one adjacent surface); B—the end of the jog (n_B is the unit normal vector at the point B to the other adjacent surface); C is the contact point in the jog (n_c is the unit vector of the contact normal line at the point C of the sector of normal lines). There is also the current point C designated in Fig. 6.5c with the coordinate v in the jog (n is the unit normal vector at the current point). Boundaries of sectors are unit normal vectors (n_A and n_B) and unit vectors of tangent lines in the jog (tangent lines are not shown in Fig. 6.5) to two lines that generate the jog. Sectors are the part of the plane with the curvilinear coordinate v that assigns the position of vectors n and τ in the sector

2. PRISMS OF NORMAL LINES AND TANGENT LINES—*two-dimensional sets* of unit normal vectors (Fig. 6.4b and 6.5c) and unit vectors of tangent lines going out of the edge, that is, of the JOG OF THE BODY SURFACE. If the edge is not the segment of the straight line, then prisms will be curvilinear with non-plane edges (as shown in Fig. 6.5c). Prisms contain two curvilinear coordinates that assign the position of vectors in prisms: u—along the edge and v in the plane perpendicular to the edge (as for sectors). PYRAMIDS OF NORMAL LINES can be at the ends of prisms of normal lines (at the point in Figs. 6.4b and 6.5c)

3. PYRAMIDS OF NORMAL LINES—*two-dimensional set* of unit normal vectors going out of the vertex (Figs. 6.4c and 6.5d), that is, out of the JOG which is the point of intersection of three surfaces of the body. Pyramids of normal lines are always three-edged. Edges of pyramids are three sectors of normal lines (in Figs. 6.4c and 6.5d these sectors are designated by digits 1, 2, 3; and each of the sectors is perpendicular to the corresponding edge of the jog on the tooth surface). Edges of pyramids of normal lines are three unit normal vectors to three intersecting surfaces of the tooth. The position of the normal line inside the pyramid of normal lines is assigned by two curvilinear coordinates v_1 and v_2. The coordinate v_1 is the angle assigning the position of the vector n in the sector with the maximum angle of the fan of normal lines; the coordinate v_2 is the angle assigning the position of the vector n in the sector with the minimum angle of the fan of normal lines (see Fig. 6.5d)

4. FAMILIES OF NORMAL LINES—one-dimensional or two-dimensional sets of unit normal vectors to the surface or line going out of the solid body. Keep in mind, that the meshing involves only smooth lines (in plane meshing) or surfaces (in spatial meshing) that are differentiable at all points. The following families are distinguished: FIELD OF NORMAL LINES (to the "usual" surface), BRUSH OF NORMAL LINES (to the "usual" profile), SECTOR OF NORMAL LINES (at the plane jog, including at the point of the edge), PRISM OF NORMAL LINES (to the edge of the jog), PYRAMID OF NORMAL LINES (to the vertex of the jog). All sets are two-dimensional, besides the brush and sector of normal lines which are one-dimensional (Figs. 6.4 and 6.5)

Table 6.5 presents additional terms for new images, concepts and parameters used for describing jogs, geometry and kinematic images and also their specific features.

Table 6.5 Additional terms for jogs and geometrical images

1. PLANE LINE—the segment of the plane curve assigned by one equation or splines with the focused normal line; the points can be present at the ends of the segment
2. TOOTH PROFILE—the connected set of segments with common points
3. PIECE OF THE SURFACE—a part of the surface described by one equation; it is limited by edges; it has orientation (the normal line goes outwards)
4. TOOTH SURFACE—the connected set of pieces of surfaces with edges and vertexes in places of connection of pieces
5. GEAR-RIM—the connected set of tooth profiles or surfaces
6. EDGE—the line of intersection of two surfaces; it has the prism of normal lines and two curvilinear coordinates; vertexes can be at the ends of the edge. The edge with the prism of normal lines is regenerated into the line with the line of normal lines if the neighboring surfaces are contacting rather than intersecting
7. POLYLINE—the connected set of segments of lines in which neighboring lines are either contacting or forming the angular jog. See in more details in [7], shortly—in the p. 3.4
8. CURVILINEAR COORDINATE—the parameter that characterizes the position of the point and unit normal vector on the segment, polyline, tooth profile, piece of the surface, tooth surface, gear-rim, point, vertex or edge. It is shortly described in p. 3.5

6.3.3 Polylines—Sets of Similar or Different Segments of Lines

We use segments of typical plane lines (straight lines, arc of circumference, involute, etc.) and we get two kinds of typical polylines (poly-arc—the set of segments of straight lines and arcs of circumferences) and the poly-involute of circumferences. Numerous issues related to assignment and application of polylines (choice of typical lines, unified parameters for them, parameters of curvature, parameters of jogs, profiles, surfaces and polylines, and other) are considered in [7].

6.3.4 Curvilinear Coordinates on Lines, Surfaces and in Jogs

Here we consider shortly the issue of curvilinear coordinates, since it is one of components of the "triad", on which the developed TKG is based: jogs + multi-parametric envelopes + special actions with curvilinear coordinates. We use their two types: continuous and discrete [7]. About half dozen types of curvilinear coordinates are distinguished: natural, unified, regulated and other [7]. The preference is given to natural curvilinear coordinates of lines, when u-coordinate is the length of the arc S of this coordinate line. It allows for composing complete tooth profiles, gear rims and sometimes [10] meshing lines of segments of *different lines*.

Figure 6.6a, b shows the most essential coordinates—unified regulated ones for the tooth profile or for the face profile of the gear rim (parameter u). The limits of measurement of u are for the tooth N1 $\{0 \leq u \leq 1\}$; for the tooth N2 $\{1 < u \leq$

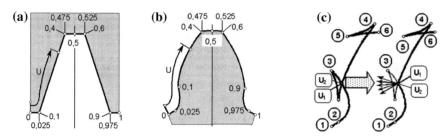

Fig. 6.6 Regulated curvilinear coordinates and their transformation: **a** on the generating element; **b** on the gearwheel; **c** removal of the loop 3 on the profile

2}; and so on. The "standard" of regulation is proposed—see Fig. 6.6a, b: (a) for the main profile $\Delta u = 0.30$ is assigned; (b) for segments of apexes and roots $\Delta u = 0.05$ for each; (c) for the transient curve and for the jog $\Delta u = 0.075$ for each. The mutual correspondence between points of the generating and generated profiles will be provided here, and also between their curvilinear coordinates (even in case of losing the conjugacy of profiles due to undercutting and removal of loop-type segments). Figure 6.6c shows that the loop on the left profile is removed and instead of it the jog with the prism of normal lines appears on the right profile. For such a profile correction the range Δu of the fan of normal lines in the jog is taken to be equal to the range Δu of the curvilinear coordinate calculated for the loop. For consequent enveloping (for example, grinding wheel → worm → worm gearwheel) it allows for inheriting initial curvilinear coordinates, for instance, of the grinding wheel on the worm gearwheel. Moreover, having assigned the point on the tooth of the worm gearwheel, one can find "conjugated" points both on the worm and the grinding wheel even when the point is located in the jog or removed with the loop (see Fig. 6.6c).

6.3.5 Local and Global Quality Parameters of Meshing

The list and systematization of quality parameters applied in the alternative TG are given in Babichev [2, 3, 7, 15]. The following sets of parameters are considered: local and global; for gears and for machine-tool meshing; geometry, force and kinematical; and so on. Note, that the "standard" of regulation of curvilinear coordinates simplifies the "standard" representation of local quality parameters obtained on tooth surfaces in gears and machine-tool meshing.

6.3.6 Multi-parametric Meshing—An Important Part of TG and TKG

Here we consider shortly multi-parametric meshing, since it is also one of the components of the "triad" which the developed TKG is based on. *Meshing of surfaces with the number of enveloping parameters greater than two (for spatial meshing) and one (for plane meshing) is called multi-parametric.* Almost all machine-tool types of meshing are multi-parametric; they are operating by kinematic generation: *motion of cutting-in of the ITS into the workpiece makes them multi-parametric. In gears the multi-parametric meshing is necessary* for analysis of methods of assembling and disassembling of gear elements. For example, if gearwheel teeth in the gear with high helix angle of the worm are cut by axial cutting-in of the hob, it can turn out that such a worm pair can be assembled only by screwing the worm into gearwheel teeth.

Properties and features of multi-parametric meshing. The generated surface is called *the envelope of the multi-parametric family of generating surfaces.* Its physical essence (in terms of machine-tool meshing) is the boundary of the area on the item where no point of the tool body can reach. Mathematicians say that the enveloping and enveloped surfaces should be in constant contact with each other. It takes place in meshing not always. It is established [1, 5] that there is such a set of envelopes in multi-parametric meshing:

- the envelope of the n-parametric family of enveloped surfaces;
- envelopes of $(n-1)$-parametric families of enveloped surfaces;
- and so on up to one-parametric enveloping families;
- stationary surfaces (envelopes of the 0-parametric family).

All these envelopes can be determined by the kinematic method and by the same software, as shown in Fig. 6.7. It presents the simulation results for gear machining by the involute shaping cutter with internal teeth. Parameters of the machine-tool meshing are: $z_0 = 15$, $z_2 = 20$, m $= 10$. The small difference (z_2-z_0) is taken to get large cuts on the gearwheel tooth profile. They are easier to be replaced and analyzed.

Figure 6.7a, b, c shows one-parametric generation (by rolling motion), Fig. d, e, f shows two-parametric generation (by joints and independent motions of rolling and approach-removal of the shaping cutter).

Features of determining the gearwheel tooth profile shown in Fig. 6.7. Firstly, there are two lines of meshing (ML) both for one-parametric and two-parametric enveloping; and they are all the closed lines with jogs. Secondly, each smooth segment on each ML gives its own smooth segment in the coordinate system of the gearwheel (on the tooth and outside the gearwheel body). Two segments of straight lines of the *upper* ML in Fig. 6.7a form segments of involutes of the right and left tooth flanks. Two segments of straight lines of the *lower* ML in Fig. 6.7a form segments of involutes in the area of secondary cutting. Arcs of circumferences on the ML with centers on the axis of the shaping cutter are the place where jogs of the cutter profile

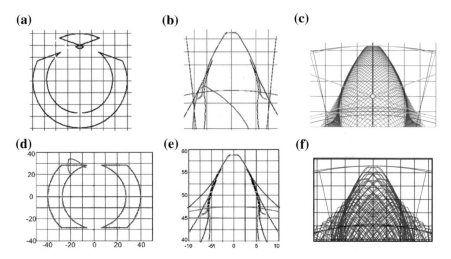

Fig. 6.7 Lines of meshing of the involute shaping cutter with the gearwheel; and root profiles of the gearwheel with internal teeth: **a, b** meshing lines at rolling and two-parametric enveloping; **c, d** root profiles and all possible cuts on the tooth (they are obtained by kinematic method); **e, f** root profiles obtained by rolling by non-differential method

form lines in the area of secondary cutting which can intersect the generated tooth profile and cut its portion.

Figure 6.7e, f show determination of the gearwheel tooth root profile by the non-differential method—by the direct enveloping. Having compared the right and the middle columns in Fig. 6.7, we state the advantages of the differential method: (1) the number of analyzed points is less by 1-3 order; (2) the visualization is much better; (3) one can evaluate quantitatively the risk or value of interference of any kind (by the least distance from the main profile to lines of possible cuts); (4) one can determine all segments of surfaces generated by enveloping—the enveloping line itself, including the cuts in areas of the secondary cutting, and also cuts from approach-removal of the tool. Additionally, differential methods allow for calculating the curvatures; and the sign of the reduced curvature can be used to reject the segments of profiles generated inside the body of the generating element. Note, that we managed to obtain the *meshing line* (Fig. 6.7) in *multi-parametric meshing* and determine the generated *line of possible cut* of the gearwheel tooth for the first time [6]. In more details the theory of multi-parametric meshing as the part of the "triad" important for TKG is considered in [6–8].

6.3.7 Theoretical Investigations and Computer-Aided Developments Made by the Alternative TG

The main results of our theoretical investigations on this theme are stated in theses [3, 7] and published in Babichev and Serebrennikov [1, 2, 4, 5, 8, 10, 15]. These investigations are focused on the development of kinematic methods of the theory of gearing applied at *computer simulation* of gear and tool operation. In [3] the methodology of the numerical analysis of gearing is developed, that does not require carrying out the analytical investigation of the geometry and kinematics of specific gears and manufacturing types of meshing. In Babichev [14] computer-aided implementation of this ideology is described. One of the elements of this ideology is the application of the generalized gearing with coordinate systems and motions that are used for all operating and manufacturing types of meshing. In Babichev [1, 5] fundamentals of the alternative theory of generation are developed and stated; they are based on new geometry and kinematic images and concepts [4]: sector, prism and pyramid (or fan, wedge and bunch) on normal lines in jogs of profiles and surfaces [8]; and acceleration of feeding-in [2]. The essential component of the alternative theory is the application of consequent and multi parametric enveloping [6] and discrete and continuous curvilinear coordinates, including natural [10][1] and regulated ones with the controllable arrangement of points along profiles [7]. The difference of the developed theory of generation from the classic one is that we consider not separate (the most important) segment of the tooth, but the whole profile or surface of the tooth (or root or even the whole gear rim) [5, 7]. The technique is developed for such complete describing of any blade tools (shaping cutters and different hobs) that allows for determining both the generated surface and dynamic angles of cutting. In this case, the ITS will have continuous smooth curvilinear coordinates [5].

Software developments made within this theme. About fifty software programs applying new images, concepts and parameters have been developed and tested. They are mainly the investigation programs purposed for working out the techniques and algorithms of analysis and synthesis of gears and machine-tool meshing. These software programs are described to the fullest extent in theses [1, 3, 7, 21]. The thesis [7] can be found in Internet.

---

[1]The natural equation of the line is the equation that relates the length of its arc calculated from the initial point on the line (natural curvilinear coordinate u) with the radius of curvature R of the line at this point. For example, for the involute: $R = \sqrt{2r_b u}$, where r_b is the radius of the basic circle; the initial point is the return point on the involute. This equation of the line does not depend on applied coordinate systems—it is the unique property of natural equations.

6.4 Development of the System of Russian and English Terms for New Concepts, Images and Parameters

The problem of translating the Russian terms of the theory of gearing into English is far from being solved. There are plenty of standards, rules, and regulations on English terms for gear elements, geometry, etc. like ISO, DIN, ASME and so on; and Russian State Standards on gear terminology. They were published long time ago and do not lead to any unambiguous correspondence of Russian and English terms for many reasons, including the constant development of this field by independent gear scientists from all over the world who publish their research results without a backward glance to the numerous regulations on designations in different countries. The question of coordinating English and Russian terms in gear science is a matter of an integral encyclopedic research of many people that can take several decades, to our opinion. In this manuscript we deal only with translation of the newly introduced terms and tend to show all the complexity of correlation of terms in different languages.

Table 6.6 presents the Russian-English dictionary of 10 terms for new images and concepts discussed in this manuscript. The table contains the type of the concept or image, the recommended term for it, the paragraph in the manuscript where it is specified, and comments to the term. Let us discuss how these terms have been chosen and specified.

6.4.1 Specification of "Accepted" Terms for New Concepts and Images

The first of concepts of Table 6.6—"acceleration of feeding-in" was proposed in Russian by V. A. Shishkov in the end of 1940s. He also called it "the velocity of approaching and removal". In 1960s in the Sverdlovsk Polytechnic Institute the term "velocity of pressing-in" was applied. The name of the second concept—"the acceleration of feeding-in" is rather logic and understandable. The method of choosing *English analogs* is shown in the column "Comments" in Table 6.6.

Terms for concepts "jog of the surface" and "jog of the line" are present in practice of Russian and English languages. They should just be recommended for application in TG.

Terms of the group of secondary images—"edge of jog", "vertex of jog", "point of jog" are inherited from geometry; and they should only be clarified when applied in TG.

Table 6.6 Russian and English terms for new images and concepts of the theory of gearing

Type	Original term in Russian	The term translated into English	Essence described in	Comments
Concepts—parameters	1. Скорость внедрения	Velocity of feeding-in	p. 2.3.2	In Russian these terms became fixed. When translated into English this new term was obtained by choosing the closest one among the synonyms (intrusion, approach, punching, etc.)
	2. Ускорение внедрения	Acceleration of feeding-in		
Concepts	3. Излом поверхности	Jog of the surface	p. 1 Table 6.3	In Russian the term "jog of the surface" or "jog of the line" became common. In English there are still no publications on this theme
	4. Излом линии (профиля зуба)	Jog of the line (tooth profile)	p. 2 Table 6.3	
Primary images	5. Сектор нормалей	Sector of normal lines	p. 1 Table 6.4	Initially other terms were used in Russian and English: "fan, wedge, bunch" (when translated from Russian, the method of calquing was used). Now the authors recalled them as "sector, prism, pyramid" for the reasons explained above. The terms are still not used in English
	6. Призма нормалей	Prism of normal lines	p. 2 Table 6.4	
	7. Пирамида нормалей	Pyramid of normal lines	p. 3 Table 6.4	
Secondary images	8. Ребро излома	Edge of the jog	p. 1 Table 6.3	These terms come from geometry. Russian terms are common and comprehensible. They have been easily translated into English by mathematical calquing of the corresponding terms
	9. Вершина излома	Vertex of the jog	p. 1 Table 6.3	
	10. Точка излома	Point of the jog	p. 2 Table 6.3	

6.4.2   Choosing of Russian and English Terms for New Primary Geometrical Images

The question is about **three primary images**: "sector, prism and pyramid of normal lines" presented in Table 6.6, pp. 5–7. Originally in the Thesis [7] and the contributed volume [8] these terms were chosen individually:

- The set of normal lines as the plane "**sector** of normal lines" was named "**fan**" [7], since it resembled in its shape the usual household fan both in the jog of lines (Fig. 6.4a) and in the jog of surfaces (Fig. 6.5); and it was for the first time translated into English by calquing as "**Fan**" in [8].
- The set of vectors "**prism** of normal lines" was named "**wedge**" [7], since it resembled in its shape the wedge, if the line of the jog was the segment of the straight line (Fig. 6.4b), and it was for the first time translated into English by calquing as "**Wedge**" in [8].
- The set of vectors "**pyramid** of normal lines" (Fig. 6.4c) was named "**bunch**" [7]; and it was for the first time translated into English by calquing in [8] as "**bunch**".

Note, that in mathematics the following concepts are applied in Russian: "bunch of planes" (meaning the set of all planes passing through one straight line) and "bunch of straight lines" [33]. The "bunch of straight lines" in this case is close in mathematics to the image previously named "**fan**", and later "**sector**".

The terms "**Fan, Wedge, Bunch**" used by the authors in [7] are complex for English speaking readers; their translation from Russian into other languages is very disputable due to many linguistic factors. For instance, "**fan**" in English is not only the household appliance or a hand fan; it is also a part of many technical devices operating with the flow of gases: a cooler, a propeller blade, etc. Also the term "fan" in English has about one hundred other meanings, including the enthusiast, the admirer, the blow dryer, etc. In order to eliminate the polysemy and misunderstanding, the authors proposed to rename previously introduced (still not common) terms into the new ones closer to geometrical. The new proposed terms are consonant in both Russian and English; they are international mathematical terms understandable for everyone. It was also desired to make them a group of similar terms based on geometrical figures. The Russian and English triples of terms are suitable here "**сектор, призма, пирамида**" in Russian and "**Sector, Prism, Pyramid**" in English. These terms are consonant in many languages and are related to volumetric bodies similar to sets of normal lines which they denote. In this case the difficulties in communication of gear geometry experts are overcome; and the translation of texts on geometry of gears and tools by the corresponding software programs will be unambiguous.

6.4.3   Refinement of English Terms for Secondary Geometrical Images and Additional Terms for Jogs and Images

The question is about terms for *secondary images*: "edge, vertex and point of jog" in Table 6.6, pp. 8–10. The essence of these terms is explained in Table 6.4. Table 6.6 gives comments on refinement of these terms.

As far as concepts-parameters and concepts (Table 6.6, pp. 1–4) are concerned, the authors translated them by means of common geometrical concepts similar in these two languages. The clarification made by drawings, figures and schemes provide these new terms to be understandable and unambiguous.

Nevertheless, the issue of translating common and new terms from the theory of gearing from Russian into English and vice versa remains limitless for many linguistic, social, technical and other reasons.

6.5   Conclusions

(1) It is shown that geometrical concepts that came to the theory of gearing from mathematics and kinematic concepts of mechanics do not provide the reliable solution of many essential problems of the theory of kinematic generation. Having done the analysis of basic concepts applied in mechanics, geometry and theory of gearing, new and updated images and concepts (first of all, *geometry and kinematic*) have been proposed to be applied in the theory of gearing.

(2) A new and supplemented system of *geometrical kinematic* parameters of motion of points of the surface of one solid body relative to the other body (including the velocity of feeding-in, acceleration of feeding-in and other) has been developed.

(3) The system has been developed that allows for representing any solid body or its generating elements as a continuous smooth surface differentiable at all its points (including those of the jogs if they are present on the generating body, cutting edges of the tool, for instance). This system is based on new *geometrical images* (sector, prism and pyramid of normal lines) and on the definite methodology of introduction and application of *curvilinear coordinates*.

(4) These and other developed concepts, images and parameters allowed for developing a new direction in the theory of kinematic generation focused on getting the computational models, adequate to the reality and aimed at creation of reliable software programs of computer-aided design.

(5) Maximum similar Russian and English terms (obtained mainly by calquing) for the proposed new concepts, images and parameters are introduced.

6.6 Financing and Acknowledgements

The work is financially supported by the project N 9.6355.2017/БЧ of the state assignment of the Ministry of Education and Science for the period 2017–2019 in Tyumen Industrial University.

The authors are thankful to the management of Tyumen Industrial University and Institute of Mechanics of Kalashnikov ISTU for the support of this research work.

We also appreciate the assistance of Ph.D., Senior Researcher Sergey A. Lagutin in the selection of English terminology and any-time support.

The authors apologize for the great number of self citations on application of new concepts and images in the theory of gearing. The reason is in the absence of other researchers currently dealing with the development of kinematic methods of TG at all and the new geometrical and kinematic concepts and images, in particular.

References

1. Babichev, D.A.: Development and approbation of the number of qualitative parameters for synthesis of spur and helical gears. Ph.D. Thesis (in Russian) (2013)
2. Babichev, D.A., Serebrennikov, A.A., Babichev, D.T.: Qualitative indexes of planar gearing operation. In: Proceedings of the 7th International Conference on Research and Development of Mechanical Elements and Systems, Zlatibor, Serbia, pp. 623–630 (2011)
3. Babichev, D.T.: Issues of investigation of geometry and kinematics of gearing. Ph.D. Thesis, Sverdlovsk, UPI (in Russian) (1971)
4. Babichev, D.T.: On basic geometrical primitives of the theory of gearing. In: Proceedings of the International Conference on Theory and Practice of Gearing, Izhevsk, pp. 469–474 (in Russian) (1996)
5. Babichev, D.T.: Basic rack surface of edge tools. In: Proceedings of the International Conference on Theory and Practice of Gearing, Izhevsk, pp. 412–421 (1998)
6. Babichev, D.T.: On application of multi-parametric envelopes at computer-aided simulation of processes of generation on operating and machine-tool gearing. In: Proceedings of the International Symposium Theory and Practice of Gearing, Izhevsk, ISTU Publication, pp. 302–315 (in Russian) (2004)
7. Babichev, D.T.: Development of the theory of gearing and generation of surfaces on the basis of new geometrical kinematic representations. In: DSc in Engineering Thesis, Tyumen, TumGNGU Publication (in Russian) (2005)
8. Babichev, D.T.: Development of kinematic method of theory of gearing to determine areas of tooth flanks produced by jogs of generating solids. In: Litvin, F.L., Goldfarb, V., Barmina, N. (eds.) Theory and Practice of Gearing and Transmissions, vol. 34, pp. 159–188. Springer (2015)
9. Babichev, D.T.: Development of geometric descriptors for gears and gear tools. In: Goldfarb, V., Trubachev, E., Barmina, N. (eds.) Advanced Gear Engineering. Mechanisms and Machine Science, vol. 51, pp. 231–254. Springer (2018)
10. Babichev, D.T., Babichev, D.A.: Optimization synthesis of tooth profile as the segment of the curve assigned by natural equation. In: Proceedings of the International Symposium Theory and Practice of Gearing, Izhevsk, ISTU Publication, pp. 301–308 (in Russian) (2014)
11. Babichev, D.T., Babichev, D.A., Lebedev, S.Yu.: Concept of gear synthesis based on assignment of instant contact areas for loaded teeth. J. Int. Rev. Mech. Eng. **12**(5), 420–429 (2018)

12. Babichev, D.T., Lagutin, S.A., Barmina, N.A.: Russian school of the theory and geometry of gearing: its origin and golden period (1935–1975). J. Front. Mech. Eng. **11**(1), 44–59 (2016)
13. Babichev, D.T., Lagutin S.A., Barmina, N.A.: Russian school of the theory and geometry of gearing. Part 2. Development of the classical theory of gearing and establishment of the theory of real gearing in 1976–2000 (2020)
14. Babichev, D.T., Plotnikova, V.S.: On development of the software complex for computer-aided numerical investigation of gearing. J. Mech. Mach. Moscow, Nauka (45), 36–43 (in Russian) (1974)
15. Babichev, D.T., Storchak, M.G.: Synthesis of cylindrical gears with optimum rolling fatigue strength. In: Production Engineering. Research and Development, vol. 9, No. 1, pp. 87–97. Springer (2015)
16. Babichev, D.T., Storchak, M.G.: Quality characteristics of gearing. In: Goldfarb, V., Trubachev, E., Barmina, N. (eds.) Advanced Gear Engineering. Mechanisms and Machine Science, vol. 51, pp. 73–90. Springer (2018)
17. Goldfarb, V.I., Tkachev, A.A.: Design of involute spur and helical gears. New approach. Izhevsk, ISTU Publication (in Russian) (2004)
18. Goldfarb, V.I., Trubachev, E.S., Lunin, S.V.: System of hobs unification for gear-wheel cutting of worm-type gears. In: Proceedings of the ASME International Conference IDENC'07, Las-Vegas, USA (2016)
19. Korostelev, L.V.: Kinematic parameters of load capacity of spatial gearing. J. Izvestiya Vuzov, Mashinostroenie **N10**, 5–15 (1964). (in Russian)
20. Lagutin, S.A.: The meshing space and its elements. J. Soviet Mach. Sci. **4**, 63–68 (reprinted from Mashinovedenie **4**, 69–75) (1987)
21. Langofer, A.R.: Enhancement of layout and operation of rolling gear-cutting hobs based on mathematical modeling of their loading. Ph.D. Thesis, Mosstankin (in Russian) (1987)
22. Litvin, F.L.: Theory of Gearing. Nauka, Moscow (in Russian) (1968)
23. Litvin, F., Fuentes, A.: Gear Geometry and Applied Theory, 2nd ed. Cambridge University Press (2004)
24. Rodin, P.R.: Fundamentals of Generation of Surfaces by Cutting. Kiev, Visha shkola (in Russian) (1977)
25. Rodin, P.R.: Fundamentals of Design of Cutting Tools. Kiev, Visha shkola (in Russian) (1999)
26. Segal, M.G.: Types of localized contact of bevel and hypoid gears. Mashinovedenie **N1**, 56–63 (in Russian) (1970)
27. Sheveleva, G.I.: Theory of Generation and Contact of Moving Bodies. Mosstankin, Moscow (in Russian) (1999)
28. Shishkov, V.A.: Formation of Surfaces by Cutting According to Generation Method. Mashgiz, Moscow (in Russian) (1951)
29. Shishov, V.P., Nosko, P.L., Fil, P.V.: Theoretical fundamentals of synthesis of gears. Lugansk, SNU n.a. V. Dal (in Russian) (2006)
30. Starzhinsky, V.E., Antonyuk, V.E., Kane, M.M., et al.: Reference Dictionary on Gears: Russian-English-German-French. Saint-Petersburg, CCP OAO "Svetoch" (2004)
31. Trubachev, E.S.: Several issues of tooth generating process by two-parametric families of generating lines. In: Litvin, F.L., Goldfarb, V., Barmina, N. (eds.) Theory and Practice of Gearing and Transmissions, vol. 34, pp. 97–116. Springer (2016)
32. Volkov, A.E., Babichev, D.T.: History of gearing theory development. In: Terminology for the Mechanism and Machine Science. Proceedings of the 25th Working Meeting of IFToMM Permanent Commission on MMS, pp. 71–102. Gomel–Saint-Petersburg (2016). ISBN 978-985-6477-45-7
33. Vygodskiy, M.Ya.: Reference Book on Higher Mathematics (in Russian) (2001)
34. Zalgaller, V.A.: Theory of Envelopes. Nauka, Moscow (in Russian) (1975)

Chapter 7
Aspects of the Kinematic Theory of Spatial Transformations of Rotations: Analytic and Software Synthesis of Kinematic Pitch Configurations

Valentin Abadjiev and Emilia Abadjieva

Abstract The present work deals with the kinematic theory of spatial rotations transformation in the context of defining a kinematic and geometrical essence of the basic building elements of the mathematical models for synthesis of hyperboloid gear sets, which are called by the authors of this study—*pitch configurations: pitch circles and pitch surfaces.* They configure one contemporary, as a content and terminology, direction from the *Theory of Gearing,* defining such basic characteristics of hyperboloid gears as: the structure and geometry of the gear system, the longitudinal and cross orientation of the active tooth surfaces of the gears, the values of the gears' modules, the forces acting in the contact zone of the gears, on the shafts and on the bearing supports of the gear mechanism, coefficient of efficiency and etc. Hence, from a methodological and application view point, it is one of the reasons, that the authors of the current study have devoted a great part of their researches on the mentioned above thematic. Activities of such type are a potential opportunity for emerging of new ideas related to the creation of innovative hyperboloid gears. When these pitch configurations are synthesized in condition of a static contact of the active tooth surfaces of the hyperboloid gears, they define so-called *geometric pitch configurations.* These configurations are subject of study "*Geometric Pitch Configurations—Basic Primitives of the Mathematical Models for the Synthesis of Hyperboloid Gear Drives*", published in Advance Gear Engineering, Mechanism and Machine Science 51, Springer. The research, oriented to the synthesis of hyperboloid transmissions, will be complete and effective, when it leads to the creation of an adequate mathematical model, describing the status of the pitch configurations in the process of spatial rotations transformation. In this work, the treated scientific

V. Abadjiev · E. Abadjieva
Department of Mechatronics, Institute of Mechanics-BAS, Acad. G. Bonchev Str., Block 4, 1113 Sofia, Bulgaria
e-mail: abadjiev@imbm.bas.bg

E. Abadjieva (✉)
Graduate School of Engineering Science, Akita University, 1-1 Tegatagakuenmachi, Akita, Akita Prefecture 010-8502, Japan
e-mail: abadjieva@gipc.akita-u.ac.jp

© Springer Nature Switzerland AG 2020
V. Goldfarb et al. (eds.), *New Approaches to Gear Design and Production,*
Mechanisms and Machine Science 81,
https://doi.org/10.1007/978-3-030-34945-5_7

concepts, are related to the content of the concepts of *kinematic pitch configura-tions*, when their geometry (shape and dimensions) and mutual position directly take into account the process (law) of spatial rotations transformation. These type characteristics of hyperboloid gears are of essential importance for their synthesis, for considering the magnitude and orientation of the sliding velocity vector in the mesh region of the active tooth surfaces.

Keywords Mathematical modelling · Synthesis · Pitch configurations · Vector analysis · Kinematic surfaces of level

7.1 Introduction

For a relatively long time in the theory of spatial gearing the terms "*primary surfaces*" [1–6] and the terms "*pitch surfaces*" and "*pitch cones*" [7–9] are used together. To a greater extent in the cited specialized literature with these two terms—primary and pitch surfaces are defined the same concepts.

In the early seventies of the last century, the well-known Russian scientist—Prof. F. Litvin gave the following definition of primary surfaces [3]:

"The primary surfaces H_1 and H_2 firmly connected with the movable links of the mechanism are called primary ones if the following conditions are fulfilled: (a) the rotation axis of the primary surface coincides with the rotation axis of the movable link; (b) the surfaces H_1 and H_2 tangent at a given point P of the fixed space, and the velocity of the relative motion of the links 1 and 2 at P lies on the common tangent to the helical lines of the surfaces H_1, H_2 and Q (author' note: Q is a family of coaxial cylinders and the vector (helical) lines of the vector field of the relative motion velocity $\bar{V}^{(12)}$ are situated on them). The second requirement means that they have a common normal at the chosen point, and the velocity vector of the relative motion $\bar{V}^{(12)}$ lies in the common tangent plane of H_1 and H_2. …, If $i_{12} = constant$ the primary surfaces could be arbitrary surfaces of revolution only if: (a) the axis of rotation $i - i$ of H_i is an axis of rotation of ith link; (b) $\bar{V}^{(12)}$ lies in the common tangent plane of H_1 and H_2.

In order to avoid any misunderstanding it is necessary to note (record) the following principles: (a) in the most common case the primary surfaces cannot be identified with the axoids; such identifications is possible only in the case of gear-pairs with parallel or intersecting axis, and is not permitted for gear sets with crossed axes of rotations; (b) the tooth surfaces Σ_1 and Σ_2 do not coincide with the primary surfaces. Although Σ_1 and Σ_2 tangent at the point P, the normal vectors $\bar{e}^{(\Sigma_1)}$ and $\bar{e}^{(H_i)}$ have different directions. The common tangent plane of Σ_1 and Σ_2 at the point P does not coincide with the common tangent plane of H_1 and H_2 but $\bar{V}^{(12)}$ belongs to each of them; c) the condition of the simultaneous tangent of H_1, H_2 and Q at the point P is possible but not obligatory…".

At the end of the 20th century, Professor F. Litvin, already as a Professor of Mechanical Engineering at the Illinois University in Chicago, keeps the concept of primary surfaces almost unchanged. In [7–9], the primary surfaces are now called *"operating pitch surfaces"*, *"pitch surfaces"* and *"pitch cones"*, in connection with their practical applicability, in the design of spatial gear mechanisms with crossed axes. Here, it is noteworthy that the mentioned geometric primitives of the mathematical models for the synthesis and design of hyperboloid gears have nothing in common with the axodes of their movable links [10–14].

In the studies [1, 3–5] algorithms for calculation of the geometric parameters of the primary surfaces of spatial gear drives with an external mating gears are offered. They are oriented to the design of hyperboloid mechanisms with an arbitrary crossed axes. In publication [15], W. Nelson treats pitch surfaces for one concrete type of spatial gear sets—the Spiroid®[1] ones. There, he uses the terms *"primary pitch cone"* (a coaxial cone, limiting the tips of the Spiroid pinion threads) and *"pitch surface"* (an envelope of the primary pitch cone in its relative motion relative the axis of the second movable link of the Spiroid gear). He searches for that pitch contact point, on the common line of contact of both pitch surfaces, which determines the most suitable spatial curve, to be used as a longitudinal line of the synthesized tooth surfaces of the Spiroid pinion. This is an approach that does not differ essentially from the above already considered ones.

The authors of the current study, in connection with solving the tasks related to the synthesis and design of spatial gear mechanisms, have devoted a big part of their researches to the problems connected with the mentioned thematic [16–21]. These studies are oriented primarily into précising the content of the two basic concepts that form the common name—*pitch configurations*: *pitch circles* and *pitch surfaces*. Here, it will be noted especially, that the exact defining of these concepts and the content included in them, gives the possibility of refining the applied mechano-mathematical models for the synthesis of spatial gear mechanisms and, on the other hand, provides potential opportunities for new ideas related to the creation of hyperboloid gears with new qualities, respectively—with new applications in the technique. It is natural, that a study in this field will be effective, when it leads to the creation of an adequate mathematical model describing the status of the pitch configurations in the process of spatial rotations transformation. In [10], the *geometric pitch configurations,* which are the basic primitives of the mathematical models for synthesis and design *upon* a *pitch contact point* of spatial transmissions with crossed axes, are terminologically and analytically defined in details. In their synthesis these configurations are treated as static characteristics of the hyperboloid mechanisms. The present study deals with new concepts concerning the content of the *kinematic pitch surfaces* and *pitch circles*. The specific ideology of these concepts has its natural reflection on the constructions of the mathematical models, used for the synthesis of the above mentioned pitch configurations, since the synthesis of these characteristics takes into account the law of spatial transformation of the synthesized transmissions, i.e. the magnitude and orientation of the sliding velocity between their active tooth surfaces are taken into account.

---

[1] Spiroid is a trademark registered by the Illinois Tool Works, Chicago, Ill.

7.2   Kinematic Model of the Process of Spatial Rotations Transformation

Transformation of rotation motions upon a preliminary given law is a basic process, the realization of which is oriented towards the predominant part of the three-link spatial gear mechanisms with crossed axes, known as *hyperboloid gear sets*. Although, some transmissions with parallel and intersecting axes of the rotating links also belong to the spatial gear mechanisms, from the theoretical point of view, the process of spatial transformation of rotations for the crossing of the axes of the movable links of the three-links transmission gear mechanism is the most common. This determines the predominant interest of researchers into them. For this reason, in creating the mechano-mathematical model, this most common case will be treated, but in aspect determined by its practical application. The study will be focused on spatial gear mechanism, transforming rotations with constant angular velocities between fixed crossed shafts (axes) [16]. On Fig. 7.1, a kinematic scheme of spatial three-link gear mechanisms, transforming rotations by means of its movable links (gears) 1 and 2 with angular velocities $\bar{\omega}_1$ and $\bar{\omega}_2$ between fixed crossed axes of rotations $1 - 1$ and $2 - 2$ (placed in the static space) of the links 1 and 2 is illustrated.

On Fig. 7.1 the illustrated symbols are as follows: $(\Sigma_1 : \Sigma_2)$—high kinematic joint; D_{12}—contact line of the geometric elements Σ_1 and Σ_2; P—contact point; $i - i(i = 1, 2)$—conjugates axes of the relative helical $(O_0, \bar{\omega}_{12}, p_{z_0})$; $\bar{a}_w$—offset between $1 - 1$ and $2 - 2$; $\bar{\omega}_i$—angular velocity of the movable link i; $\bar{V}_{12}$-relative velocity vector of the point P; $n - n$—common normal of Σ_1 and Σ_2 at point P; δ—shaft angle of axes $1 - 1$ and $2 - 2$; $\bar{\rho}_i(i = 1, 2)$—radius-vector of the contact point P; $R_{V_{c,i}}$—kinematic cylinder of level; $R_{V_{z_0}}$—zero kinematic cylinder of level; $S_{c,i}$—kinematic relative helix; $\|S_{c,i}\|$—normalized relative helix. In this case, the studied process of rotations transformation is characterized by the following conditions:

$$\omega_i = constant, \ i = (1, \ 2), \ \delta = \angle(\bar{\omega}_1, \bar{\omega}_2) = constant,$$

$$a_w = constant, \ i_{12} = \frac{\omega_1}{\omega_2} = \frac{1}{i_{21}} = constant, \tag{7.1}$$

where ω_i is a magnitude of the angular velocity vector $\bar{\omega}_i$ of rotation of the movable link i; δ—shaft angle of the axes $1 - 1$ and $2 - 2$; a_w—minimum distance between the crossed axes $1 - 1$ and $2 - 2$ (offset); i_{12} (i_{21})—velocity ratio.

The rotations transformation is realized by means of high kinematic joint $(\Sigma_1 : \Sigma_2)$, which elements are surfaces Σ_1 and Σ_2, that in some moment have a conjugate contact point P (a tangent contact point P can belong to the conjugated instantaneous contact line D_{12}). Here, it should be noted that for the actual gear pair in any moment, there is more than one pair of tooth surfaces Σ_1 and Σ_2, forming a high kinematic joint $(\Sigma_1 : \Sigma_2)$. When one of the movable links, for example $i = 1$, is put into rotation with an angular velocity $\bar{\omega}_1$, then the other link also starts to rotate $i = 2$. And if the obtained rotation of the second link is determined by the angular velocity $\bar{\omega}_2$, (for which the condition $\omega_2 = \omega_1 i_{21} = constant$ is fulfilled, that is equivalent to the last

Fig. 7.1 Structural—
kinematic scheme of spatial
rotations transformation

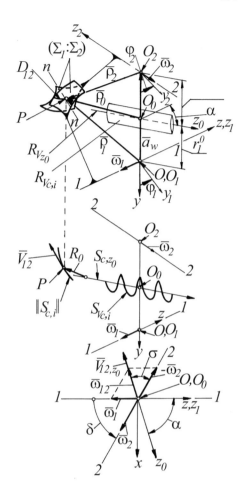

equality of the equation system (7.1)), then it can be affirmed that the illustrated in
Fig. 7.1 kinematic joint $(\Sigma_1 : \Sigma_2)$, and the studied spatial gear mechanism respectively,
are kinematically conjugated. The research is realized by introducing right-hand
coordinate systems $S(O, x, y, z)$ and $S_i(O_i, x_i, y_i, z_i)$ $(i = 1, 2)$, of which S is a
static coordinate system (connected with the fixed link (posture) of the three-links
mechanism), and S_i $(i = 1, 2)$ are the coordinate systems, firmly connected with
the movable links $(i = 1, 2)$. The current placement of S_1 and S_2 towards S is given
by the parameters of rotation (meshing) φ_i $(i = 1, 2)$. Their reading is done in
accordance with the shown on Fig. 7.1. Let's the elements Σ_1 and Σ_2 of the high
kinematic joint are given in the parametric form in the static coordinate system S [3,
16].

$$\bar{\rho}_{i,s} = \bar{\rho}_{i,s}(u_i, \vartheta_i, \varphi_i), \quad \bar{n}_{i,s} = \bar{n}_{i,s}(u_i, \vartheta_i, \varphi_i), \tag{7.2}$$

where $\bar{\rho}_{i,s}$ is a radius-vector of the contact point P, as a point from Σ_i; $\bar{n}_{i,s}$—normal vector to Σ_i at the same point; u_i, ϑ_i—independent parameters, determining the location of the point P on Σ_i.

At the point of geometric conjugation of the surfaces Σ_1 and Σ_2 the following conditions are fulfilled:

$$\bar{\rho}_{1,s}(u_1, \vartheta_1, \varphi_1 = constant) = \bar{\rho}_{2,s}(u_2, \vartheta_2, \varphi_2 = constant),$$
$$\bar{n}_{1,s}(u_1, \vartheta_1, \varphi_1 = constant) = \bar{n}_{2,s}(u_2, \vartheta_2, \varphi_2 = constant). \tag{7.3}$$

For the concrete values of the rotations' parameters φ_1 and $\varphi_2 = \varphi_1 i_{21}$, the u_1 and ϑ_1 are a pair of independent parameters, defining point P as a point from Σ_1, which has a tangent contact with the corresponding point P from Σ_2, defined by the parameters u_2 and ϑ_2. In the process of motion of the kinematic conjugated joints for the current points P, the conditions (7.3) are continually met when $\varphi_i = varia$, $i = 1, 2$. Then [3, 18]:

$$\dot{\bar{\rho}}_{1,s}(u_1, \vartheta_1, \varphi_1) = \dot{\bar{\rho}}_{2,s}(u_2, \vartheta_2, \varphi_2), \quad \dot{\bar{n}}_{1,s}(u_1, \vartheta_1, \varphi_1) = \dot{\bar{n}}_{2,s}(u_2, \vartheta_2, \varphi_2), \tag{7.4}$$

$$\ddot{\bar{\rho}}_{1,s}(u_1, \vartheta_1, \varphi_1) = \ddot{\bar{\rho}}_{2,s}(u_2, \vartheta_2, \varphi_2), \quad \ddot{\bar{n}}_{1,s}(u_1, \vartheta_1, \varphi_1) = \ddot{\bar{n}}_{2,s}(u_2, \vartheta_2, \varphi_2), \tag{7.5}$$

where $\dot{\bar{\rho}}_{i,s}$ is an absolute velocity vector of the point P, connected to the Σ_i; $\dot{\bar{n}}_{i,s}$—is an absolute velocity of the tip of the normal vector $\bar{n}_i$ in the contact point P, connected to the Σ_i; $\ddot{\bar{\rho}}_{i,s}$—an absolute acceleration of the contact point P, considered as a point from Σ_i; $\ddot{\bar{n}}_{i,s}$—an absolute acceleration of the tip of $\bar{n}_i$, at point P, joined to Σ_i.

Here and further, under conjugation of the kinematic joints and of the transmission mechanisms functioning through them, it will be understood their theoretical kinematic conjugation.

The equations systems (7.4) and (7.5) are vector conditions for conjugation of the transmission mechanism. The first vector equality (7.4) can be written as:

$$\bar{V}_{r,2} = \bar{V}_{r,1} + (\bar{V}_1 - \bar{V}_2) = \bar{V}_{r,2} + \bar{V}_{12}. \tag{7.6}$$

From (7.6) can be seen, that $\bar{V}_{12}$ is determining both for the circumferential and the relative motion of the conjugated contact points. The same equality shows more, that the vectors $\bar{V}_{12}$, $\bar{V}_{r,1}$ and $\bar{V}_{r,2}$, as well the vectors $\bar{V}_{12}$, $\bar{V}_1$ and $\bar{V}_2$ are two triplets of coplanar vectors. The first triple determines a tangential plane T_n in the conjugated contact point P of Σ_1 and Σ_2. The second triple of vectors, as it will be shown further, define plane T_m, which, under certain conditions, contains the pole of meshing of the synthesized gear mechanism. The above comment is illustrated in Fig. 7.2, from which it is obvious the existence of the vector equality:

$$\bar{n}_i . \bar{V}_{12} = 0, \tag{7.7}$$

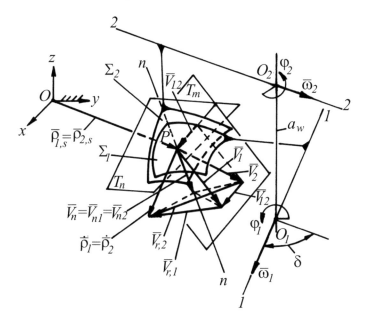

Fig. 7.2 Kinematic scheme of the conjugate action of the active tooth surfaces Σ_1 and Σ_2 of three-link hyperboloid gear drive in the contact point P

known in the Theory of Gearing as a *basic equation of meshing* [1, 3, 7–9, 16], which in the context of this study can be formulated as follows:

At each contact point of the conjugated tooth surfaces (when this point is not a node of second order, i.e., a point of undercutting), the relative velocity vector have to be placed in the common tangent plane of these surfaces at the same point.

The above theorem can be proved easily by assuming the opposite. If condition (7.7) is not fulfilled, then the vector $\bar{V}_{12}$ will have its component on the common normal $n - n$ (see Fig. 7.2). This means the surfaces Σ_1 and Σ_2 will realize a relative motion, which depending on the direction of the normal component of $\bar{V}_{12}$, will cause detachment of Σ_1 from Σ_2 or their inclination to each other, which is unacceptable. From the said up to now, the placement of the relative velocity vectors of the geometric elements Σ_1 and Σ_2 of the joint $(\Sigma_1 : \Sigma_2)$ in their common points is essential for the conjugation of high kinematic joints, which realize spatial motions transformation.

On Fig. 7.2 the shown symbols are as follows: $i - i$ $(i = 1, 2)$—axes of rotations of the movable links; a_w—offset; δ—shaft angle of the axes $1 - 1$ and $2 - 2$; T_n—common tangent plane of Σ_1 and Σ_2 in point P; $n - n$—common normal of Σ_i $(i = 1, 2)$ at point P; $\bar{\rho}_{i,s}$—radius-vector of point P, jointed to the Σ_i in the fixed space $S(O, x, y, z)$; $\dot{\bar{\rho}}_i$—absolute velocity vector of contact point P; $\bar{V}_{12}$—relative velocity vector of P; $\bar{V}_{r,i}$—relative velocity vector of the motion of point P on the tooth surface Σ_i; $\bar{V}_{n,i}$—normal components of $\dot{\bar{\rho}}_i$.

7.3 Vector Analysis of the Relative Velocity Vector

One of the essential parts of this research is dedicated to the vector analysis of the vector function of the relative (sliding) velocity vector $\bar{V}_{12}$ in the contact point P, which belongs to the geometric elements Σ_1 and Σ_2 of the conjugated high kinematic joints realizing spatial rotations transformation in accordance with conditions (7.1). Assuming that the motions transformation of this type is illustrated on Fig. 7.1, in accordance with the shown symbols, for the relative velocity vector $\bar{V}_{12}$ it can be written:

$$\bar{V}_{12} = \bar{V}_1 - \bar{V}_2 = \bar{\omega}_1 \times \bar{\rho}_1 - \bar{\omega}_2 \times \bar{\rho}_2 = \bar{\omega}_{12} \times \bar{\rho}_1 + \bar{\omega}_2 \times \bar{a}_w, \qquad (7.8)$$

where $\bar{\rho}_i$ is a radius-vector of the contact point P, considered as a point from the link i, written in the coordinate system $S(O, x, y, z)$. Due to the fact that $\omega_i = constant$ ($i = 1, 2$), $a_w = constant$ and $\delta = \angle(\bar{\omega}_1, \bar{\omega}_2) = constant$, it is obvious that the vector function of the sliding velocity vector is of the type

$$\bar{V}_{12} = \bar{V}_{12}(\bar{\rho}_1) = \bar{V}_{12}(x, y, z). \qquad (7.9)$$

7.3.1 Scalar Field of the Relative Velocity Vector $\bar{V}_{12}$

Using the illustration shown in Fig. 7.1, for the Cartesian coordinates of the vector $\bar{V}_{12}$ in the static coordinate system S, it is obtained [16]:

$$V_{12,x} = (1 - i_{21} \cos \delta)y - a_w i_{21} \cos \delta, \quad V_{12,y} = i_{21} \sin \delta z - (1 - i_{21} \cos \delta)x,$$
$$V_{12,z} = -i_{21} \sin \delta (a_w + y). \qquad (7.10)$$

In (7.10) without disturbing the community of arguments it is accepted that $\omega_1 = 1$ rad/s and $\omega_2 \le \omega_1$. Then from the last equation of (7.1) follows $0 < \omega_2 = i_{21} \le 1$. It order to have a detailed study, let's determine the orthogonal Cartesian coordinates of the relative velocity of the rotation $\bar{\omega}_{12}$, as well its magnitude. Then, in accordance with Fig. 7.1, it can be written:

$$\omega_{12,x} = -i_{21} \sin \delta, \quad \omega_{12,y} = 0, \quad \omega_{12,z} = -(1 - i_{21} \cos \delta), \qquad (7.11)$$

$$\omega_{12} = \sqrt{1 + i_{21}^2 - 2i_{21} \cos \delta}. \qquad (7.12)$$

The scalar field of the velocity of relative motion $\bar{V}_{12}$ is a scalar function in point P, considered as a point from the static space, [16], i.e.:

$$V_{12} = V_{12}(x, y, z) = \sqrt{V_{12,x}^2 + V_{12,y}^2 + V_{12,z}^2}, \tag{7.13}$$

with its definition area. During the time τ of running the process of meshing, the contact point P changes its placement in the static space, when the preliminary defined law of rotations transformation is observed, i.e.

$$\bar{\rho}_1 = \bar{\rho}_1(\tau). \tag{7.14}$$

Nevertheless the scalar field will be considered as a static one, but not depending on the time τ.

Surfaces of the type

$$V_{12}(\bar{\rho}_1) \equiv V_{12}(x, y, z) = V_c, \quad V_c = constant, \tag{7.15}$$

are *surfaces of level* of the studied scalar field and they define it geometrically.

From the analytical dependencies (7.10), (7.13) and (7.15) the general type of the surface of level is obtained:

$$a_{11}x^2 + a_{22}y^2 + a_{33}z^2 + 2a_{12}xz + 2a_{24}z + a_{44} = 0,$$
$$a_{11} = (1 - i_{21} \cos \delta)^2, \quad a_{22} = 1 + i_{21}^2 - i_{21} \cos \delta, \quad a_{33} = i_{21}^2 \sin^2 \delta, \tag{7.16}$$
$$a_{12} = -i_{21} \sin \delta (1 - i_{21} \cos \delta), \quad a_{24} = a_w i_{21}(i_{21} - \cos \delta), \quad a_{44} = a_w^2 i_{21}^2 - V_c^2.$$

The canonical type of (7.16) describes a right circular cylinder in the coordinate system $S_0(O_0, x_0, y_0, z_0)$ (Fig. 7.1):

$$x_0^2 + y_0^2 = R_0^2, \quad R_0 = \sqrt{\frac{V_c^2}{1 + i_{21}^2 - 2i_{21} \cos \delta} - \left(\frac{a_w i_{21} \sin \delta}{1 + i_{21}^2 - 2i_{21} \cos \delta}\right)^2}, \tag{7.17}$$
$$z_0 = H, \quad H \in (-\infty, +\infty).$$

Equation (7.17) shows that the surfaces of level of the defined scalar filed $V_{12} = V_{12}(x_0, y_0, z_0)$ represent families of coaxial cylinders with a family parameter V_c. Further, these *right circular cylinders of level* will be called *kinematic cylinders of level* [16]. The axis of the family coaxial cylinders of level is an axis $O_0 z_0$ of the right-hand coordinate system S_0 and it is defined in the static system S with the following equations:

$$y = -\frac{a_w i_{21}(i_{21} - \cos \delta)}{1 + i_{21}^2 - 2i_{21} \cos \delta}, \quad x = \tan \alpha.z, \quad \tan \alpha = \frac{i_{21} \sin \delta}{1 - i_{21} \cos \delta}. \tag{7.18}$$

In (7.18), α is an angle which $O_0 z_0$ concludes with the axis Oz of the static coordinate system S.

In accordance with (7.17) surfaces of level of the studied scalar field will exist, if for the choice of the parameter of this family, the following condition is fulfilled

$$V_c \geq \frac{a_w i_{21} |\sin \delta|}{\sqrt{1 + i_{21}^2 - 2i_{21} \cos \delta}}. \tag{7.19}$$

The module of sliding velocity vector $\bar{V}_{12}$ multiplied with ω_1^{-1} (where ω_1 is the magnitude of $\bar{\omega}_1$) defines the parameter V_c $\left[\text{mm} / \text{rad} \right]$. Every point, belonging to a concrete kinematic cylinder of level is characterized by this parameter, if the preliminary given velocity ratio is fulfilled. When the equality in (7.19) is realized, then system (7.18) describes *zero kinematic cylinder of level*. It contains those points of the static space, which are characterized by the smallest values of relative velocity motion's module. The locus of these points is the axis $O_0 z_0$ of S_0. The kinematic cylinders of level are noted by the symbol $R_{V_c,i}$, and the zero kinematic cylinder— with $R_{V_{z_0}}$.

7.3.2 Vector Field of the Relative Velocity Vector $\bar{V}_{12}$

Vector field of $\bar{V}_{12}$ is part of three-dimensional space, that in every point of this space is defined a vector:

$$\bar{V}_{12} = \bar{V}_{12}(\bar{\rho}_{1,s}) = \bar{V}_{12}(x, y, z). \tag{7.20}$$

Let, the vector $\bar{V}_{12}$ is presented in the coordinate system S_0, i.e.

$$\bar{V}_{12} = V_{12,x_0}.\bar{i}_0 + V_{12,y_0}.\bar{j}_0 + V_{12,z_0}.\bar{k}_0, \quad V_{12,x_0} = \sqrt{1 + i_{21}^2 - 2i_{21} \cos \delta}.y_0,$$

$$V_{12,y_0} = -\sqrt{1 + i_{21}^2 - 2i_{21} \cos \delta}.x_0, \quad V_{12,z_0} = -\frac{a_w i_{21} \sin \delta}{\sqrt{1 + i_{21}^2 - 2i_{21} \cos \delta}}, \tag{7.21}$$

where V_{12,x_0}, V_{12,y_0}, V_{12,z_0} are orthogonal Cartesian coordinates of the vector $\bar{V}_{12}$ in the coordinate system S_0; $\bar{i}_0$, $\bar{j}_0$, $\bar{k}_0$ are unit vectors of the coordinate axes of S_0.

The expressions (7.12) and (7.21) show unequivocally that for the relative angular velocity $\bar{\omega}_{12}$ for the spatial rotations transformation, the following condition (see Fig. 7.1) is fulfilled:

$$\bar{\omega}_{12} = \bar{\omega}_{12,z_0} = -\sqrt{1 + i_{21}^2 - 2i_{21} \cos \delta}.\bar{k}_0. \tag{7.22}$$

Every vector field is characterized by its own *vector lines* [16]. The vector lines of the field (7.21) represent curves and the tangents to them in every point $P(x_0, y_0, z_0)$ coincide with the direction of the vector $\bar{V}_{12}$ at the same point. They are determined by the system of differential equations (each equation is a consequence of the other two):

$$\frac{dx_0}{V_{12,x_0}} = \frac{dy_0}{V_{12,y_0}}, \quad \frac{dy_0}{V_{12,y_0}} = \frac{dz_0}{V_{12,z_0}}, \quad \frac{dx_0}{V_{12,x_0}} = \frac{dz_0}{V_{12,z_0}}. \tag{7.23}$$

The first and the third equation of the system (7.23) are presented in:

$$\frac{dx_0}{\omega_{12}y_0} = \frac{dy_0}{-\omega_{12}x_0}, \quad \frac{dx_0}{\omega_{12}y_0} = \frac{dz_0}{-\dfrac{a_w i_{21} \sin \delta}{\omega_{12}}}. \tag{7.24}$$

After solving the system (7.24) it is obtained

$$x_0^2 + y_0^2 = R_0^2 = constant, \quad z_0 + p_{z_0}t_{(1)} = C_{0(1)} = constant, \tag{7.25}$$

where $p_{z_0} = \dfrac{a_w i_{21} \sin \delta}{1 + i_{21}^2 - 2i_{21}\cos\delta}, \quad t_{(1)} = \arcsin\dfrac{x_0}{R_0}.$

Analogically, from the solution of the first and second equation from (7.23) it is received:

$$x_0^2 + y_0^2 = R_0^2 = constant, \quad z_0 = p_{z_0}t_{(2)} + C_{0(2)} = constant,$$
$$p_{z_0} = \frac{a_w i_{21} \sin \delta}{1 + i_{21}^2 - 2i_{21}\cos\delta}, \quad t_{(2)} = \arcsin\frac{y_0}{R_0}. \tag{7.26}$$

When the equations systems (7.25) and (7.26) are analyzed, it is established that the vector lines, which form vector filed of the function $\bar{V}_{12} = \bar{V}_{12}(x, y, z)$, are a family of helical lines, obtained as intersections of a family of coaxial cylinders with radiuses R_0 with a family of conoids (orthogonal Archimedean helicoids) with a helical parameter p_{z_0}. These conoids are called *kinematic conoids* and they are noted with $C_{d,i}$. The family of helical lines, described by the systems (7.25) and (7.26), (Fig. 7.3) is determined not only by the above mentioned parameters R_0 and p_{z_0},

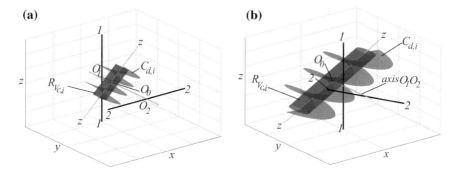

Fig. 7.3 Kinematic scheme of the vector lines generation of the vector function $\bar{V}_{12} = \bar{V}_{12}(x, y, z)$: **a** kinematic cylinder of level $R_{V_{c,i}}$ and conoid $C_{d,i}$ when $a_w = 105$ mm, $\delta = 90°$, $i_{12} = 2$, $V_c = 50$ mm/rad (external meshing); **b** kinematic cylinder of level $R_{V_{c,i}}$ and conoid $C_{d,i}$ when $a_w = 105$ mm, $\delta = 30°$, $i_{12} = 2$, $V_c = 50$ mm/rad (internal meshing)

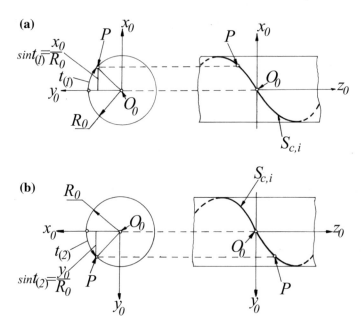

Fig. 7.4 Scheme of kinematic relative helix $S_{c,i}$ generation: **a** for the angular parameter $t_{(1)} =$ arcsin(x_0 / R_0); **b** for the angular parameter $t_{(2)} =$ arcsin(y_0 / R_0)

but also by the parameter $C_{0(1)}$ and $C_{0(2)}$, correspondingly. The vector lines of the studied vector field are called *kinematic relative helices* [16] and they are noted by the symbol $S_{c,i}$ (i is the number of the kinematic relative helix) (see Figs. 7.1 and 7.4). Obviously, when the kinematic relative helices of the sliding velocity vector field are defined, then the relative helical motion, resulting from the process of rotations transformation between crossed axes, is defined also. Focusing again on Eqs. (7.12) and (7.21), it is established that if the contact point of the conjugated tooth surfaces Σ_1 and Σ_2 is placed on this axis O_0z_0, then the relative linear and angular velocities at this point, are respectively:

$$\bar{V}_{12} = \bar{V}_{12,z_0} = -\frac{a_w i_{21} \sin \delta}{\sqrt{1 + i_{21}^2 - 2i_{21} \cos \delta}} . \bar{k}_0, \quad \bar{\omega}_{12} = \bar{\omega}_{12,z_0} = -\sqrt{1 + i_{21}^2 - 2i_{21} \cos \delta} . \bar{k}_0. \quad (7.27)$$

From (7.27) it is obvious that

$$\frac{V_{12,z_0}}{\omega_{12,z_0}} = \frac{a_w i_{21} \sin \delta}{1 + i_{21}^2 - 2i_{21} \cos \delta} = p_{z_0}. \quad (7.28)$$

Therefore, when the conjugated contact points of Σ_1 and Σ_2 are placed on the axis O_0z_0, then they participate in the same helical motion with the helical parameter p_{z_0}. Each of the points constituting the axis O_0z_0 represents an application point of $\bar{V}_{12,z_0}$, whose directrix coincides with the axis $O_0 z_0$. Hence, the axis $O_0 z_0$ is also a kinematic relative helix, which will be noted with the symbol S_{c,z_0} and will be called *minimal kinematic relative helix,* due to the fact that the "tangent" vector of the relative motion in each of its point has the smallest possible value of the magnitude for a particular spatial rotations transformation. This fact is obvious after comparing the Eqs. (7.21) and (7.27). In other words, the magnitude of the projection of vector $\bar{V}_{12}$, defined in an arbitrary point of Σ_1 and Σ_2 on the minimal kinematic relative helix, has a minimal constant magnitude and therefore, does not depend on the location of the contact point of the conjugated tooth surfaces in the static space [16].

From (7.28) it is obvious if the parameter $\delta \in (0, \pi)$, then parameter $p_{z_0} > 0$, hence (7.25) and (7.26) describe all right-hand kinematic relative helices, and when $\delta \in (\pi, \; 2\pi)$-$p_{z_0} < 0$ and therefore, (7.25) and (7.26) describe all left-hand kinematic helices.

Furthermore, for simplicity, the word *"kinematic"* from the name of the kinematic relative helix will be omitted.

7.4 Characteristics of the Relative Motion, When the Rotations Transformation Between Crossed Axes Is Realized

7.4.1 Normalized (Normed) Relative Helices

The magnitude of the relative velocity vector $\bar{V}_{12}$ is a characteristic involved in the construction of a number of important quality criteria, which are applied to the synthesis of spatial gear mechanisms and serve to evaluate the contact strength of the kinematically conjugated surfaces as well as the wear-resistance assessment of the gear drive [16]. This requires normalizing of the vector's module. For this reason, in this study, an approach, based on the symbiosis between the vector and scalar field of the vector function $\bar{V}_{12} = \bar{V}_{12}(x_0, y_0, z_0)$, is offered. This is achieved by substituting in (7.25) (or in (7.26)) the following:

$$R_0 = \sqrt{\frac{V_c^2}{1 + i_{21}^2 - 2i_{21}\cos\delta} - \left(\frac{a_w i_{21} \sin\delta}{1 + i_{21}^2 - 2i_{21}\cos\delta}\right)^2}. \qquad (7.29)$$

Then the normalized vector field of the vector function $\bar{V}_{12} = \bar{V}_{12}(x_0, y_0, z_0)$ is described by the following equation system

$$x_0^2 + y_0^2 = R_0^2, \; z_0 + p_{z_0}t = C_0, \qquad (7.30)$$

where $R_0 = \sqrt{\left(\dfrac{V_c}{\omega_{12}}\right)^2 - p_{z_0}^2}$, $p_{z_0} = \dfrac{a_w \sin\delta}{\omega_{12}^2} i_{21}$, $\omega_{12} = \sqrt{1 + i_{21}^2 - 2i_{21}\cos\delta}$.

(a)   **(b)**

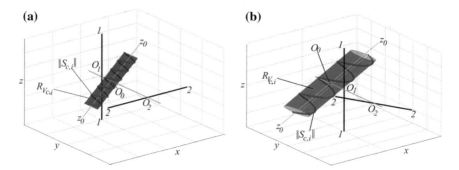

Fig. 7.5 Kinematic cylinder of level $R_{V_{c,i}}$ and normalized relative helices $||S_{c,i}||$: **a** $a_w = 105$ mm, $\delta = 90°$, $i_{12} = 2$, $V_c = 50$ mm/rad (external meshing); **b** $a_w = 105$ mm, $\delta = 30°$, $i_{12} = 2$, $V_c = 50$ mm/rad (internal meshing)

The family (7.30) of the normalized relative helices has parameters V_c and C_0. The parameter V_c illustrates the normalized magnitude of the vector $\bar{V}_{12}$. This means that in all points of the kinematic cylinder of level with radius R_0 and axis O_0z_0 (which location in the static space is determined by the system (7.18)) and chosen as contact points of the conjugated tooth surfaces Σ_1 and Σ_2 of spatial gear drive, a relative velocity $\bar{V}_{12}$ exists, and its module is determined by the expression

$$V_{12} = V_c.\omega_1. \tag{7.31}$$

Here V_{12} is the module of the normalized relative velocity in [mm/s], and ω_1- is the actual angular velocity of the movable link $i = 1$, measured in [rad/s]. The parameters C_0 defines the placement of the normalized relative helix along the kinematic cylinder of level, i.e. placement of the chosen (computational) contact points towards the axis O_1O_2 of the crossed axes $i - i$ ($i = 1, 2$). Normalized relative helices will be noted with the symbol $||S_{c,i}||$. They are illustrated on Fig. 7.5.

7.4.2 Orientation of the Normalized Relative Helices Towards the Axes of Rotation $1 - 1$ and $2 - 2$

Orientation of one normalized relative helix towards the rotation's axes $i-i$ ($i = 1, 2$) of the movable links i ($i = 1, 2$) is determined by the angle between the tangent line at a point of the helix (the directrix of the vector $\bar{V}_{12}$ at point of the helix $||S_{c,i}||$) and the axes $1 - 1$ and $2 - 2$ of the links 1 and 2, respectively. In accordance with the symbols, given in Fig. 7.6, expressions are obtained for these angles, respectively:

$$\beta_1^k = \arccos[\cos\alpha \cos\beta_0^k + \sin\alpha \sin\beta_0^k \cos t], \tag{7.32}$$

Fig. 7.6 Orientation of the normalized relative helix $||S_{c,i}||$ (of the relative velocity vector $\bar{V}_{12}$), towards the crossed axes of rotations $1-1$ and $2-2$

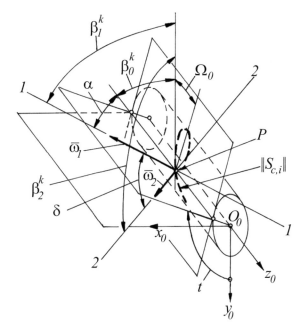

$$\beta_2^k = \arccos[\cos(\alpha + \delta)\cos\beta_0^k + \sin(\alpha + \delta)\sin\beta_0^k \cos t], \qquad (7.33)$$

$$\beta_0^k = arcctg\frac{p_{z0}}{R_0}. \qquad (7.34)$$

In expressions (7.32)–(7.34), the given symbols are: β_i^k is an angle of inclination of the tangent line to the normalized relative helix at a given point towards the axis, i - i of the link i; β_0^k is an angle of inclination of the tangent line to the normalized relative helix at the same point from it, towards the axis O_0z_0.

From Eqs. (7.32), (7.33) and (7.34) it is established, that the angles of the inclination of the minimal relative helix towards the axes $1 - 1$, $2 - 2$ and O_0z_0 are

$$\beta_1^k = \alpha, \beta_2^k = \alpha + \delta, \beta_0^k = 0, \qquad (7.35)$$

as it is obvious from Fig. 7.1. In conclusion, it should be noted that the normalized relative helices can be considered as locus of the conjugated contact points of the high kinematic joints $(\Sigma_1 : \Sigma_2)$ in the static space.

By keeping the longitudinal orientation (determined by the angles β_1^k and β_2^k) of the conjugated tooth surfaces, for a concrete contact point, the preliminary defined law for rotations transformation is achieved with a sufficient precision, ensuring also the magnitude of the sliding velocity between the synthesized teeth surfaces.

7.5 Synthesis of Kinematic Pitch Configurations

7.5.1 Synthesis of Isokinematic Quasi-hyperboloids

Let's the equations system (7.30), describing the normalization of the vector field of the vector function $\bar{V}_{12} = \bar{V}_{12}(x_0, y_0, z_0)$, is presented in the type:

$$x_0 = R_0 \sin t, \quad y_0 = R_0 \cos t, \quad z_0 = C_0 - p_{z_0} t. \tag{7.36}$$

Let us now determine an envelope of the normalized relative helix (7.36) in its rotation around the conjugated axes $1-1$ and $2-2$. To accomplish this goal, following matrix equalities for transition from the static coordinate system $S_0(O_0, x_0, y_0, z_0)$ into the coordinate systems $S_1(O_1, x_1, y_1, z_1)$ and $S_2(O_2, x_2, y_2, z_2)$, rigidly connected with the links $i = 1$ and $i = 2$, (see Fig. 7.1), are used:

$$[x_1 \ y_1 \ z_1 \ t_1]^T = M_{S_1 S_0}[x_0 \ y_0 \ z_0 \ t_0]^T, \tag{7.37}$$

where $M_{S_1 S_0} = \begin{vmatrix} \cos\varphi_1 \cos\alpha & -\sin\varphi_1 & \cos\varphi_1 \sin\alpha & r_1^0 \sin\varphi_1 \\ \sin\varphi_1 \cos\alpha & \cos\varphi_1 & \sin\varphi_1 \sin\alpha & -r_1^0 \cos\varphi_1 \\ -\sin\alpha & 0 & \cos\alpha & 0 \\ 0 & 0 & 0 & 1 \end{vmatrix}$,

$$[x_2 \ y_2 \ z_2 \ t_2]^T = M_{S_2 S_0}[x_0 \ y_0 \ z_0 \ t_0]^T, \tag{7.38}$$

where $M_{S_2 S_0} = \begin{vmatrix} \cos\varphi_2 \cos(\delta+\alpha) & -\sin\varphi_2 & \cos\varphi_2 \sin(\delta+\alpha) & -r_2^0 \sin\varphi_2 \\ \sin\varphi_2 \cos(\delta+\alpha) & \cos\varphi_2 & \sin\varphi_2 \sin(\delta+\alpha) & r_2^0 \cos\varphi_2 \\ -\sin(\delta+\alpha) & 0 & \cos(\delta+\alpha) & 0 \\ 0 & 0 & 0 & 1 \end{vmatrix}$.

The system below is obtained, after a substitution of (7.36) in (7.37) and (7.38):

$$\begin{aligned} x_i &= R_0 \sin t \cos\varphi_i \cos A_i - R_0 \cos t \sin\varphi_i + (C_0 - p_{z_0} t) \cos\varphi_i \sin A_i \pm r_i^0 \sin\varphi_i, \\ y_i &= R_0 \sin t \sin\varphi_i \cos A + R_0 \cos t \cos\varphi_i + (C_0 - p_{z_0} t) \sin\varphi_i \sin A_i \mp r_i^0 \cos\varphi_i, \\ z_i &= -R_0 \sin t \sin A_i + (C_0 - p_{z_0} t) \cos A_i, \\ t_i &= t_0 = 1, \quad A_{i=1} = \alpha, \quad A_{i=2} = \delta + \alpha. \end{aligned} \tag{7.39}$$

The upper signs in (7.39) treat the case, when $i = 1$, and the lower signs are related to the case $i = 2$. After simple transformations from (7.39), it is received

$$\frac{x_1^2}{(R_0 \cos t - r_1^0)^2} + \frac{y_1^2}{(R_0 \cos t - r_1^0)^2} - \frac{\left(z_1 + \dfrac{R_0 \sin t}{\sin\alpha}\right)^2}{\cot^2\alpha(R_0 \cos t - r_1^0)^2} = 1, \tag{7.40}$$

where $r_1^0 = \dfrac{a_w i_{21}(i_{21} - \cos \delta)}{1 + i_{21}^2 - 2i_{21}\cos \delta}$.

and

$$\frac{x_2^2}{(R_0 \cos t + r_2^0)^2} + \frac{y_2^2}{(R_0 \cos t + r_2^0)^2} - \frac{\left(z_2 + \dfrac{R_0 \sin t}{\sin(\delta + \alpha)}\right)^2}{\cot^2(\delta + \alpha)(R_0 \cos t + r_2^0)^2} = 1, \quad (7.41)$$

where $r_2^0 = \dfrac{a_w(1 - i_{21}\cos \delta)}{1 + i_{21}^2 - 2i_{21}\cos \delta}$.

Equations (7.40) and (7.41) define two families of single-parameter hyperboloids of revolution presented in the coordinate systems $S_1(O_1, x_1, y_1, z_1)$ and $S_2(O_2, x_2, y_2, z_2)$, respectively. The parameter of these families is t. These hyperboloids will be called *kinematic hyperboloids*. When (7.40) and (7.41) are used, the envelope of the two families of kinematic hyperboloids can be written, i.e. their discriminant surfaces:

$$F_1(x_1, y_1, z_1, t) = x_1^2 + y_1^2 - \left[\tan\alpha\left(z_1 + \frac{R_0 \sin t}{\sin\alpha}\right)\right]^2 - (R_0 \cos t - r_1^0)^2 = 0,$$
$$\frac{\partial F_1}{\partial t} = -\tan^2\alpha\left(z_1 + \frac{R_0 \sin t}{\sin\alpha}\right)\frac{R_0 \cos t}{\sin\alpha} + (R_0 \cos t - r_1^0)R_0 \sin t = 0 \quad (7.42)$$

and

$$F_2(x_2, y_2, z_2, t) = x_2^2 + y_2^2 - \left[\tan(\alpha + \delta)\left(z_2 + \frac{R_0 \sin t}{\sin(\alpha + \delta)}\right)\right]^2 - (R_0 \cos t + r_2^0)^2 = 0,$$
$$\frac{\partial F_2}{\partial t} = -\tan^2(\alpha + \delta)\left(z_2 + \frac{R_0 \sin t}{\sin(\alpha + \delta)}\right)\frac{R_0 \cos t}{\sin(\alpha + \delta)} + (R_0 \cos t + r_2^0)R_0 \sin t = 0. \quad (7.43)$$

Equations (7.42) and (7.43) define families of *rotational quasi-hyperboloids* pairs, which parameter $R_0 = R_0(V_c)$ is determined from (7.29). Their basic geometric characteristic is undevelopable rotational surfaces, which axes are the conjugated ones of the axis of the instantaneous relative helical motion $(O_0 z_0, \bar{\omega}_{12}, p_{z_0})$. In kinematic terms, these surfaces are characterized with the fact that as they are envelopes of concrete normalized kinematic relative helices, then in every point of their tangent contact, an instantaneous relative velocity $\bar{V}_{12}$ exists and it has one and the same magnitude $V_{12} = \omega_1 V_c$. From (7.42) and (7.43) it is easy to be established, when $R_0 = 0$ the two rotational quasi-hyperboloids are transformed into rotational hyperboloids, representing the locus (an envelope) of the normalized relative helix $||S_{c,i}|| \equiv S_{c,z_0}$.

The quasi-hyperboloids, described by the expressions (7.42) and (7.43), are called *kinematic/isokinematic pitch surfaces* and they will be noted with I_1 and I_2. And in conditions of their conjugate action they will be called *kinematic/isokinetic pitch pairs* and it will be noted with $(I_1:I_2)$ (see Fig. 7.7). On the basis of the chosen approach for generation of the surfaces I_1 and I_2, the fact that should be pointed out, is that the contact line of isokinematic pairs $(I_1:I_2) - C_i$ is a locus of the points from the normalized relative helices $||S_{c,i}||$. Hence, it should be noted especially,

(a) **(b)**

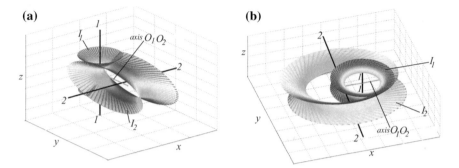

Fig. 7.7 Isokinematic pairs (I_1:I_2): **a** internally contacting isokinematic quasi-hyperboloids I_i ($i =$ 1, 2)—$a_w = 105$ mm, $\delta = 90°$, $i_{12} = 2$, $V_c = 50$ mm/rad, $t \in [-75°, +75°]$; **b** internally contacting isokinematic quasi-hyperboloids I_i ($i = 1$, 2)-$a_w = 105$ mm, $\delta = 30°$, $i_{12} = 2$, $V_c = 50$ mm/rad, $t \in [-60°, +60°]$

that these points are common ones for the corresponding kinematic surfaces of level $R_{V_c,i}$, normalized relative helices $||S_{c,i}||$ and the kinematic surfaces I_1 and I_2. Further, the next goal is to determine the analytical type of the contact line C, which in its essence is a characteristics of I_1 and I_2. Hence, the following statements should be proven:

Statement. When transforming rotations between crossed axes $1 - 1$ and $2 - 2$, the ray R_i of an arbitrary kinematic relative helix (including the normalized $||S_{c,i}||$), in its arbitrary point P_i, is a perpendicular to the plane, determined by the circumferential vectors of the point P_i, when this point is a common for the geometric elements Σ_1 and Σ_2 of the high kinematic joints (Σ_1:Σ_2), formed by the links $i = 1$ and $i = 2$ rotating around the axes $1 - 1$ and $2 - 2$.

Proof On Fig. 7.8 the rotations transformation with angular velocities $\bar{\omega}_1$ and $\bar{\omega}_2$ between the crossed axes $1 - 1$ and $2 - 2$ is shown. The ray R_i is passed through a point P_i from the relative helix $||S_{c,i}||$, so that in accordance with the above defined its characteristics, it should intersect the conjugated axes $1-1$ and $2-2$ of the relative helical motion ($O_0 z_0$, $\bar{\omega}_{12}$, p_{z_0}) in the points $O_{1,R}$ and $O_{2,R}$. According to the shown on Fig. 7.8, the following vectors are introduced: $\bar{\rho}_{1,R} = \overline{O_{1,R}P_i}$ and $\bar{\rho}_{2,R} = \overline{O_{2,R}P_i}$, so that $\overline{O_{1,R}O_{2,R}} = \bar{\rho}_{1,R} - \bar{\rho}_{2,R}$. Since, $\bar{\omega}_i$ ($i = 1$, 2) are sliding velocity vectors, then for the circumferential velocity vectors of the point P_i, considered sequentially as a point from link $i = 1$ and $i = 2$, it could be written

$$\bar{V}_1 = \bar{\omega}_1 \times \bar{\rho}_{1,R}, \quad \bar{V}_2 = \bar{\omega}_2 \times \bar{\rho}_{2,R}. \tag{7.44}$$

From (7.44), it follows that the vectors $\bar{V}_i$ ($i = 1, 2$) are perpendicular to the $\overline{O_{1,R}O_{2,R}}$, i.e. to the ray R_i of the relative helix $||S_{c,i}||$. Since R_i is a ray of the relative helix $S_{c,i}$, then it is perpendicular to $\bar{V}_{12}$ at the same point. Hence, the ray R_i of the relative helix $||S_{c,i}||$ at point P_i is perpendicular to the plane T_R, determined by the coplanar vectors $\bar{V}_{12}$ and $\bar{V}_i$ ($i = 1, 2$). Hence, the statement is proven. Here,

Fig. 7.8 Mutual placement of the kinematic relative helices $||S_{c,i}||$ and their rays R_i in an arbitrary point P_i of the vector field of the sliding velocity vector $\bar{V}_{12}$

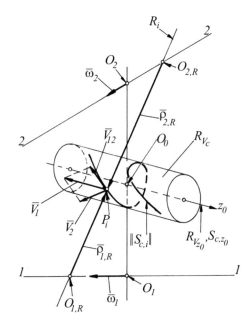

it should be mentioned, that in most commons case, plane T_R, in not a tangent one to the cylindrical surfaces, which contains the relative helix $||S_{c,i}||$.

An obvious fact is that all of the points describing the characteristic C_i of the isokinematic surfaces I_1 and I_2 have to be searched among the common points of the normalized relative helices $||S_{c,i}||$ and the corresponding rays R_i, intersecting the crossed axes $1-1$ and $2-2$ of the relative helical motion $(O_0z_0, \bar{\omega}_{12}, p_{z_0})$. Here, it should be noted, that the whole set of ∞^1 rays of the relative kinematic helices $||S_{c,i}||$ does not contain the searched points. In order a point P_i (belonging to the normalized relative helix $||S_{c,i}||$) to be a point from the searched characteristics C_i, it has be possible to pass through it a common tangent plane T_R to the I_1, I_2 and $||S_{c,i}||$, (to the corresponding kinematic surface of lever $R_{V_{c,i}}$ respectively, since the relative helix is treated as a normalized one). This means, that the characteristic C_i will be a locus of the intersections between those rays R_i with the corresponding kinematic cylinder of level $R_{V_{c,i}}$ that contains the set ∞^1 of the relative helices $||S_{c,i}||$, for which following two conditions are fulfilled: to cross the conjugated axes $1-1$ and $2-2$ and to be perpendicular to the minimal relative helix S_{c,z_0}, to the zero kinematic cylinder R_{z_0}, respectively. The realized analysis confirms that not all planes T_R, containing the point P_i and the vectors $\bar{V}_{12}$ and $\bar{V}_i$ $(i = 1, 2)$, are common tangent planes in point P_i to I_1, I_2 and to $||S_{c,i}||$ (as well as for the cylinder $R_{V_{c,i}}$ containing the normalized relative kinematic helix $||S_{c,i}||$, respectively).

Let's further write the equation for that ray R_i, which passes through point P_i from $||S_{c,i}||$ and is perpendicular to $||S_{c,i}||$ and crosses the conjugated axes $1-1$ and $2-2$ of the relative helical motion $(O_0z_0, \bar{\omega}_{12}, p_{z_0})$. For this reason, a polar plane π_i of $||S_{c,i}||$ is passed through the point P_i (i.e. the plane perpendicular to the $||S_{c,i}||$

at point P_i; perpendicular to the $\bar{V}_{12}$ at the same point P_i respectively). Then, the intersections of the $O_{1,R}$ and $O_{2,R}$ of π_i with the axes $1-1$ and $2-2$ are found, and after that a line $O_{1,R}O_{2,R} \equiv R_i$ is passed through them.

From (7.21) for the direction cosines of the vector $\bar{V}_{12}$ in $S_0(O_0, x_0, y_0, z_0)$ can be written:

$$\vartheta_{12,x_0} = \frac{y_0(1 + i_{21} - 2i_{21}\cos\delta)}{A}, \quad \vartheta_{12,y_0} = -\frac{x_0(1 + i_{21} - 2i_{21}\cos\delta)}{A},$$
$$\vartheta_{12,z_0} = -\frac{a_w i_{21}\sin\delta}{A}, \quad A = \sqrt{(x_0^2 + y_0^2)(1 + i_{21}^2 - 2i_{21}\cos\delta)^2 + (a_w i_{21}\sin\delta)^2} \tag{7.45}$$

The parametric expression of (7.45) is

$$\vartheta_{12,x_0} = \frac{R_0\cos t(1 + i_{21} - 2i_{21}\cos\delta)}{B}, \quad \vartheta_{12,y_0} = -\frac{R_0\sin t(1 + i_{21} - 2i_{21}\cos\delta)}{B},$$
$$\vartheta_{12,z_0} = -\frac{a_w i_{21}\sin\delta}{B}, \quad B = \sqrt{R_0^2(1 + i_{21}^2 - 2i_{21}\cos\delta)^2 + (a_w i_{21}\sin\delta)^2}. \tag{7.46}$$

The normal equation of the polar plane π_i, passing through the point $P_i(x_0, y_0, z_0)$, and perpendicular to the normalized relative helix $||S_{c,i}||$ is:

$$\vartheta_{12,x_0}.x_0 + \vartheta_{12,y_0}.y_0 + \vartheta_{12,z_0}z_0 + D_0 = 0, \tag{7.47}$$

and the equations of the conjugated axes $1-1$ and $2-2$ in the static coordinate system $S_0(O_0, x_0, y_0, z_0)$, according to Fig. 7.1 are respectively:

– For the axis $1-1$:

$$\cot\alpha.x_0 + z_0 = 0, \quad y_0 - r_1^0 = 0; \tag{7.48}$$

– For the axis $2-2$:

$$\cot(\delta + \alpha).x_0 + z_0 = 0, \quad y_0 + r_2^0 = 0. \tag{7.49}$$

Then for intersects of the (7.48) with (7.47), the equations system is obtained:

$$x_0 = \frac{\vartheta_{12,y_0}.r_1^0 + D_0}{\vartheta_{12,z_0}\cot\alpha - \vartheta_{12,x_0}}, \quad y_0 = r_1^0, \quad z_0 = \frac{(\vartheta_{12,y_0}.r_1^0 + D_0)\cot\alpha}{\vartheta_{12,x_0} - \vartheta_{12,z_0}\cot\alpha}, \tag{7.50}$$

and for the intersects of the (7.49) with (7.47) it is received

$$x_0 = \frac{\vartheta_{12,y_0}.r_2^0 - D_0}{\vartheta_{12,x_0} - \vartheta_{12,z_0}\cot(\delta + \alpha)}, \quad y_0 = -r_2^0, \quad z_0 = \frac{(\vartheta_{12,y_0}.r_2^0 - D_0)\cot(\delta + \alpha)}{\vartheta_{12,z_0}\cot(\delta + \alpha) - \vartheta_{12,x_0}}. \tag{7.51}$$

The ray R_i of the normalized relative helix $||S_{c,i}||$, passing through the point P_i of the polar plane π_i and crossing the conjugated axes $1-1$ and $2-2$ in points $O_{1,R}$ and $O_{2,R}$, is described by the equations system:

$$\frac{x_0 - \dfrac{\vartheta_{12,y_0} \cdot r_1^0 + D_0}{\vartheta_{12,z_0} \cot \alpha - \vartheta_{12,x_0}}}{\dfrac{\vartheta_{12,y_0} \cdot r_2^0 - D_0}{\vartheta_{12,x_0} - \vartheta_{12,z_0} \cot(\delta + \alpha)} - \dfrac{\vartheta_{12,y_0} \cdot r_1^0 + D_0}{\vartheta_{12,z_0} \cot \alpha - \vartheta_{12,x_0}}} = \frac{y_0 - r_1^0}{-a_w}$$

$$= \frac{z_0 - \dfrac{(\vartheta_{12,y_0} \cdot r_1^0 + D_0) \cot \alpha}{\vartheta_{12,x_0} - \vartheta_{12,z_0} \cot \alpha}}{\dfrac{(\vartheta_{12,y_0} \cdot r_2^0 - D_0) \cot(\delta + \alpha)}{\vartheta_{12,z_0} \cot(\delta + \alpha) - \vartheta_{12,x_0}} - \dfrac{(\vartheta_{12,y_0} \cdot r_1^0 + D_0) \cot \alpha}{\vartheta_{12,x_0} - \vartheta_{12,z_0} \cot \alpha}}. \tag{7.52}$$

The common normal of the isokinematic surfaces I_i ($i = 1, 2$) and the normalized relative helix $||S_{c,i}||$ in their common contact point P_i is searched among those rays R_i of the normalized relative helix $||S_{c,i}||$, which are perpendicular to the minimal relative helix S_{c,z_0}, i.e. of the instantaneous relative axis $O_0 z_0$. Hence, the characteristic C_i of the isokinematic surfaces I_1 and I_2 have to be searched as an intersection of the family of rays (7.52) with the family of normalized relative helices $||S_{c,i}||$ (with their envelope—the corresponding kinematic cylinder of level $R_{V_{c,i}}$, respectively) in the system S_0, when these rays cross the minimal relative helix S_{c,z_0}. For this purpose when $R_0 = 0$ is substituted in (7.46), the directorial cosines of the unit vector $\bar{V}_{12} = \bar{V}_{12,z_0}$, "tangential" to the minimal relative helix S_{c,z_0}, are obtained

$$\vartheta_{12,x_0} = 0, \quad \vartheta_{12,y_0} = 0, \quad \vartheta_{12,z_0} = -1. \tag{7.53}$$

After substituting (7.53) in (7.52), the following equations system is received:

$$\frac{x_0 + \dfrac{D_0}{\cot \alpha}}{\dfrac{-D_0}{\cot(\delta + \alpha)} + \dfrac{D_0}{\cot \alpha}} = \frac{y_0 - r_1^0}{-a_w} = \frac{z_0 - D_0}{0}. \tag{7.54}$$

In Eqs. (7.47), (7.59), (7.51), (7.52) and (7.54), D_0 is a parameter of the family of polar planes π_i, at different points P_i on the normalized relative helices $||S_{c,i}||$. Hence, D_0 is also a parameter of the family of rays R_i, including those that cross the minimal relative helix S_{c,z_0}. It is searched for the intersection between the rays' family (7.54) with the kinematic surface on level (7.17), considered here as a locus of the normalized relative helices $||S_{c,i}||$ in the coordinate system S_0. For this reason expression (7.17) is presented parametrically as

$$x_0 = R_0 \sin t, \quad y_0 = R_0 \cos t, \quad z_0 = H = constant, \quad H \in (-\infty, +\infty), \tag{7.55}$$

and the ray of the relative helix (7.54)—as an intersection between two planes, i.e.:

$$z_0 - D_0 = 0, \quad a_w \left(x_0 + \frac{D_0}{\cot \alpha} \right) = -(r_1^0 - y_0) \left(\frac{D_0}{\cot(\delta + \alpha)} - \frac{D_0}{\cot \alpha} \right). \tag{7.56}$$

Analytical type of characteristic C_i is obtained, after Eqs. (7.55) and (7.56) are solved together:

$$x_0 = R_0 \sin t, \ \ y_0 = R_0 \cos t, \ \ z_0 = D_0, \ \ D_0 = \frac{a_w}{\tan(\delta + \alpha) - \tan \alpha} \tan t. \quad (7.57)$$

On the base of analysis of the system (7.57), it can easily be found that it describes two spatial curves C_i $(i = 1, 2)$ on the kinematic cylinder of level $R_{V_{c,i}}$ having a radius R_0. They are a graphical expression of the trigonometric function:

$$z_0 = a_w \frac{(\cos \delta - i_{21})(1 - 2i_{21} \cos \delta)}{\sin \delta (1 + i_{21}^2 - 2i_{21} \cos \delta)} \tan t. \quad (7.58)$$

Hence, the characteristic C_i of the isokinematic quasi-hyperboloids I_1 and I_2 contains two curves C_1 and C_2. C_1 is placed on the lower half of the kinematic cylinder of level $R_{V_{c,i}}$, and C_2—on the upper half of $R_{V_{c,i}}$. The characteristics C_1 and C_2, of externally and internally contacting isokinematic quasi-hyperboloids I_1 and I_2 respectively, are shown in Fig. 7.9. On Fig. 7.10 are illustrated characteristics C_i $(i = 1, 2)$ on the developable cylindrical surface of the $R_{V_{c,i}}$. The cylindrical surface is interrupted along its generatrix, having equations $x_0 = 0$, $y_0 = R_0$, $z_0 = z_0$.

Depending on the geometric and kinematic parameters of the synthesized systems for the spatial rotations transformation, the kinematic cylinder of level $R_{V_{c,i}}$ can contact internally with one isokinematic surface and—externally with its conjugated or internally with the two isokinematic quasi-hyperboloids. For the first case, I_1 and I_2 are externally contacting (externally meshed), and the second case corresponds to the internally contacting (internally meshed) I_1 and I_2.

The application of the isokinematic quasi-hyperboloids as primitives in mathematical models for synthesis of hyperboloid gear mechanisms is the reason to be called, by the authors of this publication, *kinematic pitch surfaces*.

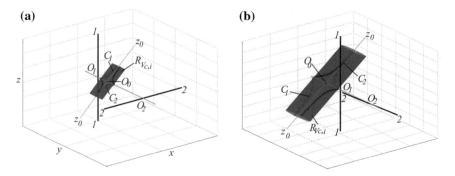

Fig. 7.9 Characteristics C_i $(i = 1, 2)$ of the isokinematic quasi-hyperboloids: **a** $a_w = 105$ mm, $\delta = 90°, i_{12} = 2, V_c = 50$ mm/rad (internal meshing); **b** $a_w = 105$ mm, $\delta = 30°, i_{12} = 2, V_c = 50$ mm/rad (external meshing)

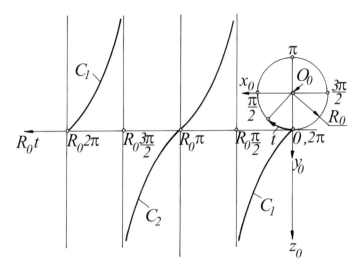

Fig. 7.10 General type of the characteristics C_i ($i = 1,\ 2$) of the isokinematic quasi-hyperboloids I_i ($i = 1,\ 2$) in the static coordinate system $S_0(O_0, x_0,\ y_0,\ z_0)$

7.5.2 Kinematic Pitch Circles

The synthesized isokinematic rotational surfaces I_1 and I_2 contact in every moment of their rotation at the characteristic C_i. The spatial curve C_i, as a locus of the common points P_i of I_1 and I_2, defines both the magnitude of sliding velocity vector $\bar{V}_{12}$ ($V_{12} = V_c.\omega_1 = constant$) and its orientation for every point P_i in the static space. As it can be seen from Fig. 7.6, the vector $\bar{V}_{12}$ concludes with the axes $i - i$ ($i = 1,\ 2$) angles β_i^k ($i = 1,\ 2$), respectively. These angles (as it is obvious from the expressions (7.32), (7.33) and (7.34)) have different values, depending on the location of the point P_i on C_i. The defined above kinematic characteristics of C_i are explained with the fact, that the curve C_i is a locus of the points, belonging to the family of normalized relative helices $||S_{c,i}||$. Here, it should be mentioned, that when the isokinematic rotational quasi-hyperboloids are transformed into revolution hyperboloids (axodes of the movable links of the spatial gear mechanism, contacting at the minimal kinematic relative helix S_{c,z_0}), the vector of the relative motion $\bar{V}_{12}$ has a minimal magnitude $V_{12} = V_{12,\min} = V_{12,z_0}$ for every point P_i of I_1 and I_2. Also its orientation, towards the axes of the movable links $i - i$ ($i = 1,\ 2$), is a constant one for every point P_i from the instantaneous helical axis $O_0 z_0$. An alternative to what has been said up to now, is the statement, that every point P_i is a common point between a normalized relative helix $||S_{c,i}||$ and its ray R_i, intersecting the rotation axes $i - i$ ($i = 1,\ 2$). Hence, every point P_i from C_i can be treated as a common point of the geometric elements Σ_1 and Σ_2 of high kinematic joints ($\Sigma_1{:}\Sigma_2$), by means the rotations transformation between crossed axes $1-1$ and $2-2$ is realized. In the treated case, every point P_i from C_i and S_{c,z_0} is a common point of two circles H_i^k ($i = 1,\ 2$), which centers lies on the axes $i - i$ ($i = 1,\ 2$), and the corresponding circles are perpendicular to these axes. In dependence on the desired geometric and kinematic parameters of the particular spatial rotations transformation, the referred pairs of circle H_i^k ($i = 1, 2$) are placed on the both sides or on one side of the plane T_R, that passes through their common point P_i and contains the coplanar vectors $\bar{V}_i$ ($i = 1, 2$) and $\bar{V}_{12}$ (this plane is perpendicular to the ray R_i of the normalized kinematic relative helix $||S_{c,i}||$ passing through the same point P_i). The diameters of these circles and their mutual position in the static space, define both the basic geometric characteristics of the structure of the three-links gear mechanism (including the basic dimension proportions of the gear blanks) and the analytical characteristics of the tooth surfaces Σ_1 and Σ_2, including the geometric-kinematic parameters defining their longitudinal orientation in the common contact point P_i. The pitch pairs H_i^k ($i = 1, 2$) will be called *kinematic pitch circles*, due to the fact that their dimensions and mutual position in the space define pitches (tooth modules) of the synthesized active tooth surfaces Σ_i ($i = 1, 2$). Their common contact point $P_1 \equiv P_2 \equiv P$ and plane T_R, passing through it, will be called a *kinematic pitch point* and a *kinematic pitch plane* respectively. On Fig. 7.11, the kinematic pitch circles, corresponding to the spatial gear mechanisms with an external and internal gearing, are illustrated, respectively. Further bellow, an algorithm, representing the solution of the task for the synthesis of kinematic pitch circles, will be shown. It is based on the main analytical dependencies, obtained on the basis of the vector analysis of the vector function of the sliding velocity vector, as well as on the algorithms that define the dimensions of the relative helices and isokinematic quasi-hyperboloids (so-called *kinematic pitch surfaces* together with the *kinematic pitch circles* define the *kinematic pitch configurations* of the spatial transformation of rotations). The study is realized, in accordance with Fig. 7.1. In this case, the coordinate systems shown there $S_i(O_i, x_i, y_i, z_i)$ ($i = 1, 2$) are treated as static ones. Then, based on the equation systems (7.42) and (7.43), the radii of the kinematic pitch circles H_i^k and their location in the systems S_1 and S_2, respectively, by means of the applicate z_i ($i = 1, 2$) of the pitch contact point $P_1 \equiv P_2 \equiv P$, can be determined, i.e.:

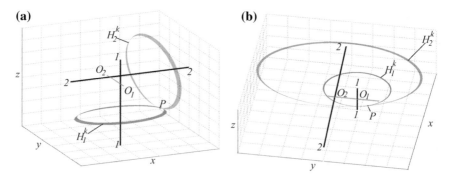

Fig. 7.11 Kinematic pitch circles H_i^k ($i = 1, 2$): **a** externally contacting H_i^k ($i = 1, 2$)—$a_w = 105$ mm, $\delta = 90°$, $i_{12} = 2$, $V_c = 50$ mm/rad, $t \in [+80°, +81°]$; **b** internally contacting H_i^k ($i = 1, 2$)—$a_w = 105$ mm, $\delta = 30°$, $i_{12} = 2$, $V_c = 50$ mm/rad, $t \in [+250°, +251°]$

$$r_1^k = \sqrt{\left[\tan\alpha\left(z_1 + \frac{R_0\sin t}{\sin\alpha}\right)\right]^2 + [R_0\cos t - r_1^0]},$$

$$z_1 = \frac{\tan t\ \sin\alpha\left(R_0\cos t - r_1^0\right)}{\tan^2\alpha} - \frac{R_0\sin t}{\sin\alpha}, \tag{7.59}$$

where $R_0 = \sqrt{\dfrac{V_c^2}{\omega_{12}^2} - p_{z_0}^2}$, $\omega_{12} = \sqrt{1 + i_{21}^2 - 2i_{21}\cos\delta}$, $p_{z_0} = \dfrac{a_w i_{21}\sin\delta}{\omega_{12}^2}$,

$$r_1^0 = \frac{a_w i_{21}(i_{21} - \cos\delta)}{\omega_{12}^2}, \quad \alpha = \arctan\frac{i_{21}\sin\delta}{1 - i_{21}\cos\delta}$$

and

$$r_2^k = \sqrt{\left[\tan(\alpha + \delta)\left(z_2 + \frac{R_0\sin t}{\sin(\alpha + \delta)}\right)\right]^2 + [R_0\cos t + r_2^0]^2},$$

$$z_2 = \frac{\tan t\ \sin(\alpha + \delta)\left(R_0\cos t + r_2^0\right)}{\tan^2(\alpha + \delta)} - \frac{R_0\sin t}{\sin(\alpha + \delta)}, \tag{7.60}$$

where $R_0 = \sqrt{\dfrac{V_c^2}{\omega_{12}^2} - p_{z_0}^2}$, $\omega_{12} = \sqrt{1 + i_{21}^2 - 2i_{21}\cos\delta}$, $p_{z_0} = \dfrac{a_w i_{21}\sin\delta}{\omega_{12}^2}$,

$$r_2^0 = \frac{a_w(1 - i_{21}\cos\delta)}{\omega_{12}^2}, \quad \alpha = \arctan\frac{i_{21}\sin\delta}{1 - i_{21}\cos\delta}.$$

The dimensions of the kinematic pitch circles H_i^k ($i = 1,\ 2$) are determined by their radii by using the systems (7.59) and (7.60), while the corresponding analytical expressions of an arbitrary crossed section of isokinematic quasi-hyperboloid with planes $z_1 = C_{z_1}$ and $z_2 = C_{z_2}$ are searched sequentially.

The cross section of I_1 is given by the equations system:

$$x_1^2 + y_1^2 = \left[\tan\alpha\left(z_1 + \frac{R_0\sin t}{\sin\alpha}\right)\right]^2 + [R_0\cos t - r_1^0]^2,$$

$$z_1 = C_{z_1} = \frac{\tan t\ \sin\alpha\left(R_0\cos t - r_1^0\right)}{\tan^2\alpha} - \frac{R_0\sin t}{\sin\alpha}. \tag{7.61}$$

The cross section of I_2 is described by the following equations system:

$$x_2^2 + y_2^2 = \left[\tan(\alpha + \delta)\left(z_2 + \frac{R_0\sin t}{\sin(\alpha + \delta)}\right)\right]^2 + [R_0\cos t + r_2^0]^2,$$

$$z_2 = C_{z_2} = \frac{\tan t\ \sin(\alpha + \delta)\left(R_0\cos t + r_2^0\right)}{\tan^2(\alpha + \delta)} - \frac{R_0\sin t}{\sin(\alpha + \delta)}. \tag{7.62}$$

In the equations system (7.61) and (7.62), the planes $z_1 = C_{z_1}$ and $z_2 = C_{z_2}$ correspond with the concrete value of the parameter t.

7.6 Computer Programs for Software Synthesis and Visualization of Kinematic Pitch Configurations

The program described below, refers to this type of software that studies the character of the kinematic pitch configurations and the geometric and kinematic primitives, which are in the basis of their synthesis and which essentially can be treated as a building element of the constructed kinematic model for synthesis of spatial gear mechanisms with crossed axes of rotation. It is based on the on the multi-paradigm numerical computing software environment MATLAB. The program contains a description and declaration of input parameters and 7 functions. The different kinematic pitch configurations, described in the study, are obtained after execution of those functions. The input parameters defined in the program are as follows:

- **Geometric input parameters**: offset—$a_w \in (0, +\infty)$ [mm]; crossed angle of the axes of rotations – $\delta \in (0, \pi)$, [rad]; angular parameter—t, [rad].
- **Kinematic input parameters**: velocity ratio—$i_{21} \in (0, 1)$; module of the given sliding velocity—$V_{12} \in (0, +\infty)$, [mm/s]; module of the angular velocity—ω_1, [rad.s^{-1}];

Instead of the last two parameters, $V_c = V_{12}\omega_1^{-1}$ [mm/rad] can be given in the program.

The program visualizes all possible pitch configurations, which are specific for hyperboloid three-link mechanism, with an arbitrary placement of the axes of rotations $i - i$ ($i = 1, 2$) of the movable links $i = 1$ and $i = 2$, which placement is determined by the offset a_w and the crossed angle δ of the mentioned axes. Structurally, this computer program consists of one main program and six executed functions. The execution of each function is realized by the input key K, which is defined in the body of the main program. For each value from 1 to 6 of the key K, the main program executes one of the functions, that graphically illustrates a concrete kinematic configuration, namely:

- When key $K = 1$, the corresponding function is called. It illustrates the axes $1 - 1$ and $2 - 2$ of the spatial mechanism, kinematic cylinders of level $R_{V_{c,i}}$ and the relative helical axis $z_0 - z_0$. When the geometric and kinematic parameters are chosen appropriately, then different placements of the kinematic cylinders of level are obtained.
- For the case when $K = 2$, the obtained graphic image, contains axes $i - i$, the offset line O_1O_2, the axis of the relative helical motion $z_0 - z_0$, kinematic cylinders of level $R_{V_{c,i}}$ and the kinematics conoids $C_{d,i}$. In this case, the vector lines of the vector function $\bar{V}_{12} = \bar{V}_{12}(x, y, z)$, representing the intersection of $R_{V_{c,i}}$ and $C_{d,i}$ (see Fig. 7.12) are illustrated.

(a) **(b)**

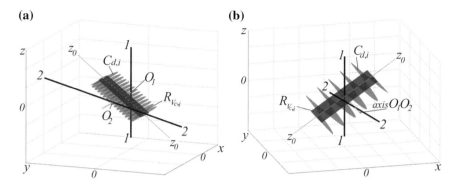

Fig. 7.12 Obtaining the vector lines of the vector function $\bar{V}_{12} = \bar{V}_{12}(x, y, z)$: **a** kinematic cylinder of level $R_{V_{c,i}}$ and left-hand kinematic conoid $S_{d,i}$ when $a_w = 50$ mm, $\delta = 225°$, $i_{21} = abs(\cos\delta)$, $V_c = 25$ mm/rad; **b** kinematic cylinder of level $R_{V_{c,i}}$ and right-hand kinematic conoid $S_{d,i}$ when $a_w = 50$ mm, $\delta = 45°$, $i_{21} = abs(\cos\delta)$), $V_c = 40$ mm/rad

- When to the key K is given value 3, it runs the function, for drawing axes of rotations $i - i$ ($i = 1, 2$), zero cylinder of level $z_0 - z_0$, axis O_1O_2, as well as kinematic cylinders of level $R_{V_{c,i}}$ together with the normalized relative helices $||S_{c,i}||$, which are placed on them (see Fig. 7.13).
- When a value 4 is given K, the executed function illustrates crossed axes $1 - 1$ and $2 - 2$, the relative helical axis $z_0 - z_0$, around which is drawn the kinematic cylinder of level $R_{V_{c,i}}$, on which the characteristics C_i of the isokinematic quasi-hyperboloids are placed. The characteristic C_1 is obtained when the parameter t changes in the interval $[-\pi/2 + \varepsilon, \pi/2 - \varepsilon]$, and C_2- when the angular parameter t belongs to $[\pi/2 + \varepsilon, 3\pi/2 - \varepsilon]$ (Fig. 7.14).
- When $K = 5$, the corresponding executive functions draws axes $1-1$ and $2-2$ and conjugated isokinematic quasi-hyperboloids (kinematic pitch surfaces) I_1 and I_2.

(a) **(b)**

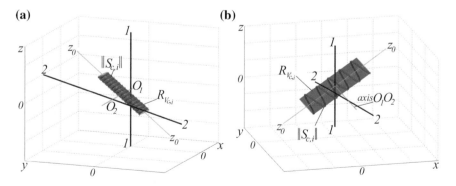

Fig. 7.13 Kinematic cylinder of level $R_{V_{c,i}}$ and normalized relative helices $||S_{c,i}||$ ($i = 1, 2$): **a** $R_0 = R_0(V_c)$ mm, $(O_0z_0, \bar{\omega}_{12}, p_{z_0})$, $i_{21} = abs(\cos\delta)$), $V_c = 25$ mm/rad; **b** $a_w = 50$ mm, $\delta = 45°$, $V_c = 40$ mm/rad

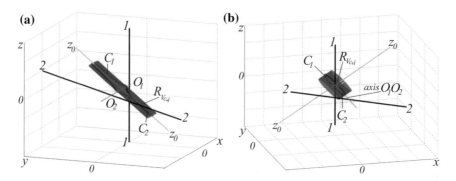

Fig. 7.14 Characteristics C_i ($i = 1,\ 2$) placed on the kinematic surfaces of level $R_{V_{c,i}}$: **a** $a_w = 50$ mm, $\delta = 225°$, $i_{21} = abs(\cos\delta)$, $V_c = 25$ mm/rad; **b** $a_w = 50$ mm, $\delta = 45°$, $i_{21} = abs(\cos\delta)$, $V_c = 40$ mm/rad

These special surfaces are obtained when each of the characteristics C_i ($i = 1,\ 2$) (Figs. 7.15, 7.16) are rotated around axes $1-1$ and $2-2$. Various graphical images, representing a structure, geometric, and kinematic alternative of the hyperboloid gear drives with external and internal gearing, are obtained for different values of the input parameters.

- When the function, called by key $K = 6$ is executed, then the axes of rotations $1-1$ and $2-2$, offset line $O_1 O_2$, which connects them and the kinematic pitch circles H_i^k ($i = 1,\ 2$) are illustrated. For the different values of the input parameters, images of the internally and externally contacting kinematic pitch circles (see Fig. 7.17) are obtained.

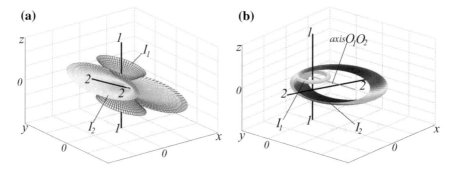

Fig. 7.15 Isokinematic pairs (I_1:I_2): **a** externally contacting isokinematic quasi-hyperboloids $I_i - \bar{V}_{12}$ mm, $\delta = 225°$, $i_{21} = abs(\cos\delta)$, $V_c = 25$ mm/rad, $t \in [-75°, +75°]$; **b** internally contacting isokinematic quasi-hyperboloids I_i ($i = 1,\ 2$)-$a_w = 50$ mm, $\delta = 45°$, $i_{21} = abs(\cos\delta)$, $V_c = 40$ mm/rad, $t \in [-75°, +75°]$

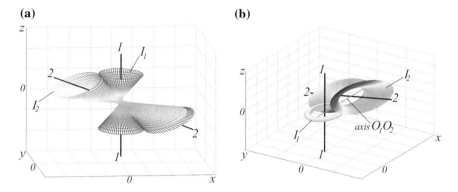

(a)   **(b)**

Fig. 7.16 Isokinematic pairs ($I_1:I_2$): **a** externally contacting isokinematic quasi-hyperboloids $I_i - a_w = 50$ mm, $\delta = 225°$, $i_{21} = abs(\cos \delta)$, $V_c = 25$ mm/rad, $t \in [110°, +260°]$; **b** externally contacting isokinematic quasi-hyperboloids I_i ($i = 1, 2$)-$a_w = 50$ mm, $\delta = 45°$, $i_{21} = abs(\cos \delta)$, $V_c = 40$ mm/rad, $t \in [110°, +260°]$

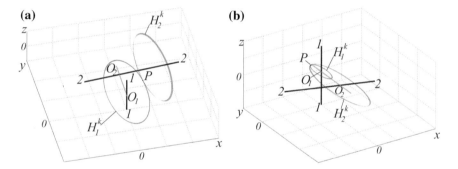

(a)   **(b)**

Fig. 7.17 Kinematic pitch circles H_i^k ($i = 1, 2$): **a** external contacting $H_i^k - a_w = 50$ mm, $\delta = 225°$, $i_{21} = abs(\cos \delta)$, $V_c = 25$ mm/rad, $t \in [+72°, +73°]$; **b** internal contacting $H_i^k - a_w = 50$ mm, $\delta = 45°$, $i_{21} = abs(\cos \delta)$, $V_c = 40$ mm/rad, $t \in [-30°, -25°]$

7.7 Conclusion

The kinematic model of the process of spatial rotations transformation between arbitrary crossed axes is illustrated in this study. It is oriented into defining the character of the relative motion of the conjugated contact point of the elements of high kinematic joints, by means of which the regular spatial transformation of rotations is realized. On the basis on the elaborated mathematical model, an analysis of the vector function of the sliding velocity vector in the contact point of the conjugated high kinematic joints is realized. These high kinematic joints represent the meshed tooth surfaces of the hyperboloid gear sets with an arbitrary crossed axes of rotations. By means of the defined scalar and vector field of the relative motion velocity, the analytical type of important features of the spatial motions transformation is obtained: kinematic cylinders of level, kinematic conoids and relative helices of the vector field of the relative motion. The analytical type of the normalized (normed) relative helices is derived on the basis of symbiosis between kinematic cylinders of level and the relative helices of the above mentioned vector field. Algorithms and computer programs for synthesis and software visualization of the kinematic pitch configurations are created, when the so called kinematic pitch configurations (isokinematic quasi-hyperboloids) and kinematic pitch surfaces are defined.

References

1. Abadjiev, V.: On the Synthesis and Analysis of Spiroid Gears. Ph. D. Thesis, Sofia (in Bulgarian) (1984)
2. Gavrilenko, V.: Methods of further development of the theory of gearing". "Mashinovedenie. Moscow **4**, 41–45 (1972)
3. Litvin, F.: Theory of Gearing. Publishing house "Nauka," Moscow (in Russian) (1968)
4. Litvin, F., Petrov, K., Ganshin, V.: The effect of geometrical parameters of Hypoid and Spiroid gears in its quality characteristics. Trans. ASME. J. Eng. Ind. **96**(1), Series B, 330–334 (1974)
5. Minkov, K.: Fundamentals of primary surfaces primary surfaces synthesis of gears with arbitrarily crossed axes. J. Theor. Appl. Mech. vol. 3, pp. 61–74, BAS, Sofia (1975)
6. Minkov, K.: Mechanical and Mathematical Modeling of Hyperboloid Gears. Sc. D. Thesis, Sofia (in Bulgarian) (1986)
7. Litvin, F.: Theory of Gearing. NASA Reference Publication 1212, AVSCOM Technical Report 88-C-035, US Government Printing Office, Washington (1989)
8. Litvin, F.: Gearing Geometry and Applied Theory. PTR Prentice Hall, A Paramount Communication Company, Englewood Cliffs, New Jersey 07632 (1994)
9. Litvin, F., Fuentes, A.: Gear Geometry and Applied Theory, 2nd edn., Cambridge University Press, Cambridge, New York, Melbourne, Madrid, Cape Town, Singapore, Sao Paolo, Delhi, Tokyo, Mexico City (2004)
10. Abadjiev, V., Abadjieva, E.: Geometric Pitch Configurations—Basic Primitives of the Mathematical Models for the Synthesis of Hyperboloid Gear Drives. In: Chapter 4 in Mechanisms and Machine Science, Advanced Gear Engineering, vol. 51, pp. 91–116, Springer (2018)
11. Figliolini, G., Angeles, J.: The synthesis of the pitch surfaces of internal and external skew-gears and their rack. ASME J. Mech. Des. **128**(4), 794–802 (2006)
12. Figliolini, G., Stachel, H., Angeles, J.: A new look at the Ball-Disteli diagram and it relevance to spatial gearing. J. Mech. Mach. Theory **42**(10), 1362–1375 (2007)
13. Phillips, J.: Freedom in Machinery, vol. 2 Screw Theory Exemplified, Cambridge University Press, Cambridge (1990)
14. Phillips, J.: General Spatial Involute Gearing, Springer (2003)
15. Nelson, W.: Spiroid gearing. Part 1—basic design practices. Mach. Des. 136–144 (1961)
16. Abadjiev, V.: Gearing Theory and Technical Applications of Hyperboloid Mechanisms. Sc. D, Thesis, Institute of Mechanics—BAS, Sofia (2007). (in Bulgarian)
17. Abadjiev, V.: Mathematical modelling for synthesis of spatial gears. In: Proceedings of the Institution of Mechanical Engineers, Part E: J. Process. Mech. Eng., vol. 216, pp. 31–46 (2002)
18. Abadjiev, V., Abadjieva, E., Petrova, D.: Pitch configurations: definitions, analytical and computer synthesis. In: Proceedings of ASME 2011 International Power Transmissions and Gearing Conference, Washington DC, USA (published on CD) (2011)
19. Abadjiev, V., Abadjieva E., Petrova, D.: Non-orthogonal hyperboloid gears. Synthesis and visualization of pitch configurations with inverse orientation. Compt. Rend. Acad. Bulg. Sci, Sofia, **64**(8), 1171–1178 (2011a)
20. Abadjiev, V., Abadjieva E., Petrova, D.: Non-orthogonal hyperboloid gears. Synthesis and visualization of pitch configurations with normal orientation. Compt. Rend. Acad. Bulg. Sci, Sofia, **64**(9), 1311–1319 (2011b)

21. Abadjiev, V., Okhotsimsky, D., Platonov, A.: Research on the Spatial Gears and Applications. Keldysh Institute of Applied Mathematics, Russian Academy of Sciences, Preprint No 89, Moscow (1997)
22. Dooner, D.: Kinematic Geometry of Gearing, WILEY, Wiley Publication (2012)

Chapter 8
Tooth Contact Analysis of Cylindrical Gears Reconstructed from Point Clouds

Ignacio Gonzalez-Perez and Alfonso Fuentes-Aznar

Abstract Non-contact metrology is being used to accelerate and improve the gear inspection process by obtaining a large amount of information on the gear tooth surfaces in form of point clouds for all teeth of the to-be-inspected gears. A new approach for tooth contact analysis and stress analysis of cylindrical gears reconstructed from point clouds is proposed in this work. It considers a whole revolution of the being-inspected pinion in mesh with a master gear to evaluate the influence of the deviations presented in all the teeth of the first in terms of contact patterns and loaded functions of transmission errors. The approach is validated throughout the generation of computationally generated point clouds whose points are obtained theoretically by the application of the modern theory of gearing. The results are then compared with those obtained with measured point clouds from manufactured gears. The application of the proposed approach to a pair of gears shows the necessity to apply a filtering process to avoid irregularities in the reconstructed gear tooth surfaces that cause high values of contact stresses and unloaded transmission errors. However, the evaluation of the loaded function of transmission errors has shown an important value as an indicator to compare the mechanical performance of the being-inspected gears.

8.1 Introduction

Integrating non-contact metrology with the application of computational tooth contact analysis of gear drives constitutes a promising technique for gear inspection and evaluation of the mechanical performance of the being-inspected gear. Besides the evaluation of the usual magnitudes currently applied in the gear inspection process,

I. Gonzalez-Perez
Universidad Politécnica de Cartagena (UPCT), Cartagena, Spain
e-mail: ignacio.gonzalez@upct.es

A. Fuentes-Aznar (✉)
Rochester Institute of Technology (RIT), Rochester NY, USA
e-mail: afeme@rit.edu

© Springer Nature Switzerland AG 2020
V. Goldfarb et al. (eds.), *New Approaches to Gear Design and Production*,
Mechanisms and Machine Science 81,
https://doi.org/10.1007/978-3-030-34945-5_8

such as profile errors, flank deviations, or pitch errors, the possibility to simulate the contact patterns and function of transmission errors from gear tooth surfaces reconstructed from point clouds adds value to the gear inspection process. The level of stresses and the peak-to-peak level of the loaded function of transmission errors constitute representative parameters of the mechanical performance of the gear drive. Reconstruction of gear tooth surfaces from point clouds and the application of tooth contact analysis as well as finite element analysis represents a step forward in the process of gear inspection.

The first machines for non-contact metrology appeared less than a decade ago. Currently, the main gear manufacturers include these types of machines to add value to their gear inspection process. Therefore, this technology is starting to be extensively applied not only for the quality assessment of gears as it is done with contact-based metrology machines, but it is being used as an extraordinary technology to reverse engineering of gear designs and for the virtual testing of manufactured gears, including the possibility to perform tooth contact analysis and finite element analysis of gear systems. The development of dedicated softwares to perform these tasks is adding value to the gear inspection process and helping to increase the application of non-contact metrology machines.

Previous publications related with the application of data provided by non-contact metrology machines to the process of gear analysis and simulation hardly exist. In [1], optical methods are considered to evaluate gear flanks from point clouds and to calculate surface deviations, thus saving time in the gear quality control. In [2], an overview of the possibilities of integration of non-contact metrology in the process of analysis and simulation of gear drives is presented. However, in that work, the modeling of the inspected gear uses the point clouds for just one teeth, and the reconstructed gear tooth surfaces for that tooth are used to form all the teeth of the gear model, missing in this way possible pitch deviation errors or the different flank deviations present at different teeth of the being-inspected gear.

In the present work, the integration of non-contact metrology with the application of computational tooth contact analysis of gear drives is proposed. All the scanned teeth are being taken into account in the reconstructed gear model since the regenerated gear tooth surfaces for each tooth of the gear model are obtained independently from the provided points clouds for all teeth of the inspected gear. A new tooth contact analysis algorithm is proposed here to obtain the contact patterns and the function of transmission errors for a whole revolution of the being-inspected gear in mesh with a master gear. Here, master gear refers to a theoretical gear in which their surfaces are obtained theoretically, thus no considering manufacturing errors. The results will show, depending on the deviations of each regenerated gear tooth surface, different contact patterns and contact paths at each tooth of the being-inspected gear. A new finite element model is proposed here as well to analyze the variation of contact stresses all over a revolution of the being-inspected gear. For that purpose, the finite element model proposed in [3] is applied with some modifications to cover the analysis for a whole revolution of the being-inspected gear.

8.2   Regeneration of Gear Tooth Surfaces

The gear tooth surfaces of the to-be-inspected gear are regenerated from the point clouds obtained with a non-contact metrology machine. Unlike the work presented in [2], here each tooth is regenerated from the point clouds provided for its left and right surface sides. Therefore, the regenerated model of the to-be-inspected gear can be used to evaluate pitch errors and the different flank deviations present at each tooth. The process of regeneration of each gear tooth surface from its corresponding point cloud is explained in detail in [2]. A NURBS (Non-Uniform Rational B-Spline) surface [4] is created from a grid of points that belong to the point cloud and that are selected using a K-D tree algorithm [5] which is a space-partitioning data structure for organizing points in a k-dimensional space.

Figure 8.1 shows a regenerated gear model comprising twenty teeth and their corresponding forty point clouds (two for each tooth). The fillets were obtained using Hermite curves as proposed in [2] to adjust the regenerated fillet geometry to some points selected from the point cloud presented in the fillet area.

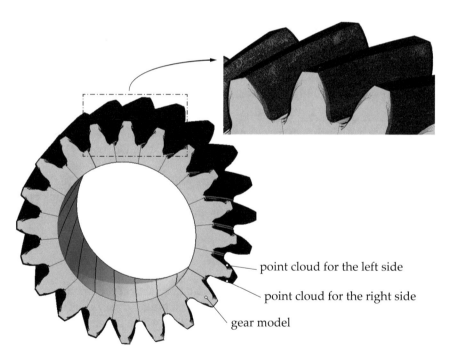

point cloud for the left side

point cloud for the right side

gear model

Fig. 8.1 Point clouds for each tooth and gear model obtained from the regenerated gear tooth surfaces

8.3 Tooth Contact Analysis

Application of tooth contact analysis (TCA) between the to-be-inspected gear and a master gear all over a revolution of the being-inspected gear represents a challenging task. Usually, application of tooth contact analysis covers two cycles of meshing to investigate the transition of load between adjacent pairs of teeth. This requires to focus the computational effort in the contact analysis of three pairs of teeth. However, application of TCA for one revolution of the being-inspected gear requires analyzing N cycles of meshing, where N is the number of teeth of the to-be-inspected gear. Therefore, the pairs of teeth used in the algorithm for tooth contact analysis change from one cycle of meshing to the following one.

For each cycle of meshing, an algorithm that minimizes the distance between the tooth surfaces of the selected three teeth of the master gear and the three teeth of the being-inspected gear has been applied here to find the contact path, the contact pattern, and the unloaded function of transmission errors. The gear tooth surfaces of the master and the being-inspected gear are considered as rigid surfaces. A virtual marking compound thickness will be considered to determine the contour of the bearing contact for each angular position of the being-inspected gear. The algorithm approaches the master gear tooth surfaces to the to-be-inspected gear tooth surfaces until contact is achieved at each angular position of the latter gear. This algorithm is independent of the contact condition of the gear tooth surfaces since it is valid for point, line, or edge contact.

The proposed algorithm for tooth contact analysis is based on the following steps:

1. For the being-inspected gear, teeth are numbered opposite to the direction of rotation that the being-inspected gear will perform during the simulated tooth contact analysis. Figure 8.2 shows the numbering from the tooth 0 to the tooth $N - 1$.
2. N cycles of meshing are established. Each cycle of meshing c, $c = \{0, 1, 2, \ldots, N - 1\}$ is defined considering the extremes values of the interval of angular position of the to-be-inspected gear, given by angle ϕ_1 as

$$\phi_1 \in \left(\frac{2\pi c}{N} - \frac{\pi}{N}, \frac{2\pi c}{N} + \frac{\pi}{N} \right] \tag{8.1}$$

   Here, the left parenthesis denotes that the extreme value of the established range for ϕ_1 is not considered. The right side square parenthesis denotes that the extreme value of the established range for ϕ_1 is considered within the analysis.
3. At each cycle of meshing c, p contact positions are considered. This yields the following expression to define the angular position of the to-be-inspected gear ϕ_1 in the cycle c for each contact position i:

$$\phi_1 = \frac{2\pi c}{N} - \frac{\pi}{N} + \frac{2\pi}{Np}(i + 1) \tag{8.2}$$

Here, $i = \{0, 1, 2, \ldots, p - 1\}$.

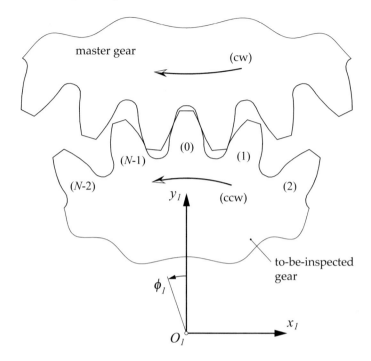

Fig. 8.2 Numbering of teeth for counter-clockwise rotation of the to-be-inspected gear

Table 8.1 Tooth selection of the being-inspected gear for application of tooth contact analysis

Interval of the cycle of meshing	Chosen tooth numbers
$\left(-\dfrac{\pi}{N}, +\dfrac{\pi}{N}\right]$	$(N-1), (0), (1)$
$\left(\dfrac{2\pi c}{N} - \dfrac{\pi}{N}, \dfrac{2\pi c}{N} + \dfrac{\pi}{N}\right]^{a}$	$(c-1), (c), (c+1)$
$\left(\dfrac{2\pi (N-1)}{N} - \dfrac{\pi}{N}, \dfrac{2\pi (N-1)}{N} + \dfrac{\pi}{N}\right]$	$(N-2), (N-1), (0)$
$^{a}c = \{1, 2, \ldots, N-2\}$	

4. Depending on the cycle of meshing c where ϕ_1 is located, the three teeth where the TCA algorithm is applied are selected. Table 8.1 summarizes the process of selection of the three teeth of interest for TCA as a function of the corresponding cycle of meshing.

5. For each value of the angular position of the being-inspected gear, ϕ_1, the cycle of meshing c is identified and the numbers of the gear teeth considered for application of TCA are also known. The TCA algorithm is based on the work [6] and has been previously applied in [7, 8]. Figure 8.3 shows schematically the process of minimization of the distance between the three teeth of the master gear and three teeth of the being-inspected gear. Here, surfaces $\Sigma_{1,c}$, $\Sigma_{1,c+1}$ and $\Sigma_{1,c-1}$, which

Fig. 8.3 Schematic view for explanation of the minimization of the distance between the master gear teeth and the being-inspected gear teeth

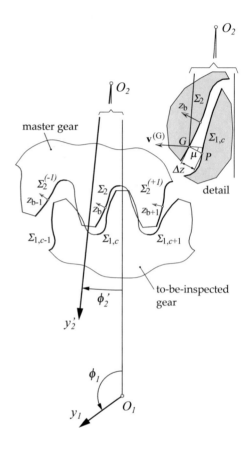

are known after the process of regeneration of the gear tooth surfaces from their corresponding point clouds, have been considered. They are different from each other. However, surfaces $\Sigma_2^{(+1)}$ and $\Sigma_2^{(-1)}$ are exactly the same surface as Σ_2, but misplaced the master gear pitch angle. The objective of the TCA method is to find the value of ϕ_2' for which the master gear tooth surfaces are in contact with any of the to-be-inspected gear tooth surfaces, which are fixed at angular position ϕ_1. Three auxiliary coordinate systems S_b, S_{b+1} and S_{b-1} are considered with their origins at the central point of each master gear tooth surface. By coordinate transformation, the master gear tooth surfaces and the being-inspected gear tooth surfaces are represented in these auxiliary systems. Minimization of the distance between tooth surfaces is carried out following a kinematic approach (see the detail in Fig. 8.3): any point G of the master gear tooth surfaces is projected on the corresponding to-be-inspected gear tooth surface in the direction given by axis z_b, obtaining point P. The distance Δz is then projected on the velocity direction of point G through the angle μ. Such projection is used to determine the angle $\Delta\phi_2'$ for each point G that approaches the master gear tooth surfaces to the to-be-inspected gear tooth surfaces as

$$\Delta\phi_2' = \frac{\Delta z \cos \mu}{|\overrightarrow{O_2 G}|}$$ (8.3)

Such a value $\Delta\phi_2'$ is subtracted from an initial value of ϕ_2'. This process is repeated until $\Delta\phi_2'$ is lower than a chosen computational threshold. The final value of ϕ_2' is transformed into the value of ϕ_2 considering the cycle of meshing c in whose interval ϕ_1 is.

$$\phi_2 = \phi_2' + \frac{2\pi \cdot c}{N_2}$$ (8.4)

6. A total of $N \times p$ angular positions are considered for one revolution of the being-inspected gear. The first and last contact positions are given by

$$\phi_{1,\text{first}} = -\frac{\pi}{N} + \frac{2\pi}{Np}$$

$$\phi_{1,\text{last}} = 2\pi - \frac{\pi}{N}$$

For each angular position ϕ_1 and once the contact has been achieved, the TCA algorithm provides a value of ϕ_2 and the bearing contact. This geometric representation of the bearing contact is determined by those points G and P whose Δz (see detail in Fig. 8.3) is equal to the chosen value of the virtual marking compound thickness. Each bearing contact is automatically represented on the corresponding tooth surface and saved to form the contact pattern on the gear tooth surfaces as the geometric representation of the bearing contacts for each value of ϕ_i.

8.4 Stress Analysis

Stress analysis is performed by the application of the finite element method. The finite element models that are applied here are based on the models presented in [3]. Such models have a reduced number of degrees of freedom since only the active portion of the gear tooth surface and the corresponding fillet have a fine mesh, whereas the body regions of the teeth have a coarse mesh. Tie-surface constraints are applied between these two regions to assure continuity in the displacements. Such a model can be suitable to analyze contact and bending stresses along a whole revolution of the being-inspected gear after some adaptations of the boundary conditions with a reduced computational time.

Figure 8.4 shows the finite element model that is being applied here. The model includes all the teeth of the gears in order to analyze one revolution of the being-inspected gear. Tie-surface constraints are applied at each interface between adjacent teeth in order to assure continuity of displacements at each interface. Nodes on the bottom side of the scanned gear and the master gear conform the corresponding rigid

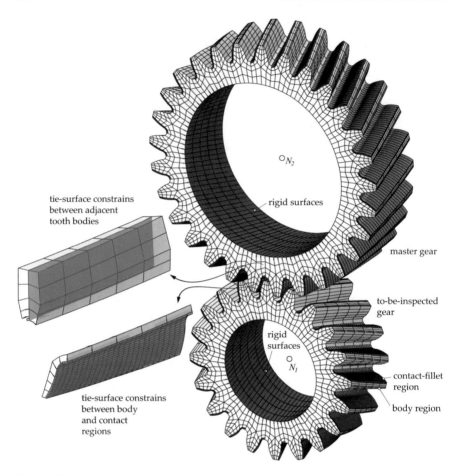

Fig. 8.4 Finite element model

surfaces, which are rigidly connected to their respective reference nodes N_1 and N_2, located at their corresponding axes of rotation. The degrees of freedom of the master gear reference node at each angular position of the master gear obtained from TCA, ϕ_2, are constrained, whereas for the reference node of the being-inspected gear, all degrees of freedom are constrained except the one representing rotation around the z axis, in which the torque is applied.

The types of finite elements applied to the finite element model are linear elements C3D8I [9] for the contact-fillet regions and quadratic elements C3D20 [9] for the body regions (see [3]). Contact iterations are defined for each pair of teeth, from pair 0 to pair $N - 1$.

8.5 Numerical Examples

Two measured gears A and B with same macro-geometry parameters but manufactured by different suppliers are being considered here in order to evaluate whether there are differences in their mechanical performance. The basic data of these two gears were presented in [2] as a result of a reverse engineering process from the point clouds and are shown in Table 8.2. In order to validate the new TCA algorithm and the proposed finite element model prior to being applied to the being-inspected gears, the following steps are being considered:

1. Design of the master gear. Any number of teeth for the master gear can be chosen because it will not affect the obtained results.
2. Generation of the point-clouds for all the teeth of a theoretical pinion based on the parameters shown in Table 8.2.
3. Application of the proposed TCA algorithm to the meshing between the regenerated surfaces of the theoretical pinion from the generated point clouds and the master gear to validate the proposed methodology and establish the reference contact patterns and peak-to-peak level of the unloaded function of transmission errors.
4. Application of the proposed finite element model for stress analysis between the theoretical pinion and the master gear to validate the proposed methodology and establish the reference values of stresses and peak-to-peak level of the loaded function of transmission errors.
5. Application of the proposed TCA algorithm to the meshing between the regenerated surfaces of the being-inspected pinions from their measured point clouds and the master gear.
6. Application of the proposed finite element model for stress analysis between the being-inspected pinions and the master gear.

Table 8.2 Basic data of the being-inspected gears

Gear parameter	Value
Tooth number, N_1	20
Normal module, m_n [mm]	3.0
Normal pressure angle, α_n [°]	20.0
Helix angle, β [°]	20.0
Helix hand	Right
Addendum coefficient	0.915
Dedendum coefficient	1.398
Profile shift coefficient	0.0546
Root radius coefficient	0.15

Table 8.3 Macro-geometry parameters of the master gear

Gear Parameter	Value
Tooth number, N_2	29
Normal module, m_n [mm]	3.0
Normal pressure angle, α_n [°]	20.0
Helix angle, β [°]	20.0
Helix hand	Left
Addendum coefficient	0.915
Dedendum coefficient	1.398
Profile shift coefficient	−0.0546
Root radius coefficient	0.38
Profile crowning [μm]	2.0 (parabolic)
Lead crowning [μm]	25.0 (parabolic)
Tip relief [μm]	15.0 (parabolic), 1.5 mm from the tip
Root relief [μm]	30.0 (parabolic), 2.0 mm from the root

8.5.1 Design of a Master Gear

A theoretical gear, denoted here as master gear, is designed to be in mesh with the theoretical pinion and the being-inspected gears, all with same macro-geometry parameters. The basic data of the master gear and the parameters of its micro-geometry tooth surface modifications are shown in Table 8.3. The micro-geometry modifications of the master gear include lead and profile crownings to localize the contact pattern on the gear tooth surfaces. The values for the micro-geometry modifications were chosen after a trial-and-error procedure where the level of the unloaded function of transmission errors, the variation of contact pressures, and the level of the loaded function of transmission errors were taken into account. In this stage, a conventional TCA algorithm, applied previously in [7, 8], was considered between the master gear and the theoretically generated pinion.

Figure 8.5 illustrates the results of TCA including the contact pattern and the contact path on the theoretical pinion tooth surfaces. Forty one contact positions were considered in this analysis with a virtual marking compound thickness of 0.0065 mm. A grid of 81×81 points was considered for tooth contact analysis. The peak-to-peak value of the unloaded function of transmission errors is about 15 arc-seconds.

Figure 8.6 shows the finite element model applied at this stage [3]. This model considers five pairs of contacting teeth since the theoretical pinion and the master gear tooth surfaces are considered without profile errors, flank deviations, or pitch errors. The model has a total of 78550 elements and 116572 nodes. Three layers of linear elements C3D8I [9] are considered in the contact region [3]. The thickness of these three layers is large enough to locate the maximum von Mises stress inside them. Quadratic elements C3D20 [9] are considered in the body regions. The applied

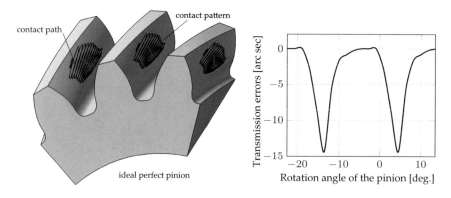

Fig. 8.5 Results of TCA for the theoretical pinion in mesh with the master gear

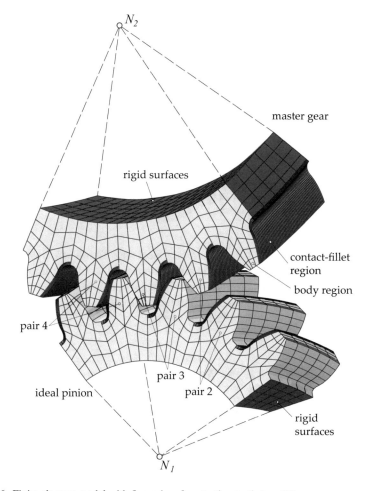

Fig. 8.6 Finite element model with five pairs of contacting teeth (see [3])

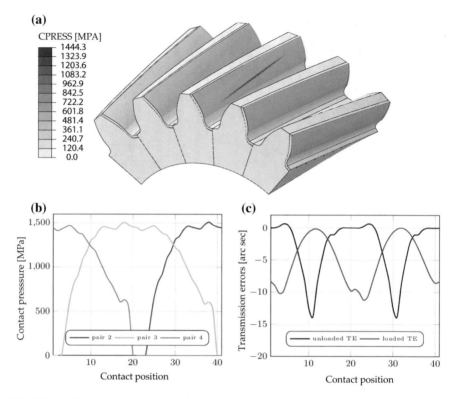

Fig. 8.7 Results of finite element analysis for the theoretical pinion in mesh with the master gear: **a** contact pressure field for a given contact position, **b** variation of contact pressure, **c** loaded and unloaded functions of transmission errors

torque to the pinion is 150 Nm. A conventional steel with Young's module of 210000 MPa and a Poisson coefficient of 0.3 is considered as material for pinion and gear. The results of stress analysis are shown in Fig. 8.7. A peak-to-peak level of 10 arc-seconds is obtained for the loaded function of transmission errors and a maximum contact pressure of about 1500 MPa is observed.

8.5.2 Generation of Point-Clouds for the Theoretical Pinion

The point-clouds for all the teeth of the theoretical pinion are now computationally generated considering 1000 points in lead direction and 500 points in profile direction (including the active part and the fillet). This means that a total of 500,000 points are being generated for each side of the tooth. All these points are generated theoretically on the gear tooth surfaces by application of the modern theory of gearing [10]. The idea is to validate the proposed algorithm for TCA and the proposed finite element model that will be applied later for the being-inspected gears. If the obtained results are the same as the ones illustrated in Figs. 8.5 and 8.7, the validation will be successful.

8.5.3   Application of the Proposed TCA Algorithm to the Meshing Between the Regenerated Theoretical Pinion from Its Point-Clouds and the Master Gear

A total of $p = 21$ contact positions were considered at each cycle of meshing. Since the pinion has $N = 20$ teeth, a overall total of $N \times p = 420$ contact positions are being considered. A virtual marking compound thickness of 0.0065 mm is considered at this stage. Figure 8.8a shows the contact pattern and the contact path at each tooth of the regenerated theoretical pinion. This set of contact patterns are obtained independently for each cycle of0 meshing. Figure 8.8b shows the unloaded function of transmission errors with a peak-to-peak level of about 15.0 arc-seconds. The results are similar to those obtained for the theoretical pinion in mesh with the master gear (see Fig. 8.7) and with them, the applied methodology was validated.

8.5.4   Application of the Proposed Finite Element Model to the Stress Analysis Between Regenerated Theoretical Pinion from Its Point-Clouds and the Master Gear

The finite element model shown in Fig. 8.4 is applied here to analyze the stresses along a whole revolution of the theoretical pinion. The model has a total of 378010 elements and 561835 nodes. A conventional steel with Young's module of 210000 MPa and a Poisson coefficient of 0.3 is considered as material for pinion and gear. A torque of 150 Nm is applied to the pinion. Figure 8.9a shows the contact pressure distribution for a chosen contact position. Figure 8.9b shows the variation of the contact pressure along the 420 contact positions, covering a whole revolution of the regenerated theoretical pinion. A maximum contact pressure of about 1500 MPa is obtained. Figure 8.9c shows the unloaded and loaded functions of transmission errors. Similar peak-to-peak values to those shown in Fig. 8.7 are obtained. The results of TCA and stress analysis show that the process of regeneration of the pinion tooth surfaces from point clouds and the proposed algorithm for TCA as well as the proposed finite element model work satisfactory to achieve the goals of this work.

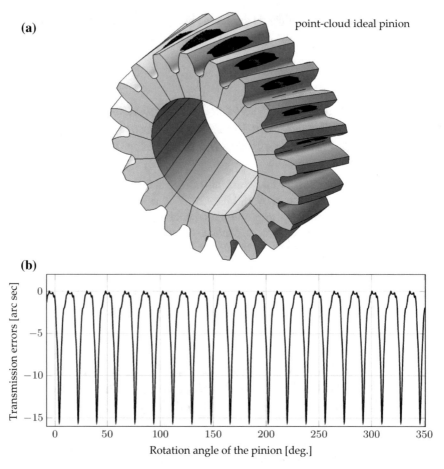

Fig. 8.8 Results of the proposed TCA algorithm applied to the meshing between the regenerated theoretical pinion from its point-clouds and the master gear

8.5.5 Application of the Proposed TCA Algorithm and Finite Element Model to the Meshing and Stress Analyses Between the Being-Inspected Gears and the Master Gear

The proposed TCA algorithm is applied now to the meshing between the master gear and the being-inspected gears A and B. A grid surface 41×41 and a virtual marking compound thickness of 0.030 mm have been considered. The virtual marking compound thickness is increased to account for the precision of measurement of the point clouds by the non-contract metrology machines that can be up to 20 μm. Figure 8.10 shows the contact patterns and the contact paths in gears A and B. Due to the irregu-

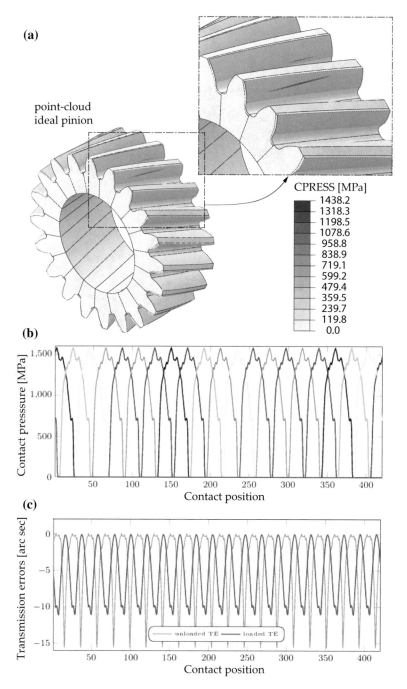

Fig. 8.9 Finite element analysis results from the meshing between the regenerated theoretical pinion from its point-clouds and the master gear: **a** contact pressure field for a given contact position, **b** variation of contact pressure, **c** loaded and unloaded functions of transmission errors

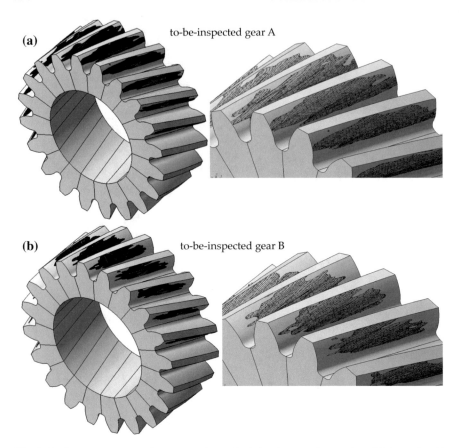

Fig. 8.10 Contact patterns and contact paths on the being-inspected gears A and B in mesh with the master gear

larities of the being-inspected gear tooth surfaces, the shapes of the contact patterns are not as uniform as the ones shown in Fig. 8.8a for the regenerated theoretical pinion from their computationally obtained point-clouds. The peak-to-peak values of the unloaded functions of transmission errors may not be very representative of the quality of the gear tooth surfaces due to the precision of measurement of the point clouds and the possibility to chose noisy points during the process of regeneration of the gear tooth surfaces if a filtering process is not applied to the point clouds.

Application of the finite element model shown in Fig. 8.4 allows the variation of the contact pressure along a whole revolution of the being-inspected gears to be determined. Figure 8.11 shows the contact pressure distribution for one of the 420 contact positions analyzed. Due to the irregularities of the being-inspected gear tooth surfaces, the pressure is not uniformly distributed along the contact pattern. High values of contact pressure are obtained for some contact points, reaching a maximum value as high as 4026 MPa for the being-inspected gear A and as high as

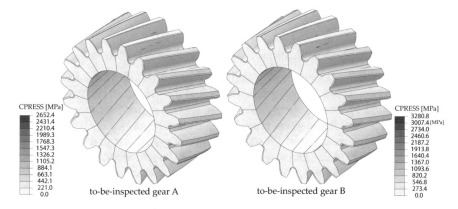

Fig. 8.11 Contact pressure for a given contact position for the being-inspected gears A and B

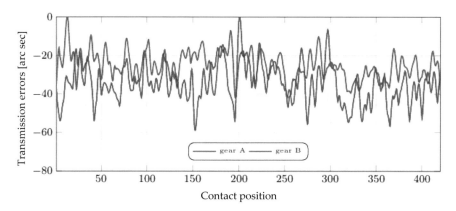

Fig. 8.12 Loaded functions of transmission errors for the being-inspected gears A and B

6245 MPa for the being-inspected gear B. Such values might not be representative of the mechanical performance of gears A and B.

The loaded functions of transmission errors are shown in Fig. 8.12 for gears A and B. The loaded functions of transmission errors take into account profile and flank deviations, as well as the pitch errors. A peak-to-peak value of 46 arc-seconds is observed for the loaded function of transmission errors of gear A in mesh with the master gear whereas a value of 59 arc-seconds is observed for the loaded function of transmission errors of gear B, also in mesh with the same master gear. These values can be representative of the relative quality of gears A and B since the deformations of the gear tooth surfaces under load mitigate possible irregularities of the regenerated surfaces from their corresponding point clouds.

8.6 Conclusions

The performed research work allows the following conclusions to be drawn:

1. A new approach for tooth contact analysis of gear drives that considers one revolution of the pinion and takes into account all the gear tooth surfaces has been proposed. The proposed algorithm has been applied for tooth contact analysis of measured manufactured gears in mesh with a master gear.
2. A new finite element model for stress analysis of gear drives that considers one revolution of the pinion and takes into account all the gear tooth surfaces has been proposed. The proposed model has been applied for stress analysis of measured manufactured gears in mesh with a master gear.
3. The proposed approaches have been validated by their application to the analysis of a regenerated theoretical pinion from computational generated point clouds. The obtained results are similar to those provided by the existing approaches wherein only two cycles of meshing are analyzed.
4. Application of the proposed approaches for TCA and stress analysis of real measured gears reflect some deficiencies in the regeneration of the gear tooth surfaces from the provided point clouds. Irregularities of the gear tooth surfaces cause high values of contact stresses and high peak-to-peak values of the unloaded function of transmission errors. The application of a filtering process is suggested before the application of the computational approaches for TCA and stress analysis of real measured gears and it will be investigated as a continuation of this work.

Acknowledgements The authors express their deep gratitude to the Spanish Ministry of Economy, Industry and Competitiveness (MINECO), the Spanish State Research Agency (AEI) and the European Fund for Regional Development (FEDER) for the financial support of research project DPI2017-84677-P.

References

1. Peng, B.Y., Ni, K., Goch, G.: Areal evaluation of involute gear flanks with three-dimensional surface data, paper 17FTM08, 1–14 (2017)
2. Fuentes-Aznar, A., Gonzalez-Perez, I.: Integrating non-contact metrology in the process of analysis and simulation of gear drives. AGMA Fall Technical Meeting, paper 18FTM21, 1–15 (2018)
3. Gonzalez-Perez, I., Fuentes-Aznar, A.: Implementation of a finite element model for gear stress analysis based on tie-surface constraints and its validation through the Hertz's theory. J. Mech. Des. **140**, 023301-1-13 (2018)
4. Piegl, L., Tiller, W.: The NURBS book. Springer, Berlin (1995)
5. Samet, H.: Foundations of Multidimensional and Metric Data Structures. Elsevier, San Francisco (2006)
6. Sheveleva, G.I., Volkov, A.E., Medvedev, V.I.: Algorithms for analysis of meshing and contact of spiral bevel gears. Mech. Mach. Theory **42**, 198–215 (2007)
7. Fuentes, A., Nagamoto, H., Litvin, F.L., Gonzalez-Perez, I., Hayasaka, K.: Computerized design of modified helical gears finished by plunge shaving. Comput. Methods Appl. Mech. Eng. **199**, 1677–1690 (2010)

8. Gonzalez-Perez, I., Fuentes-Aznar, A.: Analytical determination of basic machine-tool settings for generation of spiral bevel gears and compensation of errors of alignment in the cyclo-palloid system. Int. J. Mech. Sci. **120**, 91–104 (2017)

9. Systemes, Dassault: Abaqus/Standard Analysis User's Guide. Dassault Systemes Inc., Waltham, MA (2019)

10. Litvin, F.L., Fuentes, A.: Gear Geometry and Applied Theory, 2nd edn. Cambridge University Press, New York (2004)

Chapter 9
Tooth Surface Stress and Flash Temperature Analysis with Trochoid Interference of Gears

Akio Ueda

Abstract Typical damage to gears is pitting and breakage of teeth, so it was traditionally designed based on these two damage and tooth surface shape design was done to avoid edge contact. However, in recent years, a design method that minimizes the tooth surface modification amount to maximize the load capacity and a tooth surface modification method that can secure a contact region even in a low load region are adopted. However, damage due to trochoid interference occurs outside the geometric meshing range and can't be calculated by the strength calculation formula (ex. gear strength standard). And, with such a tooth surface modification, since tooth edge contact is accompanied in a high load region, it is necessary to consider edge contact stress analysis of the tooth due to trochoid interference. For this reason, this software (CT-FEM Opera iii [1]) adopts the finite element method (FEM) as a method capable of edge contact analysis and analyzes it. With the occurrence of edge contact, a large stress is generated in that portion and the flash temperature becomes high temperature, so temperature analysis of this portion is important. In the stress analysis of the teeth, it is necessary to analyze considering the tooth surface modification, the tooth profile error, the pitch error, and the posture error of the axis in consideration of tooth deformation and load sharing. At the time of engagement of the teeth, the curvature of the tooth profile, the sliding speed, the tooth surface roughness and the lubricant greatly change the friction coefficient and the oil film thickness depending on the tooth surface position. Therefore, tooth surface stress, edge contact stress, and flash temperature are also greatly affected. In this paper, the damaged picture of the experiment gear and the edge contact stress analysis result are shown, and they are in good agreement. And the analysis results of flash temperature, friction coefficient distribution, oil film thickness, power loss, transmission error, and tooth bending stress are shown.

Keywords Gear · Trochoid interference · Edge contact · Flash temperature · Pitting

A. Ueda (✉)
AMTEC Inc. Prior Tower 4305, 1-2-30, Benten, Minato-Ku, 552-0007 Osaka, Japan
e-mail: ueda@amtecinc.co.jp

© Springer Nature Switzerland AG 2020
V. Goldfarb et al. (eds.), *New Approaches to Gear Design and Production*,
Mechanisms and Machine Science 81,
https://doi.org/10.1007/978-3-030-34945-5_9

9.1 Introduction

In recent years, tooth surface modification has become more complex in gear design, so when analyzing the tooth surface stress of gears, it is necessary to set the tooth surface mesh more finely. Therefore, when trying to analyze with 3D-FEM detailed model, it is an unrealistic analysis method to use at the design stage due to problems of model creation and analysis time. However, it is possible to analyze stress easily in a short time even for gears with large tooth surface modification by adopting an analysis method that combines the 3D-FEM model and the tooth surface membrane element [2]. In this paper, the stress analysis is performed based on the measured tooth surface data of the manufactured gear, and the analysis is performed over the entire tooth surface considering the shaft mounting error and pitch deviation. Then, the friction coefficient of the tooth surface, oil film thickness, flash temperature, power loss, etc. are analyzed from the tooth surface roughness, lubricating oil, sliding speed, and material thermal conductivity.

In addition, the analysis of the edge can be used to understand the phenomenon of trochoid interference from the stress and flash temperature distribution at the edge contact. The results of the edge contact analysis and the damage photograph from the gear experiment were compared and verified. As a result, we obtained an analytical result that closely matched the actual tooth surface damage. In addition, the analysis results when the optimal tooth surface modification without trochoidal interference is given are shown.

9.2 Overview of Edge Contact Stress Analysis

Tooth surface stress analysis calculates the amount of tooth contact [3] from the load acting during tooth meshing, and analyzes the Hertzian stress (σ_{Hi}) at each coordinate (x_i, y_i, z_i) divided into meshes as shown in Fig. 9.1. Figure 9.2 shows an example of tooth surface stress analysis. In addition, when performing edge contact analysis [4] to analyze the tooth edge contact area, the mesh roughness at the location where the tooth edge contacts as shown in Fig. 9.3 is further refined. Then, the friction

Fig. 9.1 Load and the stress distribution

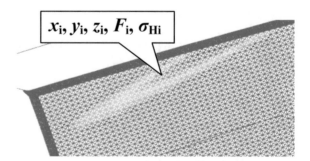

$x_i, y_i, z_i, F_i, \sigma_{Hi}$

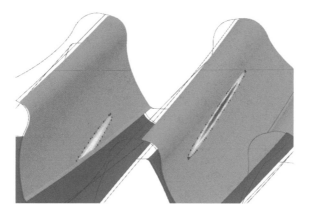

Fig. 9.2 Stress distribution on surface

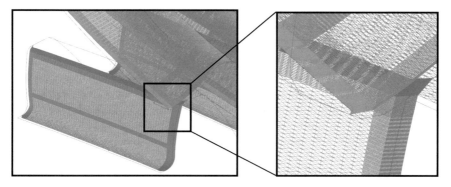

Fig. 9.3 Tooth flank film elements for edge contact analysis which have finer meshes not only near edges at tooth tip and tooth side but also on the region where edges of the mating gear would contact on the usable flank

coefficient, oil film thickness, flash temperature, and calorific value on all coordinates are analyzed.

To analyze the flash temperature, the maximum tooth surface roughness Rz is used, and the friction coefficient (Matsmoto's Equation) [5] of the tooth surface and the contact end is calculated according to Eqs. (9.1)–(9.3). The flash temperature generated on the mesh coordinates of the gingival membrane element is calculated based on AGMA formula [6].

$$f = f_{\mathrm{L}}(1 - \alpha) + f_{\mathrm{S}} \cdot \alpha \qquad (9.1)$$

$$\alpha = 0.5 \log D \qquad (9.2)$$

$$D = (R_{\mathrm{Z1}} + R_{\mathrm{Z2}})/h_0 \qquad (9.3)$$

f coefficient of friction
f_L coefficient of hydrodynamic friction, $f_L = 0.01$
f_s coefficient of boundary lubrication friction, $f_s = 0.11$
α boundary part load ratio
D lubrication condition ($1 < D$)
R_z maximum height of the surface roughness
h_0 EHL minimum oil film thickness

9.3 Gear Specifications and Tooth Surface Measurement Data

The material of the gear to be examined is SCM420 (JIS G 4053) material, and the tooth surface hardness is HRc 60–62. As shown in Figs. 9.4 and 9.5, the gear specifications are helical gears with $m_n = 4$, $z_1 = 15$, $z_2 = 43$, $\alpha_n = 20°$, $\beta = 28°$, and the total meshing ratio is $\varepsilon_\gamma = 2.30$. Figure 9.6 shows the tooth profile, and Fig. 9.7 shows the sliding ratio. As shown in Figs. 9.8 and 9.9, tooth surface modification is a helical gear with a simple tooth profile and tooth modification commonly used.

As shown in Fig. 9.10, the ground tooth profile was measured using a gear measuring machine (Osaka Seimitsu Kikai: CLP-35), measuring 17 tooth profiles and tooth trace directions. When this measurement data file (csv) is read by software [1], it can be displayed as shown in Fig. 9.11. However, tooth surface modification is not given to the pinion, and it is a theoretical tooth profile, but only a single pitch deviation (f_{pt} = 17.9 μm) is given in order to grasp the effect of stress due to the pitch deviation.

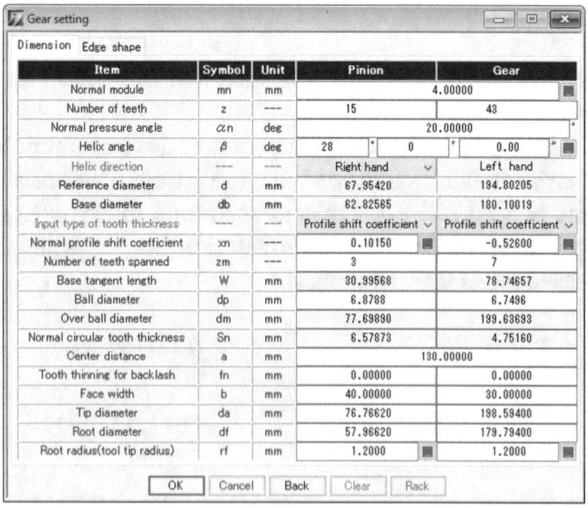
Gear setting

Dimension | Edge shape

Item	Symbol	Unit	Pinion	Gear
Normal module	mn	mm	4.00000	
Number of teeth	z	---	15	43
Normal pressure angle	αn	deg	20.00000	
Helix angle	β	deg	28 ° 0 ' 0.00 "	
Helix direction	---	---	Right hand	Left hand
Reference diameter	d	mm	67.95420	194.80205
Base diameter	db	mm	62.82565	180.10019
Input type of tooth thickness	---	---	Profile shift coefficient	Profile shift coefficient
Normal profile shift coefficient	xn	---	0.10150	-0.52600
Number of teeth spanned	zm	---	3	7
Base tangent length	W	mm	30.99568	78.74657
Ball diameter	dp	mm	6.8788	6.7496
Over ball diameter	dm	mm	77.69890	199.63693
Normal circular tooth thickness	Sn	mm	6.57873	4.75160
Center distance	a	mm	130.00000	
Tooth thinning for backlash	fn	mm	0.00000	0.00000
Face width	b	mm	40.00000	30.00000
Tip diameter	da	mm	76.76620	198.59400
Root diameter	df	mm	57.96620	179.79400
Root radius(tool tip radius)	rf	mm	1.2000	1.2000

OK | Cancel | Back | Clear | Rack

Fig. 9.4 Gear specification

Fig. 9.5 Tooth tip chamfer setting

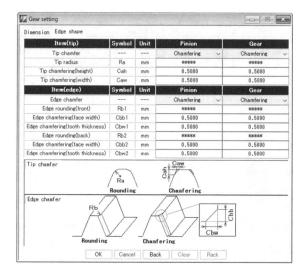

Fig. 9.6 Tooth profile

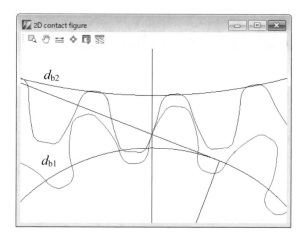

9.4 Tooth Surface Stress Analysis (Edge Contact Analysis Disabled)

Here, only the involute tooth surface is analyzed. In Fig. 9.12, torque, elastic modulus and Poisson's ratio are set, and the mesh of the tooth surface sets the number of divisions in the tooth tip portion, tooth surface portion and tooth width direction. In Fig. 9.13, the shaft crossed error ($\varphi_1 = 0.02°$) and the parallelism error ($\varphi_2 = -0.006°$) set by the gear unit measurement are set. Then, the tooth meshing contact range angle ($\theta = 64.107°$) is divided into 80 parts for analysis. As a result of analysis, the maximum tooth surface stress is $\sigma_{Hmax} = 2434$ MPa as shown in Fig. 9.14. This analysis result is a tooth surface analysis, and no edge contact analysis is performed.

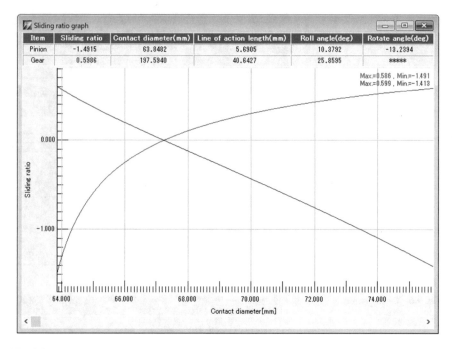

Fig. 9.7 Sliding ratio

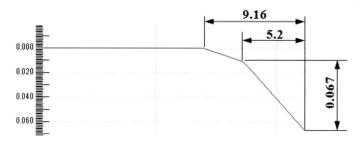

Fig. 9.8 Tooth modification (gear)

Fig. 9.9 Lead modification (gear)

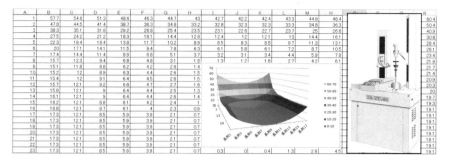

Fig. 9.10 Tooth surface measured data (gear, actual measurement)

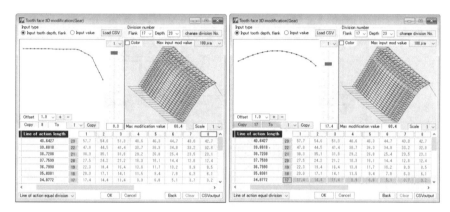

Fig. 9.11 Tooth surface deviation data (gear)

Fig. 9.12 Setting of a torque and so on

Tooth face item setting

Edge contact analysis setting

○ Edge analysis(edge curvature setting)   ● No edge analysis

Item	Symbol	Unit	Pinion	Gear
Maximum curvature	ρ	mm	---	---
Curvature modification range	h	mm	---	---

Analysis teeth number   Analysis point

○ 1   ○ 3   ● 5   ● root + face + tip   ○ Only tooth face

Item	Symbol	Unit	Pinion	Gear
Face width center position	bm	mm	0.0000	0.0000
Torque	T	N·m	1400.0000	4013.3333
Normal forth at transverse plane	F	N	44567.8	
Modulus of elasticity	E	MPa	205800.0	205800.0
Poisson's ratio	ν	---	0.3000	0.3000
Root division number	Nh1	---	20	20
Involute division number	Nh2	---	40	40
Tip, edge division number	Nh3	---	10	10
Face width division number	Nb	---	40	40

Pitch error(μm)

| Pinion | 0.0 | 0.0 | 17.9 | 0.0 | 0.0 |
| Gear | 0.0 | 0.0 | 0.0 | 0.0 | 0.0 |

Positive : weak contact
Negative : strong contact

OK   Cancel   Default   Back   Clear

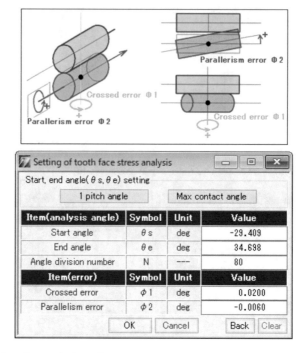

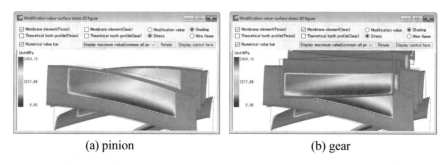

Fig. 9.13 Setting of an analysis angle

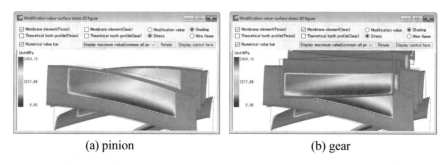

| (a) pinion | (b) gear |

Fig. 9.14 Tooth surface stress, $\sigma_{Hmax} = 2434$ MPa

In Fig. 9.15, set the necessary items for flash temperature calculation. The coefficient of friction is R_Z in Eq. (9.3), and Ra is used to calculate the probability of scuffing. Select 60(W/m · K) from the material selection table in Fig. 9.15 for the thermal conductivity of the gear material SCM420. The lubricant used is ISO VG150, and the temperature of the forced lubricant is 75 °C.

Under the conditions shown in Fig. 9.15, (a) flash temperature, (b) friction coefficient, (c) oil film thickness, (d) frictional heating value, PV value, etc. are calculated. Figures 9.16, 9.17, 9.18 and 9.19 show the analysis results.

Flash temperature item setting

Item(tooth face)	Symbol	Unit	Pinion	Gear
Max tooth surface stress	σHmax	MPa	1798.013	
Torque	T	N·m	1400.0	4013.3
Power	P	kW	201.586	
Rotation speed	n	min-1	1375.000	479.651
Tangential velocity	v	m/s	4.841	
Surface roughness(max height)	Rz	μm	2.500	2.500
Surface roughness(average)	Ra	μm	0.400	0.400
Item(material)	**Symbol**	**Unit**	**Pinion**	**Gear**
Material	----	----	Chromium ste	Chromium ste
Density	ρ	g/cm^3	7.870	7.870
Specific heat	CM	J/(kg℃)	461.000	461.000
Thermal conductivity	λ	W/mK	60.000	60.000
Gear temperature	GTc	℃	70.000	
Item(lubricant)	**Symbol**	**Unit**	**Value**	
Oil temperature	OTc	℃	75.000	
Lubricant type	----	----	Mineral oil(other)	
ISO grade	----	----	ISO VG 150	
Kinematic viscosity(40℃)	----	mm²/s	150.000	
Average scuffing temperature	MTc	℃	227.000	
Standard deviation temperature	SD	℃	37.000	
Absolute viscosity	μ0	cP	22.757	
Viscosity pressure	α	MPa-1	0.01880	

OK Cancel Back Clear Material

Material selection(flash temperature)

Material	Temp. (℃)	Density (g/cm^3)	Specific heat (J/Kg℃)	Thermal conductivity (W/mK)	Pinion	Gear
Pure iron	20	7.87	461	67	○	○
Cast iron	20	7.28	461	48	○	○
Carbon steel(less 0.5C)	20	7.83	461	53	○	○
Carbon steel(1C)	20	7.80	461	45	○	○
Carbon steel(1.5C)	20	7.75	461	36	○	○
Chromium steel(1Cr)	20	7.87	461	60	◉	◉
Chromium steel(2Cr)	20	7.87	461	52	○	○
Chromium steel(5Cr)	20	7.83	461	38	○	○
Chromium steel(10Cr)	20	7.79	461	31	○	○
Chromium steel(23Cr)	20	7.68	461	22	○	○
Nickel steel(10Ni)	20	7.95	461	26	○	○
Nickel steel(20Ni)	20	7.99	461	19	○	○
Nickel steel(30Ni)	20	8.07	461	12	○	○
Nickel steel(40Ni)	20	8.17	461	10	○	○
Nickel steel(50Ni)	20	8.27	461	14	○	○
Nickel chrome steel18Cr8Ni	20	7.82	502	16	○	○
Nickel chrome steel20Cr15Ni	20	7.85	470	15	○	○
Nickel chrome steel25Cr20Ni	20	7.86	463	13	○	○
Nickel chrome alloy 40Ni 15CrNi	20	8.07	461	12	○	○
Nickel chrome alloy 80Ni 15CrNi	20	8.52	461	17	○	○

OK Cancel

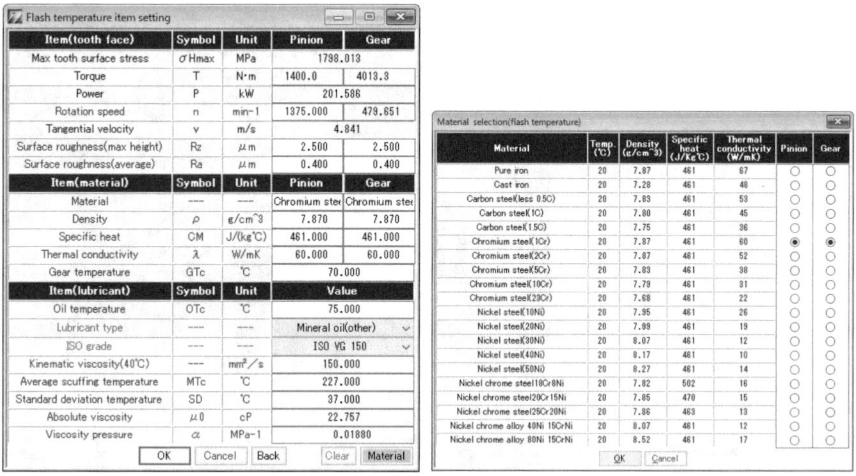

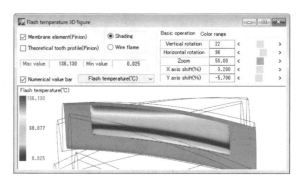

Fig. 9.15 Setting of gear speed and so on

Fig. 9.16 Flash temperature

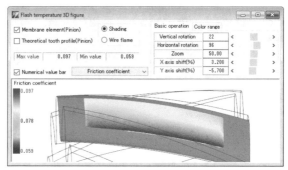

Tf_{max}=136℃

Fig. 9.17 Friction coefficient

μ_{max}=0.097

Fig. 9.18 Oil film thickness

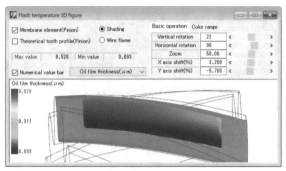

$$\lambda_{min} = 0.093\,\mu m$$

Fig. 9.19 Heat capacity

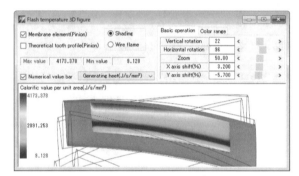

$$Q_{kmax} = 4173\,(J/s/mm^2)$$

The flash temperature must be set correctly because it is greatly affected by the tooth surface roughness (Rz), the thermal conductivity of the material, the type of lubricant, and the lubricant temperature. When the gear pair under consideration has no pitch error (shaft angle error is the same as in Fig. 9.13), the maximum tooth surface stress decreases to $\sigma_{Hmax} = 2022$ MPa, and the maximum flash temperature also decreases to $Tf_{max} = 89.7\,°C$.

9.5 Edge Contact Stress Analysis

The tooth surface stress distribution shown in Fig. 9.14 is the stress analysis set in Fig. 9.12, so the stress analysis of the tooth tip and the side edge is not performed. Therefore, here, the stress on the tooth surface and the edge contact part and the generated flash temperature are analyzed.

In the tooth profile inspection graph of Fig. 9.20, the radius of roundness of the tooth tip is 0.3 mm in terms of the line of action. When this tooth profile is plotted as shown in Fig. 9.21, the connection between the tooth tip C chamfer surface and

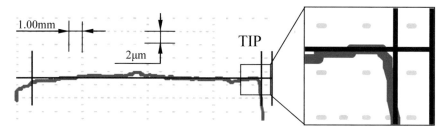

Fig. 9.20 Radius of the tooth tip corner (inspection graph of the pinion)

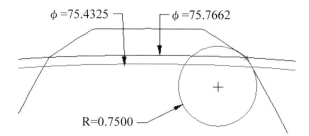

Fig. 9.21 Contact radius of the edge

the involute surface has a radius R = 0.75 mm. This edge radius R = 0.75 mm is set in Fig. 9.22, and the tooth surface stress and edge contact stress are analyzed. Figure 9.23 shows the results of edge contact stress analysis.

Figure 9.23 shows the results of edge contact stress analysis. Figures 9.24, 9.25, 9.26 and 9.27 show the results of analysis such as flash temperature. As a result of edge contact analysis, the tooth surface stress is $\sigma_{Hmax} = 8.0$ GPa, and the flash temperature is high at $Tf_{1max} = 1108\ °C$ at the end of the pinion meshing end point.

9.6 Comparison of Analysis Results and Experimental Results

Comparison of the pinion tooth surface damage photograph [7] after the experiment in Fig. 9.28 and the analysis result in Fig. 9.29 shows the maximum flash temperature ($Tf_{max} = 1108\ °C$) at the tooth tip at the pinion end position. In addition, the damage of the tooth tip (C chamfer) is in good agreement between the experimental results and the analysis results of the flash temperature distribution.

Comparing the stress distribution of the gear damage photograph in Fig. 9.30 and the analysis result in Fig. 9.23b (Fig. 9.32), the result agrees well with the analysis result of the trochoid damage position of the tooth.

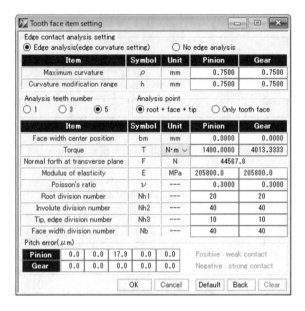

Fig. 9.22 Setting of a torque and so on (edge contact analysis)

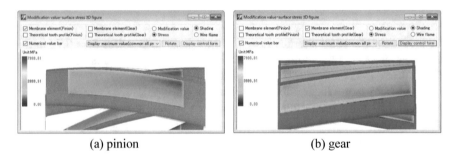

(a) pinion   (b) gear

Fig. 9.23 Tooth surface stress, $\sigma_{Hmax} = 8000$ MPa

Fig. 9.24 Flash temperature

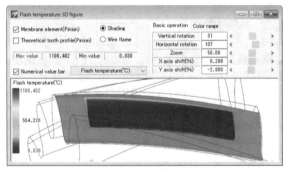

$Tf_{max} = 1108\,°C$

Fig. 9.25 Friction coefficient

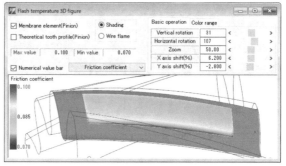

$$\mu_{max}=0.100$$

Fig. 9.26 Oil film thickness

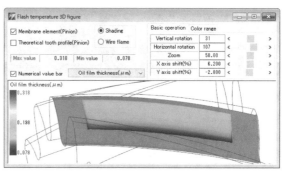

$$\lambda_{min}=0.078\mu m$$

Fig. 9.27 Heat capacity

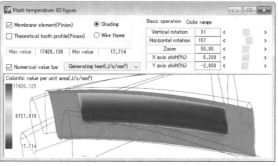

$$Q_{kmax}=17426(J/s/mm^2)$$

The damage in which part of the damaged portion at the gear engagement end tooth root in Fig. 9.31 melted is in good agreement with the temperature distribution of the flash temperature $Tf_{max} = 1108\ °C$ in Fig. 9.24 of the analysis results. In addition, the stress distribution in Fig. 9.32 agrees well with the damaged state of the

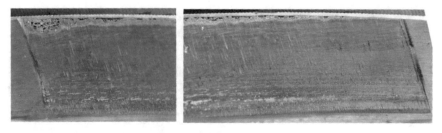

Fig. 9.28 Tooth surface damage photograph(pinion tooth number $= 14$, $N_P = 1 \times 10^6$)

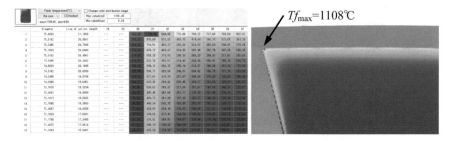

Fig. 9.29 Flash temperature, pinion, $Tf_{max} = 1108\ °C$

Fig. 9.30 Tooth surface damage photograph of the gear, No. $= 14$, $N_P = 1.0 \times 10^6$

Damage to the tooth root at the end of gear engagement

Fig. 9.31 Tooth surface damage photograph of the gear, No. $= 14$, $N_P = 1.0 \times 10^6$

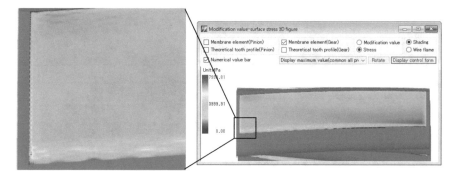

Fig. 9.32 Edge contact stress analysis (gear), $\sigma_{Hmax} = 8000$ MPa

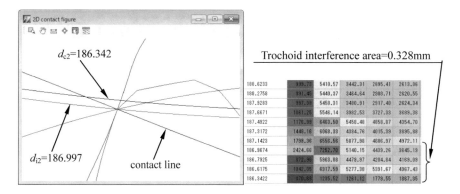

Fig. 9.33 Meshing position of the tooth

tooth side surface. Furthermore, when the center of the tooth surface is observed, no damage occurs and the pattern after grinding remains.

When the meshing position of the teeth is confirmed, the minimum geometric gear diameter is $d_{i2} = 186.997$ mm, but the minimum contact diameter at which trochoidal interference occurs is $d_{c2} = 186.342$ mm as shown in Fig. 9.33. Therefore, the distance at which trochoidal interference occurs is $L_{tr} = 0.328$ mm.

9.7 3D-FEM Stress Analysis

Figures 9.34 and 9.35 show the root stress (maximum value of maximum principal stress σ_{1max}) and tooth profile displacement. In the analysis results, the influence of the pitch deviation ($f_{pt} = 17.9\ \mu$m) given in Fig. 9.12 appears in the stress distribution and the tooth profile displacement.

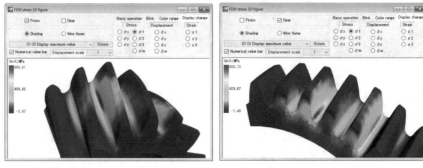

(a) pinion, σ_{1max}=884MPa   (b) gear, σ_{1max}=861MPa

Fig. 9.34 FEM analysis, bending stress

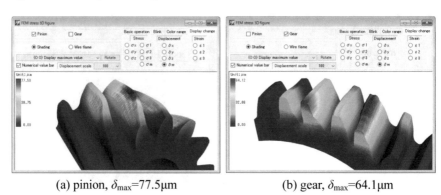

(a) pinion, δ_{max}=77.5μm   (b) gear, δ_{max}=64.1μm

Fig. 9.35 FEM analysis, tooth displacement (×100)

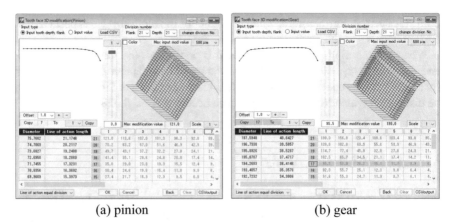

(a) pinion   (b) gear

Fig. 9.36 Tooth surface deviation data

9.8 Optimal Tooth Surface Modification

Optimal tooth surface modification is a function that determines tooth surface modification that does not cause edge contact in consideration of tooth deformation and load sharing when a load is applied to the tooth. That is, since the stress distribution acts evenly over the entire tooth surface, a tooth surface modification that produces the minimum tooth surface stress is generated.

The software [1] generated the optimal tooth surface modified tooth profile that does not generate trochoidal interference under the same conditions as the gears studied, including gear specifications, pitch deviation ($f_{pt} = 17.9\ \mu$m), and assembly error (Fig. 9.13). The generated tooth profile is shown in Fig. 9.2.

As a result of analyzing the end of the tooth profile surface shape generated here, the tooth surface stress was $\sigma_{Hmax} = 2754$ MPa as shown in Fig. 9.37, and no trochoidal interference occurred (Figs. 9.36, 9.38, 9.39, 9.40 and 9.41).

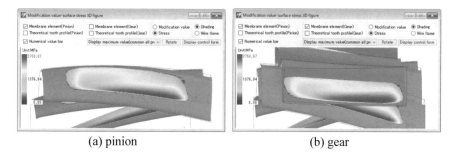

      (a) pinion   (b) gear

Fig. 9.37 Tooth surface stress, $\sigma_{Hmax} = 2754$ MPa

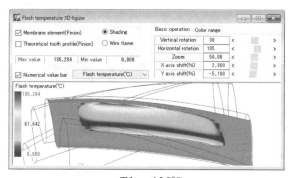

$$Tf_{max} = 135\ ^{\circ}\text{C}$$

Fig. 9.38 Flash temperature

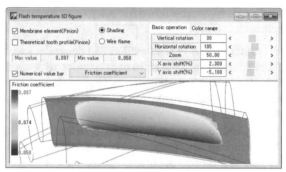

$$\mu_{\max}=0.097$$

Fig. 9.39 Friction coefficient

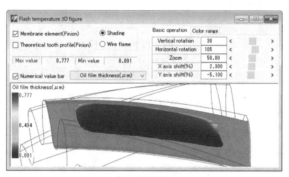

$$\lambda_{\min}=0.091\mu m$$

Fig. 9.40 Oil film thickness

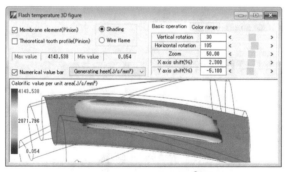

$$Q_{k\max}=4144(J/s/mm^2)$$

Fig. 9.41 Heat capacity

When the optimum tooth surface modification is performed with the pitch deviation $f_{tp} = 0$, the optimum tooth surface modification shown in Fig. 9.42 can be obtained. The tooth surface stress (edge contact analysis) decreases to $\sigma_{Hmax} = 1955$ MPa as shown in Fig. 9.43 (Figs. 9.44 and 9.45).

The summarized analysis results are shown in Figs. 9.46 and 9.47. In (C), the tooth surface stress and flash temperature due to trochoid interference are extremely large, but in (B) with the optimum tooth surface modification, both are greatly reduced. Furthermore, even if there is an axial angle error, if the pitch error is set to $f_{pt} = 0$, the tooth surface stress will decrease to $\sigma_{Hmax} = 1955$ MPa as shown in (A) even if the edge contact analysis is performed.

(A)   Optimal tooth surface modification 2

Figure 9.45, 1955 MPa

(B)   Optimal tooth surface modification 1

Figure 9.37, 2754 MPa

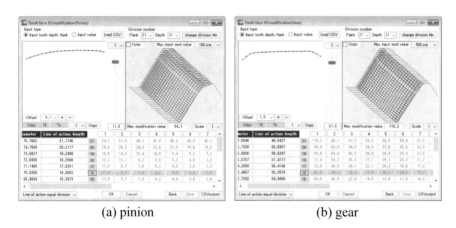

(a) pinion   (b) gear

Fig. 9.42 Tooth surface deviation data

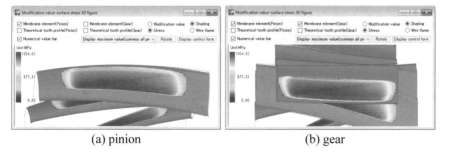

(a) pinion   (b) gear

Fig. 9.43 Tooth surface stress, $\sigma_{Hmax} = 1955$ MPa

Fig. 9.44 Flash temperature

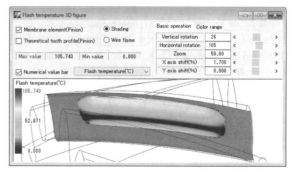

$$Tf_{max} = 106\,°C$$

Fig. 9.45 Friction coefficient

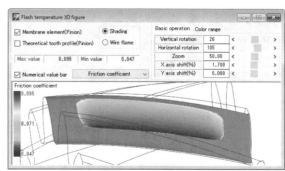

$$\mu_{max} = 0.095$$

Fig. 9.46 Comparison of the tooth surface stress

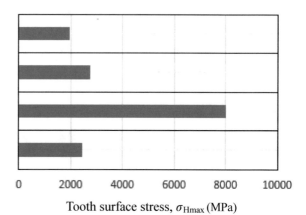

Tooth surface stress, σ_{Hmax} (MPa)

Fig. 9.47 Comparison of the flash temperature

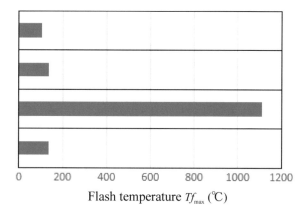

Flash temperature Tf_{max} (℃)

(C) Edge contact stress analysis

   Figure 9.23, 8000 MPa

(D) Tooth surface stress analysis

   Figure 9.14, 2434 MPa

(A) Optimal tooth surface modification 2

   Figure 9.44, 106 °C

(B) Optimal tooth surface modification 1

   Figure 9.38, 135 °C

(C) Edge contact stress analysis

   Figure 9.24, 1108 °C

(D) Tooth surface stress analysis

   Figure 9.16, 136 °C

9.9 Conclusion

(1) Although it was not possible to grasp by tooth surface analysis, the phenomenon of trochoid interference could be grasped by edge contact analysis.

(2) In a gear with general tooth surface modification, the temperature rises to $Tf_{max}=$ 1108 °C, which is close to the melting temperature of iron, as shown in Fig. 9.32, due to trochoidal interference.

(3) This temperature is convincing from the experimental results (Fig. 9.31).

(4) In the optimum tooth surface modification, the tooth surface profile in which the stress generated on the tooth surface uniformly acts can be generated by software [1] in consideration of the tooth deflection accompanying the load.
(5) Transmission error analysis when a load is applied is also possible but omitted.

References

1. AMTEC INC.: CT-FEM Opera iii Software Manual, pp. 10–15 (2014)
2. Moriwaki, I.: Finite element analysis of gear tooth stress with tooth flank film elements. In: VDI-2005 International Conference on Gears, pp. 39–53 (2005)
3. Kubo, A., Umezawa, K.: On the power transmitting characteristics of helical gears with manufacturing and alignment errors. JSME **43**(371), pp. 2771–2783 (1977)
4. Moriwaki, I., Ueda, A., Iba, D.: Edge contact analysis for power transmission gears using tooth-flank-film elements. JSME **84**(863) (2018)
5. Matsumoto, S.: Estimation formula of friction coefficient of rolling-sliding contact surface under mixed lubrication condition. Tribologist **56**(10), pp. 632–638 (2011)
6. AGMA2001-C95.: Fundamental rating factors calculation methods for involute spur and helical gear teeth, pp. 46–47 (1995)
7. Kubo, A.: JGMA X-project, report, p. 58 (2012)

Chapter 10
Estimation of Bearing Capacity and Wear Resistance of Spur Gear Meshing Taking into Account Tooth Profile Correction and Sliding Friction Coefficient

Myron V. Chernets, Serge V. Shil'ko and Victor E. Starzhinsky

Abstract The authors present a method for studying the wear kinetics during sliding friction, based on the fatigue mechanism of frictional failure under the influence of friction forces. Based on it, a method for calculating gears has been developed, which allows one to predict contact stresses in gearing, linear wear of teeth and transmission durability. Using this method, the influence of the friction coefficient on the durability and wear of a cylindrical spur gear drive of a locomotive when the conditions of contact due to wear of teeth in a corrected two-one-two-pair meshing are investigated. The advantage of this method in comparison with the known ones is the possibility of taking into account the wear of teeth, as an operational factor, a gear shift coefficient, as well as teeth engagement conditions (a pairness of meshing). It has been established that an increase in the coefficient of sliding friction in the engagement within 0.05 … 0.11 (2.2 times) results in a non-linear reduction in the transmission durability by 5.46 times. Regardless of the magnitude of the friction coefficient, the highest gear durability is achieved with a high-altitude correction $x_1 = -x_2 = 0.1$. A further increase in the gear shift coefficients has a positive effect on reducing the maximum contact pressures, but at the same time leads to a reduction in the durability of the gear for all studied values of the friction coefficient. The highest durability of the x-gear pair at $x_1 = -x_2 = 0.1$ exceeds its value in comparison with the x-zero gear pair by 1.22 … 1.32 times. It has been established that at optimal values of the gear shift coefficients, almost identical wear occurs at the entrance to the two-pair and single-pair meshing, as well as at the exit from the single-pair one.

M. V. Chernets
Lublin University of Technology, Faculty of Mechanical Engineering, Nadbystrycka St., 36, 20-618 Lublin, Poland
e-mail: m.czerniec@pollub.pl

S. V. Shil'ko (✉) · V. E. Starzhinsky
State Scientific Institution "V.A. Belyi Metal-Polymer Research Institute of National Academy of Sciences of Belarus", Kirov St, 32a, 246050 Gomel, Belarus
e-mail: shilko_mpri@mail.ru

V. E. Starzhinsky
e-mail: star_mpri@mail.ru

© Springer Nature Switzerland AG 2020
V. Goldfarb et al. (eds.), *New Approaches to Gear Design and Production*,
Mechanisms and Machine Science 81,
https://doi.org/10.1007/978-3-030-34945-5_10

Keywords Spur gear · Sliding friction coefficient · Profile correction · Gear
service life · Contact pressure · Tooth wear · x-zero gear pair · x-gear pair

10.1 Introduction

The use of lubricants is an effective method for reducing frictional forces in tribojoints such as toothed gears. If selected properly, they can significantly decrease
tooth wear and increase gear service life. One of the fundamental factors describing friction conditions in a given tribological system is a sliding friction coefficient,
f. Consequently, investigation of the effect of this parameter on the service life of
closed gears operating at boundary or mixed friction is of vital practical importance.
The literature of the subject does not contain mention of the effect of changing f on
gear service life determined with relevant computational methods Brandão et al. [3],
Brauer and Andersson [4], Drozdov [10], Flodin and Andersson [11], Flodin and
Andersson [12], Grib [13], Kahraman et al. [14], Flodin and Andersson [15], Pasta
and Mariotti Virzi [17], Pronikov [18], Shil'ko et al. [22].

Furthermore, there are a wide opportunities for increasing of service life by controlling of friction coefficient in gear meshing made of polymer materials modified by
micron size disperse particles and short fibres [23]. For designing wear resistance of
these composite gears, the effective two-level method has been developed in Shil'ko
et al. [20–22], Shil'ko [19]. The method allows us to determine the stress-strain state,
compliance, strength and wear resistance of composite gears, as well as to find an
optimum composition of the material on the basis of requirements for the service life
of gearing without geometrical corrections of the wheels. It is proposed directly in
the micromechanical models to determine the mechanical and tribological characteristics of disperse-filled plastics taking into account their structure and characteristics
of the matrix, filler and interphase components. Thereafter, such characteristics as
an initial data for gear drive calculation are used.

Authors use the well-known Archard law of wear that assumes a linear relationship
between the linear wear h, the sliding path L and the contact pressures p ($h = cLp$)
occurring in abrasive wear. However, it is hard to expect the occurrence of this kind
of wear in real operating conditions for oil-lubricated closed gears. A characteristic
of the above Archard law as well of other, later developed equations of similar type
($h = cLp^m$, $h = cp/t$, $h = cp^m/t$, where c, m are the parameters of wear) is
that they do not directly make use of the wear-affecting sliding friction coefficient f.
Based on the experimental results reported in the literature for different conditions,
the sliding friction coefficient in tribomechanical systems can range from 0.01 to 1.0
when contact pressures are maintained constant. Therefore, contact pressures can be
directly used to determine the contact strength of gear teeth; they cannot however be
used to estimate the wear resistance of gear teeth.

Taking this obvious fact into consideration in Andreykiv et al. [2], Chernets [5]
a law of friction described by a specific friction force, τ, that was determined in
accordance with Coulomb's law of friction ($h = cL(\tau - \tau_0)^m$ Andreykiv et al.

[2], Chernets [5], where $\tau = fp$, τ_0 is the threshold value of the specific friction force that does not cause wear). A modified version of this law was later applied in studies Aleksandrov and Kovalenko [1], Kovalenko and Teplyi [16] investigating sliding bearings ($h = cL\tau$ Aleksandrov and Kovalenko [1], $h = cL\tau^m$ Kovalenko and Teplyi [16]). Nevertheless, previous methods for estimating the wear of toothed gears in Brandão et al. [3], Brauer and Andersson [4], Drozdov [10], Flodin and Andersson [11, 12], Grib [13], Kahraman et al. [14], Kolivand and Kahraman [15], Pasta and MariottiVirzi [17], Pronikov [18], with the exception of Chernets and Yarema [8], Chernets et al. [9], Chernets and Chernets [6, 7], do not contain mention of a solution that would determine the sliding friction coefficient taking into account of tooth engagement conditions and wear.

Using a method based on the mechanism of wear that was developed by the authors Chernets and Yarema [8], Chernets et al. [9], Chernets and Chernets [6, 7], this paper presents results of a study of the effect of the sliding friction coefficient on the wear and service life of toothed gears, taking into account of operational (tooth wear) and constructive (profile correction) parameters, as well as tooth engagement conditions.

10.2   Problem Solution Method

To investigate the kinetics of tooth wear in engagement, a mathematical model of a sliding process is applied. The model which is described with a system of linear differential equations Andreykiv et al. [2], Chernets and Chernets [6, 7].

$$\frac{1}{v}\frac{dh_k}{dt} = \Phi_k^{-1}(\tau), k = 1; 2, \tag{10.1}$$

where h is the linear wear of the material of tribojoint elements; t is the time of wear; v is the sliding velocity; $\Phi(\tau)$ is the function of wear resistance of the material; $\tau = fp$ is the specific friction force; f is the sliding friction coefficient; p is the maximum contact pressure; k is the number of elements in the tribosystem.

Experimental values of the wear-resistance function are approximated by the relation

$$\Phi_k(\tau) = C_k\left(\frac{\tau_S}{\tau}\right)^{m_k}, \tag{10.2}$$

where C_k, m_k are the indicators of resistance to wear of tribological pair materials; $\tau_S = R_{0,2}/2$ is the shear strength of a material; $R_{0,2} = 0, 7R_m$ is the conventional yield strength of a material under tension; R_m is the tensile strength of a material.

The wear resistance function $\Phi_i(\tau_i)$ of a gear teeth material is determined in the following way

$$\Phi_i(\tau_i) = L/h_i.$$

where h_i is the linear wear of material samples; L is the sliding path; $i = 1, 2, 3$ are loading steps.

Taking into account relation (10.2) after separation of variables and system integration (10.1) on condition that $\tau = fp = $ const, the following dependence will arise

$$t_k = \frac{C_k}{v}\left(\frac{\tau_S}{\tau}\right)^{m_k} h_k. \tag{10.3}$$

Then, the function of linear wear of gear teeth at an arbitrary point j of the working tooth flank surface over a period t_j of their interaction has been determined as

$$h_k = \frac{v t_k}{C_k}\left(\frac{\tau}{\tau_S}\right)^{m_k}. \tag{10.4}$$

In accordance with the above method in Chernets and Chernets [6, 7], the sliding friction coefficient f has a direct effect on the linear wear of the gear teeth. One can also observe that it has an indirect effect on other parameters of wear. The linear wear of the gear teeth h'_{kjn} at any point j of the profile in the tooth engagement time t'_{jh} is determined using the following formula Chernets and Chernets [6, 7]:

$$h'_{kjn} = \frac{v_j t'_{jh}(fp_{jh\max})^{m_k}}{C_k(0.35R_m)^{m_k}}, \tag{10.5}$$

where $j = 0, 1, 2, 3, \ldots, s$ are the contact points of tooth profiles; $j = 0$, $j = s$ are the first and last point of tooth engagement, respectively; $t'_{jh} = 2b_{jh}/v_0$is the time of tooth wear at displacement along the profile of a j-th contact point over the contact area width $2b_{jh}$; $v_j = v$ is the sliding velocity at a j-th point of the gear profile; $p_{jh\max}$ is the maximum contact pressure (at tooth wear) at a j-th contact point; $v_0 = \omega_1 r_1 \sin \alpha$ is the velocity of contact point travel along the tooth profile; ω_1 is the angular velocity of the pinion; r_1 is the pitch radius of the pinion; $\alpha = 20°$ is the pressure angle.

Tooth wear causes an increase in the curvature radii of tooth profiles, which leads to a decrease in the initial maximum contact pressures $p_{j\max}$ and the contact area width $2b_j$ at every j-th point of contact. The values of $p_{jh\max}$ and $2b_{jh}$ are calculated in accordance with the modified Hertz equations, where a radius of curvature, ρ_{jh}, is introduced:

$$p_{jh\max} = 0.564\sqrt{N'\theta/\rho_{jh}}, \quad 2b_{jh} = 2.256\sqrt{\theta N'\rho_{jh}}, \tag{10.6}$$

where $N' = N/b_w$; $N = 9550P/r_1n_1 \cos \alpha$ is the engagement force; P is the power on the driving shaft (pinion); w is the number of engaged tooth pairs; $\theta = (1 - v_1^2)/E_1 + (1 - v_2^2)/E_2$ is elastic constant; E, v are the Young modulus and Poisson's ratio of toothed gear materials, respectively; n_1 is the number of revolutions of the drive shaft; $\rho_{jh} = \frac{\rho_{1jh}\rho_{2jh}}{\rho_{1jh}+\rho_{2jh}}$ is the equivalent radius of curvature of the tooth profiles; ρ_{1jh}, ρ_{jh} are the changeable radii of curvature of the pinion and gear tooth profiles, respectively.

With operation, according to Chernets and Yarema [8], Chernets et al. [9] the initial curvature radii ρ_{1j}, ρ_{2j} of the gear profiles and the equivalent radius ρ_j increase due to tooth wear.

Operational gear wear makes the curvature radii ρ_{1jh}, ρ_{2jh} increase, which leads to a decrease in the initial values of p_{jmax}. In previous studies, a method that takes into account a change of the initial curvature radii ρ_{1j}, ρ_{2j} due to wear and the equivalent radius ρ_j have been proposed in Chernets et al. [9], Chernets and Chernets [6].

The tested traction spur gear used in locomotive engines is characterized by double—single—double tooth engagement. The angles of transition from a double tooth engagement $(\Delta\varphi_{1F_2})$ to a single tooth engagement and, again, to a double tooth engagement $(\Delta\varphi_{1F_1})$ in the spur gear with profile correction are determined in the following way in Chernets and Chernets [6, 7]:

$$\Delta\varphi_{1F_2} = \varphi_{10} - \varphi_{1F_2}, \quad \Delta\varphi_{1F_1} = \varphi_{10} + \varphi_{1F_1}, \tag{10.7}$$

where $\varphi_{1F_1} = \tan\alpha_{F_1} - \tan\alpha_w$, $\varphi_{1F_2} = \tan\alpha_{F_2} - \tan\alpha_w$ are the angles, corresponding to beginning of engagement and disengagement correspondingly; $\varphi_{10} = \tan\alpha_{10} - \tan\alpha_w$ is the angle of gear position, corresponding to the first point of engagement;

$$\tan\alpha_{F_1} = \frac{r_{w1}\sin\alpha_w - (p_b - e_1)}{r_1\cos\alpha}, \quad \tan\alpha_{F_2} = \frac{r_{w2}\sin\alpha_w - (p_b - e_2)}{r_2\cos\alpha};$$

$$\tan\alpha_{10} = (1+u)\tan\alpha_w - \frac{u}{\cos\alpha_w}\sqrt{(r_{2s}/r_{w2})^2 - \cos^2\alpha_w};$$

$p_b = \pi m \cos\alpha_w$ is the base pitch of tooth engagement; $e_1 = \sqrt{r_{1s}^2 - r_{b1}^2} - r_{w1}\sin\alpha_w$; $e_2 = \sqrt{r_{2s}^2 - r_{b2}^2} - r_{w2}\sin\alpha_w$; α_w is the operating pressure angle; u is the gear ratio; m is the module. Other symbols with signs 1 for pinion and 2 for gear are the following: $r_{1,2} = mz_{1,2}/2$ are the reference radii; $z_{1,2}$ are the tooth numbers; $r_{b1,2} = r_{1,2}\cos\alpha$ are the base radii; $r_{a1,2} = r_{1,2} + m$ are the tip radii for x-zero gear pair; $r_{1,2s} = r_{a1,2} - r$; $r = 0.2m$ is the radius of the gear tooth fillet for x-zero gear pair.

The angle $\Delta\varphi_{1E}$ describing the end of tooth engagement is:

$$\Delta\varphi_{1E} = \varphi_{10} + \varphi_{1E}, \tag{10.8}$$

where $\varphi_{1E} = \tan\alpha_E - \tan\alpha_w$, $\alpha_E = \arccos(r_{b1}/r_{1s})$.

The equation should include a change of gear and pinion tip radii for x-gear pair:

$$r_{a1} = r_1 + (1+x_1)m, \quad r_{a2} = r_2 + (1+x_2)m, \tag{10.9}$$

where $x_1 = -x_2$ are the shifting coefficients for the pinion and gear respectively.

The sliding velocity of engaged teeth is calculated as in Chernets and Chernets [6, 7]:

$$v_j = \omega_1 r_{b1} (\tan \alpha_{1j} - \tan \alpha_{2j}), \tag{10.10}$$

where

$$\alpha_{1j} = \arctan(\tan \alpha_{10} + j\Delta\varphi), \quad \alpha_{2j} = \arccos[(r_{w2}/r_{2j})\cos \alpha_w], \quad r_{2j} = \sqrt{a_w^2 + r_{1j}^2 - 2a_w r_{1j} \cos(\alpha_w - \alpha_{1j})}, r_{1j} = r_{w1} \cos \alpha_w / \cos \alpha_{1j}.$$

Here $\Delta\varphi$ is the angle of rotation of the pinion teeth from a point of initial contact (point 0) to point 1, and so on; a_w is the center distance.

The gear service life $t_{B\min}$ for a given number of gear revolutions n_{1s} or n_{2s} when the maximum permissible tooth wear is reached is determined by Chernets and Chernets [6]:

$$t_{b\min} = n_{1s}/60n_1 = n_{2s}/60n_2, \quad n_{2s} = n_{1s}/u. \tag{10.11}$$

10.3 Numerical Solution

Let us now consider the case of a toothed gear used in electric locomotives. This tribocontact problem is solved using the following data: $z_1 = 24$ is the pinion tooth number; $b_w = 230$ mm is the width of the pinion; $P = 670$ kW; $K_g = 1.6$; $m = 16$ mm; 4; $n_1 = 400$ rpm; $h_{1*} = 1.4$ mm is the maximum permissible wear of the pinion teeth; $h_{2*} = 2.0$ mm is the maximum permissible wear of the gear teeth; $f = 0.05$, 0.07, 0.09; 0.11 is boundary or mixed friction; boundary lubrication with oil of the traction spur gear; $x_1 = -x_2 = 0, 0.1, 0.2, 0.3, 0.4$; $a_w = 960$ mm. As regards the gear material, pinion is made of 20CN3A steel after carburizing or nitrocarburizing down to a depth ranging from 1.6 to 2.4 mm, 58 ± 3 HRC; $C_1 = 5.5 \times 10^6$, $m_1 = 1.9$ are the characteristics of wear resistance; while the wheel is made of 55F steel subjected to volume hardening with high-temperature tempering, 280–321 HB; $C_2 = 0.4 \times 10^6$, $m_2 = 2.2$; $E = 2.1 \times 10^5$ MPa, $v = 0.3$.

The results are given in Figs. 10.1, 10.2, 10.3, 10.4 and 10.5. Figure 10.1 illustrates the relationship between the minimal gear life $t_{B\min}$, the friction coefficient f and the profile shift coefficients $x_1 = -x_2$ at the contact point on the gear teeth profile where the maximum permissible wear occurs faster.

It has been found that gear life is the longest when $x_1 = -x_2 = 0.1$, irrespective of the applied value of f. Compared to the gear without profile correction, the service life of the gear with profile correction is longer by 1.245 ($f = 0.05$), 1.321 ($f = 0.07$); 1.218 ($f = 0.09$) and 1.217 times ($f = 0.11$), respectively. In this case, when $x_1 = -x_2 = 0.1$, decreasing the sliding friction coefficient by 2.2 times causes an increase in the gear life by 5.46 times.

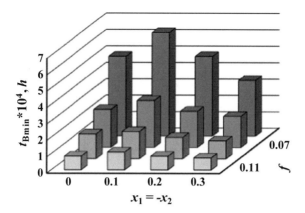

Fig. 10.1 Minimal service life of gear

Variations in the initial contact pressures $p_{j\max}$ for different profile shift coefficients during the pinion rotation are shown in Fig. 10.2a. Figure 10.2b illustrates the initial contact pressures change when the maximum permissible wear h_{2*} is reached at one of the contact points on the gear teeth profile. In the initial stage of double tooth meshing (Fig. 10.2b, left), the pressures $p_{j\max}$ undergo a significant change due to wear for all tested values of the profile correction coefficients, whereas a change of the sliding friction factor has practically no effect on the relationship plotted in the figure. On the other hand, in the case of single tooth meshing (central part of the figure), the quantitative changes of $p_{j\max}$ are much more varied and depend on the value of $x_1 = -x_2$.

Consequently, it can be claimed that a change of the sliding friction factor has practically no effect on the qualitative and quantitative variations in contact pressures; however, it does have a significant impact on gear life.

For example, in Fig. 10.3 gear life versus profile shift coefficients are plotted for the boundary friction ($f = 0.05$).

The above plot confirms the results given in Fig. 10.1 that the application of profile correction is justified when $x_1 = -x_2 \langle 0.15$ (the optimum being $x_1 = -x_2 = 0.1$).

The linear wear h_{kj} of gear teeth profiles at selected points (angles of rotation) for boundary friction and profile shift coefficients is plotted in Figs. 10.4 and 10.5. It has been found that the gear teeth reach the maximum permissible wear faster than the pinion teeth. The applied values of shifting coefficients determine the point of contact where the maximum permissible wear is reached. In the case of the x-zero gear pair, this occurs from the beginning of single tooth engagement (central part of the plot), and a similar value of wear rate can also be observed from the beginning of double tooth engagement (left part of the plot). For the gear with optimal shifting coefficients $x_1 = -x_2 = 0.1$, the maximum permissible wear rate is obtained at the end of single tooth engagement, although a similar wear rate can be observed from the beginning of both double and single tooth engagement. Similar wear patterns

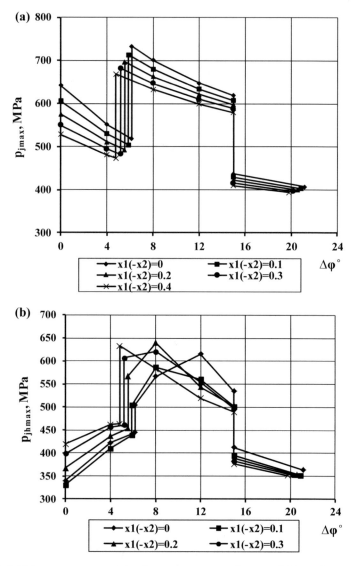

Fig. 10.2 Initial contact pressures (**a**) and contact pressures after wear (**b**) in tooth engagement

can be observed for other tested values of the sliding friction coefficient, leading to a proportional decrease in gear life.

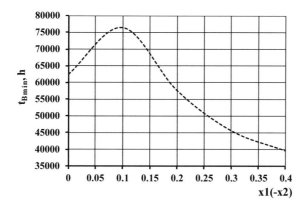

Fig. 10.3   Minimal gear service life versus profile shift coefficients

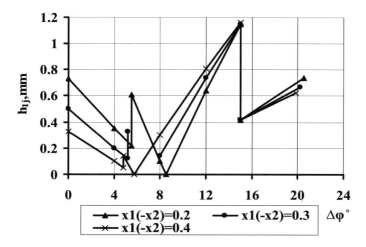

Fig. 10.4   Linear wear of pinion teeth versus angle $\Delta \varphi$ and profile shift coefficients

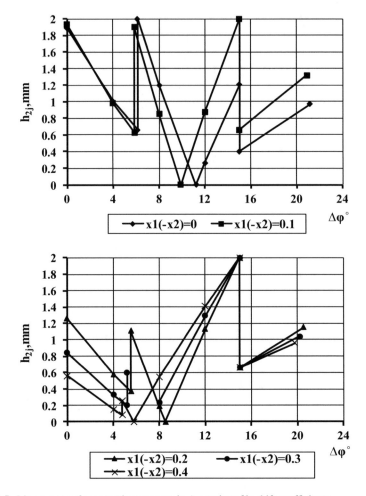

Fig. 10.5 Linear wear of gear teeth versus angle $\Delta\varphi$ and profile shift coefficients

10.4 Conclusions

1. A decrease in the sliding friction coefficient leads to a nonlinear increase in the service life of toothed gears for the tested range of profile shift coefficients (Fig. 10.1).
2. The service life of the gear with profile correction is the longest irrespective of the value of the sliding friction coefficient when the profile shift coefficients are set to $x_1 = -x_2 = 0.1$ (Fig. 10.1).
3. Change of the sliding friction coefficient in a range from 0.11 to 0.05 (by 2.2 times) leads to a nonlinear increase in the service life of the gear by 5.46 times.
4. The service life of the x-gear pair is the longest when $x_1 = -x_2 = 0.1$. Compared to the gear x-zero gear pair, the service life of the x-gear pair is longer by 1.22 ($f = 0.11; 0.09$), 1.32 ($f = 0.07$) and 1.25 times ($f = 0.05$), respectively.
5. The sliding friction coefficient has no observable effect on contact pressures. The contact pressures change with changing the profile shift coefficients (Fig. 10.2).
6. When the optimal boundary friction takes place at profile shift coefficients $x_1 = -x_2 = 0.1$, the wear of the gear teeth at the end of single tooth meshing is almost identical to that observed for the beginning of both double and single tooth meshing (Figs. 10.4 and 10.5).

References

1. Aleksandrov, V.M., Kovalenko, E.V.: On the problem of wear in a shaft-sleeve system. Frict. Wear **3**, 1016–1025 (1982)
2. Andreykiv, A.E., Panasyuk, V.V., Chernets, M.V.: On the problem of wear of materials under dry friction. physicochemical mechanics of materials. Mater. Sci. **2**, 51–57 (1981). (in Russian)
3. Brandão, J.A., Martinsa, R., Seabra, J.H.O., Castro Manuel, J.D.: Calculation of gear tooth flank surface wear during an FZG micropitting test. Wear **311**(1–2), 31–39 (2014)
4. Brauer, J., Andersson, S.: Simulation of wear in gears with flank interference—a mixed FE and analytical approach. Wear **254**, 1216–1232 (2003)
5. Chernets, M.V.: Determination of run-in time for two cylindrical elements. physicochemical mechanics of materials. Mater. Sci. **3**, 101–104 (1981). (in Russian)
6. Chernets, M.V., Chernets, Y.: Evaluation of the strength, wear, and durability of a corrected cylindrical involute gearing, with due regard for the tooth engagement conditions. J. Frict. Wear **1**, 71–77 (2016)
7. Chernets, M., Chernets, Y.: The simulation of influence of engagement conditions and technological teeth correction on contact strength, wear and durability of cylindrical spur gear of electric locomotive. J. Eng. Tribol. **231**(1), 57–62 (2017)
8. Chernets, M.V., Yarema, R.Y.: Generalized value method of teeth correction influence on resource, wear and contact durability of cylindrical involute gears. Mater. Sci. **4**, 56–58 (2011)
9. Chernets, M.V., Yarema, R.Y., Chernets, Y.: A method for the evaluation of the influence of correction and wear of the teeth of a cylindrical gear on its durability and strength. part 1. service live and wear. Mater. Sci. **3**, 289–300 (2012)
10. Drozdov, Y.: To the development of calculation methods on friction wear and modeling. Chapter in *Wear Resistance*. Nauka, Moscow, 120–135. (in Russian) (1975)
11. Flodin, A., Andersson, S.: Wear simulation of spur gears. Tribotest J. **5**(3), 225–250 (1999)

12. Flodin, A., Andersson, S.: A simplified model for wear prediction in helical gears. Wear **249**(3–4), 285–292 (2001)
13. Grib, V.: Solution of Triboengineering Problems by Numerical Methods. Science, Moscow (1982). (in Russian)
14. Kahraman, A., Bajpai, P., Anderson, N.E.: Influence of tooth profile deviations on helical gear wear. J. Mech. Des. **127**(4), 656–663 (2005)
15. Kolivand, M., Kahraman, A.: An ease-off based method for loaded tooth contact analysis of hypoid gears having local and global surface deviations. J. Mech. Des. **132**(7), 0710041–0710048 (2010)
16. Kovalenko, E.V., Teplyi, M.I.: Contact problems in the nonlinear wear of coated bodies. part 1. Frict. Wear **3**, 440–448 (1983)
17. Pasta, A., Mariotti Virzi, G.: Finite element method analysis of a spur gear with a corrected profile. J. Strain Anal. **42**, 281–292 (2007)
18. Pronikov, A.: Reliability of Machines. Mashinostroenie, Moscow (1978). (in Russian)
19. Shil'ko, S.V.: Calculation of stress-strained state and wear resistance of gears made of dispersed-filled polymer materials. In: Algin, V.B., Starzhinsky, V.E. (eds.) Chapter 3 in Gears and Transmissions in Belarus: Design, Technology, Estimation of Properties, pp. 116–136 (in Russian) Minsk, Belaruskaya Navuka (2017)
20. Shil'ko, S.V., Starzhinsky, V.E., Petrokovets, E.M., Chernous, D.A.: Two-level calculation method for tribojoints made of disperse-reinforced composites: part 1. J. Frict. Wear **34**(1), 65–69 (2013)
21. Shil'ko, S.V., Starzhinsky, V.E., Petrokovets, E.M., Chernous, D.A.: Two-level calculation method for tribojoints made of disperse-reinforced composites: part 2. J. Frict. Wear, **35**(1), 40–47 (2014)
22. Shil'ko, S.V., Starzhinsky, V.E., Petrokovets, E.M.: Methods and results of composite gears design. In: Chapter 16 in Gears, pp. 341–368. Springer (2015)
23. Starzhinsky, V.E.: Polymer gears. In: Wang, O.J., Chung, Y.-W. (eds.) Chapter in Enciclopedia of Tribology, vol. 6, pp. 2592–2602. Springer Science+Business Media, LLC, New York

Chapter 11
On Possibility of Cutting Bevel Gearwheels by Hobs

Evgenii S. Trubachev

Abstract The formation of spiral teeth of bevel gears is one of the most complicated areas for simulation, synthesis and analysis of processes in the gear technology in general. The practical technique of production is also complex and specific with regard to equipment, tools and personnel. The main purpose of this manuscript is to present an alternative, simpler and cheaper method of spiral teeth generation with respect to production preparation. The paper gives reasons for the possibility of generating the spiral teeth of bevel gearwheels by two helical surfaces with the constant pitch. The idea of generation comes from the evident exterior similarity of gearwheels of non-orthogonal spiroid gears and bevel gears. The paper presents mathematical models of operating and machine-tool surface generating gearing schemes. The following stages of gearing synthesis are introduced: synthesis of the machine-tool meshing of the pinion; synthesis of the bevel gear scheme; synthesis of the machine-tool meshing of the gearwheel in accordance with local conditions stated at points that are arbitrarily chosen on opposite tooth flanks; and analysis of contact localization. Due to similarity of the problem statement, there exists a big enough number of parameters to control the tooth geometry. The paper presents the corresponding algorithm for the choice of machine-tool setting parameters, methods of calculation of flanks and transient tooth surfaces, and evaluation of the level of contact localization. Calculation examples, conclusions on their results and recommendations on the choice of parameters are given. In particular, the main features of the method are specified: narrowness of the area of possible solutions; acceptability and even request of a certain undercut of concave tooth flanks; the tooth spiral angle that is greater than that of traditional bevel gears; preferability of application of the proposed method for gear ratios 1…1.4. The obtained results give prospects of development of the cost-efficient technique of tooth machining of spiral bevel gears within low series production.

Keywords Bevel gear · Tooth cutting · Hobs

E. S. Trubachev (✉)
Institute of Mechanics, Kalashnikov Izhevsk State Technical University, Studencheskaya Str. 7, P.O. Box 426069, Izhevsk, Russia
e-mail: truba@istu.ru

© Springer Nature Switzerland AG 2020
V. Goldfarb et al. (eds.), *New Approaches to Gear Design and Production*,
Mechanisms and Machine Science 81,
https://doi.org/10.1007/978-3-030-34945-5_11

11.1 Introduction

The technique of production of spiral bevel gears is traditionally considered to be one of the most complicated areas of the "gearing" branch. This opinion reflects those difficulties to be overcome by an enterprise that decides to master this technique, namely they are related to the necessary availability of:

– expensive and specialized gear cutting equipment applied practically only for production of the mentioned gears and requiring highly-skilled setting and re-setting;
– specific assembled tools requiring the whole complex of maintenance—production, sharpening, assembling and setting;
– properly qualified employees at all stages of production and its preparation from gear designing engineers to machine-tool service technicians.
    Exactly these difficulties became the main reason for evaluation of the possibility to generate the spiral teeth of gear pairs by means of two cylindrical helicoid generating worms. The obvious consideration to the benefit of this alternative is its simplicity for manufacturers as compared with the traditional technique:
– gear milling machines are common, relatively cheap, their setting is comparatively simple and evident;
– tools that represent the generating worms (hobs, flying cutters and shaving cutters) are also relatively simple for manufacture, control, sharpening and setting.

11.2 Several Preconditions

One can say that the idea itself is plain to see: teeth of spiroid gearwheels reproduced by a helicoid generating surface look very similar to spiral teeth of bevel or hypoid gearwheels—Fig. 11.1.

Despite of its obviousness and background to its simple practical implementation, the method of generating the spiral teeth of bevel gearwheels by means of a helicoid generating surface did not get any extensive use. Let us dare to suppose that the main reason of this is the poor investigation of this issue and, jumping ahead a bit, extremely narrow ranges of parameters with more or less acceptable solutions. The decisive precondition to turn to this theme became the long-term and generally theoretically and practically successful experience of design and production of spiroid gears with the localized contact [4, 18]. We have developed and implemented into practice the method of calculation of setting parameters for generation of spiroid gearwheel teeth [12, 14]; this method allows to vary both the level of contact localization and possible combinations of parameters of generating worms. The main idea of the method implies that geometry parameters of the generating worm and its arrangement relative to the gearwheel workpiece are selected in accordance with the conditions:

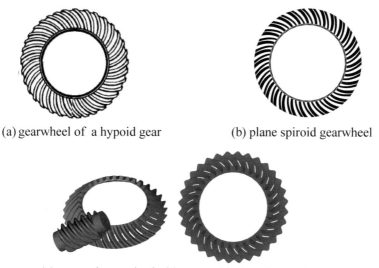

(a) gearwheel of a hypoid gear   (b) plane spiroid gearwheel

(c) non-orthogonal spiroid gear and its bevel gearwheel

Fig. 11.1 External view of gearwheels with spiral teeth

– of meshing of three surfaces [6]—conjugated surfaces of the worm and gearwheel and the sought generating surface—at arbitrarily assigned points:

$$\begin{aligned} \mathbf{n}\mathbf{v}_{wg} &= \mathbf{n}(\mathbf{v}_w - \mathbf{v}_g) = 0, \\ \mathbf{n}\mathbf{v}_{hg} &= \mathbf{n}(\mathbf{v}_h - \mathbf{v}_g) = 0, \end{aligned} \Rightarrow \mathbf{n}\mathbf{v}_{hw} = 0, \tag{11.1}$$

where $\mathbf{n}$ is the common normal vector to the pointed three surfaces, $\mathbf{v}_{h(w,g)}$ is the velocity of the generating worm motion (operating worm, gearwheel) in meshing, $\mathbf{v}_{wg}$, $\mathbf{v}_{hg}$ and $\mathbf{v}_{hw}$ are relative velocities of elements in the imaginary joint meshing;

– of constant pitch and profile of surfaces of the operating and generating worms-helicoids for which the following relations are valid that are invariant with respect to the choice of a coordinate system [13]:

$$\tan \gamma = -\frac{\mathbf{n}\mathbf{e}_t}{\mathbf{n}\mathbf{k}}, \tag{11.2}$$

$$\tan \alpha_x = -\frac{\mathbf{n}\mathbf{e}_r}{\mathbf{n}\mathbf{k}}. \tag{11.3}$$

where α_x is the axial angle of the helicoid profile, γ is its helix angle, $\mathbf{e}_t$, $\mathbf{e}_r$ and $\mathbf{k}$ are unit vectors of tangential, radial and axial directions with respect to the axis of the helicoid.

Note, that the synthesized bevel gearwheels must have the differently directed spirals of teeth; that is why, differently directed generating worms should be chosen in order to form their teeth—one right worm and one left worm. Their other parameters—diameters, axial modules, angles and radii of the profile—can also be different.

11.3 General Scheme of Calculation. Setting Parameters

One can say that the general scheme inherits the well established in practice consequence of spiroid gear design; and already at the first stage of considering the issue (the stage of studying the possibility of generation itself) it allowed for applying the available calculation modules of the program software SPDIAL + developed at the Institute of Mechanics of Kalashnikov ISTU [17] for the initial evaluation of the very possibility of generating the bevel gear tooth geometry by the considered method. The calculation scheme is as follows.

1. *Calculation of the non-orthogonal machine-tool meshing of the generating worm and the bevel pinion* (in its essence it is the calculation of the geometry of a non-orthogonal spiroid gear); it determines the angle δ_1 of the pinion cone and it includes:

    1.1. the choice of parameters of the machine-tool gearing scheme (Fig. 11.2):

Fig. 11.2 Scheme of the machine-tool meshing of the generating worm and the bevel gearwheel

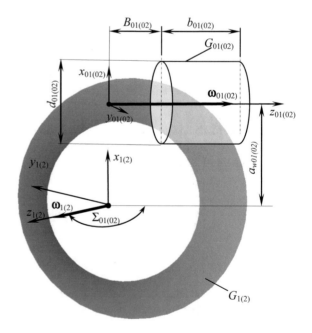

- machine-tool center distance a_{w01};
- machine-tool interaxial angle Σ_{01};
- machine-tool gear ratio i_{01} determined by the ratio:
  - of the gearwheel tooth number z_1;
  - of the number of threads of the generating worm z_{01};
- diameter of the generating worm d_{01};
- external and internal diameters of the gearwheel d_{e1} and d_{i1};
- axial module of the generating worm for the pinion m_{x01};

1.2. the choice of parameters of the generating worm of the pinion, including:

- profile angles $\alpha_{01R,L}{}^{1}$;
- profile radii $\rho_{01R,L}$;
- thread thickness s_0;
- thread height h_0;

1.3. the calculation of pinion tooth flanks as the enveloping surfaces of a one-parameter family of surfaces of the generating worm for the pinion in its relative motion.

2. *Choice of parameters of the bevel gear scheme*, including:

- interaxial angle Σ_{12} or angle δ_2 of the gearwheel cone in accordance with the condition:

$$\delta_1 + \delta_2 = \Sigma_{12}; \tag{11.4}$$

- number of gearwheel teeth $z_{(2)}$ or gear ratio i_{12} in accordance with the condition:

$$i_{12} = \frac{z_{(1)}}{z_{(2)}} = \cos \Sigma_{12} - \sin \Sigma_{12} / \tan \delta_1. \tag{11.5}$$

3. *Calculation of linearly conjugated flanks of bevel gearwheel teeth* as the enveloping surfaces of a one-parameter family of surfaces of the bevel pinion in its relative motion (according to M. G. Segal [8, 9] the enveloping surfaces of gearwheel teeth obtained at this stage will be further called reference ones).

4. *Calculation of the machine-tool setting for gearwheel teeth forming* by a generating worm, including:

4.1. choice of design points on reference surfaces;
4.2. choice of the assigned setting parameters:
- machine-tool interaxial angle Σ_{02};
- number of threads z_{02}, axial module m_{x02}, radii of the generating worm profile $\rho_{02R,L}$;

---

[1]Hereinafter indices R and L are related to parameters of right and left flanks of teeth and threads, correspondingly.

4.3. calculation of the remaining parameters of setting:
- machine-tool interaxial distance a_{w02};
- diameter of the generating worm d_{02}.

5. *Evaluation of the modification field* (MF)—distances between points of two surfaces of the gearwheel: reference and formed by the generating worm.

The paper describes further the main statements of the mathematical model that we have developed and applied in order to achieve the main aim of the work— to determine the possibility of generating the curvilinear teeth by the considered method.

11.4 Scheme of Machine-Tool Gearing, Pitch Surfaces of Elements

The scheme of machine-tool gearing, applied systems of coordinates and parameters of the scheme (Fig. 11.2) are similar to the corresponding elements of calculation applied when designing spiroid gears at arbitrary arrangement of axes of the gear elements. The transform matrix of coordinate systems $S_{1(2)}$ and $S_{01(02)}$ is as follows:

$$\mathbf{M}_{01(02)\to 1(2)} = \begin{pmatrix} 1 & 0 & 0 & a_{w01(02)} \\ 0 & \cos \Sigma_{01(02)} & -\sin \Sigma_{01(02)} & 0 \\ 0 & \sin \Sigma_{01(02)} & \cos \Sigma_{01(02)} & 0 \\ 0 & 0 & 0 & 1 \end{pmatrix}. \tag{11.6}$$

Pitch surfaces in the machine-tool setting—the cylinder $G_{01(02)}$ coaxial to the generating worm (for instance, the reference one) and the cone $G_{1(2)}$ coaxial to the gearwheel—are contacting each other at the mean diameter of the gearwheel. The condition of contacting is [3]:

$$\begin{cases} x_{01(02)}^2 + y_{01(02)}^2 - r_{01(02)}^2 = 0, \\ x_{01(02)}z_{01(02)} + y_{01(02)}a_{w01(02)} \cot \Sigma_{01(02)} = 0, \end{cases} \tag{11.7}$$

where the first equation is the equation of the pitch cylinder of the generating worm (the radius $r_{01(02)}$ can be taken to be equal to, for example, the reference one); and the second one is actually the condition of contacting.

Though surfaces of tooth roots generated by tips of worms and the surface equally distant from thread roots by the value of radial clearance are strictly saying not equidistant to each other and are not conical (they have curvilinear generating lines), these deviations can nevertheless be neglected in most cases of calculation without any harm to the gear under design; and that is why, teeth of gearwheels formed by generating worms will be considered to be uniform.

11.5 Tooth Flanks—Enveloping Surfaces of One-Parametric Family of Generating Helicoids

The family of helicoid generating surfaces resulted by the helicoid rotation can be described by equations in the coordinate system $S_{01(02)}$:

$$\begin{cases} x_{01(02)} = r_{01(02)} \cos(\vartheta_{01(02)} + \varphi_{01(02)}), \\ y_{01(02)} = r_{01(02)} \sin(\vartheta_{01(02)} + \varphi_{01(02)}), \\ z_{01(02)} = f_{01(02)}(r_{01(02)}) + p_{\gamma 01(02)} \vartheta_{01(02)}, \end{cases} \quad (11.8)$$

where $r_{01(02)}$, $\theta_{01(02)}$ are parameters—curvilinear coordinates, $p_{\gamma 01(02)} = m_{\times 01(02)} z_{(01),(02)}/2$ is the helix parameter, $\varphi_{01(02)}$ is the angle of rotation, $f_{01(02)}(r_{01(02)})$ is the function of the axial profile.

Projections of normal vectors to surfaces (11.8) are subject to the relations:

$$\mathbf{n}_{01(02)} : \begin{cases} n_{01(02)x} = -y_{01(02)} p_{\gamma 01(02)} + x_{01(02)} r_{01(02)} tg\, \alpha_{x01(02)}, \\ n_{01(02)y} = x_{01(02)} p_{\gamma 01(02)} + y_{01(02)} r_{01(02)} tg\, \alpha_{x01(02)}, \\ n_{01(02)z} = -r_{01(02)}^2. \end{cases} \quad (11.9)$$

Points of contact of the generating worm and its generated gearwheel are determined by the system of equations [16]:

$$\begin{cases} \mathbf{n}_{01(02)} \mathbf{v}_{01(02)} = r_{01(02)}^2 p_{\gamma 01(02)} \left(i_{01(02)} \cos ec\, \Sigma_{01(02)} - \cot \Sigma_{01(02)} \right) \\ + p_{\gamma 01(02)} \left(y_{01(02)} z_{01(02)} - x_{01(02)} a_{w01(02)} \cot \Sigma_{01(02)} \right) \\ - \left(x_{01(02)} z_{01(02)} + y_{01(02)} a_{w01(02)} \cot \Sigma_{01(02)} \right) r_{01(02)} \tan \alpha_{x01(02)} - r_{01(02)}^2 \left(x_{01(02)} + a_{w01(02)} \right) = 0, \\ x_{01(02)}^2 + y_{01(02)}^2 - r_{01(02)}^2 = 0, \end{cases}$$

$$(11.10)$$

where projections of the vector $\mathbf{v}_{01(02)}$ are determined for rotation of elements around stationary skew axes in the coordinate system $S_{01(02)}$ by a traditional method [7]. The solution—coordinates of contact points—should be transformed by means of (11.6) to the coordinate system $S_{1(2)}$ and then to the system $S_{c1(2)}$ rigidly connected with the rotating gearwheel; this is the way to determine the points of tooth flanks obtained at one-parameter enveloping by a helicoid generating surface:

$$\mathbf{M}_{1(2) \to c1(2)} = \begin{pmatrix} \cos \varphi_{1(2)} & \sin \varphi_{1(2)} & 0 & 0 \\ -\sin \varphi_{1(2)} & \cos \varphi_{1(2)} & 0 & 0 \\ 0 & 0 & 1 & 0 \\ 0 & 0 & 0 & 1 \end{pmatrix}, \quad (11.11)$$

where the angle $\varphi_{1(2)}$ of gearwheel rotation at cutting is:

Fig. 11.3 Regular net of
points of the bevel pinion
(bevel gearwheel) tooth

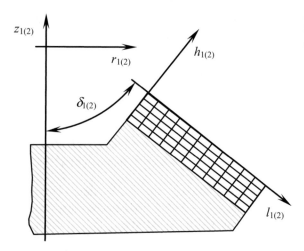

$$\varphi_{1(2)} = \varphi_{01(02)}/i_{01(02)} = (\arctan(y_{01(02)}/x_{01(02)}) - (z_{01(02)} - f_{01(02)}(r_{01(02)}))/p_{y01(02)})/i_{01(02)} \quad (11.12)$$

The solution (11.8) is determined in nodes of the so-called regular net of points
obtained by sections of the generated tooth flank by cones equidistant to cones of tips
and faces of teeth (Fig. 11.3). When moving from one node of the net to the other,
coordinates of the tooth height and length are regularly changing (with the assigned
step):

$$h_{1(2)} = (z_{1(2)} - z_{1(2)O_{2lh}}) \sin \delta_{1(2)} - (r_{1(2)} - r_{1(2)O_{2lh}}) \cos \delta_{1(2)},$$
$$l_{1(2)} = (z_{1(2)} - z_{1(2)O_{2lh}}) \cos \delta_{1(2)} + (r_{1(2)} - r_{1(2)O_{2lh}}) \sin \delta_{1(2)}. \quad (11.13)$$

11.6 Conjugated Surfaces of Gearwheel Teeth

In accordance with the described above calculation scheme, the pair of different name
flanks for one of elements (to be definite—a pinion) is determined at the step 1. Their
conjugated surfaces of the second element (gearwheel) are determined at the step 3.
Pinion tooth flanks are generating surfaces here; and they are assigned discretely (in
nodes of the regular net). Point out, that real tooth flanks of the gearwheel will be
formed by the generating worm; the conjugated surfaces of the gearwheel are not
actually reproduced—they are necessary in calculations only as reference ones to
determine parameters of the real generation process.

Figure 11.4 shows the arrangement of coordinate systems applied in search of
conjugated flanks of gearwheel teeth. The coupling matrix of these systems is as
follows:

Fig. 11.4 Scheme of the
bevel gear and coordinate
systems of the pinion and
gearwheel

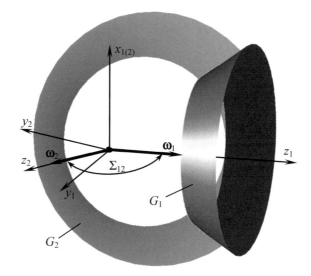

$$\mathbf{M}_{1\rightarrow 2} = \begin{pmatrix} 1 & 0 & 0 & 0 \\ 0 & \cos\,\Sigma_{12} & -\sin\,\Sigma_{12} & 0 \\ 0 & \sin\,\Sigma_{12} & \cos\,\Sigma_{12} & 0 \\ 0 & 0 & 0 & 1 \end{pmatrix}. \tag{11.14}$$

When searching the solution (11.8), normal vectors (coinciding with normal vectors to the generating helicoid surface) are determined for each node of the pinion tooth flank. Let us write the obtained normal vectors $\mathbf{n}_1$ independently to a specific angular position of the pinion, namely, in projections on its axial (index z), radial (r) and tangential (t) directions:

$$\begin{aligned} \mathbf{n}_1 &= \mathbf{n}_{1t} + \mathbf{n}_{1r} + \mathbf{n}_{1z} = n_{1t}\mathbf{e}_{1t} + n_{1r}\mathbf{e}_{1r} + n_{1z}\mathbf{e}_{1z} \\ &= (\mathbf{n}_1\mathbf{e}_{1t})\mathbf{e}_{1t} + (\mathbf{n}_1\mathbf{e}_{1r})\mathbf{e}_{1r} + (\mathbf{n}_1\mathbf{e}_{1z})\mathbf{e}_{1z}, \end{aligned} \tag{11.15}$$

where $\boldsymbol{e}_{1t} = \{-y_1/r_1;\ x_1/r_1;\ 0\}$ and $\boldsymbol{e}_{1r} = \{x_1/r_1;\ y_1/r_1;\ 0\}$ are unit vectors of tangential and radial directions, correspondingly. When the pinion is rotated in a fixed coordinate system S_1, the field of normal vectors $\tilde{\mathbf{n}}$ is formed (Fig. 11.5):

$$\tilde{\mathbf{n}} : \begin{cases} n_{1x} = (-y_1 n_{1t} + x_1 n_{1r})/r, \\ n_{1y} = (x_1 n_{1t} + y_1 n_{1r})/r, \\ n_{1z} = n_{1z} \end{cases} \tag{11.16}$$

and the equation of conjugated meshing of the pinion and gearwheel will be as follows [16]:

Fig. 11.5 Axial, radial and
tangential components of the
normal vector

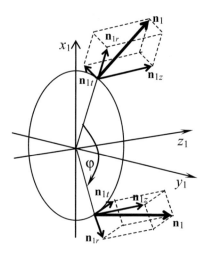

$$\tilde{n}v_{12} = n_{1t}r_1^2(i_{12}\mathrm{cosec}\ \Sigma_{12} - \cot\ \Sigma_{12}) + n_{1t}(y_1z_1 - x_1a_{w12}\cot\ \Sigma_1)$$
$$- n_{1r}(x_1z_1 + y_1a_{w12}\cot\ \Sigma_{12}) + r_1n_{1z}(x_1 + a_{w12}) = 0, \qquad (11.17)$$

where projections of the vector v_{12} are determined for rotation of elements around
stationary intersecting axes in the coordinate system S_1 [7].

In case of the fixed point on the enveloping surface of the pinion with the fixed
parameters r_1 and z_1, the Eq. (11.17) comprises linearly two unknown values that
vary at rotation of the element—coordinates x_1 and y_1. Therefore, similar to the first
equation of the system (11.10), when considered jointly with the expression $x_1^2 + y_1^2 = r_1^2$, the Eq. (11.17) can be reduced to the square equation, thus providing a
comparatively simple algorithm of searching the points of conjugated surfaces of
gearwheel teeth.

Similar to solution of (11.10), the solution of (11.17) is within nodes of the net of
the gearwheel tooth flank with regularly varied parameters h_2, l_2 (Fig. 11.3).

11.7   Calculation of Root Fillet and Undercut Areas

Surfaces of transient and undercut areas on tooth flanks can become considerable;
and their calculation is necessary. The technique of this calculation can be similar and
performed within general procedures of calculation of tooth flanks – both the surfaces
formed by generating worms and reference conjugated surfaces of the gearwheel
generated by enveloping of pinion tooth flanks. For this purpose we applied the
method proposed by Prof. D. T. Babichev [1], in accordance with it the edges of
generating surfaces are represented as channel generating surfaces of an infinitely

Fig. 11.6 Normal vectors to the upper channel surface of the generating worm

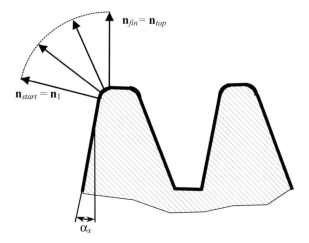

(or finitely—it is technically insignificant) small radius; their normal vectors can be represented as the wedge described around the edge—the generating line. One of normal vectors of the wedge becomes contact at generation of each point of the transient or undercut area. For example, the wedge of normal vectors for the upper edge of the generating worm can be described by the same Eq. (11.9) as the normal vector to the flank of the generating worm flank, if the profile angle α_{x1} is varied from the value equal to the profile angle of the flank up to 90° (the profile angle for which the helical surface is degenerated into the cylinder of worm tips – Fig. 11.6); and the equation of meshing for the upper edge is nominally coinciding with the first equation of the system (11.10). The issue of generation of transient and undercut areas of surfaces formed by a helicoid generating worm is considered in more details in [16].

Similar to the whole tooth flank, the line of points (point edge) of the pinion tooth is assigned discretely in the accepted scheme, i.e., by the number of points each of them possessing the normal vector $\mathbf{n}_1$ to the flank. The normal vector $\mathbf{n}_{top1}$ to the surface of the pinion tooth points depends on coordinates of the points as follows:

$$\mathbf{n}_{top1} : \begin{cases} n_{top1x} = x_1, \\ n_{top1y} = y_1, \\ n_{top1z} = r_1 \tan \delta_1. \end{cases} \tag{11.18}$$

When the wedge of normal vectors is generated for each point of the top edge, the normal vector $\mathbf{n}$ to the channel should "run" through a number of positions from $\mathbf{n}_1$ to $\mathbf{n}_{top1}$:

$$\mathbf{n} = \mathbf{n}_1 + \lambda(\mathbf{n}_{top1} - \mathbf{n}_1), \tag{11.19}$$

where the parameter λ takes the values from 0 to 1. After substitution of $\mathbf{n}$ in (11.12) instead of $\mathbf{n}_1$ and the consequent transformation (11.14), the equation of meshing (11.17) can also be applied for calculation of points of the transient and undercut areas of the conjugated tooth of the bevel gearwheel.

11.8 Choice of Setting Parameters for Generation of Gearwheel Teeth

As a matter of fact, the generating worm should be properly "fit" into the obtained coordinate system S_2 of the reference surface of the gearwheel tooth. For this purpose the design point M is chosen on each of different name flanks (in the first approximation—in the central part of the tooth flank; the peculiarities of the choice will be considered below) and the normal vector $\mathbf{n}_2$ is re-established in it. The choice of the coordinate system for the algebraic form of vector expressions (11.1)–(11.3) strongly determines the success in search for solution. Axes arrangement for the generated gearwheel and generating worm, shown in Fig. 11.7, is very likely the most common one. In addition to the enumerated above setting parameters, it is characterized by the design parameters:

– Δz_2—displacement of the machine-tool interaxial line from the point that is chosen as the base one for the generated gearwheel (it is « 0 » of the coordinate system S_2 in which the nets of points of reference surfaces are assigned);

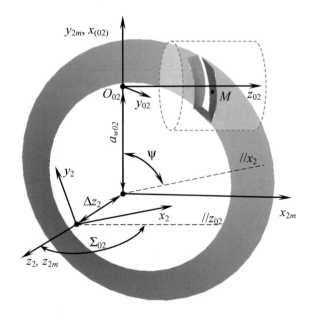

Fig. 11.7 Arrangement of coordinate systems for calculation of the setting for gearwheel generation

– $\psi_{R,L}$—the angle between the machine-tool interaxial line and the plane $x_2 O_2 z_2$ of the system S_2 in which the point M is assigned.

Though two latter parameters are not used at setting, they are nevertheless necessary to establish the relation of the coordinate system in which the design point is assigned with the coordinate system of the generating worm—therefore, they directly determine the tool parameters.

Conditions (11.1)–(11.3) should be stated for the design points on both flanks. Let us agree, that first of all we are dealing with the most practically convenient case of cutting by the two-side method with the two-side cutter head, when both the pinion and the gearwheel are produced by one generating surface at one and the same adjustment and for one and the same tool travel, that is, principally the same as it takes place at cutting of traditional worm and spiroid gearwheels (note here, that application of one-flank cutting can also be of interest, first of all because it will allow for better controlling the degree of contact localization and (or) parameters of generating worms). As for the scheme of two-flank cutting, parameters Σ_{02}, a_{w02} and Δz_2 should be common (the same) for different name flanks, and the parameter ψ can basically take its own value (different from the value for the other flank) for each of the flanks.

Matrices of transformations of coordinates shown in Fig. 11.7 are as follows:

$$\mathbf{M}_{02 \rightarrow 2m} = \begin{pmatrix} 0 & \cos \Sigma_{02} & \sin \Sigma_{02} & 0 \\ 1 & 0 & 0 & a_{w02} \\ -\sin \Sigma_{02} & \cos \Sigma_{02} & 0 & 0 \\ 0 & 0 & 0 & 1 \end{pmatrix}. \tag{11.20}$$

$$\mathbf{M}_{2m \rightarrow 2} = \begin{pmatrix} \sin \psi & \cos \psi & 0 & 0 \\ -\cos \psi & \sin \psi & 0 & 0 \\ 0 & 0 & 1 & -\Delta z_2 \\ 0 & 0 & 0 & 1 \end{pmatrix}. \tag{11.21}$$

Parameters of arrangement of coordinate systems can be chosen in accordance with conditions of the first order of conjugation [10] similar to (1) that are stated at the chosen design points of the reference surfaces:

$$\mathbf{n v}_{02} = 0, \tag{11.22}$$

where $\mathbf{n}$ is the normal vector re-established at the design point chosen on the reference surface, $\mathbf{v}_{02}$ is the vector of the relative velocity at the machine-tool meshing of the gearwheel:

$$\mathbf{v}_{02} = \mathbf{v}_0 - \mathbf{v}_2, \tag{11.23}$$

$$\mathbf{v}_2 : \begin{cases} v_{2x} = -y_2, \\ v_{2y} = x_2, \\ v_{2x} = 0, \end{cases} \tag{11.24}$$

$$\mathbf{v}_0 = \boldsymbol{\omega}_0 \times (\mathbf{r}_2 - \overline{O_2 O_0}) : \begin{cases} v_{0x} = -(z_2 + \Delta z_2)\cos\psi \sin\Sigma_{02} + (y_2 - a_{w02}\sin\psi)\cos\Sigma_{02}, \\ v_{0y} = -(z_2 + \Delta z_2)\sin\psi \sin\Sigma_{02} + (x_2 - a_{w02}\cos\psi)\cos\Sigma_{02}, \\ v_{0x} = (y_2 - a_{w02}\sin\psi)\sin\psi \sin\Sigma_{02} + (x_2 - a_{w02}\cos\psi)\cos\psi \sin\Sigma_{02}, \end{cases}$$
$$\tag{11.25}$$

$$\boldsymbol{\omega}_0 : \begin{cases} \omega_{0x} = \sin\psi \sin\Sigma_{02}, \\ \omega_{0y} = -\cos\psi \sin\Sigma_{02}, \\ \omega_{0x} = \cos\Sigma_{02}, \end{cases} \tag{11.26}$$

The vector $\overline{O_2 O_{02}}$ connecting the origins of coordinates S_2 and S_{02} (Fig. 11.7) is as follows:

$$\overline{O_2 O_0} : \begin{cases} x_{2O_{02}} = a_{w02}\cos\psi, \\ y_{2O_{02}} = a_{w02}\sin\psi, \\ z_{2O_{02}} = -\Delta z_2, \end{cases} \tag{11.27}$$

Expressions (11.23)–(11.27) are made up for the coordinate system S_2 in which the reference surfaces are determined; they contain the sought parameters Σ_{02}, a_{w02}, Δz_2 and ψ. Taking these expressions into account, we will get the Eq. (11.22) in the algebraic form:

$$A a_{w02} + B\Delta z_2 + C = 0, \text{where}$$
$$A = n_x\sin\psi \cos\Sigma_{02} - n_y\cos\psi \cos\Sigma_{02} - n_z\sin\Sigma_{02},$$
$$B = -n_x\cos\psi \sin\Sigma_{02} - n_y\sin\psi \sin\Sigma_{02},$$
$$C = (-z_2\cos\psi \sin\Sigma_{02} - y_2\cos\Sigma_{02})n_x + (-z_2\sin\psi \sin\Sigma_{02} + +x_2\cos\Sigma_{02})n_y +$$
$$+ (x_2\cos\psi \sin\Sigma_{02} + y_2\sin\psi \sin\Sigma_{02})n_z - i_{02}\mathbf{n}\mathbf{v}_2. \tag{11.28}$$

The parameter ψ (parameters ψ_R and ψ_L for the left and right flanks) can be determined by using the property (11.2) written in the coordinate system S_2:

$$\tan\gamma = \frac{p_{\gamma 02}}{r_{02}} = \frac{m_{x02}z_{(02)}}{2r_{02}} = -\frac{\mathbf{n}\mathbf{e}_{t02}}{\mathbf{n}\mathbf{k}_{02}} = -\frac{\mathbf{n}\mathbf{v}_0}{r_{02}\mathbf{n}\mathbf{k}_{02}}$$
$$= -\frac{\mathbf{n}\mathbf{v}_2}{r_{02}\mathbf{n}\mathbf{k}_{02}} \quad \rightarrow \quad a\cos\psi + b\sin\psi + c = 0, \tag{11.29}$$

where

$$a = -n_{y2}\cos\Sigma_{02},$$
$$b = n_{x2}\sin\Sigma_{02},$$
$$c = n_{z2}\cos\Sigma_{02} + \frac{\mathbf{n}\mathbf{v}_2}{p_{\gamma 02}}.$$

The Eq. (11.29) is reduced to the square one with respect to the chosen trigonometric function $\sin\psi$ or $\cos\psi$ and, along with (11.28), it allows for determining the parameters of arrangement a_{w02}, Δz_2 (therefore, to determine completely the relations of coordinates (11.20), (11.21)) at the assigned design point and assigned values of the machine-tool interaxial angle Σ_{02} and axial module m_{x02} of the generating worm for the gearwheel. Transformation of coordinates of the assigned design point and normal vector into the system S_{02} allows for determining the radius of this point relative to the axis of the generating worm (correspondingly, the worm diameter) and profile angles in accordance with (11.3).

11.9 Choice of the First Approximation

Calculation practice showed that solution of the Eq. (11.29) exists in a narrow area of the space of parameters of the machine-tool setting for cutting the gearwheel. And ranges providing the necessary longitudinal modification of teeth are even narrower. Success in search for solution is strongly determined by the choice of the correct first approximation—the assigned parameters Σ_{02} and m_{x02}. Note, that the principal difference for the inherited sequence of spiroid gear design is the absence of such an operating meshing with the worm which is completely determined by the beginning of synthesis of the machine-tool meshing at spiroid gear analysis and which parameters can be the first approximation to the corresponding assigned parameters. When calculating the machine-tool setting for the bevel gearwheel in the absence of the operating worm, it is necessary to determine somehow the assigned setting parameters in the first approximation. We propose the approach which is based on the initial fitting of the generating worm for the gearwheel (for instance, to be specific, with the left direction of threads) to the generating worm of the pinion (correspondingly, with the right direction) as it is schematically shown in Fig. 11.8c. The following conditions are fulfilled here:

(i). generating worms have the same diameters;
(ii). worm axes are parallel, their cylinders are contacting along the straight line;
(iii). the line of contact of pitch surfaces of worms and the line of contact of pitch surfaces of bevel gearwheels have the common point M_p which is the pitch point in each of the machine-tool meshing.[2]

Let us determine the first of the sought parameters (machine-tool interaxial angle Σ_{02}) by unit vectors of axes $\mathbf{k}_{02}$, $\mathbf{k}_2$—they are unit vectors of axes z_{02} of the generating worm and z_2 of the gearwheel, correspondingly:

$$\Sigma_{02} = 180 - \arccos(\mathbf{k}_{02}\mathbf{k}_2), \qquad (11.30)$$

Unit vectors $\mathbf{k}_{02}$ and $\mathbf{k}_{01}$ coincide by the condition *ii*:

---

[2] As a rule, the pitch (M_p – Fig. 11.8) and the design (M – Fig. 11.7) points do not coincide.

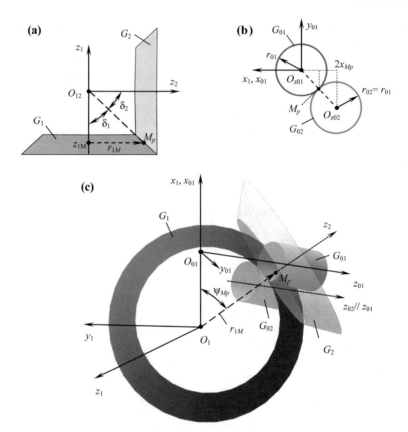

Fig. 11.8 Contact of pitch surfaces at the point M_p: **a** contact of pitch surfaces of bevel gearwheels; **b** contact of pitch surfaces of generating worms; **c** mutual arrangement and contact of four pitch surfaces

$$\mathbf{k}_{02} = \mathbf{k}_{01} : \begin{cases} k_{02x} = 0, \\ k_{02y} = -\sin \Sigma_{01}, \\ k_{02z} = \cos \Sigma_{01}. \end{cases} \tag{11.31}$$

The unit vector $\mathbf{k}_2$ in the coordinate system S_1 is as follows:

$$\mathbf{k}_2 : \begin{cases} k_{2x} = \cos \psi_{Mp} \sin \Sigma_{12}, \\ k_{2y} = -\sin \psi_{Mp} \sin \Sigma_{12}, \\ k_{2z} = \cos \Sigma_{12}. \end{cases} \tag{11.32}$$

where ψ_{Mp} is the angle between the plane $x_1 O_1 z_1$ and the axis z_2 of the gearwheel.

Let us determine this angle by choosing the point M_p on the surface G_{01} (Fig. 11.8) and keeping in mind that this point should belong to the axis of the machine-tool meshing of the pinion [5] determined by equations:

$$\begin{cases} x_{011,11}^2 + x_{011,11}[a_{w01} - p_{\gamma01}((z_1/z_{01})\,\mathrm{cosec}\,\Sigma_{01} - \cot\Sigma_{01})] + a_{w01}\,p_{\gamma01}\cot\Sigma_{01} = 0, \\ x_{01}z_{01} + y_{01}a_{w01}\cot\Sigma_{01} = 0. \end{cases}$$

(11.33)

Solving (11.30) with equations of the pitch surface (the first equation of the system (11.7)), we get:

$$\begin{cases} x_{01Mp} = \dfrac{-[a_{w01} - p_{\gamma01}((z_{(1)}/z_{(01)})\mathrm{cosec}\,\Sigma_{01} - \cot\Sigma_{01})] - \sqrt{[a_{w01} - p_{\gamma01}((z_{(1)}/z_{(01)})\,\mathrm{cosec}\,\Sigma_{01} - \cot\Sigma_{01})]^2 - 4a_{w01}\,p_{\gamma01}\cot\Sigma_{01}}}{2}, \\[2mm] y_{01Mp} = \sqrt{r_{01}^2 - x_{01Mp}^2}, \\[2mm] z_{01Mp} = \dfrac{-y_{01Mp}a_{w01}\cot\Sigma_{01}}{x_{01Mp}}. \end{cases}$$

(11.34)

Using the matrix (11.6) one can turn to the coordinate system S_1 and determine the angle ψ_{Mp}:,

$$\psi_{Mp} = \arctan\left(-\frac{y_{1Mp}}{x_{1Mp}}\right).$$

(11.35)

The first approximation for the second targeted parameter—the axial module m_{x02}—can be determined presuming that M_p is the pitch point in the machine-tool meshing of the gearwheel (condition *iii*); and at this point the vector of the relative velocity in the machine-tool meshing of the gearwheel should become collinear with the tangent line to the helical line of the generating worm for the gearwheel. It is customary to call the corresponding axial module the ideal one [2]:

$$m_{x02} = m_{x02id}(M_p) = \frac{r^2(x_{02Mp} + a_{w02})\sin\Sigma_{02}}{z_{(2)}(r_{Mp}^2((z_{(2)}/z_{(02)}) - \cos\Sigma_{02}) - z_{02Mp}y_{02Mp}\sin\Sigma_{02} - x_{02Mp}a_{w02}\cos\Sigma_{02}},$$

(11.36)

where x_{02Mp}, y_{02Mp}, z_{02Mp} are coordinates in the system S_{02} for the point M_p which we determined by (11.34) and (11.35) in systems S_{01} and S_1. Axes z_2 and z_{02} contain our known unit vectors ($\mathbf{k}_2$, $\mathbf{k}_{02}$, correspondingly); it is also of little difficulty to determine the radius-vectors $\mathbf{r}_{k2}$ and $\mathbf{r}_{k02}$ of points through which these axes are passing (Fig. 11.8a, b):

$$\mathbf{r}_{k2} = \overline{O_1O_2} : \begin{cases} r_{k2x} = 0, \\ r_{k2y} = 0, \\ r_{k2z} = z_{1M} + r_{1M}\cot(\pi/2 + \delta_2). \end{cases}$$

(11.37)

$$\mathbf{r}_{k02} : \begin{cases} r_{k02x} = 2x_M + a_{w01}, \\ r_{k02y} = 2y_M\cos\Sigma_{01}, \\ r_{k02z} = 2y_M\sin\Sigma_{01}. \end{cases}$$

(11.38)

Therefore, arrangement of axes of all elements should be considered completely determined; and derivation of the transformation formulas $S_1 \rightarrow S_{02}$ that we need to calculate (11.35) the first approximation of the module m_{x02} can be omitted due to its trivialness.

11.10 Several Results and Features of Calcualtions

The described scheme of calculation and the mathematical model are implemented in the program modules of the software SPDIAL + [17]. The main results to assess the design solutions are modification fields (p. 5 of the considered above general scheme of calculation).

During the first numerical investigation we limited ourselves by orthogonal gears with the gear ratio 1. The Table 11.1 presents some obtained results.

The calculation procedure and the obtained results give ground for the following conclusions.

1. The process of choosing the parameters is iterative and it always implies correction of two machine-tool meshing: both for the pinion and gearwheel.
2. The areas in the space of parameters where solutions (11.26) exist simultaneously for different name flanks are very narrow; and the acceptable ranges of parameters for which it is possible to get the longitudinal localization of contact

Table 11.1 Basic parameters of the considered gears and modification fields (Calculations are made by eng. T. A. Pushkareva)

Parameter	No of the gear		
	1	2	3
$z_{01}: z_1/z_{02}: z_2$	1:36/1:36	2:36/2:36	3:38/3:38
z_{2id}^{a}	36.18	35.84	37.99
d_{i1}/d_{i2}	90.0/96.2	100.0/106.7	108.0/115.6
d_{e1}/d_{e2}	117.0/119.4	130.0/132.1	140.4/143.1
m_{x01} /m_{x02}	1.46/1.46	1.75/1.75	2.0/2.0
α_{n01} /α_{n02}	20.5/20.4	21.0/20.9	21.0/21.0
d_{a01}/d_{a02}	62.0/66.1	66.0/65.0	63.0/68.5
a_{w01}/a_{w02}	45.0/45.8	50.0/49.9	54.0/56.3
Σ_{01}/Σ_{02}	128.5/128.3	128.0/128.8	127.5/128.1
Modification field (L)			
Modification field (R)			

[a] z_{2id} is the ideal number of gearwheel teeth calculated by angles of reference cones

at least in the small area are even narrower. It forces to search for solution by means of extremely small steps of variation of the assigned parameters. Thus, it is unacceptable to change separately the module m_{x02} relative to values shown in the Table 11.1 even by 0.005...0.01 mm or separately the angle Σ_{02} by 0.2°...0.5°.

3. Profile angles for different name flanks of generating worms should be achieved to be made equal or almost equal. Unlike the orthogonal spiroid gearing, it is possible, since the limiting profile angles determined by elimination of undercut for the concave tooth flank are abruptly decreased at machine-tool interaxial angles considerably exceeding 90°.

4. Profile localization of contact is provided exceptionally only by assigning the necessary curvatures of worm profiles.

5. The main difficulties of calculation are related to providing the longitudinal localization of contact. To our opinion, this feature is the continuation of the abrupt decrease in the risk of undercut in the non-orthogonal gearing. Approximation of undercut within variation of parameters promotes the variation of longitudinal curvature of the tooth and, therefore, essentially changes the pattern of modifications. The main influential parameters here are diameters, profile angles and axial modules of generating worms. They should be chosen in such a way that the undercut could be close by or even (in most of our considered cases) present to some (small) extent. Localization on the convex tooth flank of the gearwheel is provided by approximation or admission of undercut of the concave flank of the pinion; and, vice versa, localization on the convex tooth flank of the pinion is provided by approximation or admission of undercut of the concave flank of the gearwheel.

6. The favorable pattern of longitudinal modifications is promoted by the choice of modules of generating worms equal to or a little greater than the value of ideal modules for mean face sections of reference cylinders of worms.

7. For each of generating worms the profile that produces the convex tooth flank in the longitudinal direction should be made concave; and the profile that produces the concave tooth flank in the longitudinal direction should be made straight or convex. It is easier here to reach the compromise between the undercut (dimensions and the shape of the undercut area) and the correct localization of contact.

8. When iterations result in approaching the correct solution, the pattern of tooth modification has the form of a "fish"; the main drawback here is the minimum modifications at the edge—"fish tail" (its residuals can be seen on modification fields shown in Table 11.1). Therefore, the design points (centers of the localized contact) should be shifted to the tooth toe.

9. Due to relatively large transient and undercut areas at the tooth toe, the design point should be shifted to the point on the concave (subjected to undercut) tooth flank of the gearwheel and to the root (corresponding to point of the concave flank of the pinion) on the convex flank.

10. When the gear ratio is 1, single-type and "mirror-like" modification fields of different name flanks should be achieved; it minimizes the tooth undercut and

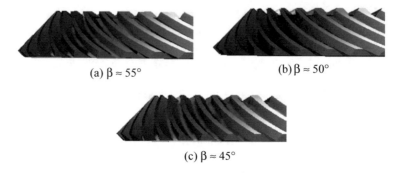

(a) β ≈ 55°   (b) β ≈ 50°

(c) β ≈ 45°

Fig. 11.9 Pinion teeth for gears corresponding to numbers in Table 11.1: **a** N1; **b** N2; **c** N3

    provides approximately equal parameters of gear-cutting tools for the pinion and gearwheel.

11. Apparently, the proposed generation method is possible for bevel gears with the gear ratio not too much different from 1. We managed to get the solutions within the range of 1.0…1.4. The possibility to widen the range requires an additional investigation. Moreover, the upper limit of the range is determined by the limit of rotation of the gear-milling machine-tool support stand.

12. Gearwheels generated by the proposed method have a little larger helix angle β (45°…55°) as compared to traditional bevel gearwheels with curvilinear teeth (Fig. 11.9). The decrease in the angle β is promoted by the increase of the number of threads of generating worms and decrease in the ratio of the gearwheel diameter and machine-tool interaxial distance (Fig. 11.9).

11.11   Conclusion

The first calculation results for bevel gears with curvilinear teeth produced by helicoid generating surfaces give ground to state that the cutting process and correct meshing of such teeth is possible. The first experience of cutting pairs of the gearbox for changing the direction required about one month of preparation and the cutting time for one gearwheel having the diameter 63 mm by the high-speed steel flying cutter was about 20 min. Therefore, the proposed generation method has, to our opinion, the prospects of implementation first of all in the low-series production, since it implies application of the common and cheap equipment (universal gear-milling machine-tools) and relatively simple in production gear-cutting tools (hobs or flying cutters).

    The development of this method is related to solving a number of problems:

– studying the possibilities of one-side cutting of one or each of the gearwheels generated by two settings of one tool for opposite tooth flanks;

– studying the possibility of cutting and preparation of production by means of assembled running cutter heads [15] that reproduce multi-thread generating worms.
– studying the possibilities of cutting the gearwheels with non-orthogonal bevel gears, including those with small interaxial angles [11] by the proposed method.

References

1. Babichev, D.T.: Development of kinematic method of theory of gearing to determine areas of tooth flanks produced by jogs of generating solids. In: Honor of Professor Faydor, L., Litvin, V., Goldfarb, N., Barmina (eds.) Chapter in Theory and Practice of Gearing and Transmissions, vol. 34, pp. 159–188. Springer, Berlin (2016)
2. Goldfarb, V.I., Ezerskaya, S.V.: To the choice of the value of the helical parameter in the orthogonal spiroid gear with the cylindrical worm. J. Izvestiya Vuzov, Mashinostroyeniye **2**, 184–186 (1975). (in Russian)
3. Goldfarb, V.I., Russkikh, A.G.: Computer-aided synthesis of gear structure at arbitrary arrangement of axes. In: Proceedings of 6th International Conference on the Theory of Machines and Mechanisms, pp. 65–70. Czechoslovakia (in Russian) (1992)
4. Goldfarb, V.I., Trubachev, E.S.: System of hob unification for cutting the gearwheels of worm-type gears. In: Proceedings of ASME 2007 International Design Engineering Technical Conferences and Computers and Information in Engineering Conference. Las-Vegas, USA (2007)
5. Korostelev, L.V., Baltadzhi, S.A., Lagutin, S.A.: Conjugated lines of meshing of general-type worm gears. J. Mashinovedeniye **5**, 49–56 (1978). (in Russian)
6. Lagutin, S.A.: Local synthesis of general type worm gearing and its applications. In: Proceedings of 4th World Congress on Gearing and Power Transmissions, vol.1, pp. 501–506. Paris (1999)
7. Litvin, F.L.: Theory of Gearing. Nauka (in Russian), Moscow (1968)
8. Lopato, G.A., Kabatov, N.F., Segal, M.G.: Bevel and Hypoid Gears with Circular Teeth. Mashinostroenie (in Russian), Leningrad (1977)
9. Segal, M.G.: To Determination of Boundaries of the Contact Pattern of Bevel and Hypoid Gears, vol. 4, pp. 61–68. Moscow, Mashinovedeniye (in Russian) (1972)
10. Sheveleva, G.I.: Theory of Generation and Contact of Moving Solids. Moscow, Stankin (in Russian) (1999)
11. Syzrantsev, V., Kotlikova, V.: Mathematical and program provision of design of bevel gearing with small shaft angle. In: Proceedings of International Conference on Gearing, Transmissions and Mechanical Systems, pp. 13–18. Nottingham (2000)
12. Trubachev, E., Savelyeva, T., Pushkareva T.: Practice of design and production of worm gears with localized contact. In: Goldfarb, V., Trubachev, E., Barmina, N. (eds.) Chapter in Advanced Gear Engineering. Mechanisms and Machine Science, vol. 51, pp. 327–344. Springer, Berlin (2018)
13. Trubachev, E.S.: Vector field of normal vectors and its application to investigation of geometry of spiroid gearing with the helicoid worm. In: Chapter in Problems of Design of Mechanical Engineering and Information Technology Products, pp. 3–14. Izhevsk, 3–14 (in Russian) (1999)
14. Trubachev, E.S.: Method of parameters calculation for the machine-tool gearing with the helicoid generating surface. In: Chapter in Advanced Information Techniques. Problems of Investigation, Design and Production of Gears, pp. 163–169. Izhevsk (in Russian) (2001)
15. Trubachev, E.S.: New Possibilities of Gear Cutting by Rolling Cutter Heads. In this book (2020)

16. Trubachev, E.S., Beresneva, A.V., Monakov, A.V.: Calculation of point coordinates of the surface generated by a helicoid. In: Proceedings of the Scientific Seminar of the Educational Scientific Center of Gears and Gearbox Engineering, Problems of Improving Gears, pp. 124–134. Izhevsk-Moscow (in Russian) (2000)
17. Trubachev, E.S., Oreshin, A.V.: CAD system for spiroid gears. J. Inform. Math. **1**(3), 159–165 (2003). M.: Publishers of physical mathematical literature (in Russian)
18. Trubachev, E.S., Savelyeva, T.V.: Optimization task at computer-aided design of spiroid gears based on single-thread unified hobs. J. Inform. Math. **1**(5), 121–130 (2005). M.: Publishers of physical mathematical literature (in Russian)

Chapter 12
New Possibilities of Tooth Cutting by Running Cutter Heads

Evgenii S. Trubachev

Abstract The traditional method of worm and spiroid gearwheel tooth machining is both convenient and understandable for manufacturers and hard-to-improve with regard to increasing its production efficiency, control of contact localization, terms and cost of production preparation, and limitations of making gears with the relatively small gear ratios. The manuscript considers the possible alternative—tooth cutting by multi-tooth flying (running) cutter that is proposed to be made assembled, equipped with carbide inserts and called the running cutter head. In order to provide this possibility, it is proposed to calculate the machine-tool setting by the method well acknowledged to be applied for designing machine-tool settings for hobs with the small number of threads. It is noted, that the most general relative arrangement of the machine-tool and operating meshes is used in this method; it gives possibility to change parameters and the level of surface modification within a wide range. In order to provide the application of the multi-cutter running head, it is proposed to increase the number of threads of the generating worm as compared to the gear worm and increase the tool diameter. The performed numerical calculations showed that for the case of the worm gear this approach can be applied with minimum limitations; and for the case of the spiroid one the limitations are unfavorable tooth modifications and tooth undercut. To overcome the first limitations, it is proposed to apply the tool feed with the variable velocity. It is easily arranged for CNC machine-tools; and it allows for applying the method not only for multi-thread gears, but for the single-thread ones (therefore, for any type). The second limitation can be lowered by choosing the parameters of the conjugate gear with a certain reserve for the undercut. Due to a number of advantages, the proposed method can become competitive, especially, for worm-type gears with the relatively small gear ratios and for bevel gears produced in low series.

Keywords Worm gear · Spiroid gear · Tooth cutting · Cutter head · Contact localization

E. S. Trubachev (✉)
Institute of Mechanics, Kalashnikov Izhevsk State Technical University, Studencheskaya str. 7, P.O. Box 426069, Izhevsk, Russia
e-mail: truba@istu.ru

© Springer Nature Switzerland AG 2020
V. Goldfarb et al. (eds.), *New Approaches to Gear Design and Production*,
Mechanisms and Machine Science 81,
https://doi.org/10.1007/978-3-030-34945-5_12

12.1   Introduction

Tooth hobbing became such a customary technique of machining of worm and spiroid gearwheels that it is considered to be almost the only method of at least primary (and most commonly final) cutting. Moreover, practically anytime when the manufacturer is aimed at producing a worm or spiroid gear, his first thought is often: "Do I have (or somewhere around) the necessary hob and if no, where can I order or make it and how much will it cost?". By the way, in most practical cases (of course, when we talk about a small series or a single case) the absence of the hob and impossibility to get it in some way is the reason to ask another manufacturer to produce a gear (though this version is not always simply implementable, especially for gears with specific parameters) or worse, to refuse from the technical solution or interesting order. Of course, other methods are known that are applied for a relatively long time—cutting by a flying cutter, stamping, pressing and casting and newly developed ones. They are somehow related to special conditions of production:

- single production—cutting by a flying cutter;
- mass production of polymeric or powder gearwheels—stamping, pressing and casting;
- relatively low-loaded gears—casting of polymeric or cast-iron gearwheels (comparatively accurate for the method of casting but inaccurate for the gear).

The method of localized contact analysis firstly proposed in [7, 8] allowed for defusing a problem. It permitted to relatively simply and quickly apply one hob—the tool that requires a special technique of production and operation—for cutting several worm or spiroid gears differing essentially from each other by dimensions and gear ratios [4, 11], providing at the same time the contact localization both in the height (profile) and longitudinal direction of the tooth. The main practical benefits achieved by this possibility are as follows:

- practically one-order decrease in the range of the tool in the multi-range production. Here, only 8 dimensions of one- and two-thread spiroid hobs provide the production of 15 dimension types of spiroid gears with more than 70 strongly different gear ratios—from 6 to 90;
- simplification of spiroid and worm hobs due to decreasing the number of their threads (the technique of production and operation of multi-thread hobs is very sensitive to the action of manufacturing errors);
- reduction of gear sensitivity to the action of errors—correspondingly, the possibility of a certain widening of manufacturing tolerances.

In the other implementation version, the technique allows for selecting the available hob for production of necessary gear pairs; here we deal with two or even several possible versions—available hobs. However, there still exists the principal tooth machining technique—worm gear hobbing—along with its relatively low production efficiency of tooth machining and the relatively expensive, time consuming, and specialized process of hob production. Moreover, for gear ratios under 8 (that are small for worm and spiroid gears) the situation is common when it is impossible to avoid applying inconvenient in production and operation hobs with the number of threads equal to 3 and above.

The present paper considers the principal question: is it possible to develop the technique of worm and spiroid gearwheel cutting that is competitive to the traditional tooth hobbing; or, to be specific, can the cutting process by a flying (running) cutter become such a competitive technique; and if yes, under what conditions? The considered method is known for a long time; and it has rather limited application mainly due to the low production efficiency. The proposed in the manuscript improvements give ground not only to overcome this disadvantage, but also to exclude the drawbacks of traditional tooth hobbing my means of hobs. Implementation of this method is urgent for design and series production of spiroid and worm gearwheels and also, perhaps, for low series and single production of bevel gears.

12.2   Tooth Machining Technique by Means of a Flying (Running) Cutter and Its Possible Development

The technique of worm and spiroid gearwheel tooth cutting by a flying cutter is traditionally applied in a single production by combining relatively low time costs and other resources for production preparation but, at the same time, providing a relatively low production efficiency. Even a visual comparison of this method (Fig. 12.1b) with the traditional one that is applied in series production (tooth hobbing by a hob—Fig. 12.1a) shows evidently the reasons of the following features:

– the flying cutter is an essentially simpler tool, where its mandrel part is constant for several versions of the thread geometry of worms having similar dimensions and its cutting part is changeable and produced when necessary. In this case, the hob is significantly more complicated and it is principally unchangeable in operation (not accounting for a slight variation of dimensions at resharpening);
– the flying cutter represents in its essence one cutting tooth (as a rule, the hob has several dozens of teeth); the load on it and its total pass of cutting are relatively big, thus forcing to relieve the cutting mode;

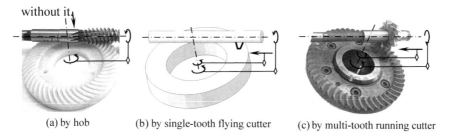

(a) by hob   (b) by single-tooth flying cutter   (c) by multi-tooth running cutter

Fig. 12.1   Traditional methods of edge machining of worm and spiroid gearwheel teeth

– the tool travel of the flying cutter (a little greater than the length of the corresponding generating worm including its several axial pitches—$n\pi m_{x0}$) within tooth
  cutting is considerably greater than the length of the cutting pass of the hob (equal
  or even equal to the profile height—usually 2.0–$2.5m_{x0}$).

In case of multi-thread gears with relatively small gear ratios (this case should
be considered as a special one, first of all, from the production point of view) the
situation is different:

– the flying (running) cutter becomes the multi-tooth one (Fig. 12.1c) [2];
– its structural layout is abruptly complicated, its production time becomes comparable with that of the hob (at least, single-thread one).

Therefore, one drawback is defused, but at the same time, its virtually unique
advantage is defused. These two negative trends can be overcome if the cutter is
made assembled and equipped with carbide inserts. In this case, to our opinion, it is
more correct to talk not on the cutter, but on a specific type of a cutting head, further
called the running cutter head. Of course, the assembled structure of a tooth cutting
tool, including the assembled layout of hobs, is not something peculiar in its essence.
However, note that the flying (running) cutter is itself more convenient (production
efficient) for implementation of this idea—at least, at its greater dimensions—its
conditions of adjacency and machining of seats for cutters-threads are not so strict
as for the hob.

The problem that is the first and determining the principal possibility of generation is the issue of obtaining the required tooth geometry that provides the contact
localization in the gear along with the absence (or, at least, the allowable low degree)
of undercut and sharpening of teeth. Let us show the possibility of positive solution
of this issue by applying the mentioned above method of calculation of parameters
for the setting with the helicoid generating surface. Our task also implies the solution
not only for multi-thread gears in which the comparatively big number of teeth of
the running head is practically natural, but also for single-thread ones.

12.3   Method of Calculation of the Setting Parameters and the Idea of Its New Application

The method of calculation of the setting parameters for the case of tooth generation
by means of a helicoid generating surface firstly proposed in [7, 8] found rather wide
application for tooth cutting of spiroid [4], worm [11], and lately bevel [9] gearwheels
by hobs. This method became so common in practice of design and production of
spiroid gears that it is impossible to imagine the efficient multi-product manufacture
of spiroid gearboxes for pipeline valves [3] without it.

Let us give the description and characteristics of possibilities of this method. The
common meshing of three elements is considered at the assigned design point M:
gearwheel (the subscript is 2) and two worms—operating (worm of the gear, the

subscript is 1) and generating (the subscript is 0). The scheme of arrangement of these elements and the corresponding coordinate systems are shown in Fig. 12.1 (the generating worm is conditionally not shown not to overload the figure). Equations of helical surfaces of worms generated by the helical motion of axial profiles $f_{1,2}(r_{1,2})$ are:

$$\begin{cases} x_{0,1} = r_{0,1}\cos(\vartheta_{0,1} + \varphi_{0,1}), \\ y_{0,1} = r_{0,1}\sin(\vartheta_{0,1} + \varphi_{0,1}), \\ z_{0,1} = f_{0,1}(r_{0,1}) + p_{\gamma 0,1}\vartheta_{0,1}, \end{cases} \quad (12.1)$$

where $r_{0,1}$, $\theta_{0,1}$ are parameters—curvilinear coordinates of the helical surface (the radius and polar angle), $p_{\gamma 0,1} = 0.5 m_{x0,1} z_{(0),(1)}$ is the helix parameter ($m_{x0,1}$, $z_{(0),(1)}$ are the axial module and the number of worm threads).

Normal vectors **n** to surfaces of worms are:

$$\mathbf{n}_{0,1} : \begin{cases} n_{0,1x} = -y_{0,1}p_{\gamma 0,1} + x_{0,1}r_{0,1}\tan\alpha_{x0,1}, \\ n_{0,1y} = x_{0,1}p_{\gamma 0,1} + y_{0,1}r_{0,1}\tan\alpha_{x0,1}, \\ n_{0,1z} = -r_{0,1}^2, \end{cases} \quad (12.2)$$

where $\tan\alpha_{x0} = f'_{0,1}(r_{0,1})$.

Two properties of normal vectors (12.2) to helicoid surfaces are used in calculations, the properties being invariant with regard to the choice of the coordinate system [7]:

$$\tan\gamma = -\frac{\mathbf{ne}_t}{\mathbf{nk}}, \quad (12.3)$$

$$\tan\alpha_x = -\frac{\mathbf{ne}_r}{\mathbf{nk}}, \quad (12.4)$$

where γ is the helix angle, $\mathbf{e}_t$, $\mathbf{e}_r$ and $\mathbf{k}$ are unit vectors of the tangential, radial and axial directions with regard to the axis of the helical surface.

The equation of meshing of each of the worms with the gearwheel (the gear worm with the conjugated one and the generating worm with the formed modified one) is:

$$\mathbf{n}_{0,1}\mathbf{v}_{02,12} = r_{0,1}^2 p_{\gamma 0,1}\left(i_{02,12}\cos ec\ \Sigma_{0,1} - \cot\ \Sigma_{0,1}\right) + p_{\gamma 0,1}\left(y_{0,1}z_{0,1} - x_{0,1}a_{w0,1}\cot\ \Sigma_{0,1}\right)$$
$$- \left(x_{0,1}z_{0,1} + y_{0,1}a_{w0,1}\cot\ \Sigma_{0,1}\right)r_{0,1}\tan\alpha_{x0,1} - r_{0,1}^2\left(x_{0,1} + a_{w0,1}\right) = 0, \quad (12.5)$$

where $\Sigma_{0,1}$ is the interaxial angle; $a_{w0,1}$ is the centre distance, $i_{02,12} = z_{(2)}/z_{(0),(1)}$ is the gear ratio ($z_{(2)}$ is the number of gearwheel teeth).

When designing a worm-type gear, the synthesis of the conjugated gearing is preceding the synthesis of the machine-tool gearing of the gearwheel; that is, in its essence, the position of the generating worm axis should be aligned to the assigned conjugated operating gearing. Besides the mentioned unknown parameters Σ_0 and a_{w0} there are two more parameters of such an "alignment" (Fig. 12.2): linear Δz_2 and

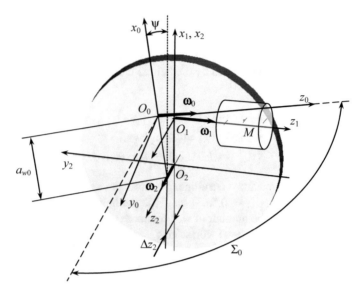

Fig. 12.2 Relative arrangement of coordinate systems of the conjugate and machine-tool meshing for the case of spiroid gear

angular ψ displacements of the machine-tool interaxial line relative to the interaxial line of the gear along and around the gearwheel axis. As it is known, the straight line (in this case it is the axis of the generating worm) can by unambiguously assigned by four independent parameters. Therefore, the considered arrangement of the machine-tool gearing relative to the gear and the introduced set of parameters (at least their number) are, probably, the most common.

The transform matrix of coordinate systems $S_0(x_0, y_0, z_0)$ and $S_1(x_1, y_1, z_1)$ is as follows:

$M_{10} =$

$$
\begin{bmatrix}
\cos\psi & -\cos\Sigma_1\sin\psi & \sin\Sigma_1\sin\psi & a_{w1}\cos\psi - a_{w3} \\
\cos\Sigma_0\sin\psi & \cos\Sigma_1\cos\Sigma_0\cos\psi + \sin\Sigma_1\sin\Sigma_0 & -\sin\Sigma_1\cos\Sigma_0\cos\psi + \cos\Sigma_1\sin\Sigma_0 & a_{w1}\cos\Sigma_0\sin\psi - \Delta z_2\sin\Sigma_0 \\
-\sin\Sigma_0\sin\psi & -\cos\Sigma_1\sin\Sigma_0\cos\psi + \sin\Sigma_1\cos\Sigma_0 & \sin\Sigma_1\sin\Sigma_0\cos\psi + \cos\Sigma_1\cos\Sigma_0 & -a_{w1}\sin\Sigma_0\sin\psi - \Delta z_2\cos\Sigma_0 \\
0 & 0 & 0 & 1
\end{bmatrix}
$$

$$(12.6)$$

The choice of parameters Σ_0, a_{w0} Δz_2 and ψ is not arbitrary; it should comply with conditions (12.3) and (12.5). In accordance with Lagutin [5] and taking into account that at common contact points $M_{R,L}$ $\mathbf{n}_1 = \mathbf{n}_1 = \mathbf{n}_2$ и $\mathbf{n}_0\mathbf{v}_0 = \mathbf{n}_1\mathbf{v}_1 = \mathbf{n}_2\mathbf{v}_2$ (note, that design points are assigned in the known conjugate meshing and, therefore, contact normal vectors $\mathbf{n}_1$ are known), we will get these conditions in the convenient for calculations form [7, 8]. To be exact, we will get according to (12.3):

$$
\tan\gamma_{0R,L} + \frac{\mathbf{n}_{1R,L}\mathbf{e}_{t0R,L}}{\mathbf{n}_{1R,L}\mathbf{k}_{0R,L}} = \frac{p_{y0}}{r_{0R,L}} + \frac{\mathbf{n}_{1R,L}\mathbf{v}_{0R,L}}{r_{0R,L}\omega_0\mathbf{n}_{1R,L}\mathbf{k}_{0R,L}}
$$

$$= \left(\frac{p_{\gamma 0}}{z_{(0)}} + \frac{\mathbf{n}_{1R,L} \mathbf{v}_{1R,L}}{z_{(1)} \mathbf{n}_{1R,L} \mathbf{k}_{0R,L}} \right) \frac{z_{(0)}}{r_{0R,L}} = 0,$$

or, after substitutions of the known projections of vectors in the coordinate system S_1 and the solution with respect to the unknown $\sin \psi$:

$$A_{1R,L} \sin^2 \psi_{R,L} + B_{1R,L} \sin \psi_{R,L} + C_{R,L} = 0, \tag{12.7}$$

where

$$A_{R,L} = a_{R,L}^2 + b_{R,L}^2, \ B_{R,L} = 2a_{R,L} c_{R,L}, \ C_{R,L} = c_{R,L}^2 - b_{R,L}^2,$$

$$a_{R,L} = -n_{x1R,L} \sin \Sigma_0,$$

$$b_{R,L} = (n_{z1R,L} \sin \Sigma_1 - n_{y1R,L} \cos \Sigma_1) \sin \Sigma_0,$$

$$c_{R,L} = (n_{y1R,L} \sin \Sigma_1 - n_{z1R,L} \cos \Sigma_1) \cos \Sigma_0 + r_{0R,L}^2 \frac{p_{\gamma 1}}{p_{\gamma 0}} \frac{z_{(0)}}{z_{(1)}}.$$

We will obtain from (12.5):

$$\mathbf{n}_{1R,L} (\mathbf{v}_{1R,L} - \mathbf{v}_{2R,L}) = \mathbf{n}_{1R,L} (\mathbf{v}_{1R,L} - \mathbf{v}_{0R,L}) = A_{R,L} a_{w0} + B_{R,L} \Delta z_2 + C_{R,L} = 0 \tag{12.8}$$

where

$$A_{R,L} = n_{x1R,L} \cot \Sigma_0 \sin \psi_{R,L} + n_{y1R,L} (\sin \Sigma_1 + \cos \Sigma_1 \cot \Sigma_0 \cos \psi_{R,L})$$
$$+ n_{z1R,L} (\cos \Sigma_1 - \sin \Sigma_1 \cot \Sigma_0 \cos \psi_{R,L})$$

$$B_{R,L} = -n_{x1R,L} \cos \psi_{R,L} + n_{y1R,L} \cos \Sigma_1 \sin \psi_{R,L} - n_{z1R,L} \sin \Sigma_1 \sin \psi_{R,L}$$

$$C_{R,L} = -n_{z1R,L} [p_{\gamma 1}(z_{(0)}/z_{(1)}) \operatorname{cosec} \Sigma_0 - y_{1R,L} \sin \psi_{R,L}]$$
$$+ (n_{z1R,L} p_{\gamma 1} - n_{y1R,L} a_{w1})(\sin \Sigma_1 \cos \psi_{R,L} + \cos \Sigma_1 \cot \Sigma_0)$$
$$+ [n_{x1R,L} z_{1R,L} - n_{z1R,L}(x_{1R,L} + a_{w1})](\cos \Sigma_1 \cos \psi_{R,L} - \sin \Sigma_1 \cot \Sigma_0)$$
$$- n_{y1R,L} z_{1R,L} \sin \psi_{R,L}$$

In the latter expressions the subscripts R and L are related to elements of right and left tooth flanks, correspondingly, and their meshing (for a traditional worm gear with the symmetrical profile this subdivision is more often nominal one; while the spiroid gearing is asymmetrical in all its respects along with the evident exterior difference: right flanks are convex tooth surfaces and left flanks are concave ones).

In case of two-side cutting of teeth the set of the sought parameters includes:

– five parameters characterizing the arrangement of the generating worm with respect to the gear in the common meshing: Σ_0, a_{w0}, Δz_2 are common

for two opposite flanks and two parameters ψ_R and ψ_L are usually different, since the design points M_R and M_L in the general case come into contact at different phases of both operating (φ_0) and machine-tool (φ_1) meshing $\left(\varphi_{0R,L}z_{(0)} - \varphi_{1R,L}z_{(1)} = \psi_{R,L}z_{(2)}\right)$;
– parameters of the generating worm $z_{(0)}$, m_{x0}.

Therefore, there is the sum total of 7 unknown parameters. Conditions (12.7), (12.8) stated in pairs for opposite flanks (4 equations in total) are establishing the relations between parameters, thus reducing the number of unknown ones—those that can be controlled to influence the geometry of the generated teeth. It is proposed in the developed algorithm [8] to consider parameters Σ_0, $z_{(0)}$, m_{x0} as the independent variable ones (assigned within the search of the solution) and determine the rest parameters by calculations of (12.7), (12.8). Moreover, the modification pattern can be varied by a certain correction of gear parameters and (or) changing the position of design points [11]. The set of independent variable parameters should also include radii $\rho_{x0R,L}$ of the axial profile of the generating worm that influence mainly the level of profile (height) contact localization. The solution of (12.7), (12.8) completely determines the position of coordinate systems S_1 and S_0 and, therefore, allows for determining the coordinates of points $M_{R,L}$ in S_0, the diameter d_0 of the generating worm and the angles $\alpha_{x0R,L}$ of its axial profile in accordance with (12.4).

Thus, due to the mentioned generality of the problem statement, there are a lot of parameters for controlling the contact localization—sufficient to find the solution for very different cases. Summarizing the description of the mathematical model, let us enumerate these parameters:

– 5 independent parameters of synthesis of the machine-tool setting Σ_0, $z_{(0)}$, m_{x0}, ρ_{x0R}, ρ_{x0L};
– 4 coordinates of design points $M_{R,L}$—two for each point: along the height and length of the tooth;
– parameters of the conjugated gear that can also be corrected in the iterative mode (usually they are: $z_{(2)}$, d_1, m_{x1}, α_{x1R}, α_{x1L}, $\rho_{x1R,L}$).

The customary and practically appropriate method is the reduction of the setting calculation to single- and double-thread generating worms that differ from multi-thread ones by their simplicity and cheapness of their implementation as the corresponding hobs. However, the practice of application of the method for worm and spiroid gearwheel production showed certain drawbacks and limitations of focusing on low-thread hobs [4, 11]:

– at relatively high gear ratios (the conditional boundary is 15) there is a challenging problem of controlling the level of longitudinal modification of right (convex) flanks of spiroid gearwheel teeth;
– better contact localization is usually provided at a certain increase in the diameter of the multi-thread worm as compared to the diameter of the low-thread hob and the decrease in the axial diameter of the worm as compared to the traditionally recommended one; all this totaled is limiting the possibilities of increasing the gear efficiency and reduction of forces acting in meshing.

The latter problem can be overcome by applying the multi-tooth running cutter head with the number of cutters **greater** (at least not less in case of smaller gear ratios) **than the number of the gear worm threads**. Actually, it is the new proposal which is opposite to the mentioned above traditional approach of decreasing the number of threads of the generating worm. Let us show it by the example of three-thread worm and spiroid gears for which the single-thread and six-thread worms are taken as the generating tool. The basic data obtained within the analysis (this analysis and all the consequent ones are performed by means of the software "SPDIAL+" [10]) is shown in Table 12.1.

Of course, the presented combinations of parameters are not the only ones; a similar or another correct modification can be obtained for other parameters of the gear and machine-tool settings. However, examples show a rather specific relation of diameters of the gear worm and generating worm (highlighted in bold and underlined in Table 12.1). Evidently, at greater number of threads of the latter, its diameter and,

Table 12.1 Versions of gearwheel tooth machining parameters for different numbers of threads of the generating worm

Parameter		Gear	Machine-tool setting for gearwheel cutting	
			$z_{(0)} = 1$	$z_{(0)} = 6$
Spiroid gear	i	46:3	46:1	46:6
	a_w, mm	60	62.78	53.60
	Σ, °	90	91.90	89.10
	d, mm	**36.00**	**28.50**	**52.0**
	m_x, mm	3.000	2.970	3.090
Modification field [a]	R	–		
	L	–		
Worm gear	i	46:3	46:1	46:6
	a_w, mm	138,6	131.85	147.87
	Σ, °	90	82.40	99.50
	d, mm	**51.2**	**40.50**	**70/2**
	m_x, mm	5.00	4.740	5.550
Modification field*		–		

[a]in projection on the axial plane of the gearwheel

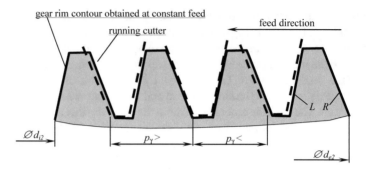

Fig. 12.3 The effect of application of the variable feed

therefore, the diameter of the running cutter head naturally become greater than the diameter of the gear worm. Correspondingly, it becomes possible to:

– choose a smaller diameter of the gear worm, thus increasing the gear efficiency;
– to simplify the structural layout of the assembled running head.

Obviously, in case of a spiroid gear that has unequal properties of meshing of opposite flanks the modifications of teeth at the increase in the number of generating worm threads are also different: on left flanks they lead to an extremely localized contact and on right flanks—to the edge contact at tooth faces. However, when the generating worm is reproduced at the helical motion of generating lines—cutting edges of cutters of the running cutter head—there is a rather simple (at least for CNC machines) possibility to implement the variable pitch of the generating worm by changing the tangential feed of the head during the tooth cutting. The effect of this approach is shown in Fig. 12.3 in a simplified way; it shows how modifications can be redistributed as required: to increase the fall on edges of right flanks and to decrease it on left flanks.

The coordinate z of points of the generating worm with the linearly variable pitch can be determined in accordance with the relation obtained similarly in Georgiev and Goldfarb [1] for the other interesting case—a spiroid gear with the worm having an ideally variable pitch:

$$z_0 = z_{0\text{ref}} + f_0(r_0) + \int_{\vartheta_{0\text{ref}}}^{\vartheta_0} \left(p_{\gamma 0\text{ref}} + k_p \vartheta_0\right) d\vartheta_0$$

$$= z_{0\text{ref}} + f_0(r_0) + p_{\gamma 0\text{ref}}(\vartheta_0 - \vartheta_{0\text{ref}}) + k_p \frac{(\vartheta_0 - \vartheta_{0\text{ref}})^2}{2} \qquad (12.9)$$

where $z_{0\text{ref}}$ is the coordinate of the medium (taken to be the reference one) face section of the worm in which the worm helical line has the helix parameter assigned at synthesis; the corresponding reference angular parameter of the helix surface is $p_{\gamma 0\text{ref}} = z_{(0)} m_{x0}/2; f_0(r_0)$ is the coordinate of the worm axial profile at the radius r of the considered point; k_p is the factor (velocity) of feed variation with variation of the angular parameter θ_0 of the helix surface. This expression describes approximately the generating surface of the variable pitch that is reproduced by cutting edges of the running cutter (the cutter profile in general does not belong to the axial plane); however, it can be implemented in practice (the error in the worst cases does not exceed 0.5 mcm) for calculation of coordinates of points of the generated tooth surfaces (flank and transient).

The following algorithm of actions is proposed when calculating the setting with the variable feed of the running cutter head.

1. Choice of the conjugated gear including determination of conjugated surfaces.
2. Choice of design points on conjugated tooth flanks (in the first approximation—in midpoints of these surfaces).
3. Choice of assigned parameters of the machine-tool setting (including the principally important for our case number of threads $z_{(0)}$ and m_{x0} (p_{y0})).
4. Calculation of the rest parameters of the machine-tool setting by (12.7), (12.8).
5. Assessment of local characteristics of the contact at design points. If the results are satisfactory—move to p. 6, otherwise return to p. 3 or p. 1.
6. Calculation of modified surface with application of (12.1), (12.5) at $p_{y0} = p_{y0ref}$ = const. Assessment of the modification field—deviations of the generated surface from the conjugate one which is called the theoretical reference surface in accordance with M. G. Segal [6]. If modification at opposite faces of the tooth is satisfactory (usually symmetrical)—move to p. 7, otherwise—return to p. 2.
7. Assignment of the variable feed (the factor k_p).
8. Calculation of modified surface with application of (12.1), (12.5), (12.9) at p_{y0} =var. Assessment of modifications at a variable feed. If the results are satisfactory—the end of the algorithm, otherwise—return to p. 7 or p. 2.

12.4 Examples of Gear Calculation for Multi-thread Running Cutter Heads

Table 12.2 presents the basic results of calculations for the gear of a series manufactured spiroid gearbox RZA-S-2000 [3] for strongly different gear ratios—8 and 46.

The main conclusions on the performed calculations are the following.

1. It is principally possible to provide the localized contact and control the level of localization for both single-thread and multi-thread spiroid gears by applying the multi-thread generating worm (multi-cutter running head) having the increased diameter.
2. The increase in the diameter of the running cutter head is limited by appearance of undercut of the concave tooth flank; that is why, when choosing the gear

Table 12.2 Calculation results for machine-tool settings for cutting of spiroid gearwheels by running cutter heads

1	i		46:1	48:6
2	a_{w1}, mm		60	39.88
3	d_1, mm		36.5	3.4
4	m_{x1}, mm		2.75	
5	a_{w0}, mm		51.58	59.0
6	Σ_0, °		88.80	89.10
7	d_0, mm		46.2	51.2
8	$z_{(0)}$		4	8
9	m_{x0}, mm		2.804…2.836	3.410…3.430
10	Modification field	R		
		L		

Table 12.3 Calculation results for machine-tool settings for cutting of worm gearwheels by running cutter heads

1	i	48:1	48:6
2	a_{w1}, mm	138.6	
3	d_1, mm	51.2	
4	m_{x1}, mm	4.6	4,7
5	a_{w0}, mm	151.8	141.2
6	Σ_0, °	99.00	97.50
7	d_0, mm	89.73	67.0
8	$z_{(0)}$	4	8
9	m_{x0}, mm	4.88	5.19
10	Modification field	run-in run-out	run-in run-out

parameters one should not admit the probability of this phenomenon by leaving the reserve for variation of parameters.

3. In order to obtain the localized contact on both flanks for the number of threads of the gear worm and generating worm not less than 4, the variable feed of the running cutter head should be practically always applied. For all the rest gears the application of the variable feed can also be desirable, since it allows for more effectively controlling the head diameter and the level of the longitudinal contact localization.

Table 12.3 presents the results of calculations for the worm cylindrical gear with the center distance 138.6 mm and similar gear ratios.

The main conclusion on calculations made for the worm gear coincides with the first enumerated above conclusion for the spiroid one—generation of tooth surfaces that provides the localized contact for the considered method is possible. Of course, due to the symmetry of a worm gear, application of the variable feed of the running cutter head is not so urgent, probably, except for certain special cases as gears with two pitches or gears with the asymmetric profiles that, similar to spur and helical ones, can provide the benefit in operation characteristics. In the worm gear there are greater possibilities of the increase in the generating worm diameter. However, they are limited by the risk of appearance of under-profiled segments of flanks that can not be covered by a simple increase in the height of the tool profile without its sharpening—Fig. 12.4.

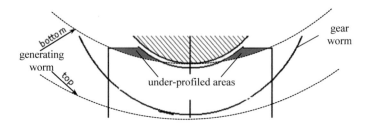

Fig. 12.4 Under-profiling of the worm gearwheel tooth at a big difference in diameters of the gear worm and generating worm

12.5 Prospects of Implementation of the Method

To our opinion, at least two versions of the structural layout of running cutter heads are possible:

1. The casing of the cutter head with slots for carbide inserts that are changeable; it is mounted on the mandrel that is common for several similar dimension types of heads—Fig. 12.5a.[1]
2. The casing with slots is the solid unit with the mandrel—Fig. 12.5b see Footnote 1.

The first version provides greater flexibility in production and the second one—higher rigidity of the tool. In any case, to our opinion, there are evident advantages of the considered method as compared to the traditional one based on the hob application with regard to the tool production and operation:

– smaller number of manufacturing operations is required; the manufacturing process as a whole takes several-fold less time;
– in order to perform operations, it is possible to use a more common equipment (universal machining centers instead of specialized relieving and sharpening machine-tools).

Running heads are also more convenient in operation: cutting properties of the tool are recovered not by resharpening, but by replacement of carbide inserts.

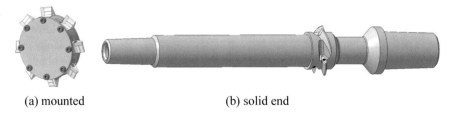

(a) mounted   (b) solid end

Fig. 12.5 Versions of principle structural layout of the running cutter head

---

[1]Layout versions are developed by eng. K. V. Bogdanov.

The possibility of application of carbide inserts in the assembled layout of the running cutter head provides conditions for the crucial exaggeration of the tooth machining mode, especially the cutting speed. Similar to the technique of tooth machining of bevel and hypoid gearwheels it is reasonable to single out groups of cutters that machine opposite flanks of teeth, thus making it possible to apply typical plates instead of special ones for each required profile, simplifying the chip removal, increasing the stability of cutter loading and, as a consequence, promoting conditions of increasing the accuracy and production efficiency of machining.

Summarizing the present manuscript, let us answer the question stated at its beginning: tooth cutting by means of the running cutter head can be a competitive technique as compared to the traditional tooth hobbing of worm and spiroid gearwheels due to greater possibilities of the choice of gear parameters and control of the localized contact, comparable production efficiency, less time consumption and cheapness of production preparation, higher flexibility and simpler operation of tools. Especially beneficial areas of implementation of the considered method can be production of multi-thread worm and spiroid gears with small gear ratios and bevel gears with spiral teeth produced by generating worms [9].

References

1. Georgiev, A.K., Goldfarb, V.I.: To investigation of the orthogonal spiroid gear with the cylindrical worm having threads of the ideally variable pitch. J. Mech. Mach. M. Nauka **45**, 91–99 (1974)
2. Georgiev, A.K., Maltsev, Yu.I., Mansurov, I.I.: To the question of the choice of rational design and geometrical parameters of running cutters for cutting the gearwheels of main spiroid gears by gear turning. In: Prospects of Development and Application of Spiroid Gears and Gearboxes. Proceedings of the All-Union Scientific Technical Meeting, Izhevsk Mechanical Institute, Izhevsk, pp. 39–45 (1979)
3. Goldfarb, V.I., et al.: Spiroid Gearboxes of Pipeline Values. Veche, Moscow (2011)
4. Goldfarb, V.I., Trubachev, E.S., Savelieva, T.V.: Unification of hobs in spiroid gears. VDI Berichte, N 1904 II, pp. 1755–1759 (2005)
5. Lagutin, S.A.: Local synthesis of general type worm gearing and its applications. In: Proceedings 4th World Congress on Gearing and Power Transmissions, Paris, vol. 1, pp. 501–506 (1999)
6. Lopato, G.A., Kabatov, N.F., Segal, M.G.: Bevel and Hypoid Gears with Circular Teeth. Mashinostroenie, Leningrad (in Russian) (1977)
7. Trubachev, E.S.: Vector field of normal vectors and its application to investigation of geometry of spiroid gearing with the helicoid worm. In: Problems of Design of Mechanical Engineering and Information Technology Products, pp. 3–14. Izhevsk (in Russian) (1999)
8. Trubachev, E.S.: Method of parameters calculation for the machine-tool gearing with the helicoid generating surface. In: Advanced Information Methods. Problems of Investigation, Design and Production of Gears, pp. 163–169. Izhevsk (in Russian) (2001)
9. Trubachev E.S.: On Possibility of Cutting bevel Gearwheels by Hobs (2020)
10. Trubachev, E.S., Oreshin, A.V.: CAD system for spiroid gears. J. Inf. Math. **1**(3), 159–165. Publishers of Physical Mathematical Literature (in Russian) (2003)
11. Trubachev, E., Savelyeva, T., Pushkareva, T.: Practice of design and production of worm gears with localized contact. In: Goldfarb, V., Trubachev, E., Barmina, N. (eds.) Advanced Gear Engineering. Mechanisms and Machine Science, vol. 51, pp. 327–344. Springer (2018)

Chapter 13
Biplanetary Gear Trains and Their Analysis Through the Torque Method

Kiril Arnaudov, Ognyan Alipiev and Dimitar P. Karaivanov

Abstract The paper shows the ability to analyze an alternative method different from the classical ones—analytical and graphical. This method uses no angular and peripheral velocities, as the classical methods do, but the torques. It is characterized by simplicity and clarity. The last is especially important for the designer, since nothing is more dominant in the engineer's mind than the need, and hence its ambition for clarity and ease of usage.

Keywords Calculation method · Efficiency · Planetary gear trains · Ratio · Torque

13.1 Introduction

The biplanetary gear trains are sophisticated engineering products, their application area being extremely large. If the number of the elementary, i.e. single-carrier planetary gear trains that are more or less used is about a score or so, then the number of the complex compound multi-carrier planetary gear trains built of them on combination principle is calculated as several thousands. A particularly complex variety of the common planetary gear trains are the so called biplanetary gear trains [14, 19]. The use of two (or more) carriers which are joined so that the planets realize rotation not toward two axes but toward three (or more) axes is characteristic for them. Thus they make sophisticated movements and describe complicated trajectories that are found to be useful for certain type of machines. The complicated construction and

K. Arnaudov
Bulgarian Academy of Sciences, Institute of Mechanics, Acad. G. Bonchev St., Bl. 4,
1113 Sofia, Bulgaria
e-mail: k_arnaudov@abv.bg

O. Alipiev
University of Ruse, 8 Studentska St., 7017 Ruse, Bulgaria
e-mail: oalipiev@uni-ruse.bg

D. P. Karaivanov (✉)
University of Chemical Technology and Metallurgy, 8 Kl. Ohridsky Blvd., 1756 Sofia, Bulgaria
e-mail: dipekabg@yahoo.com

© Springer Nature Switzerland AG 2020   311
V. Goldfarb et al. (eds.), *New Approaches to Gear Design and Production*,
Mechanisms and Machine Science 81,
https://doi.org/10.1007/978-3-030-34945-5_13

Table 13.1 Characteristic of the analytical and graphical methods

Willis analytical method

Advantages:	Shortcomings:
– Accuracy	– Absence of whatever visualization
– Universality	– Growing volume of calculations particularly in compound gear trains
	– Growing danger of errors
	– Determination of the speed ratio alone

Kutzbach-Smirnov graphical method

Advantages:	Shortcomings:
– Visualization, most of all in the simple single-carrier planetary gear trains	– Loss of visualization in the complex compound planetary gear trains
	– In principle the accuracy of the graphical methods is not high
	– Determination of the speed ratio alone

sophisticated kinematics of the biplanetary gear trains determine their considerably more difficult analysis compared to the common planetary gear trains.

13.2 Present State

Generally, there are different methods for analysis of the planetary gear trains, both simple elementary and complex compound ones. Two of them are the most frequently used. These are Willis [17] analytical method and Kutzbach-Smirnov [12, 13] graphical method. They have proved their qualities and capabilities in the course of long years. And yet, besides advantages they have also shortcomings which are manifested particularly negatively right in the analysis of the complex compound multi-carrier planetary gear trains. A comparison of the advantages and shortcomings of the said two methods are presented in Table 13.1.

13.3 Aim of the Paper

The limited capabilities of the classical methods—analytical and graphical as well as the growing difficulties in their use particularly for the analysis of the complex compound multi-carrier planetary gear trains determine the necessity of another alternative method which should possess more capabilities and advantages compared to the used up to now classical methods. Such is the method using not the angular velocities, the rotation frequency n or the periphery velocity v as it is with the analytical and graphical methods but the torques T. In this alternative method the focus is on the visualization as nothing dominates more the engineer way of thinking than the necessity and from there the pursuit of visualization and easy use of certain method.

The aim of the present paper is to show the capabilities of precisely the alternative torque method for analysis of the especially complex gear trains such as the biplanetary gear trains.

13.4   Essence and Capabilities of the Torque Method

Torque method is described in a number of publications [1–5]. It is built on the following known points of mechanics as well as on some small but practical, very useful innovations.

Known points of mechanics

(1) Equilibrium of the three ideal external torques (without taking into account the losses in the gear train, i.e. at efficiency $\eta = 1$), two of them being unidirectional ($T_{D\text{min}}$ and $T_{D\text{max}}$), and the third torque T_{Σ} opposite to the other two and its algebraic value equal to their sum (Fig. 13.1)

$$\sum T_i = T_{D\text{min}} + T_{D\text{max}} + T_{\Sigma} = 0. \tag{13.1}$$

(2) Equilibrium of the three real external torques (with taking into account the losses, i.e. $\eta < 1$)

$$\sum T_i' = T_{D\text{min}}' + T_{D\text{max}}' + T_{\Sigma}' = 0. \tag{13.2}$$

(3) Sum of the ideal powers ($\eta = 1$), from where the gear train speed ratio i is determined

$$\sum P_i = P_A + P_B = T_A \cdot \omega_A + T_B \cdot \omega_B, \tag{13.3}$$

$$i = \frac{\omega_A}{\omega_B} = -\frac{T_B}{T_A}, \tag{13.4}$$

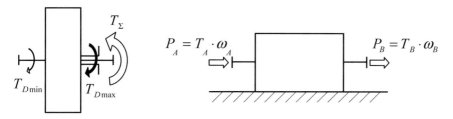

Fig. 13.1 External torques, angular velocities, and powers of a planetary gear train

where

P_A and P_B are the input and output powers;
T_A and T_B are the input and output torques;
ω_A and ω_B are the input and output angular velocities.

(4) Sum of the real powers ($\eta < 1$), from where the gear train efficiency η is determined

$$\sum P'_i = P'_A + P'_B = \eta \cdot T'_A \cdot \omega_A + T'_B \cdot \omega_B \tag{13.5}$$

$$\eta = -\frac{\frac{T'_B}{T'_A}}{\frac{\omega_A}{\omega_B}} = -\frac{\mu}{i}. \tag{13.6}$$

Innovations (Fig. 13.2)

(1) Wolf symbol [18] was used but modified, not in its original form [1–5]. In contrast to the original symbol in the modified symbol the three external gear train

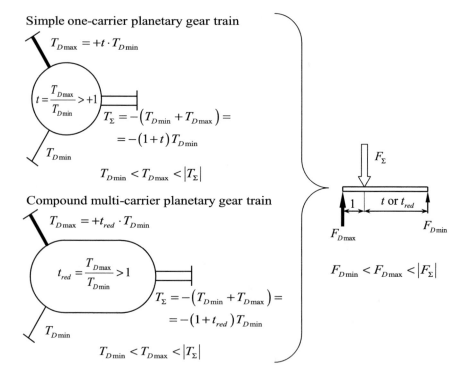

Fig. 13.2 Modified Wolf symbol, torque rations t and t_{red}, and external torques T_{Dmin}, T_{Dmax}, and T_Σ

shafts, whether this is a simple one-carrier or complex compound multi-carrier (including biplanetary), were designated in the following way. The shafts of the unidirectional torques T_{Dmin} and T_{Dmax} were designated by single line in conformity with the size of the torques—the first shaft by thin line and the second—by thick one. The shaft of the greatest opposite torque T_Σ was designated by double line [18].

(2) The following torque ratio was introduced using the unidirectional ideal external torques

$$t = \frac{T_{Dmax}}{T_{Dmin}} > +1. \tag{13.7}$$

This new relation provides unchanging connection between the three external torques, undependably on the gear train type and operational mode [1, 3, 5]. For the complex compound multi-carrier planetary gear trains this torque ratio was designated as reduced, i.e. t_{red} (Fig. 13.2).

(3) The modified Wolf symbol and the innovative torque ratio t_{red} finds the most visualized expression in the lever analogy. Generally, this is the simplest and visualized way in which the things can be presented.

The sequence of applying the torque method for the compound two-carrier planetary gear trains is best described in Arnaudov and Karaivanov [6–8], in Troha et al. [15] it is applied for studying two-speed planetary gear trains and in Karaivanov et al. [7, 9, 10] for studying of multicarrier change-gears (vehicle gearboxes).

The simple algorithm makes this method suitable for creation software and obtaining results by changing the geometrical parameters of the component gear trains, respectively the torque ratio t, within certain limits. In this way, it is convenient to make multifactor optimization (selecting the appropriate structure scheme and its parameters). For two-carrier planetary gear trains, this has done in Karaivanov [11] and Troha et al. [16], where the torque ratios of the two component gear trains t_I and t_{II} change at a certain step in the range from 2 to 12. The criteria for optimization in these works are dimensions, efficiency, and reduced to input shaft clearances.

13.5 Sequence of Biplanetary Gear Train Analysis

Using the above said points the biplanetary gear train analysis was made in the following sequence.

(1) It was started with the torque of one of the three external gear train shafts; it was expedient to start from the shaft of the lowest torque, not being difficult to find which it was. It was most convenient this torque to be chosen $T_{Dmin} = +1$, that being not obligatory.

(2) Using the speed ratios through the number of teeth of each engaged couple of gearwheels, step by step all torques were calculated reaching the last gearwheel of the chain. The direction of operation of the individual torques was also determined.

(3) The torque of the chain last gearwheel no doubt was greater than the initial torque $T_{Dmin} = +1$. Whether this last torque was the torque T_{Dmax} or T_Σ was determined by its direction of operation, its algebraic sign, respectively. If they coincide with those of the initial torque T_{Dmin}, then the last torque is T_{Dmax}. In the opposite case it is T_Σ.

(4) From the equilibrium condition of the external torques (13.1) the third torque was determined—T_Σ or T_{Dmax}.

(5) Knowing the three external torques T_{Dmin}, T_{Dmax}, and T_Σ, as well as the function that each of them fulfills as input torque T_A, output torque T_B or reactive torque T_C, it was easy to determine by formula (13.4) every speed ratio which the gear train could realize.

13.6   Examples

Figure 13.3 presents the solution of the first example that is relatively the simplest one. The figure shows the consecutive determination of all torques. In the torque method this services the purpose not only for analyzing the gear train kinematics but also for determination of the loading of every gearwheel which is necessary for its strength calculation. This is an additional substantial advantage of the torque method. It is clearly seen in the figure that the external reactive torque T_C has the same direction as the input torque $T_A \equiv T_{Dmin} > 0$, which means that $T_C \equiv T_{Dmax} > 0$. The third external torque $T_B \equiv T_\Sigma$, which is output torque, was determined by the condition for equilibrium of the external torques (13.1), as it is shown in the figure. It is also seen in the figure the determination of the reduced torque ratio t_{red} and the speed ratio i.

For comparison, the analytical solution according to Willis is as follows. In this simple case it is still comparatively simple.

$$i_{1S_1(6)} = \frac{\omega_A}{\omega_B} = 1 - i_{16(S_1)}, \tag{13.8}$$

where

$$i_{16(S_1)} = i_{12} \cdot i_{3S_2(4)} \cdot i_{56} \tag{13.9}$$

and the individual speed ratios are

$$i_{12} = -\frac{z_2}{z_1}; \, i_{3S_2(4)} = 1 - i_{34(S_2)} = 1 + \frac{z_4}{z_3}; \, i_{56} = +\frac{z_6}{z_5}.$$

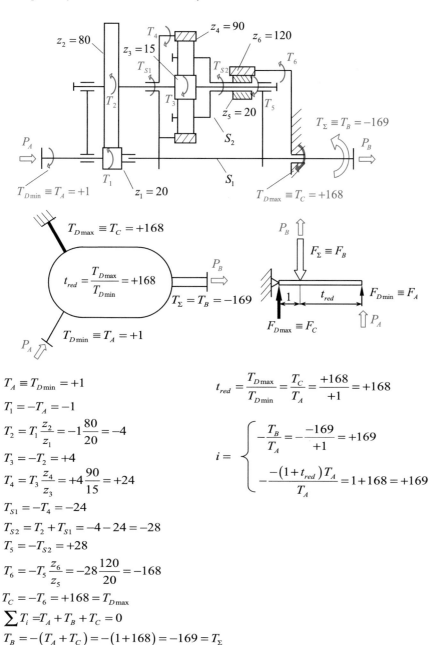

$$T_A \equiv T_{D\min} = +1$$

$$T_1 = -T_A = -1$$

$$T_2 = T_1 \frac{z_2}{z_1} = -1\frac{80}{20} = -4$$

$$T_3 = -T_2 = +4$$

$$T_4 = T_3 \frac{z_4}{z_3} = +4\frac{90}{15} = +24$$

$$T_{S1} = -T_4 = -24$$

$$T_{S2} = T_2 + T_{S1} = -4 - 24 = -28$$

$$T_5 = -T_{S2} = +28$$

$$T_6 = -T_5 \frac{z_6}{z_5} = -28\frac{120}{20} = -168$$

$$T_C = -T_6 = +168 = T_{D\max}$$

$$\sum T_i = T_A + T_B + T_C = 0$$

$$T_B = -(T_A + T_C) = -(1 + 168) = -169 = T_\Sigma$$

$$t_{red} = \frac{T_{D\max}}{T_{D\min}} = \frac{T_C}{T_A} = \frac{+168}{+1} = +168$$

$$i = \begin{cases} -\dfrac{T_B}{T_A} = -\dfrac{-169}{+1} = +169 \\[4mm] -\dfrac{-(1+t_{red})T_A}{T_A} = 1 + 168 = +169 \end{cases}$$

Fig. 13.3 Determination of the individual torques and the speed ratio of the biplanetary gear train

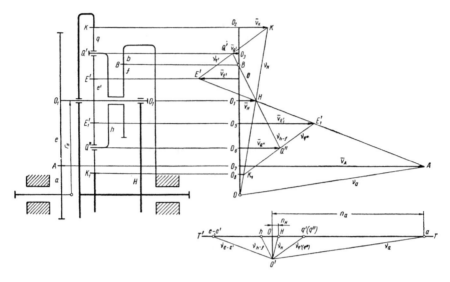

Fig. 13.4 Graphical solution of the biplanetary gear train from Fig. 13.3

The final formula by this method is

$$i_{1S_1(6)} = 1 - i_{12}(1 - i_{34(S_2)})i_{56}$$

$$= 1 + \frac{z_2}{z_1}\left(1 + \frac{z_4}{z_3}\right)\frac{z_6}{z_5} = 1 + \frac{80}{20}\left(1 + \frac{90}{15}\right)\frac{120}{20} = 169$$

The result is the same as by the torque method.

Figure 13.4 presents the graphical solution contained in Tkachenko [14]. It is easy to estimate that besides some visualization it could not be relied on accuracy. In this comparatively simple example the obtained knotwork could be somehow yet accepted.

Figure 13.5 presents the solution of the so called closed biplanetary gear train. It is characteristic for this gear train that the gearwheels 6 and 8, forming the immovable component C, are connected and close the kinematic chain. The determination of the individual torques, the torque ratio t_{red} and the speed ratio i is carried out in the same manner as in the previous case and is shown in the figure.

The analytical solution according to Willis is as follows:

For the basic planetary mechanism (carrier S_1 movable central gearwheel 1 and fixed central gearwheel 8)

$$i_{AB} = \frac{\omega_A}{\omega_B} = i_{1S_1(8)} = 1 - i_{18(S_1)} = 1 - i_{12} \cdot i_{3S_2(S_1)} \cdot i_{78}$$

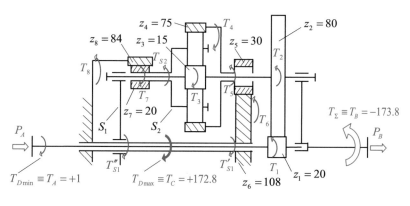

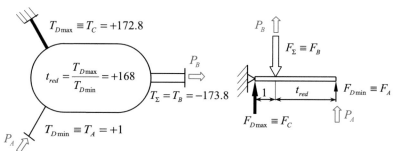

$T_A \equiv T_{D\min} = +1$

$T_1 = -T_A = -1$

$T_2 = T_1 \dfrac{z_2}{z_1} = -1\dfrac{80}{20} = -4$

$T_3 = -T_2 = +4$

$T_4 = T_3 \dfrac{z_4}{z_3} = +4\dfrac{75}{15} = +20$

$T_5 = -T_4 = -20$

$T_6 = T_5 \dfrac{z_6}{z_5} = -20\dfrac{108}{30} = -72$

$T'_{S1} = -T_6 = +72$

$T_{S2} = T_2 + T_5 = -4 - 20 = -24$

$T_7 = -T_{S2} = +24$

$T_8 = -T_7 \dfrac{z_8}{z_7} = -24\dfrac{84}{20} = -100.8$

$T''_{S1} = -T_8 = +100.8$

$T_C = T'_{S1} + T''_{S1} = +72 + 100.8 = +172.8 = T_{D\max}$

$\sum T_i = T_A + T_B + T_C = 0$

$T_B = -(T_A + T_C) = -(1 + 172.8) = -173.8 = T_\Sigma$

$t_{red} = \dfrac{T_{D\max}}{T_{D\min}} = \dfrac{T_C}{T_A} = \dfrac{+172.8}{+1} = +172.8$

$i = \begin{cases} -\dfrac{T_B}{T_A} = -\dfrac{-173.8}{+1} = +173.8 \\[4mm] -\dfrac{(1 + t_{red})T_A}{T_A} = 1 + 172.8 = +173.8 \end{cases}$

Fig. 13.5 Determination of the individual torques and the speed ratio of the closed biplanetary gear train

where $i_{3S2(S1)}$ is determined by the closed differential mechanism (planetary gear train 3-S_2-4 and closing chain 5-6-8-7) obtained after fixation of the carrier S_1 and discharge of the gearwheels 8 and 6.

From

$$\frac{\omega_3 - \omega_{S_2}}{\omega_4 - \omega_{S_2}} = i_{34(S_2)}$$

it was obtained

$\omega_3 = \omega_4 \cdot i_{34} + \omega_{S_2}(1 - i_{34})$, $\frac{\omega_3}{\omega_{S_2}} = \frac{\omega_4}{\omega_{S_2}} i_{34} + 1 - i_{34}$, respectively, where $\frac{\omega_3}{\omega_{S_2}} = i_{3S_2(S_1)}$, and $\frac{\omega_4}{\omega_{S_2}} = \frac{\omega_5}{\omega_7} = i_{57}$.

Then

$$i_{3S_2(S_1)} = i_{57} \cdot i_{34} + 1 - i_{34},$$

and for the speed ratio of the considered gear train it was obtained

$$i_{AB} = \frac{\omega_A}{\omega_B} = i_{1S_1(8)} = 1 - i_{12} \cdot i_{3S_2(S_1)} \cdot i_{78} = 1 - i_{12}(i_{57} \cdot i_{34} + 1 - i_{34})i_{78},$$

where the individual speed ratios were

$$i_{12} = -\frac{z_2}{z_1} = -\frac{80}{20} = -4; \quad i_{34} = -\frac{z_4}{z_3} = -\frac{75}{15} = -5;$$

$$i_{57} = i_{56} \cdot i_{87} = -\frac{z_6}{z_5}\frac{z_7}{z_8} = -\frac{108}{30}\frac{20}{84} = -\frac{6}{7}; \quad i_{78} = \frac{z_8}{z_7} = \frac{84}{20} = 4\frac{1}{5},$$

and ultimately

$$i_{1S_1(8)} = 1 - i_{12}(i_{57} \cdot i_{34} + 1 - i_{34})i_{78} = 1 + 4\left[\left(-\frac{6}{7}\right)(-5) + 1 - (-5)\right]4\frac{1}{5} = 173\frac{4}{5}$$

It was seen that in this more complicated case Willis method is significantly more sophisticated compared to torque method, let alone the absence of visualization.

Figure 13.6 presents also the graphical solution of this example, contained in Tkachenko [14]. Here also the said in the first case is valid concerning the obtained network of lines and visualization. Naturally, accuracy cannot be mentioned here, too.

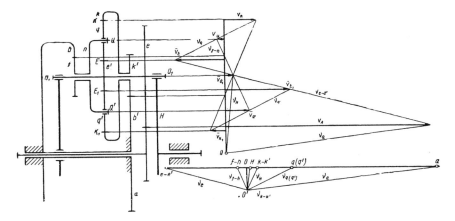

Fig. 13.6 Graphical solution of the biplanetary gear train from Fig. 13.5

13.7 Conclusions

The above examples demonstrated the qualities and capabilities of the torque method for analysis of planetary gear trains in general and particularly of the complex compound multi-carrier planetary gear trains including also the biplanetary ones.

These capabilities are as follows:

(1) Simplicity, maximum visualization and easy applicability of the method which combines in itself accuracy and visualization existing separately in the analytical and graphical methods.

(2) Universality of the method—applicability both to the simple one-carrier planetary gear trains and to the complex multi-carrier planetary gear trains including the biplanetary ones.

(3) Achievement of more aims, that it to say—determination of not only the speed ratio as it is with the analytical and graphical methods but also additional possibility for determination of the torques of the individual gear train components—gearwheels, shafts—as necessary prerequisite for their reliable calculation. This paper did not reveal but by the torque method the efficiency could be also determined very easily and visually.

(4) Decreased danger of errors as a result of the simplicity, visualization and easy applicability of the method.

(5) Owing to the simplicity, visualization and easy applicability of the method it is suitable both for the practice—for engineers and for the educational process—for the students.

References

1. Arnaudov, K.: Einfaches Verfahren zur Ermittlung des Übersetzungsverhältnisses zusammengesetzter Planetengetriebe. VDI-Berichte **1230**, 313–328 (1996)
2. Arnaudov, K., Karaivanov, D.: Engineering analysis of two-carrier planetary gearing through the lever analogy. In: Proceedings of the International Conference on Mechanical Transmissions, Chongqing, China, China Machine Press, pp. 44–49 (2001)
3. Arnaudov, K., Karaivanov, D.: Systematik, Eigenschaften und Möglichkeiten von zusammengesetzten Mehrsteg-Planetengetrieben. Antriebstechnik **5**, 58–65 (2005)
4. Arnaudov, K., Karaivanov, D.: Higher Compound Planetary Gear Trains. VDI-Berichte, 1904-1 pp. 327–344 (2005)
5. Arnaudov, K., Karaivanov, D.: The Complex Compound Multi-Carrier Planetary Gear Trains—A Simple Study. VDI-Berichte, 2108-2, pp. 673–684 (2010)
6. Arnaudov, K., Karaivanov, D.: The torque method used for studying coupled two-carrier planetary gear trains. Trans. FAMENA, **37**(1), 49–61 (2013)
7. Arnaudov, K., Karaivanov, D.: Torque Method for Analysis of Compound Planetary Gear Trains. Beau Bassin [Mauritius], LAP Lambert Academic Publishing (2017)
8. Arnaudov, K., Karaivanov, D.: Planetary Gear Trains. Boca Raton [FL, USA], CRC Press (2019)
9. Karaivanov, D., Petrova, A., Ilchovska, S., Konstantinov, M.: Analysis of complex planetary change-gears through the torque method. Mach. Technol. Mater. **X**(6), 38–42 (2016)
10. Karaivanov, D., Velyanova, M., Bakov, V.: Kinematic and power analysis of multi-stage planetary gearboxes through the torque method. Mach. Technol. Mater. **12**(3), 102–108 (2018)
11. Karaivanov, D.: Structural analysis of coupled multi-carrier planetary gear trains—from lever analogy to multi-objective optimization. In: Proceedings of the 3rd International Conference on Manufacturing Engineering (ICMEN), Chalkidiki, Greece, pp. 579–588 (2008)
12. Kutzbach, K.: Mehrgliedrige Radgetriebe und ihre Gesetze. Maschinenbau **22**, 1080 (1927)
13. Smirnov, L.P.: Investigation of rotation movement through "Velocity Triangles. Vestnik Inzhenerov i Tehnikov **8** (in Russian) (1938)
14. Tkachenko, V.: Planetary Mechanisms. Optimal Design. Harkov, HAI (in Russian) (2003)
15. Troha, S., Lovrin, N., Milovančević, M.: Selection of the two-carrier shifting planetary gear train controlled by clutches and brakes. Trans. FAMENA **36**(3), 1–12 (2012)
16. Troha, S., Petrov, P., Karaivanov, D.: Regarding the optimization of coupled two-carrier planetary gears with two coupled and four external shafts. Mach. Build. Electr. Eng. **LVIII**(1), 49–55 (2009)
17. Willis, R.: Principles of Mechanism. John W. Parker, London (1841)
18. Wolf, A.: Die Grundgesetze der Umlaufgetriebe. Braunschweig, Friedrich Viewegn Sohn (1958)
19. Zyplakov, J.S.: Biplanetary Mechanisms. Mashinostroenie, Moscow (in Russian) (1966)

Chapter 14
Optimization of Planocentric Gear Train Characteristics with CA-Tools

Gorazd Hlebanja, Miha Erjavec, Simon Kulovec and Jože Hlebanja

Abstract The planocentric gearboxes are characterized by the coaxial input and output shafts and large transmission ratios, which can be achieved based on a ring gear with internal gearing in combination with usually two planet gears with external gearing, where the difference in the numbers of teeth between the gear ring and the planet gears is small, preferably amounting one. The proposed solution is based on the S-shaped tooth flank geometry. Since the required gear train characteristics impose strict limitations (e.g. near zero backlash) on the assembled product, it is necessary to simulate its tolerance behaviour. So, a professional or proprietary developed CA tools are necessary to deal with such analyses, e.g. KissSoft. Properties of a such gear train can be successfully modified in this way. A sophisticated testing rig was developed to verify actual gear train characteristics, and to test its short-term and long-term behaviour. Backlash, stiffness, kinematic error, and dynamic behaviour of produced gear trains are measured in this way.

Keywords Planocentric gear train · Backlash · Tolerance analysis · S-gearing

G. Hlebanja (✉)
University of Novo Mesto, FME, Na Loko 2, 8000 Novo mesto, Slovenia
e-mail: gorazd.hlebanja@siol.net; gorazd.hlebanja@podkriznik.si

G. Hlebanja · M. Erjavec · S. Kulovec
Podkrižnik d.o.o., Lesarska cesta 10, 3331 Nazarje, Slovenia
e-mail: miha.erjavec@podkriznik.si

S. Kulovec
e-mail: simon.kulovec@podkriznik.si

J. Hlebanja
University of Ljubljana, FME, Aškerčeva c. 6, 1000 Ljubljana, Slovenia
e-mail: joze.hlebanja@siol.net

© Springer Nature Switzerland AG 2020
V. Goldfarb et al. (eds.), *New Approaches to Gear Design and Production*,
Mechanisms and Machine Science 81,
https://doi.org/10.1007/978-3-030-34945-5_14

14.1 Introduction

Planocentric gear boxes are in technical use for many decades due to their main characteristic, which is reduction of rotational speed and accordingly increased torque in the smallest available volume. This type of device is used in aircraft, marine, robotics, and many other industries. The expected efficiency may be as high as 90% or more and gear ratios can achieve up to 160:1 in a single step configuration. Basic arrangements of this type are described in renowned references, e.g. Radzevich [1] and Niemann and Winter [2]. The available industrial solutions include *Sumitomo* cyclo gearboxes [3], *Spinea* drives [4], *Nabtesco* [5], *Onvio* [6] and many others. Gearings are usually cycloidal or lantern. Some solutions combine planetary gear train and a cycloidal stage with three eccentrics having origins in the centers of planets. Some other devices incorporate a threefold eccentric rotating three planets. Such gearboxes are more compact but also more complex. All producers claim near zero or zero backlash. There is also a strong patent activity all over the world, also in Russia and China. This indicates importance of the field. Some patented solutions comprise gear boxes of Sumitomo and Jtekt Corporation [7–9], etc. The expected characteristics are low hysteresis for accurate positioning, low lost motion, compact design, high torsional stiffness, low inertia, high efficiency, overload capacity, easy assembly, and lifetime lubrication. Some robot producers develop and produce in-house patented gearings, e.g. *Stäubli's* JCS reduction gear boxes [10], which improve the precision, dynamics, and service life of its robots and lower their maintenance.

Additionally, some additional features can be built into a device, such as accurate output shaft positioning and output torque sensorics, which enable incorporation of such devices in collaborative robot's arm joints.

The paper presents kinematic circumstances of the proposed planocentric gear train. Some historical notes show basic mechanisms used in a contemporary gear train. The produced prototypes have been tested and results helped in improvements of new series. The comprehensive analysis in KissSoft helped in the design of new prototype, conforming to the requirements.

14.2 Kinematic Circumstances of a Planocentrcic Gear Train

The planocentric gearbox has coaxial input and output shafts, and large transmission ratios can be achieved based on a gear ring with internal gearing in combination with usually two planet gears with external gearing, where the difference in the numbers of teeth between the gear ring z_r and the planet gears z_p rules the gear ratio, Eq. (14.1). The difference in ring and planet numbers of teeth should be small, preferably amounting one.

$$i = \frac{z_p - z_r}{z_p} \tag{14.1}$$

Fig. 14.1 Scheme of double cage, double bearing planocentric gear box

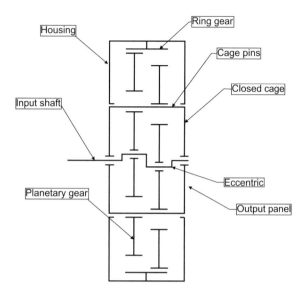

The planet gears are mounted on an eccentric shaft, where bearings separate the planet gears from the eccentric. The planet gears wobble around the gear ring, that is, they reverse for one tooth in each revolution of the eccentric. The wobbling movement is in accordance with a hypocycloidal movement where the generating circle with the radius of the eccentric is rolling on the kinematic circle of the ring gear. And at the same time, the planet gear kinematic circle rolls in the inner side of the ring gear kinematic circle, which is simultaneous with the rotation of the eccentric. In this way the planetary gears develop rotation superimposed on the wobble. So, the input rotation of the eccentric is transformed into the reduced output rotation of the cage with the pins according to the gear ratio in the reverse direction of the input shaft in the same axis. And the gear ring is fixed to the housing. The planocentric gear train arrangement with a closed cage is illustrated in Fig. 14.1.

The eccentric driven planocentric gear train can be regarded as a simple mechanism with two links. The first link size is the radius of the eccentric and its joint indicates its position. The second one connects the eccentric with a point on the planet gear (a rigid body), e.g. the contact point. The eccentric link rotates and induces movement of the chosen point on the planet gear, which is restricted by a following rule:

$$r_v = r_p \frac{\varphi_p}{\varphi_g} \qquad (14.2)$$

r_v and r_p are the radii of the kinematic circles of the gear and the planet gear, respectively. If the ring gear rotates for φ_v the planet rotates for φ_p.

A simple algorithm can be used to define movement of the planet based on the rotation of the eccentric and limited by Eq. (14.2) (Fig. 14.2).

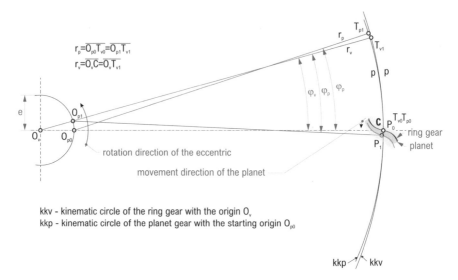

Fig. 14.2 Planetary gear movement in accordance with hypocycloid generated on the kinematic circle of the ring gear

- T_{p0} and T_{v0} coincide with C. P_0 is a point on the planet also coinciding with C.
- T_{p1} and T_{v1} are calculated according to Eq. (14.2). It is true: $p = \pi\, m = p = \pi\, m = \widehat{CT_{v1}} = \widehat{CT_{p1}}$.
- Eccentric turns for φ_v to the new point O_{p1}. kkp rolls on kkv in such a way that T_{p1} coincides with T_{v1}. So, tangents and normals of kkv and kkp coincide in T_{v1}.
- The normal of the planet in this point runs through O_v and O_{p1}.
- Since the planet is a rigid body, the right leg of the angle φ_p rotates around O_{p1} in CW direction for the difference $\Delta\varphi = \varphi_v - \varphi_p$.
- The procedure is continuous, but it can be numerically calculated by an adequate number of steps.

The above procedure can be formalized. Thus, succesive points on the ring gear kinematic circle T_{vi} are defined as follows:

$$x_{Tvi} = r_v \cos \varphi_{vi} \text{ and } y_{Tvi} = r_v \sin \varphi_{vi}. \tag{14.3}$$

Similarly, successive position points O_{pi} of the eccentric are

$$x_{Opi} = e \cos \varphi_{vi} \quad \text{and} \quad y_{Opi} = e \sin \varphi_{vi}. \tag{14.4}$$

And coordinates of the moving point P_i on the planet gear are

$$x_{Pi} = x_{Opi} + r_p \cos \Delta\varphi_i \quad \text{and} \quad x_{Pi} = y_{Opi} + r_p \sin \Delta\varphi_i. \tag{14.5}$$

The planet gear tooth movement into a new ring gear tooth space is illustrated in Fig. 14.3 by 20 iterations. So, each point and planet gear position in Fig. (14.3) are based on successive rotations of the eccentric for 18°.

14.3 Historical Background

Potential advantages of planocentric gear boxes, namely a high-speed reduction ratio combined with high output torque in a relatively small volume, were a driving force for a new development already twenty years ago. So, the modified lantern gears were developed for use in planocentric gear drives with eccentric [11] and patented [12] The internal gear pair during meshing is illustrated in Fig. 14.4, clearly indicating the value of eccentricity.

The teeth of the ring gear are designed as semi-circular extremities, whereas the planetary gears are designed with corresponding semi-circular spaces, adapted in size for a tolerance. Design aims were focused in automatic production lines and CNC-machinery. The planocentric lantern gear box design was robust. The transmission of rotation from the planet gears through a pin composition to the output shaft was provided by a single sided cage with double bearing output shaft. Small series were produced with various gear ratios up to 100 and with various modules, down to $m = 0.5$ mm.

A lantern gear train assembly is represented in Fig. 14.5. Gear boxes of this type have been produced in various sizes and configurations to serve primarily wood cutting automation machines. Materials for gears included among others self-lubricating plastics for planet gears, despite through hardened steel with tensile strength of 900 N/mm² prevailed. The specific load k_{lim} for continuous working conditions in the latter case amounted up to 125 N/mm², which implied the output torque of 100 Nm for the illustrated gear box with the gear ratio 100 and the module 1 mm and $z_{ring} = 101$ and $z_{planet} = 100$.

14.4 Gear Tooth Flank Geometry

Beside semicircular, many other gear flank geometries have been proposed for planocentric gear trains, all in search for an optimal geometry.

The involute gear shape can be used to compose a planocentric gear box. Park et al. [13] conducted a research with the goal to produce a robotic gear box in order to replace an existing cycloidal planocentric gearbox. The key point of this model was also regarded to be economic in manufacture due to little influence of a manufacturing and assembly errors.

However, due to possible gear interference, such gears exhibit rather high pressure angle α_w around 30°, and Δz cannot amount to less than 5, or 4 based on the condition that the numbers of teeth of pinion and ring gear are rather high, 167 and 171 in this

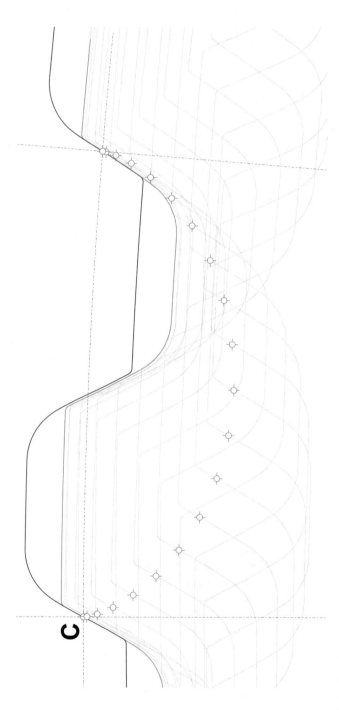

Fig. 14.3 Planetary gear movement in accordance with hypocycloid generated on the kinematic circle of the ring gear

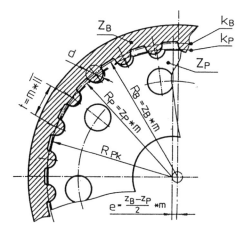

Fig. 14.4 Gear ring and planet gear during meshing

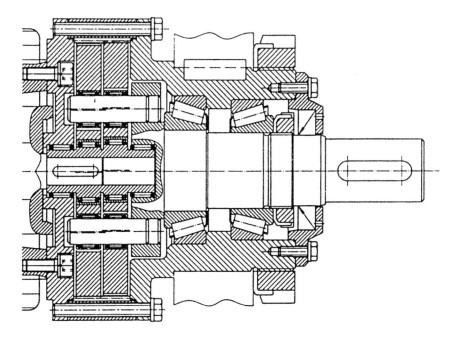

Fig. 14.5 Assembly of a robust planocentric lantern gear train

case. This also means essentially lower gear ratio (Eq. 14.1), $i^{-1} = 41.75$ for the above data.

Another attempt was made with a rectilinear tooth profile where a tooth is defined by two lines enclosing an angle, by an inside or outside circle, by the root circle and fillet arcs [14]. The power is transmitted by the arc at the pinion tooth tip which slides over the linear tooth part of the gear ring. The problem is that such a gear composition does not follow the law of gearing. If a high-speed rotation is required, then such a gear arrangement can develop high noise and torque fluctuation.

Yet another tooth design is trapezoidal [15], where the contact of teeth is surface-like. However, the efficiency of such gear box may be poor, due to lack of rolling, amount of sliding and (non)conformity to the law of gearing.

A proposed solution is based on S-shaped tooth flank geometry for the meshing ring gear and planetary gears of the planocentric gear box. General ideas about S-gears have been described in several papers, e.g. [16–18].

The S-gear configuration has several advantages, the most important being the following:

1. Convex-concave contact in the vicinity of the meshing start and meshing end point;
2. Low amount of sliding during meshing which is due to the curved path of contact;
3. Evenly distributed flank load, which is due to similar sizes of addendums and dedendums of both meshing gears.

The other features include better lubrication due to high relative velocities and amount of rolling. In the case of internal-external gear pair some features may become less pronounced. Additionally, the path of contact is less curved, which is on behalf of smaller dedendum and addendum heights. However, S-gear shape is ruled by two parameters, the height parameter a_p and the curve exponent n. And by optimizing these two parameters one can shape this type of gearing in such a way to allow the gear and planet teeth number difference to be only one. So, it is possible design gear boxes with the diameter small diameters and high reduction ratios. The S-gearing for planocentric gear trains is illustrated in Fig. 14.6.

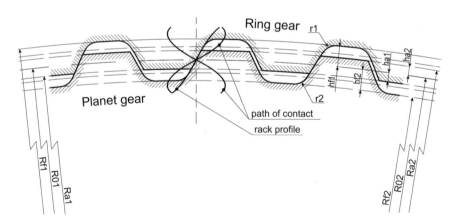

Fig. 14.6 S-gear flank geometry for the planocentric gear box

14.5 Planocentric Gear Box Prototypes

A small series of prototypes should be produced and assembled to test characteristics of a such gear box. The gear box is similar in function to those already mentioned. It contains an input shaft with eccentrics. As a motor rotates the shaft the eccentric rotates two planetary gears which wobble on the ring gear. The planetary gears are positioned in such a way that they enclose 180° for the sake of symmetry. Arrangements with three or more planetary gears are possible, which would impose high manufacturing skills regarding eccentrics. For each rotation of the eccentric the planetary gear proceeds for one tooth on the ring gear in the opposite direction. A cage consists of a supporting ring and output ring (serving also as the output shaft) that are connected by pins in an interference fit. The cage is rotated by planetary gears, having appropriate holes in which connecting pins with bearings comply. The cage is fixed to the input shaft by bearings at the extremities and in a similar manner to the housing with the ring gear. In this way a compact low volume gear box is achieved. It was decided that the device having module 1 mm, reduction ratio 80 ($z_v = 81$ and $z_p = 80$) and outer diameter around $\phi 100$ is the most interesting. The required maximal working torque is 120 Nm. The rendered image of this prototype is represented in Fig. 14.7 (from the input side) and the actual device in Fig. 14.8 (from the output side).

Fig. 14.7 Rendered image of the prototype planocentric gear box

Fig. 14.8 Prototype planocentric gear box

Gear trains were tested, when assembled. Performed tests were backlash, hysteresis and stiffness, kinematic error, vibrations and noise, and durability tests. Special testing rigs were prepared or manufactured for testing of this type and similar devices. After that, the devices were disassembled, and components inspected on a CMM (Computerized measurement machine) and optically. Some typical testing results are presented below.

Since the device is intended for precision industry and robotics, its backlash and stiffness characteristic become crucial. The characteristic should be symmetric, regardless of any initial position of the planets and rotation direction, and the backlash in very narrow limits, <1 arcmin. As Fig. 14.9 indicates, the specimen 04 reveals backlash of 9 arcmin and stiffness characteristic which becomes considerably stiffer with increasing load. Stiffness of this device in the zone 3–50 Nm is 10 Nm/arcmin, in the zone 50–100 Nm 30 Nm/arcmin and 16Nm/arcmin in the zone 3–100 Nm. This gear train has actually higher values of stiffness, since the working torque limit amounts to 120 Nm. Similar gear train of *Spinea* (Spinea TS110) with gear ratio 89 has stiffness of 22 Nm/arc min. Sumitomo Cyclogear drive A15 with ratio 89 is a bit stronger drive has stiffness values as follows: 15 Nm in 3–50%, 28 Nm/arcmin in 50–100%, and 20 Nm/arcmin in the segment 3–100%.

Kinematic error of a gear train is defined as a deviation of the actual angular position from the theoretical angular position:

$$\Delta\varphi = i \cdot \varphi_{inp} - \varphi_{out} \tag{14.6}$$

Rotation of the input shaft (φ_{inp}) is measured with a 16-bit incremental encoder, whereas the output shaft (φ_{out}) rotation is measured by a built in absolute optical encoder with a high resolution. The optical encoder and the reading head are mounted with some tolerance, which reflects in a sinusoid carrying actual error signal. Several measurements of the kinematic error of the specimen 04, Fig. 14.10, yield maximal

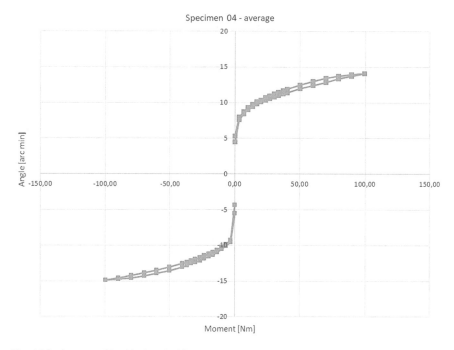

Fig. 14.9   Average of backlash and stiffness measurement of a specimen 04

Fig. 14.10   Kinematic error of a specimen 04

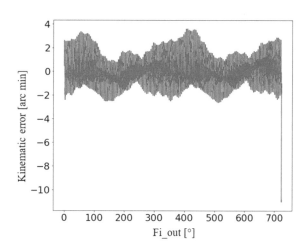

error limit of 6 arcmin.

Figure 14.11 shows the ring gear and the planets of the specimen 01 after disassembly. The specimen was submitted to high torques and speeds. The planet gears were made of 42CrMo4 and the ring of 25CrMo4, all gears plasma nitrided to HV700. The gears were carefully examined by an optical microscope. Gear teeth did not show

Fig. 14.11 Ring gear and planets of a specimen 01 after disassembly

any wear or damages. Initial wear appeared in some planet teeth tips and at certain locations in teeth tips. The reason is in meshing errors, which was discovered by measuring teeth spaces of the planets and the ring gear with a CMM.

In general, it could be concluded that first prototypes did not yet meet all prescribed requirements. So, it is necessary to improve the design and the quality of manufacturing so, that all measures are within prescribed tolerances. And with regard, So, all components should be inspected by the CM machine. Regarding gearing, the nominal circumferential backlash should be as low as possible, i.e. less than 10 μm, having manufacturing tolerance in the range of 5 μm. The circumferential arc of 10 μm for the radius 40.5 mm gives the angle of 0.85 arcmin for geometrically precise circumstances.

14.6 Influences of Tolerances

This imposes the necessity to analyze influences of tolerances in varying circumstances. Such an analysis could disclose additional measures which would enable a convergence of the design towards accomplishing design requirements. As a primary tool for this task, the KissSoft [19] system was employed. KissSoft is software for effective, high-quality tool for calculating machine elements, reviewing these calculations, determining component strength, and documenting safety factors and product life parameters, incorporating currently valid standards (DIN, ISO, AGMA).

However, KissSoft uses the prevailing involute gear geometry. So, the first aim is to adapt geometry in such a way to reflect S-gear geometry. The current *KissSoft* User manual [20] includes possibility of a progressive profile modification (p. 343 of the named manual), which can be used as a modification in the addendum and the dedendum of a gear tooth, and is defined as follows:

$$\Delta_{ad} = 2 \cdot C_{ad} \cdot \left(\frac{d - d_k}{d_t - d_k} \right)^{f_{ad}/5} \quad \text{and} \quad \Delta_{dd} = 2 \cdot C_{dd} \cdot \left(\frac{d - d_k}{d_v - d_k} \right)^{f_{ad}/5} \quad (14.7)$$

Δ_{ad} and Δ_{dd} stand for a profile modification function in addendum and dedendum. C_{ad} and C_{dd} are modifying tip relief (or corresponding active dedendum modification) and f_{ad} and f_{dd} power coefficients. If a coefficient amounts to 5 the relief is linear. d_t, d_v, d_k, and d, are diameters of the tip circle, dedendum circle, kinematic circle and current circle. One can adapt the involute flank addendum and dedendum to S-gear flank. Such a modification is justified since addendum and dedendum heights are rather small, between 0.2 and 0.25 m.

So, a tolerance analysis of key elements and their mutual influences on functional properties of the gear train becomes possible. A theoretical analysis of various tolerance combinations of the gearing and their effects is presented was conducted. And subsequently influences of tolerances in bearings and assembly of connected shafts was studied, which can have a negative influence on the function of the gear box in certain working conditions. To prevent the planned function of the gear box, the backlash should be kept below 1 arcmin = $0.016°$.

The aim is also to create a method of manufacturing and assembly, which would ensure the required precision and at the same time the tooth thickness tolerance would not be less than class h25, DIN3967 and the axis distance tolerance lesser than js6 (± 0.003) is not allowed in the KissSoft system.

14.6.1 Gearing Tolerances (Analysis of Tolerance Influences)

Figure 14.12 shows a computational shape of the used S-gears in the form of deviations from the involute gears. The planet gear was additionally modified with crowning amounting to 2 μm, as to prevent possible numerical singularities. This has a positive effect on a simulation. And in manufacturing, a worker at the machine is thus warned that the tooth cannot be concave and that the lead line should be straight or at least convex.

Initial state
The lowest backlash is $0°$, but this can only be achieved if gears are manufactured on the higher tooth thickness limit, and at the same time the eccentricity is higher than nominal. For gears and eccentricity in the mean tolerance values the backlash amounts already to $0.062°$.

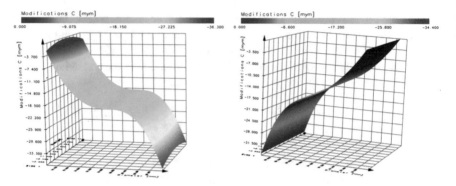

Fig. 14.12 S-gear data as they deviate from involute gearing (planet gear—left, ring gear—right)

Gears in production can be sorted in two classes based on the tooth thickness measurements. Values are collected in Table 14.1.

If a planet gear which is in the lower half class of the tooth thickness and a ring gear in the upper half class of the tooth thickness (Table 14.1, column 2), the backlash below the limit 0.016° cannot be achieved. The situation is almost identical with a combination thicker planet teeth and thinner ring gear teeth (column 3). If both, the planet and the ring gear tooth thickness are in the upper class, the limit 0.016° can be exceeded as well (column 4). So, one can conclude, that the chosen gear tooth thickness tolerances and the nominal axis distance with a selected tolerance do not lead to a gear box assembly complying to a basic functional requirement.

Table 14.1 Collected backlash limits for sorted gears in the nominal range (The effect of the angle of a planet rotation is insignificant.)

Backlash	Planet lower/ring upper	Planet upper/ring lower	Planet upper/ring upper
Highest	0.096	0.097	0.067
Mean	0.063	0.064	0.034
Lowest	0.029	0.031	0.000

Table 14.2 Collected backlash limits for sorted gears with shifted tolerance range (The effect of the angle of a planet rotation in the highest backlash range becomes notable and the backlash increases with the angle.)

Backlash	Planet upper/ring lower	Planet lower/ring upper	Planet lower/ring lower
Highest	0.011–0.012	0.0078–0.0082	0.036–0.037
Mean	0.000	0.000	0.004
Lowest	0.000	0.000	0.000

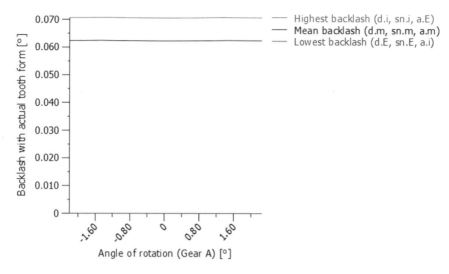

Fig. 14.13 Backlash (eccentricity 0.500js6; DIN3967 h25)

Shifted tolerance grade of the tooth thickness and increased axis distance
It is necessary to shift a tolerance grade of the tooth thickness or change the axis distance. Several iterations of the procedure described in the above sub-chapter lead to an acceptable solution.

So, the nominal axis distance increases from 0.500 to 0.520 mm and at the same time the tooth thickness tolerance is shifted for 0.020 mm towards thicker teeth for both, the planets and the ring gear (Fig. 14.13).

The highest backlash is reached for the lowest tooth thickness and axis distance values and its value (Fig. 14.14) is still above the allowable limit. Therefore, the method of discrimination of gears in two sub-groups based on tooth thickness will be used again. Values are collected in Table 14.2.

If a planet gear tooth thickness is from the upper part and the ring gear from the lower part of the tolerance grade the backlash is within prescribed limit (Table 14.2—column 2). The same is true for the reversed situation (Table 14.2—column 3). A logical conclusion is that if both, the planet and the ring gear, tooth thicknesses are in the upper part of the tolerance zone, an advantageous result is expected. If both gears are chosen from the lower part of the tolerance zone (Table 14.2—column4), the result is similar as the results in mean value of the tolerance zone (Fig. 14.14), therefore unsatisfactory.

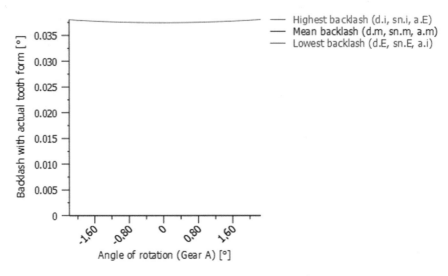

Fig. 14.14 Backlash (eccentricity 0.520js6; teeth tolerance shifted for 0.020 mm)

14.6.2   Contact Analysis

The load considered in contact analyses was prescribed working torque of 120 Nm at the output shaft. Several simulations were carried out with varying the axis distance and tooth thickness deviations. In this context, a transmission error as a function of the angle of rotation of the planet gear, a system stiffness in the contact zone and a contact pressure were simulated. And finally, a simulation of meshing of a planet gear with the ring gear is provided. The aim of the contact analysis is to discover possible interferences due to changes in the axis distance and tooth thickness values.

Several simulations were performed for both tolerance limits, that is for the eccentric and for the tooth thickness:

1.   Axis distance 0.523 mm, tooth thickness deviation −0.020 mm (Fig. 14.15).
2.   Axis distance 0.517 mm, tooth thickness deviation −0.020 mm (Fig. 14.16).
3.   Axis distance 0.517 mm, tooth thickness deviation +0.020 mm (Fig. 14.17).
4.   Axis distance 0.523 mm, tooth thickness deviation +0.020 mm (Fig. 14.18).

Transmission error is small, always in the range of less than 1 arcsec, the diagrams are presented exaggerated. The contact stiffness is in the range from 700 to 720 N/μm. And the contact pressure is in the range between 1100 and 1170 N/mm², which implies usage of heat-treated alloy steels. Furthermore, the meshing scheme shows that around five teeth pairs are always in contact. This was demonstrated also with a CAD kinematic analysis. It was also discovered that individual contact zones are rather small. This implies that only a short zone in the vicinity of the kinematic circles is actually active. And more important, no interferences resulted from these simulations.

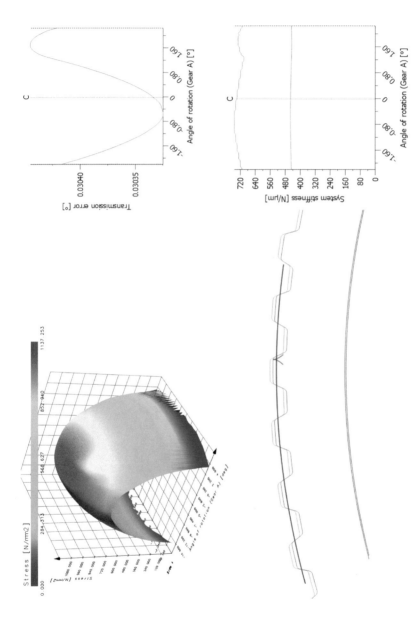

Fig. 14.15 Simulation with axis distance 0.523 mm, tooth thickness deviation −0.020 mm

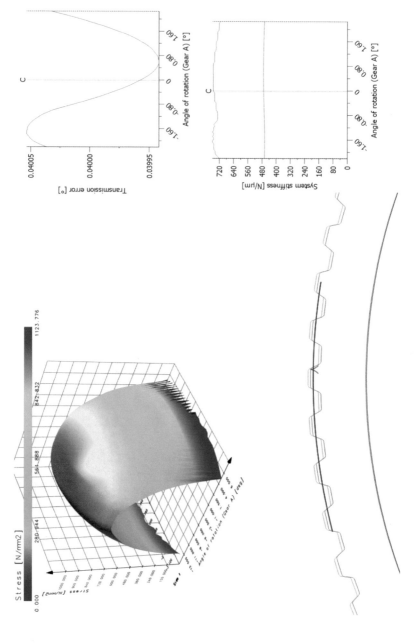

Fig. 14.16 Simulation with axis distance 0.517 mm, tooth thickness deviation −0.020 mm

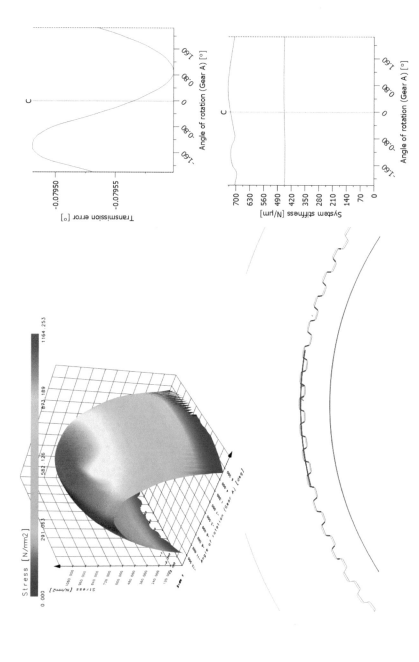

Fig. 14.17 Simulation with axis distance 0.517 mm, tooth thickness deviation +0.020 mm

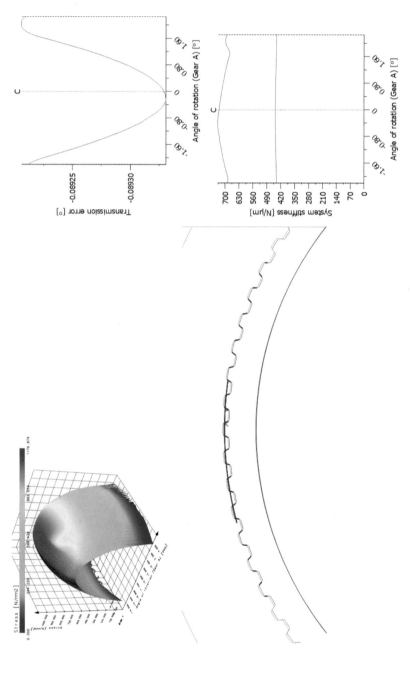

Fig. 14.18 Simulation with axis distance 0.523 mm, tooth thickness deviation +0.020 mm

14.6.3 Influence of Bearing Tolerances and Carriers on the Position of the Gear Train

The influence of carriers has not been considered in so far made calculations. That is why it is necessary to examine axis distance deviations emerging during operation under load (120 Nm). So, a model, clarifying deviations of a planet from its nominal position, was developed in KissSoft. The nominal position in the model amounts to 0 mm, which is due to easier simulation.

The situation illustrated in Fig. 14.19, shows an analysis for the case where for all bearing positions and clearances mean tolerance values are assumed. The axis distance changes for 0.007 mm in a direction towards increasing the backlash. So, if the eccentric is produced on the lowest tolerance limit 0.517 mm, the resulting eccentric link radius becomes 0.510 mm, which increases the backlash.

Assuming a possibility that bearing locations (shafts and housings bores) are made in such a way that these increase the deviation of the axis distance, as illustrated in Fig. 14.20, the axis distance deviation amounts to 0.015 mm and the axis distance 0.505 mm.

Such deviations with an additional displacement of the prescribed eccentricity on the other hand imply possible interference since the tolerances are narrow. So, the above analysis appears to be important in the context of functionality and detection of possible collisions between the ring gear and planet gears teeth tips. Figure 14.21

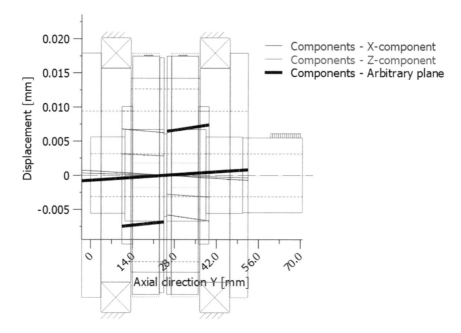

Fig. 14.19 Radial deviation of the planet gear position from the prescribed axis distance

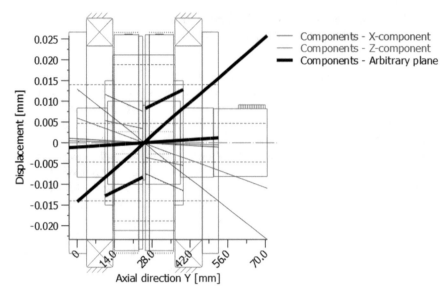

Fig. 14.20 Radial deviation of the planet gear position with increased axis distance deviations

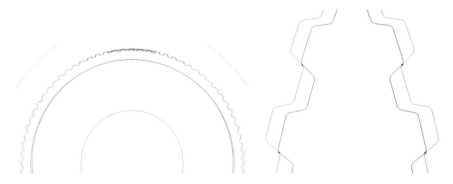

Fig. 14.21 Meshing gear and planet (left) details showing collisions (right)

(left) shows the ring gear and the planet with the already described S-gear geometry ($z_p = 80$, $z_v = 81$). However, Fig. 14.21 (right) clearly shows collisions between the planet and ring gear teeth tips which is located in zones around $-70°$ and $+70°$ from the pitch point. It is necessary to avoid such interferences, and since a near zero backlash can only be achieved with a bit bigger eccentricity (0.520 mm) and with very narrow tooth tolerances the tooth tip rounding must be increased.

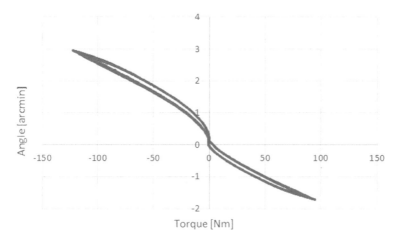

Fig. 14.22   Stiffness and backlash characteristic

14.7   A New Prototype

The next series prototype has been produced, based on the above findings. After CMM inspection of the components, the device was assembled, and its backlash and stiffness characteristic measured by two different methods. First, the testing system induced continuously controlled and measured torque in the range from −120 Nm to +120 Nm, Fig. 14.22. The other method is classical, inducing the output torque by a lever and weights. The resulting backlash was in the range of 0.25 arcmin regardless of measurement method.

14.8   Conclusions

The paper describes development of a planocentric gear train based on the S-gearing principles. The gear design enables high gear ratios. However, the produced prototypes exhibited high backlashes and to high vibrations. Design improvements contributed to better performance, but the design aims were not yet met. So, it was necessary to employ analysis of tolerances, contact analysis and radial deviations of the planet gear positions. Careful examination of the findings led to the improved gearing design—with narrow tooth thickness tolerance and bigger tooth tip rounding. The analysis also showed that planet gears and ring gear should be sorted according to their tooth tolerance position. Gears can be fully loaded and at the same time no interference appears. And the most important, the backlash of the gear train developed based on the tolerance analysis is about 0.25 arcmin (Fig. 14.22). Whereas, preliminary results for the stiffness of the system amount to app. 35 Nm/arcmin.

The new series prototypes will be tested in next weeks to disclose their important characteristics, e.g. kinematic error, vibrations and durability tests. A serial production can be prepared based on testing results analysis and a design of a family of products.

Acknowledgements The investment is co-financed by the Republic of Slovenia and the European Union under the European Regional Development Fund, no. SME 2/17-3/2017 and C3330-18-952014

References

1. Radzevich, S.P.: Dudley's Handbook of Practical Gear Design and Manufacture, Second edn. CRC Press, Taylor & Francis Group, Boca Raton (2012). ISBN 978-1-4398-6602-3 (eBook—PDF)
2. Nieman, G., Winter, H.: Maschinenelemente, Band II: Getriebe allgemein, Zahnradgetriebe Grundlagen, Stirnrad Getriebe. Springer, Berlin, Heidelberg, New York. ISBN 3-540-11149-2
3. Sumitomo Drive Technologies: Fine Cyclo®—Zero Backlash Precision Gear-boxes, Catalog #991333. www.sumitomodrive.com (2018). Accessed 1 Sept 2019
4. Spinea TwinSpin: High Precision Reduction Gears, Ed. I/2017. https://www.spinea.com/en/products/twinspin/index (2017). Accessed 1 Sept 2019
5. Nabtesco.: Precision Reduction Gear RV^{TM} (2018)—E Series/C Series/Original Series CAT.180420. https://www.nabtesco.de/en/downloads/product-catalogue/. Accessed 1 Sept 2019
6. Onvio.: Zero Backlash Speed Reducers. www.onviollc.com (2005). Accessed 1 Sept 2019
7. Wakida, M., et al.: Speed change gear and manufacturing method therefor. Ass. JTEKT Corp., US Patent Application publication, US2011/0245030 A1 (2011)
8. Takuya, H., et al.: Planetengetriebevorrichtung und Verfahren zur Herstellung einer Planetengetriebevorrichtung. Anmelder: Sumitomo Heavy Industries, DE Offenlegungsschrift, DE 10 2012 023 988 A1 (2013)
9. Egawa, M.: Reducer with internally meshing planetary gear mechanism and device incorporating reducer. Ass.: Sumitomo Heavy Industries, US Patent Application publication, US 2006/0025271 A1 (2006)
10. Stäubli.: TS2 Robot Range. https://www.staubli.com/en/file/21027.show (2019). Accessed 1 Sept 2019
11. Hlebanja, J., Hlebanja, G.: Efficiency and maximal transmitted load for internal lantern planetary gears. In: Fawcet, J.N. (ed.) International Gearing Conference, pp. 117–120 (1994)
12. Hlebanja, J., Hlebanja, G.: Patent No. 9300152, Planetary gear train. Slovenian Intellectual Property Office (SIPO) (1994)
13. Park, M.-W., et al.: Development of speed reducer with planocentric involute gearing mechanism. J. Mech. Sci. Technol. **21**(2007), 1172–1177 (2007)
14. Kim, J.H.: Analysis of planocentric gear. Agri. Biosys. Eng. **7**(1), 13–17 (2006)
15. Nam, W.K., Oh, S.-H.: A design of speed reducer with trapezoidal tooth profile for robot manipulator. J. Mech. Sci. Technol. **25**(1), 171–176 (2011)

16. Hlebanja, G.: Specially shaped spur gears: a step towards use in miniature mechatronic applications. In: Miltenović, V. (ed.) 7th International Science Conference on Research and Development of Mechanical Elements and Systems—IRMES 2011, Zlatibor, Serbia, Apr 2011. Proceedings, Niš, pp. 475–480 (2011)
17. Hlebanja, G., Hlebanja, J.: Contribution to the development of cylindrical gears. In: Dobre, G., Vladu, M.R. (eds.) Power Transmissions: Proceedings 4th International Conference, Sinaia, Mechanisms and Machine Science, vol. 13, pp. 309–320. Springer, Dordrecht (2013). ISSN 2211-0984
18. Hlebanja, G., Kulovec, S., Hlebanja, J., Duhovnik, J.: S-gears made of polymers. Ventil **20**(5), 358–367. ISSN 1318-7279. 10. 2014 (2014)
19. KissSoft.: Design software for mechanical engineering applications. https://www.kisssoft.ag/english/home/index.php (2019a) Accessed 23 Sept 2019
20. KissSoft.: KissSoft Release 2019 User Manual. KISSsoft AG—A Gleason Company, Bubikon (2019)

Chapter 15
Analysis of Complex Planetary Gears by Means of Versatile Graph Based Approaches

Józef Drewniak, Stanisław Zawiślak and Jerzy Kopeć

Abstract In the paper, compound and pseudo-compound planetary gears are analyzed by means of graph-based models of planetary gears. The differences of the structures of compound and pseudo-compound planetary gears will be shown. The applied graph-based methods of analysis of planetary are e.g. linear mixed (modified Hsu's) graphs and contour graphs. Some classical approaches are also utilized to show consistency of results. The detailed analyses made for some exemplary layouts of compound planetary gears and variants of their constructional specifications show usefulness of the proposed graph-based approach to planetary gear analysis because it is systematic and algorithmic.

Keywords Linear mixed · Contour graphs · Graph-based models of planetary gear · Compound and pseudo-compound planetary gears · Gear ratios

15.1 Introduction

Planetary gears are used widely. An importance of these mechanisms consists in their compactness, achievement of high ratios with simultaneously moderate size as well as due to the fact that they constitute the base for automatic gear boxes [24] achieving DOF = 1 in every case. The last mentioned idea consist in converting of a planetary gear into an automatic gear box by adding a system of brakes, clutches and a control system which is a modern constructional solution used currently by many leading automotive producers.

J. Drewniak · S. Zawiślak (✉) · J. Kopeć
Faculty of Mechanical Engineering and Computer Science, University of Bielsko-Biała, 2 Willowa Street, 43-309 Bielsko-Biała, Poland
e-mail: szawislak@ath.bielsko.pl

J. Drewniak
e-mail: jdrewniak@ath.bielsko.pl

J. Kopeć
e-mail: jkopec@ath.bielsko.pl

In the present paper we analyze two special, complex planetary gears by means of versatile graph based methods as well as traditional approaches for comparison. These planetary gears are of special layouts where some internal connections are arranged. The aim of the paper is to show usefulness of the graph-based analyses whereas relatively less known graphs are utilized i.e. contour and modified Hsu graphs. Nowadays, bond graphs are considered as a standard tool, however planetary gears are not modeled by means of this tool too frequently, therefore also this approach is valuable to propagate which is done in other papers of the current authors. The comparisons to the classical Willis method confirm correctness of calculations and moreover allow for highlight the advantages of proposed approaches. The comparison of graph-based methods has almost never been considered—just the opposite—usually in a single paper one particular graph-based method had been discussed. Moreover, the contour graph based approach for modeling of gears is relatively less known, but it deals with geometrical quantities which can allow for checking the correctness of dimensional relationships of constructional parts of planetary gears.

15.2   Related Work

Several tools and methods are usually used for modeling of planetary gears. There are e.g.: methods of block diagrams [20], loops [2, 5, 9], special codes [6]. However, the most widely known approach to versatile and advance modeling of mechanical system in general and in particular gears (especially including planetary gears)—is usage of different graphs e.g.: bond-graphs [1], contour graphs [3, 4, 11, 12], flow graphs [14, 22], linear graphs [7, 10, 15, 21], network vector graphs [13], etc. Recently, several review type of books and papers dedicated to the afore mentioned problems have been published (i.e. books—[8, 19, 25] and paper—[23]).

The graph-based methods of planetary gears analysis and synthesis can be used for checking the correctness of a classical approach (i.e. Willis method) but they have also additional advantages e.g. general approach, easiness in introducing constraints, usage of versatile quantities related to gear kinematics which allow for a deeper insight into a gear layout and its operational behavior via virtual engineering analyses. It just has a very important meaning, especially within a conceptual phase of planetary gear design.

It is worth to add that mixed graph is a special type of linear graphs where edges and arcs could appear.

Among others, the following tasks have been performed based on the graph models of gears, especially planetary gears: kinematical analysis, layout analysis, forces analysis, synthesis, isomorphism identification [17], searching for redundant wheels, searching for degenerate structures, artificial intelligence based design [16, 17], ratio and ranges analysis [18] (other topics listed—see review: [23]) etc. Like it can be stated upon the above list of the graph-based methods, these methods are—just recently—under constant development. Moreover, the discussed approach is very useful, effective, handy and flexible. Additionally, it provides a way of an algebraic encoding which opens a possibility for AI and computer aided approaches.

15.3 Theoretical Bases of Graph-Based Modelling of Planetary Gears

Underneath, some rules of assignment of Hsu and contour graphs to planetary gears are briefly summarized. The modelling rules are relatively rarely explained which causes a narrow spectrum of users despite simplicity and effectiveness of these methods.

15.3.1 Linear Modified Hsu Graphs

Several linear graphs have been assigned to planetary gears for tens of recent years. The essence of the approaches consists in the different rules of assignment which are formulated and applied in every case. It seems that Hsu approach is especially useful. We proposed slight but essential modifications which can be recognized as an important upgrade i.e.: usage of Freudenstein's f-cycles idea for these (Hsu's) graphs as well as introducing a path symbolizing a passage of rotational movements throughout a planetary gear (Figs. 15.2a and 15.3a). The start point of the path is described by an arrow 'in' (input) and the end point by an arrow 'out' (output). Therefore, the path starts in an element introducing rotational movement to the gear from an engine. Consequently, the path terminates in an element (shaft) passing the movement outside the gear. In general, there is a possibility of different numbers of inputs and outputs depending of DOF of a particular planetary gear. Additionally, path could turn into network of paths due to passing movement via different ways. We had also introduced the rule that the edges incident to the vertex representing a braked element are drawn by means of double lines. It causes an immediate need for adding respectable equation (that the adequate rotational velocity is equal to 0). It allows also for determination of temporary redundant elements (geared wheels). It means that such elements which are outside the considered path as well as which are not adjacent to double lines—are temporary redundant in case of analyzing of automated gear boxes.

The rules of graph assignment are roughly given. Therefore the initial remarks should be formulated: names of gear elements and relevant graph elements are the same. Sometimes the description of one part is made via double marks e.g. an element has toothing (a) but it is simultaneously a carrier (b) for another row of geared wheels. In such a case, a graph vertex is denoted as a/b.

The rough formulated rules for building a Hsu-graph of a planetary gear—are as follows:

- discretization: choosing elements being involved in passing of rotational movement across a gear. Considering them as vertices with the same description as the gear part (which is modelled). The other elements and phenomena are neglected (e.g. rings, seals, vibrations, fatigue etc.);
- the rules of descriptions are: "a gear element ↔ a graph vertex" and marking them in a traditional way for graphs (a name near a particular vertex) are preserved. However other analyses e.g. of forces, velocities and accelerations can be made also by other graph methods (e.g. bond and contour graphs) however it is beyond the scope of the present paper,
- creation of a linear mixed graph—different relations between pairs of gear elements are represented by edges different types. A pair of the elements 'being in gear (engaged)' is represented via a stripped-line edge, a pair of elements 'arm and planetary gear wheel' is represented via a continuous-line edge, all the pairs of elements rotating around the main gear axis (rotational pairs) are exceptionally represented together via a shaded polygon e.g. (see Figs. 15.2a and 15.3a). This is only for visual reasons. In fact, it is a clique spanned (induced) on all vertices of a polygon;
- cross-domain transfer of knowledge "from machine building to graph theory" which is the essence of the proposed approach. It consists in an assignment of motion equation to a so-called f-cycle, which can be distinguished in the graph. Such f-cycles represent single basic planetary gears embedded in the entire mechanism, the equation of motion is just the equation of rotational velocities. It worth to underline that, the name f-cycle was introduced by Freudenstein. Additionally, this notion should not be matched to general graph theory meaning (as a base cycles in a particular linear space of graph cycles).

New approach in comparison to original Hsu idea consists in an introducing of graph edges drawn as the double lines starting from vertices which represent one or more fixed elements of the gear as well as introducing of a directed path or set of paths (converting edges into appropriate arcs) from an input to an output. Another new idea is marking inputs and outputs. In consequence, the graph converts from a simple graph into a mixed graph.

15.3.2 Contour Graph of Planetary Gear

A contour graph idea is relatively less known. Besides the books of Marghitu, this approach is mainly (even solely) propagated by the authors of the present paper. We use here the nomenclature introduced in the afore cited Marghitu's books (Figs. 15.2c and 15.3c), especially for marking of graph vertices i.e. numbers inside circles and the name (notion) 'contour'—which is just a cycle. The rough formulated rules are as follows:

- modelling consists is approximately the same ideas as used by Marghitu. Additionally, a vertex representing the support system (bearing system) is here also considered i.e. vertex denoted by 0 is introduced;
- the main difference is: that here we consider a directed graph (digraph) built of contours. We start a particular contour from the vertex described as 0. We draw an arc to the chosen vertex representing an element which rotates around the main axis. We continue drawing arcs to the consecutive vertices representing the elements onto which rotational motion is passed one by one (passing through consecutive planets, arm and/or external toothed wheels). We terminate the contour after returning to the 0 vertex. So, we additionally establish a direction of every contour, as well. If the same vertices (representing the same elements) are in other contour in reverse sequence then two arcs in opposite directions are placed between these two vertices;
- the novelty proposed is that the orientations of the contours could be changed by a user by adding needed arcs in such a manner that between some vertices are only arcs in one direction and between some other pairs of vertices—two arcs of reverse directions are drawn. It does not overcome any mechanical rules but it allows even for checking the correctness, solving the systems for two different orientations of some or every contour—the same results are obtained. A change of orientation causes that the indices are assigned in a reverse order. It—in turn—causes that the appropriate values have opposite signs—so changing the orientation of a contour is equivalent to multiplying a respectable equation by '-1' which does not change the solution of the equation system;
- cross-domain transfer of knowledge "from machine building to graph theory": e.g. assignment of several equations to a contour. The complete maximal set of independent contours (in the light of graph theory) has to be distinguished. It is equivalent to the fact that the system of equations is solvable. Independence (linear independence) is here considered for the special linear space of contours (cycles) according to graph theory and algebra rules. Proves are given in the afore mentions books. Moreover it was proved that such a system is solvable!

15.4 Gears' Layouts and Assigned Graphs

In the paper, we consider two planetary gears which are at first glimpse similar. However, like it is shown via presented considerations and calculations—they are essentially different. The first one can be recognized as structurally closed (compound) whereas the second one is structurally similar but it is not compound in the light of statics and it has not inner circulating energy circle. So we propose to call the second one as a pseudo-compound planetary gear. We would like to show that just the graph-based analyses can enable discover and pinpoint the aforementioned differences. The considered gears have three row serial structure which is not too frequently analyzed however graph-based approach allows for systematic analysis of these designs.

The variants of operation are achieved via braking particular parts. Such approach is well known in design of automatic gear boxes and it also is connected with changing the degree of freedom of the gear.

Comparative kinematical analyses of the planetary gears, which functional schemes are presented in Fig. 15.1, have been performed. The considered planetary gears are presented in Fig. 15.1a, b—compound gear and Fig. 15.1c, d—pseudo-compound, respectively. The elements (links) of the gears are depicted via Latin numbers i.e.: 1, 2 ... or h_1, h_2 ... (usually for carriers) but also I, II ... for inputs or outputs. The last mentioned descriptions can be changed if we consider other inputs or outputs.

The assumed numbers of teeth are as follows: $z_1 = z_4 = z_7 = 18$; $z_2 = z_5 = z_8 = 45$; $z_3 = z_6 = z_9 = 108$ (-108). The assumed moduli for stages are as follows: $m_I = m_1 = m_2 = m_3 = 1$, $m_{II} = m_4 = m_5 = m_6 = 1.5$ and $m_{III} = m_7 = m_8 = m_9 = 2$.

The difference in the signs, in the case of the last mentioned quantities, is connected with the notation conventions for number of teeth on inner ring i.e.: "+" for graph based methods and '−' for Willis approach, respectively. However, in turn, there are special rules of assignment sign "±" in gear kinematical equations derived

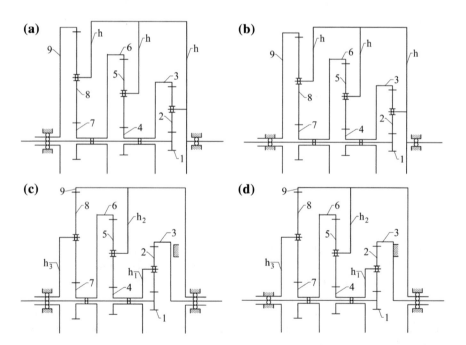

Fig. 15.1 Planetary gears: **a** compound—general scheme, **b** compound with input (1), output (h) and braked (9) elements, **c** pseudo-compound **d** pseudo-compound with input (1), output (9) and braked (3) elements

upon a graph-based method. These are directly connected with inner of outer meshing of geared wheels which all together make that booth approaches are equivalent.

The mechanism (Fig. 15.1a) has 2 DOF (two degrees of freedom). So, braking of a particular element allows us for consideration of variants with different single input and output elements (Fig. 15.1b). Detailed considerations are given in the next chapters.

The DOF of the first considered gear (Fig. 15.1a) can be calculated by means of so called Gruebler's equation:

$$M = 3 \cdot n - 2 \cdot c_5 - c_4 = 3 \cdot 8 - 2 \cdot 8 - 6 = 2 \tag{15.1}$$

Where, n—the number of moving links, c_5—the number of joints of class 5 (full joints), c_4—the number of joints of class 4 (half joints)

The number of independent contours related to the contour graph:

$$N = c - n = 14 - 8 = 6 \tag{15.2}$$

where, c—the number of joints ($c = c_5 + c_4$).

However, the system has one DOF in case when the element 9 is braked:

$$M = 3 \cdot n - 2 \cdot c_5 - c_4 = 3 \cdot 7 - 2 \cdot 7 - 6 = 1 \tag{15.3}$$

In this case, number of independent contours remains the same:

$$N = c - n = 13 - 7 = 6 \tag{15.4}$$

In general, there are three elementary (basic) planetary gears inside both considered planetary gears. The difference between their layouts consists in the fact that in the first case all three elementary gears are embedded on the same arm (carrier) $h = h_1 = h_2 = h_3$. However, in the second case, three distinct arms are considered (h_1, h_2, h_3).

The selected graph-based models of these gears are presented, utilized and compared underneath in this paper. All approaches are applied for modeling of the considered planetary gears, especially kinematical analysis. The graphs representing the compound gear are shown in Fig. 15.2 and the graphs of the pseudo-compound gear are given in Fig. 15.3, respectively.

The most important feature of this approach is that based on these graphs we can encode the facts about a gear in an algebraic way. One exemplary f-cycle and one exemplary contour (Fig. 15.2b, d) were distinguished from two initial graphs (Fig. 15.2a, c), respectively.

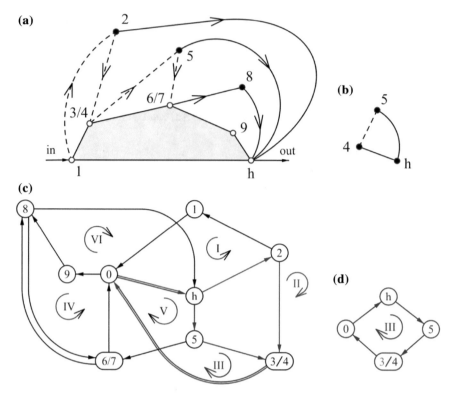

Fig. 15.2 Graphs representing the planetary compound gear (see Fig. 15.1): **a** linear modified Hsu graph of the gear; **b** exemplary f-cycle; **c** contour graph of the same gear; **d** exemplary contour

15.4.1 Compound Gear

Firstly we consider a compound planetary gear which functional scheme is presented in Fig. 15.1a. The calculations of kinematical analysis are performed for every of two graph based methods as well as for Willis's approach. Both, graph methods allow for derivation of the systems of equations describing kinematics of considered gears, however different rotational velocities are taken into account:

– absolute, in case of linear graphs,
– relative, in case of contour graph, so finally they have to be converted into absolute ones.

We consider a particular case when element 9 is braked as well as links 1 and $h = h_1 = h_2 = h_3$ are input and output, respectively.

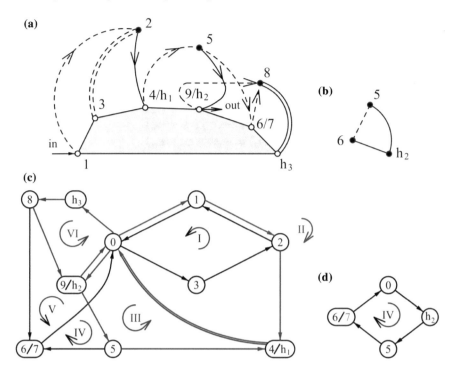

Fig. 15.3 Graphs representing the planetary pseudo-compound gear (presented in Fig. 15.1b): **a** linear modified Hsu graph of the gear; **b** exemplary f-cycle; **c** contour graph of the same gear, **d** exemplary contour

15.4.2 Linear Graph Representing the Considered Compound Planetary Gear

The kinematical analysis of a planetary gear can be performed, among other, based on its linear graph. The linear Hsu's graph for the gear presented in Fig. 15.1a is shown in Fig. 15.2a.

The f-cycles, which can be distinguished in the graph assigned to the considered planetary gear, have the following codes: $(8, 9)h$, $(7, 8)h$, $(5, 6)h$, $(4, 5)h$, $(2, 3)h$ and $(1, 2)h$. The cycle $(5, 6)h$ is presented in Fig. 15.2b together with its code. We place here a description 6 instead of 6/7 because planetary wheel 5 is engaged with a gear wheel with an internal toothing 6. We can observe that the edge $(6/7, h)$ is not explicitly drawn in the graph (Fig. 15.2a). On contrary, it is hidden in the afore-mentioned polygon. In the light of mechanics, an f-cycles represents an elementary gear inside the gear as a whole.

In general, a code of an f-cycle consists of descriptions of the ends (end vertices) of stripped line edge (one end is a planetary wheel) and an arm which description is placed outside the brackets. It has been assumed that the sequence of description in brackets is in increasing or lexicographic order. Every f-cycle code generates one equation describing the kinematics of the singled out mechanism.

Therefore, the entire system of equations describing kinematics of the planetary gear, which can be written upon these codes, has the following form:

$$\begin{cases} (\omega_8 - \omega_h) = +N_{98} \, (\omega_9 - \omega_h) \\ (\omega_7 - \omega_h) = -N_{87} \, (\omega_8 - \omega_h) \\ (\omega_5 - \omega_h) = +N_{65} \, (\omega_6 - \omega_h) \\ (\omega_4 - \omega_h) = -N_{54} \, (\omega_5 - \omega_h) \\ (\omega_2 - \omega_h) = +N_{32} \, (\omega_3 - \omega_h) \\ (\omega_1 - \omega_h) = -N_{21} \, (\omega_2 - \omega_h) \end{cases} \tag{15.5}$$

where: $\pm N_{ij} = z_i / z_j$—local ratio, sign depends on external ($-$) or internal meshing ($+$),

ω_i—absolute rotational speed of i element of the gear.

We can see that a single equation of the above system is equivalent to the Willis formula, however its lower indices are assigned in an algorithmic manner.

Additional equation has to be added—it represents constraints i.e.: it is relevant to the fact that the element 9 is braked i.e. $\omega_9 = 0$.

After some rearrangements we can obtain the formula:

$$\omega_1 - \omega_h = N_{21} \cdot N_{32} \cdot N_{54} \cdot N_{98} \cdot N_{87} \cdot N_{65} \cdot \omega_h \tag{15.6}$$

Therefore, we finally have the value of ratio for the assumed teeth numbers (in this case all teeth numbers are positive):

$$i_{ih} = \frac{\omega_1}{\omega_h} = 1 + N_{21} \cdot N_{32} \cdot N_{54} \cdot N_{98} \cdot N_{87} \cdot N_{65} = 217 \tag{15.7}$$

Underneath, the same gear is analyzed via other graph-based method.

15.4.3 Contour Graph Representing the Considered Compound Planetary Gear

The contour graph for the considered planetary gear is presented in Fig. 15.2c. Six independent contours can be considered inside the contour graph. The contours have been chosen in such a way that every contour starts and ends in the support system (bearings embedded in a gear housing). It causes that relative velocities according to the support system can be easily converted into their absolute equivalents. In fact, just these quantities are searched upon solution of the derived system of equations.

The contours can be described in the following forms:

$$\begin{cases} \text{(i)} & 0 \to h \to 2 \to 1 \to 0 \\ \text{(ii)} & 0 \to h \to 2 \to 3/4 \to 0 \\ \text{(iii)} & 0 \to h \to 5 \to 3/4 \to 0 \\ \text{(iv)} & 0 \to 9 \to 8 \to 6/7 \to 0 \\ \text{(v)} & 0 \to h \to 5 \to 6/7 \to 0 \\ \text{(vi)} & 0 \to 9 \to 8 \to h \to 0 \end{cases} \tag{15.8}$$

For example, the contour (iii) is presented in Fig. 15.2d, denoted as: $0 \to h \to 5 \to 3/4 \to 0$. In detailed description we place the edge description (5, 4) because the planetary wheel 5 is in gear (engaged) with the geared wheel 4 of external toothing. Based upon these descriptions of the contours, their equations can be systematically written. Here we consider only the equations connected with kinematics of the gear but some similar equations for forces and force moments could also be assigned. The original equations deal with the vector quantities. However, due to the fact that all considered rotational velocities (as vectors) act along the main gear axis—the vector notation can be omitted. Sign plus '+" means compatibility with the input rotations, '−' relates to the reverse vector direction. We do not consider bevel wheels. Moreover, the cross product of vectors can be converted into product of scalar quantities due to the fact that an angle between a radius vector and a rotational velocity vector is always equal to 90°, in case of a planetary gear with cylindrical geared wheels. Moreover, for some velocities the radiuses are equal to zero, therefore such summands are omitted in the respectable equations (i.e. in the odd-numbered equations). In case of kinematical analysis, we assign two equations to one description of every contour— because here some unwanted quantities have to be introductory considered and then eliminated.

The discussed system of equations (rewritten into scalar version) is as follows:

$$\begin{cases} \omega_{h0} + \omega_{2h} + \omega_{12} + \omega_{01} = 0 \\ \omega_{h0} + \omega_{2h} + \omega_{32} + \omega_{03} = 0 \\ \omega_{h0} + \omega_{5h} + \omega_{45} + \omega_{04} = 0 \\ \omega_{90} + \omega_{89} + \omega_{78} + \omega_{07} = 0 \\ \omega_{h0} + \omega_{5h} + \omega_{65} + \omega_{06} = 0 \\ \omega_{90} + \omega_{89} + \omega_{h8} + \omega_{0h} = 0 \end{cases} \tag{15.9}$$

$$\begin{cases} \omega_{2h} \times (r_1 + r_2) + \omega_{12} \times r_1 = 0 \\ \omega_{2h} \times (r_1 + r_2) + \omega_{32} \times (r_1 + 2r_2) = 0 \\ \omega_{5h} \times (r_4 + r_5) + \omega_{45} \times r_4 = 0 \\ \omega_{89} \times (r_7 + 2r_8) + \omega_{78} \times r_7 = 0 \\ \omega_{5h} \times (r_4 + r_5) + \omega_{65} \times (r_4 + 2r_5) = 0 \\ \omega_{89} \times (r_7 + 2r_8) + \omega_{h8} \times (r_7 + r_8) = 0 \end{cases} \tag{15.10}$$

where:

ω_{ij}—vector of relative rotational velocity of i link in relation to movement of j link,

r_i—means a pitch radius vector of i point (from wheel center to meshing point or carrier axis, etc.).

Therefore in consequence, we have 12 equations. The code of a contour allows us for systematic assignment of indices of particular quantities. The fifth and sixth equations are assigned to the third contour i.e. $0 \rightarrow h \rightarrow 5 \rightarrow 3/4 \rightarrow 0 \equiv (0, h, 5, 3/4, 0)$. Therefore the sequence of indices in theses equations is as follows: $((h, 0); (5, h); (4, 5); (0, 4))$. Moreover, we have as previously $\omega_{90} = 0$. Since the geometrical configuration of gear axes, i.e. parallel to the main axis of symmetry of the planetary gear as well as straight angle between angular velocities and respective radii it is possible to abandon the vector notation. We assume that the wheel are cylindrical, so true is the relationship $d_i = 2r_i = m \cdot z_i$ and adequate modules are reduced. Therefore, solving the system, we obtain the needed ratio:

$$i_{ih} = \frac{\omega_1}{\omega_h} = \frac{1}{r_1}\left[2r_1 + 2r_2 + \frac{r_1 + 2r_2}{r_4}\left(2r_5 + \frac{2r_4 r_8 + 4r_5 r_8}{r_7}\right)\right] = 217 \quad (15.11)$$

The same result was also obtained via Hsu graph method. The moduli were not used explicitly in case of mixed graph method and in contour method were reduced.

15.4.4 Pseudo-Compound Planetary Gear

Now, we can analyze the pseudo-compound planetary gear (Fig. 15.3).

15.4.5 Linear Graph for Pseudo-Compound Planetary Gear

The f-cycles, which can be distinguished from the graph of pseudo-compound planetary—presented in Fig. 15.3a, have the following codes: $(2, 3)h_1, (1, 2)h_1, (5, h_1)h_2$, $(5, 6)h_2, (8, h_2)h_3$ and $(7, 8)h_3$.

The system of equations describing kinematics of the planetary gear, which can be written upon these codes, has the following form:

$$\begin{cases} (\omega_2 - \omega_{h1}) = +N_{32}(\omega_3 - \omega_{h1}) \\ (\omega_1 - \omega_{h1}) = -N_{21}(\omega_2 - \omega_{h1}) \\ (\omega_5 - \omega_{h2}) = -N_{45}(\omega_{h1} - \omega_{h2}) \\ (\omega_5 - \omega_{h2}) = +N_{65}(\omega_6 - \omega_{h2}) \\ (\omega_8 - \omega_{h3}) = +N_{98}(\omega_{h2} - \omega_{h3}) \\ (\omega_7 - \omega_{h3}) = -N_{87}(\omega_8 - \omega_{h3}) \end{cases} \quad (15.12)$$

Additional equations have to be added. It is relevant to the fact that the elements 3 and h3 are braked $\omega_3 = 0$, and $\omega_{3h} = 0$.

The solution of the system has the following form:

$$i_{ih2} = \frac{\omega_1}{\omega_{h2}} = (1 + N_{21} N_{32}) \left[1 + \frac{N_{65}}{N_{45}} (1 + N_{87} N_{98}) \right] = 301.00 \qquad (15.13)$$

The same gear will be also analyzed via a contour-graph approach.

15.4.6 Contour Graph for Pseudo-Compound Planetary Gear

The contour graph for the considered planetary gear is presented in Fig. 15.3c. Contour no (iv) is presented in Fig. 15.3d. The code of this contour is: $0 \rightarrow h_2 \rightarrow 5 \rightarrow 6 \rightarrow 0 \equiv (0, h_2, 5, 6, 0)$, therefore the sequence of indices in the Eqs. 15.7 and 15.8 is as follows: $((h_2, 0);\ (5, h_2);\ (6, 5);\ (0, 6))$. Six independent contours have to be considered. In this case, the contour codes have the following forms:

$$\begin{cases}
\text{(i)} & 0 \rightarrow 3 \rightarrow 2 \rightarrow 1 \rightarrow 0 \\
\text{(ii)} & 0 \rightarrow 1 \rightarrow 2 \rightarrow h_1 \rightarrow 0 \\
\text{(iii)} & 0 \rightarrow h_2 \rightarrow 5 \rightarrow 4 \rightarrow 0 \\
\text{(iv)} & 0 \rightarrow h_2 \rightarrow 5 \rightarrow 6 \rightarrow 0 \\
\text{(v)} & 0 \rightarrow h_3 \rightarrow 8 \rightarrow 7 \rightarrow 0 \\
\text{(vi)} & 0 \rightarrow h_3 \rightarrow 8 \rightarrow 9 \rightarrow 0
\end{cases} \qquad (15.14)$$

Based upon these descriptions of the contours—their equations can be systematically written. Here we consider only the equations connected to kinematics of the gear. Therefore, we assign two equations to one description of a particular contour:

$$\begin{cases}
\omega_{30} + \omega_{23} + \omega_{12} + \omega_{01} = 0 \\
\omega_{10} + \omega_{21} + \omega_{h1,2} + \omega_{0,h1} = 0 \\
\omega_{h2,0} + \omega_{5,h2} + \omega_{45} + \omega_{04} = 0 \\
\omega_{h2,0} + \omega_{5,h2} + \omega_{65} + \omega_{06} = 0 \\
\omega_{h3,0} + \omega_{8,h3} + \omega_{78} + \omega_{07} = 0 \\
\omega_{h3,0} + \omega_{8,h3} + \omega_{98} + \omega_{09} = 0
\end{cases} \qquad (15.15)$$

$$\begin{cases}
\omega_{23} \times (r_1 + 2r_2) + \omega_{12} \times r_1 = 0 \\
\omega_{21} \times r_1 + \omega_{h1,2} \times (r_1 + r_2) = 0 \\
\omega_{5,h2} \times (r_4 + r_4) + \omega_{65} \times r_4 = 0 \\
\omega_{5,h2} \times (r_4 + r_4) + \omega_{65} \times (r_4 + 2r_5) = 0 \\
\omega_{8,h3} \times (r_7 + r_8) + \omega_{78} \times r_7 = 0 \\
\omega_{8,h3} \times (r_7 + r_8) + \omega_{98} \times (r_7 + 2r_8) = 0
\end{cases} \qquad (15.16)$$

and, as previously, additional equations were taken into account: $\omega_3 = 0, \omega_{3h} = 0$.
Solving this system of equations, we obtain the same solution as previously:

$$i_{i,h2} = \frac{\omega_1}{\omega_{h2}} = \frac{2(r_1 + 2r_2)}{r_1\, r_4}\left[2\,r_4 + 2\,r_5 + \frac{(r_4 + 2r_5)(r_7 + 2\,r_8)}{r_7}\right] = 301 \quad (15.17)$$

The same final results were obtained for all applied methods for both considered layouts of planetary gears. It confirms the correctness of the calculations. It is worth to highlight that these approaches used different set of variables, but finally the ratios are expressed via numbers of teeth.

An idea of cross-domain knowledge transfer is connected with usage of all algebraic structures connected immanently with a particular graph e.g.: matrices, vector spaces of cycles and cuts, paths, polynomials, subgraphs as well as adequate algorithms for generation of the mentioned notions etc. So, the mechanical properties are turn into graph properties, mechanical equations (laws) are expressed by means of the related graph objects. It causes that the graph-based approaches are general, algorithmic and systematic. The indices in all above listed equations are arranged in an algorithmic manner. The completeness of the system consists in consideration of all stripped edges of the appropriate graph i.e. all f-cycles built upon consecutive stripped line edges.

15.5 Classical Approach to Gear Analysis for Compound Gear

The calculations are performed for the case of the compound gear, for the case when $n_9 = 0$ i.e. the same as previously via graph-based approach. The formulas for local ratios based on Willis method are as follows:

$$i_{13}^{h} = \frac{n_1 - n_{h1}}{n_3 - n_{h1}} = \frac{z_3}{z_1}, \quad i_{46}^{h} = \frac{n_4 - n_{h2}}{n_6 - n_{h2}} = \frac{z_6}{z_4}, \quad i_{79}^{h} = \frac{n_7 - n_{h3}}{n_9 - n_{h3}} = \frac{z_9}{z_7} \quad (15.18)$$

The additional conditions are connected to internal unification of some parts as well as the assumed input rotational velocity:

$$n_4 = n_3, \quad n_7 = n_6, \quad n_{h1} = n_{h2} = n_{h3} = n_h = \frac{n_1}{i_{1h}^{9}} = \frac{1500}{217} = 6,91 \text{ rev}/\min \quad (15.19)$$

Notion i_{1h}^{9} is related to the ratio from element 1 to h, in case when element 9 is stopped. Experienced engineer can perform the following recalculations:

$$n_7 - n_h = -n_h \cdot \frac{z_9}{z_7}, \quad n_6 = n_h \cdot \left(1 - \frac{z_9}{z_7}\right), \quad n_4 - n_h = n_h \cdot \frac{z_6}{z_4} \cdot \frac{z_9}{z_7} \quad (15.20)$$

$$n_4 = n_h \cdot \left(1 - \frac{z_9}{z_7} \cdot \frac{z_6}{z_4}\right), \quad n_1 - n_h = -n_h \cdot \frac{z_3}{z_1} \cdot \frac{z_6}{z_4} \cdot \frac{z_9}{z_7}, \quad n_1 = n_h \cdot \left(1 - \frac{z_3}{z_1} \cdot \frac{z_6}{z_4} \cdot \frac{z_9}{z_7}\right) \quad (15.21)$$

Finally, the total ratio (from 1 to h in case when 9 is stopped) is equal to:

$$i_{1\,h}^{9} = \frac{n_1}{n_h} = 1 - \frac{z_3}{z_1} \cdot \frac{z_6}{z_4} \cdot \frac{z_9}{z_7} = 1 - \frac{-108}{18} \cdot \frac{-108}{18} \cdot \frac{-108}{18} = 217. \tag{15.22}$$

So, the same result was achieved like for graph-based method which allows for mutual checking and comparison of the methods.

15.6 Conclusion

Two graph-based methods were successfully applied for an analysis of special planetary gears. Both methods are relatively new, both were updated and enhanced in comparison to initial versions.

References

1. Coudert, N., Dauphin-Tanguy G., Rault, A.: Mechatronic design of an automatic gear box using bond-graphs. In: Proceedings of 1993 International Conference on Systems, Man and Cybernetics, Le Touquet, 2, pp. 216–221 (1993)
2. Ding, H.F., Zhao, Z., Huang, Z.: The establishment of edge-based loop algebra theory of kinematic chains and its application. Eng. Comput. **26**(2), 119–127 (2009)
3. Drewniak, J., Zawiślak, S.: Linear-graph and contour-graph-based models of planetary gears. J. Theor. Appl. Mech. **48**(2), 415–433 (2010)
4. Drewniak, J., Zawiślak, S.: Synthesis of planetary gears by means of artificial intelligence approach—graph based modeling. Solid State Phenom. **164**, 243–248 (2010)
5. Huang, Z., Ding, H.F.: Loop theory and applications to some key problems of kinematic structure of kinematical chains. Front. Mech. Eng. China **4**(3), 276–283 (2009)
6. Hwang, W.-M., Huang, Y.-L.: Connecting clutch elements to planetary gear trains for automotive automatic transmissions via coded sketches. Mech. Mach. Theory **46**(1), 44–52 (2011)
7. Kamesh, V.V., Rao, K.M., Rao, A.B.S.: An innovative approach to detect isomorphism in planar and geared kinematic chains using graph Theory. J. Mech. Des. **139**(12), 122301 (2017)
8. Kaveh, A.: Structural Mechanics: Graph and Matrix Methods, 3rd edn. Research Studies Press (Wiley), Exeter, U.K., (2004)
9. Kecskemethy, A., Krupp, T., Hiller, M.: Symbolic processing of multiloop mechanisms dynamics using closed-form kinematics solutions. Multibody Sys. Dyn. **1**, 23–45 (1997)
10. Laus, L.P., Simas, H., Martins, D.: Efficiency of gear trains determined using graph and screw theories. Mech. Mach. Theory **52**, 296–325 (2012)
11. Marghitu, D.B., Crocker, M.J.: Analytical Elements of Mechanisms. Cambridge University Press, Cambridge, U.K. (2001)
12. Marghitu, D.B.: Kinematic Chains and Machine Components design. Academic Press, London as well as Elsevier, San Diego, USA (2005)
13. McPhee, J.J.: On the use of linear graph theory in multibody system dynamics. Nonlinear Dyn. **8**(1–2), 73–90 (1996)
14. Nagaraj, H.S., Hariharan, R.: Flow-graph techniques for epicyclic gear train analysis. Int. J. Control **17**(2), 263–272 (1973)

15. Nitu, I., Cononovici, S.B., Bogdan, R.C.: Kinematic study of the planetary mechanisms using the directed graphs. In: Eight IFToMM International Symposium on Theory of Machines and Mechanisms—SYROM 2001, pp. 325–332. Bucharest, III, (2001)
16. Ping, Y., Ningbo, L.: Approach on complex neural-genetic algorithm modeling for isomorphism identification in conceptual design of mechanisms. Int. J. Comput. Syst. Sci. Eng. **6**, 61–69 (2009)
17. Ping, Y., Pei, Z., Ningbo, L., Yang, B.: Isomorphism identification for epicyclic gear mechanism based on mapping property and ant algorithm. Eng. Comput. **23**, 49–54 (2007)
18. Salgado, S.R., Del Castillo, J.M.: Analysis of the transmission ratio and efficiency ranges of the four-, five-, and six-link planetary gear trains. Mech. Mach. Theory **73**, 218–243 (2014)
19. Tsai, L.-W.: Mechanism Design: Enumeration of Kinematic Structures According to Function. CRC, Boca Raton, USA (2001)
20. Tsai, M.C., Huang, C.C., Lin, B.J.: Kinematic analysis of planetary gear systems using block diagrams. J. Mech. Des. **132**(6), (2010)
21. Uematsu, S.: An application of graph theory to the kinematic analysis of planetary gear trains. Int. J. Jpn. Soc. Precis. Eng. **31**(2), 141–146 (1997)
22. Wojnarowski, J., Lidwin, A.: The application of signal flow graphs—the kinematic analysis of planetary gear trains. Mech. Mach. Theory **10**(1), 17–31 (1975)
23. Xue, H.-L., Geng, L., Yang, X.-H.: A review of graph theory application research in gears. Proc. Inst. Mech. Eng. C J. Mech. Eng. Sci. **230**(10), 1697–1714 (2016)
24. Zawiślak, S.: Graph-Based Methodology as An Artificial Intelligence Tool for Mechanical Engineering Design. Bielsko-Biała, Poland, p. 296 (2010)
25. Zawiślak, S., Rysiński, J. (eds.): Graph-Based Modelling in Engineering. Springer, Cham (2017)

Chapter 16
Simulation of the Teeth Profile Shaping During the Finishing of Gears

Michael Storchak

Abstract The main process of workpiece transformation into a finished product is the shaping of the working profile of the gear teeth. The processes occurring in the contact of the tool-wheel pair determine the service properties of the treated gear wheel and the technical and economic parameters of the technological system for processing these products. During this period, the specified volume of material is removed and the accuracy of the gear crown and the quality parameters of the wheel teeth surface layer, which uniquely determine the service properties of the transmission, are finally formed. The article presents the results of the development of an imitation model for the shaping of the wheel tooth profile, developed based on the fundamental concepts of the gearing theory and analytical geometry. The model describes the characteristics of the geometric contact in the machine pair. This provides a simulation of the process for machining allowance and changing the contact area in a machine pair at the current machining time. The simulation is performed taking into account the removed volume of material, the conditions for moving the contact area along the surface of the machined wheel teeth, etc. The result of simulation of the shaping model is the vector of the corresponding objective functions, used to optimize the design parameters. The simulation model of shaping is a part of the information system for parametric synthesis of technological systems for finishing of gears. This model is also used as an independent unit for the synthesis of new methods of finishing processing and tool designs presented in the article.

Keywords Computer-aided engineering · Simulation model · Shaping · Gear honing · Objective function

M. Storchak (✉)
Institute for Machine Tools, University of Stuttgart, Stuttgart, Germany
e-mail: michael.storchak@ifw.uni-stuttgart.de

© Springer Nature Switzerland AG 2020
V. Goldfarb et al. (eds.), *New Approaches to Gear Design and Production*,
Mechanisms and Machine Science 81,
https://doi.org/10.1007/978-3-030-34945-5_16

16.1   Introduction

The processes occurring in the contact of the tool-wheel pair determine both the technical and economic parameters of the technological system and the service properties of the gear. One of the main processes of the workpiece transformation into the finished product is the formation of the working teeth profile. During this period, a given volume of material is removed and the accuracy of the gear ring and the quality parameters of the wheel teeth surface layer are finally formed: the altitude and stepping parameters of microgeometry and the physical-mechanical parameters. These parameters uniquely define the service properties of the gear. Therefore, the development of a simulation model of shaping the profile of the gear teeth is a central step in the creation of an information system necessary for the synthesis of technological systems for the gear finishing. A simulation model of profile shaping describes a geometric contact in a machine pair. The model allows at any time of processing to follow the process of removing the allowance, changing the contact area in a pair of tool-wheel, taking into account the volume of material to be removed, the conditions for moving the contact area on the surface of the wheel teeth to be machined, etc. The result of the shaping model functioning should be the vector of the corresponding objective functions used to optimize the design parameters in conjunction with similar vectors obtained using other models of the information system. However, a simulation model of shaping can also be used as an independent development used to synthesize new methods of finishing processing and tool designs. The shaping model is developed based on the main provisions of the gearing and shaping theory.

16.2   Methodology of Shaping Model Developing

Contact conjugative surfaces of the tool and the machined gear wheel is possible with a point, linear and surface tangency. To implement a surface touch in the processes of gear finishing at a given level of industrial development is extremely difficult. With a point contact, the movement of the tool and the wheel mating surfaces is determined by two parameters of their position, and with a linear contact, by one [1]. As a rule, the rotation angle φ of one of the machine pair links, most often of the enveloped link, is taken as one parameter, and the second is the linear movement of one of the links of the pair—ψ. In practice, it is impossible to implement two independent parameters, since the process of formation takes place in time. If we exclude time from the equations describing two-parameter shaping, then we arrive at an equation where one of the parameters of the enveloping depends on the other. Thus, in machine engagement, the surface of one of the machine pair links is the envelope of the one-parameter family of producing surfaces of the other in relative motion or consists of a set of such surfaces. The greater the difference in the change rates of the enveloping parameters, the closer to the one-parameter envelope surface will be the real surface of the product profile. A considerable amount of investigates is devoted to the study of shaping processes, the basis of which is the use of analytical and graph-analytical methods [1–3]. With all the essential inherent advantages, one common drawback of analytical and graph-analytical methods of ideal shaping of complex profile periodic geometrically connected surfaces is the need to find specific solutions for different types of surfaces of instrumental bodies, methods of tool feeding and the required shapes of the workpiece surface. Despite the considerable practical value, sufficient clarity and applicability, the solutions obtained by the methods described cover a relatively small subset of theoretically possible options. This sharply limits the possibilities for synthesizing new technological systems and processing methods. The impossibility of a general analytical solution to the problem of shaping surfaces is due to the substantial nonlinearity of the equations describing the real surface of the tool and workpiece. This nonlinearity is generated primarily by the fact that the surfaces under consideration, as a rule, consist of a set of sections, each of which is easily described by an analytic (continuous and having final derivatives) function, but the condition of continuity of derivatives is violated on the lines of conjugation of surfaces. These shortcomings are devoid of methods related to the group of simulation methods (SM). In application to the field of processing of various products, the methods of SM allow numerical shaping in the presence of an arbitrary number of mutually independent parameters depending on time. The simulation methodology assumes the existence of limited structural equivalence between the model and the object being modeled. The essence of the SM methods is that the object under study is considered as a moving object that exists in time. This object has a set of its own parameters selected by the criteria of materiality and linear independence, also called the "object state vector":

$$V = V_n = \left[a_0, a_1, \ldots, a_{(n-1)} \right], \tag{16.1}$$

where n is the dimension of the object.

The object state vector is represented in simulation as the set of numerical values. Variation of the state vector in time is called "motion of the system":

$$V(t) = V_n(t). \tag{16.2}$$

Movement in the absence of interaction with other objects is called the object's "own motion". The movement caused by the interaction with other objects in the absence of proper movement is called "forced movement". In the presence of interaction and proper motion of an object, one can speak of "disturbed motion." Since the components of the state vector are subject to the requirement of linear independence, one can also introduce the formal concept of an "object state space" [2] as the space of domains of variation of vector component values:

$$M = M_n = \left[A_0, A_1, \ldots, A_{(n-1)} \right], \tag{16.3}$$

where, A_i is the area of change of the i-th component.

The formality of this definition is because there is no suggestion about a fixed dimension of the state space. Moreover, it is necessary to postulate the following: *the number of components of the state vector can be known as one of the motion parameters, i.e. change over time.* In this way:

$$n = n(t), \ V(t) = V(n(t), t), \ M(t) = M(n(t), t). \tag{16.4}$$

The existence of a postulate on the variable dimension of the state space is crucial for the application of the methods of SM in the problems of forming surfaces. The first stage of the computational experiment is the determination of the components of the state space and the construction of motion models. This stage of work is, as a rule, heuristic in nature and cannot be formalized. However, as a component of the state vector, quantities that are usually familiar to a person skilled in the art are usually used. This task is reduced to the choice of their linearly independent subset. Analytical relations characteristic of the subject domain also act as models of motion. The motion of an object can be continuous ($t = t$) or discrete ($T = T_i$) in time, at the same time, the computational process is discrete in nature. As a result, in the case of object motion continuity, mathematical models $M(t)$ constructed for it with continuous time, describing proper, forced and disturbed motion, should be replaced with adequate mathematical models $M(T_i)$ [4] with discrete time:

$$V(T_i) = V(t)|t = T_i. \tag{16.5}$$

Solutions at discrete time points must match. In the discrete-time model, solutions for:

$$V(T)|T_i < T < T_{(i+1)} \tag{16.6}$$

may be missing.

For the software implementation of SM it is necessary to choose a classification set that combines the involved objects, which have common properties, on which it is possible to perform the necessary operations within this set. Such a classification set is the base class of objects. The base class of objects is the basic concept for simulation modeling of the shaping processes. The base classes of objects used in the simulation of machining operations should allow the construction of the eigenvectors of the systems under study and specify their movement. In our case, such classes are Point— a two-dimensional (three-dimensional) point, Matrix—a matrix of a generalized coordinate transformation and Polygon—a list of points used to define contours [5]. Conducting simulation calculation experiment is reduced to a universal algorithm. The flowchart of the simulation algorithm is shown in Fig. 16.1.

To modeling the formation of the gear teeth profile, the following statements are postulated:

- as the interacting objects are the workpiece and the tool;
- models of the workpiece and the tool are their flat sections;

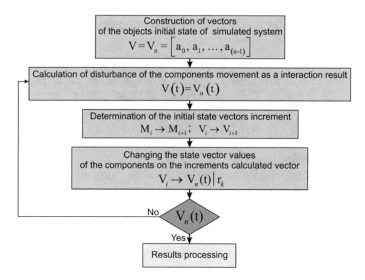

Fig. 16.1 The flowchart of the simulation algorithm for shaping the teeth profile

- state vector of an object of class (type) "flat section" consists of the following components:

  - two spatial coordinates of the anchor center (XC, YC) (t);
  - a list of relative (to the anchor center) coordinates $(X, Y)_i$(t), $i = 1 \ldots N(t)$ of the vertices of the polygon bounding the section contour;
  - logical direction of bypass C;
  - angular coordinate α;
    those, sections are specified in the form of a list of points coordinates connected by straight line segments.

- approximation of the curvilinear sections of the instrument is carried out taking into account the specified maximum deviation of the straight segment (secant) from the theoretical profile by setting the appropriate number of points of the profile section;
- motion occurs discretely with a constant time increment equal to dT. The choice of the discrete value depends on the tool geometrical characteristics and the specific research task;
- own movements of the attachment centers and angular coordinates of the tool and the workpiece occur in space independently, in accordance with the kinematic scheme of the machine on which the processing is carried out. Forced movement centers are absent;
- forced and own movements of the tool contour are absent;
- own movements of the workpiece contour are absent;
- forced movements of the workpiece contour are the result of processing. The increment $d(dX, dY)_i(T_j), T_j = dT \cdot j$ calculated as the intersection of the areas of

space limited by the tool and the workpiece on the j-th simulation cycle is a chip section.

Considering these assertions, at the last step of the generalized algorithm, the workpiece profile (surface) is defined as the envelope of the producing surfaces of the tool in relative motion. This takes into account macro and micro deviations, the cross section of the material removed layer, the geometric contact area in the tool-wheel pair, taking into account the material removed layer (previous cuts) and the total cross section area of the chip. The latter can be used as a measure of the energy intensity of the shaping process.

For the processes of shaping complex profiles with geometrically and functionally related surfaces, such as gears, it is necessary to distinguish the basic classes inherent only in this type of products. These specific classes describe the generating contour, the tool body, the kinematic processing scheme, the main and auxiliary movements of the tool and the workpiece, their geometric parameters, processing modes, etc. The evaluation of the accuracy of the simulation modeling of shaping gears was performed by comparing the envelope surfaces obtained by the simulation method and the well-known analytical method.

To determine the desired characteristics, we use known relations from the theory of differential geometry and gearing [1]. Consider the general case of the tool location and the processed gear wheel. The casting center (pole) of the tool and the gear wheel lie at an arbitrary point in space—Fig. 16.2 [6].

Fig. 16.2 Scheme of the relative screw movement reduction of the tool and gear wheel mating surfaces

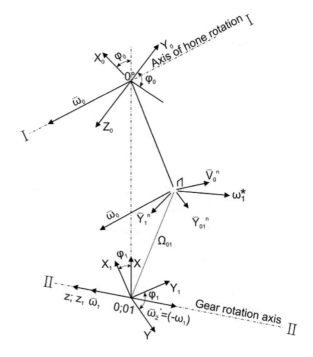

The system of spatial rectangular Cartesian coordinates $X_1 O_1 Y_1 Z_1$ is associated with the gear wheel, the tool – with system $X_0 O_0 Y_0 Z_0$, and with the fixed stand—system XOYZ. The first two systems are mobile the third one is fixed. The counterclockwise motion for an observer looking from the positive Z-axis direction is taken as the positive direction of rotation of the links. The profile unfolding from right to left is taken as the right gear wheel tooth profile. The positive direction of the crossing angle of the axes γ is the rotation of the Y- and Z-axes in a clockwise direction for an observer looking from the positive direction of the X-axis. The definition of the envelope surface is made using the kinematic method. In accordance with this method, the velocity vector of relative motion at the point of contact—v_1 should lie in a plane tangent to the conjugative surfaces [1, 7]:

$$n_1 \cdot v_1 = 0, \tag{16.7}$$

where, n_1 is the normal vector to the lateral surface of the gear wheel teeth.

The profile of the lateral surface of the wheel teeth is given by the radius-vector in parametric form:

$$r_1 = r_1(u, v), \tag{16.8}$$

where, u and v are the parameters of the surface, the envelope to which is the desired one.

If the lateral surface of the wheel teeth is a right involute helicoid, then a column vector represents its radius vector [1, 2]:

$$r_1 = \begin{vmatrix} r_{b1} \cdot cosv + u \cdot cos\lambda_{b1} \cdot \sin v \\ r_{b1} \cdot sinv - u \cdot cos\lambda_{b1} \cdot v \\ p \cdot v - u \cdot sin\lambda_{b1} \end{vmatrix}, \tag{16.9}$$

where, p—screw parameter; λ_{b1}—elevation angle of the helix.

In the left involute helicoid, the signs in front of the v parameter will be changed. For r_1, v_1 and n_1 are determined, and after substituting the latter in (16.8) and performing the transformations, we obtain the engagement equation:

$$f_1(u, v, \varphi_1) = 0, \tag{16.10}$$

where, φ_1 is the angle of rotation of the machined gear wheel.

Relationship (16.10) together with (16.9) for a specific φ_1 determine the equation of the contact line, and for different values of φ_1—the equation of the engagement surface in the moving wheel coordinate system, that is, the lateral surface of the wheel teeth. To determine the gearing surface in the fixed coordinate system and the moving system associated with the tool (i.e. the lateral surface of the tool teeth), it is necessary to make the transition from the gear wheel system to the fixed system or tool system. The transition is carried out by transforming the coordinates by multiplying the orthonormal transition matrix from the left by the radius vector (16.9) with (16.10) taken into account:

$$r_2 = M_{21} \cdot r_1, \quad r_2 = M_{02} \cdot r_1, \tag{16.11}$$

where, r_2 is the radius vector of the engagement surface in the fixed system; r_0 is the radius vector of the tool teeth lateral surface; M_{21} and M_{02} are orthonormal transition matrices from the gear wheel system to the fixed coordinate system and from it to the tool system, respectively:

$$M_{21} = \begin{vmatrix} cos\varphi_1 & -sin\varphi_1 & 0 & 0 \\ sin\varphi_1 & cos\varphi_1 & 0 & 0 \\ 0 & 0 & 1 & 0 \\ 0 & 0 & 0 & 1 \end{vmatrix} \tag{16.12}$$

$$M_{02} = \begin{vmatrix} cos\varphi_0 & cos\gamma \cdot sin\varphi_0 & -sin\gamma \cdot sin\varphi_0 & a \cdot cos\varphi_0 \\ -sin\varphi_0 & cos\gamma \cdot sin\varphi_0 & -sin\gamma \cdot sin\varphi_0 & -a \cdot cos\varphi_0 \\ 0 & sin\gamma & cos\,\gamma & 0 \\ 0 & 0 & 0 & 1 \end{vmatrix}, \quad \varphi_0 = \frac{\varphi_1}{i_{10}} \tag{16.13}$$

where, φ_0—the angle of tool rotation; i_{10}—gear ratio.

Thus, the engagement surface of a fixed coordinate system is determined by the expression:

$$\bigcup_{k=1}^{m} \{f_1(u_k, v_k, \varphi_{1k}) = 0, r_1\}, \quad r_1 \subseteq \mathbb{R}^3, \tag{16.14}$$

where $\mathbb{R}^3$ is three-dimensional space,

and the lateral surface of the tool teeth:

$$\bigcup_{k=1}^{m} \{f_1(u_k, v_k, \varphi_{0k}) = 0, r_0\}, \quad r_0 \subseteq \mathbb{R}^3 \tag{16.15}$$

Equation (16.18) defines the lateral surface of the tool teeth or the producing tool surface. To describe the surface of a workable tool, its lateral surface (16.15) or engagement surface (16.14) should be limited to the size, shape of the teeth tips surface and the butt surfaces. Thus, the producing surface of an operable tool or the active engagement surface can be interpreted as the intersection of the coordinates set of the engagement surface (16.14) with the set of coordinates of the bounding surfaces:

$$\bigcap_{k=1}^{m} \{r_{ks3}, r_{ks1}, r_{ks0}, r_{kb1}, r_{kb0}\}, \quad r_k \subseteq \mathbb{R}^3, \tag{16.16}$$

where, rks3, rks, rkb, are the radius vectors of the engagement surfaces, the teeth tips and the butt surfaces, respectively.

One of the main parameters of the ruled contact of the tool with the gear wheel is the length of the contact line, which is determined in accordance with (16.16) from the known relationship for the length of the curve [4]:

$$S = \int_{v_1}^{v_2} \sqrt{\left(\frac{dx}{d\varphi_1}\right)^2 + \left(\frac{dx}{d\varphi_1}\right)^2 + \left(\frac{dx}{d\varphi_1}\right)^2} \, d\varphi_1 \qquad (16.17)$$

where, v_1, v_2—the smallest and largest value of the parameter v of Eq. (16.8), defining the boundaries of the engagement surface.

The dependences given allow determining the engagement surface, identifying the lateral surface of the tool teeth and the length of the contact lines. To consider the characteristics of the linear contact with two-profile gearing in a pair of tool-wheel, one must enter the second (left) profile. This is done by changing the sign to the opposite before the parameter v in (16.8) and shifting the beginning of the involute helicoid in the direction of the wheel rotation, by an amount determined by the thickness of the wheel teeth to be machined. According to the developed software for determining the parameters of enveloping at the ruled contact of the tool with the gear wheel being processed, a software-implemented algorithm has been developed which has a hierarchical structure. The developed simulation model of shaping on the basis of numerical modeling provides identification of the gear profile to be machined and tracking of its shaping at each fixed point in time. This allows one to study the patterns of profile formation of the gear wheels teeth and chip formation parameters. On this basis, the vector of objective functions is determined, which is used later to optimize the design parameters of the technological system, and new processing methods and tools for their implementation are synthesized. The developed model makes it possible to numerically simulate the process of mutual enveloping of the tool teeth profiles and the gear wheel being machined at their linear contact. It provides the study of the line contact characteristics regularity—the length of the contact lines, the total length of the contact lines, the size and location of the engagement field in space, and the change in the length of the contact lines in the engagement phase. These characteristics are used as objective functions, and also allow you to determine the area of search for new designs of tools that provide ruled contact with the gear wheel being processed.

16.3 Analysis of Objective Functions of the Model

Numerical modeling of the wheels teeth profile to be machined is performed in the butt sections of the wheel by means of rack and helical profiles [2]. The analysis of graphic materials allows to establish the regularity of the relative movement of the tool and the wheel during its shaping. Consecutive positions of the producing surface of the tool determine the cross section of the cutting layer. The dimensions of the cross section of the layer cut by the working surface of the tool depend on the shape of the cutting edge and the law of its movement between two successive cuts. Characteristic for loading cutting edges or cutting surfaces is the variability of sections of cuts in the process of relative movement, changing the thickness of the slice and uneven loading of cutting surfaces even for tools with a symmetric tooth profile. The ability to determine the section of the cut at any given time or at a given point of the machined surface allows you to determine the contour area of contact of the tool working surface and the gear wheel at a given moment of shaping, that is, taking into account the previous material cuts. The shape of the heat source acting on the contact surface is taken in the form of a rectangular strip elongated perpendicular to its movement [8], and the distribution of thermal energy over the contact area is uniform. However, the uniform distribution of thermal energy over the contact area, which corresponds to such a scheme, can only be taken as a first approximation [9], in cases where the length of the platform is significantly greater than the width. By numerous experiments, it was found that the maximum temperature is located on the periphery of the profiling surface, i.e. the shape of the source is significantly different from the rectangle and segment. Numerical simulation of shaping the gears teeth profile with regard to previous sections allows you to determine the shape of the contact area and its dimensions. The contact area is located on the surface of the transition from the machined surface of the gear wheel to the untreated one. In the case of grinding gears with the "Maag" null-degree method, it is a shape bounded by a curvilinear triangle abc, when the disc circles are displaced, by an amount of the rolling path S_0 (I and II—the position of the axes of symmetry of the circles)— Fig. 16.3. The volume of material removed decreases to the axis of the circle and the point a of the circle lying on its axis is profiling.

According to the dimensions of the section of the cut, its area and volume, the area of contact of the tool with the wheel, the energy intensity of the finishing process is estimated. The integral characteristic of energy intensity can be the total area of the

Fig. 16.3 Contour contact area during gear grinding by the zero-degree method "Maag"

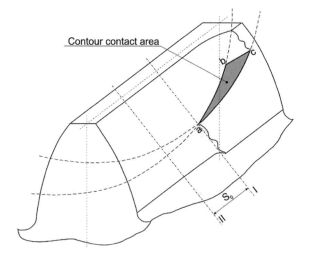

cut over all simultaneously working cutting surfaces or cutting edges (or its linear prototype—the total length of the cut), as well as the total area of the cut during the formation of the tooth profile or a given part of it.

16.4 Analysis of Linear Contact Objective Functions

The following are taken as independent geometrical parameters of the machine tool when calculating the characteristics of a ruled contact: machine gear angle α_{nw01}, number of tool teeth—z_0, angle of inclination of the tool or wheel teeth line, or crossing angle of their axes $\beta(\gamma)$. The location in space, the shape and dimensions of the engagement field uniquely determine all the geometrical parameters of the contact. The engagement field in a fixed coordinate system is a closed part of space that extends along the Z-axis—Fig. 16.4. In moving coordinate systems, the engagement field bounded by the butt surfaces of the tool and the wheel to be machined and the surfaces of their teeth tips is the lateral surface of the gear wheel teeth (in the gear wheel system) and the tool (in the tool system). The position of the engagement field

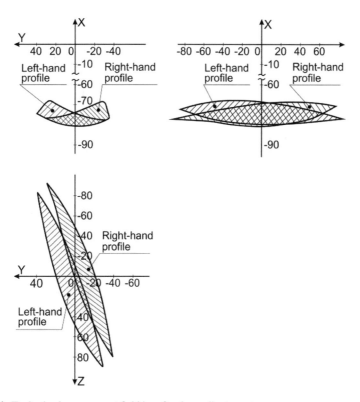

Fig. 16.4 Tool-wheel engagement field in a fixed coordinate system

is expediently analyzed by the angle of rotation of the wheel φ_k to when the tool comes into contact with the wheel—φ_{kin} and out—φ_{kout}. The difference ($\varphi_{kout}-\varphi_{kin}$) determines the dimensions of the engagement field, and the absolute values of φ_k—the position of the engagement field in space. By changing the initial parameters of the machine gear, for example, α_{nw01} and β (γ), it is possible to control the shape and location of the engagement field in space—Fig. 16.5.

One of the main characteristics of a tool-wheel line contact is the parameters of contact lines: their length, total length, change in total length, and the difference in total length of contact lines along the right and left profiles. Since the boundaries of the engagement field are the locus of the points of the contact lines ends, the parameters of the contact lines are uniquely determined by the shape, size, and location in the engagement field. Consequently, the parameters of contact lines are determined by the values of independent parameters: α_{nw01}, β (γ), z_0, as well as a certain combination of these values. Therefore, their magnitude and nature of change can be controlled using independent parameters of the machine engagement.

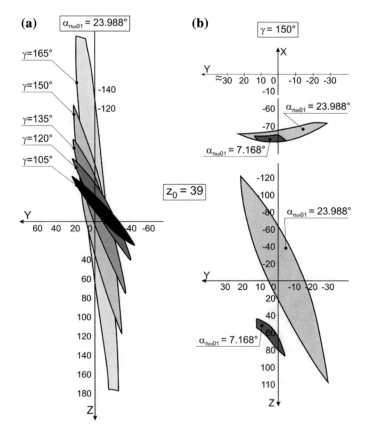

Fig. 16.5 Resizing and positioning in the space of the engagement field projection depending on the crossing angle γ (**a**) and on the engagement angle α_{nw01} (**b**)

With increasing φ_k to the length of the contact lines changes according to a parabolic law, taking the greatest value approximately in the middle of the engagement phase and the minimum at the beginning and end of the engagement—Fig. 16.6.

The total length of the contact lines is the main characteristic that determines the stability of the machining process, since it is proportional to such process characteristics as the contact area, the engagement factor, the force acting in the engagement, the stiffness of the engagement, etc. With a change in the independent parameters of the machine gear γ, α_{nw01} and z_0, the total length of the contact lines changes in the same way as the change in the length of the contact line. When limiting the field of engagement with the surface of the tool teeth tips and the wheel being machined and

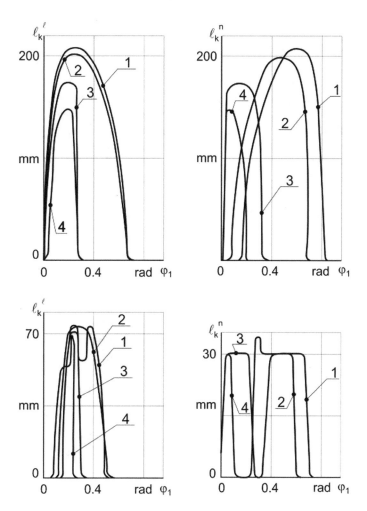

Fig. 16.6 Change the length of the contact lines on the angle φ_k (1—$\alpha_{nw01} = 23.483°$; 2–$\alpha_{nw01} = 20.483°$; 3—$\alpha_{nw01} = 11.483°$; 4—$\alpha_{nw01} = 8.056°$; $z_0 = 43$; $\beta_0 = 15°$)

their butt surfaces, the total length of the contact lines varies significantly, both along the right and along the left profiles. The greatest influence on them has a change α_{nw01}.

The considered objective functions reflect the interrelation of independent parameters and properties of gear wheels. It is not possible to determine the desired functions explicitly due to their complex description and the absence of an analytical solution. Therefore, the objective functions are determined numerically at a given point in time, as the result of the functioning of the developed simulation model.

As a result of the analysis, it has been established that the desired values are most affected by α_{nw01}, $\gamma(\beta_0$ or $\beta_1)$, z_0, the relative location of the tool and the wheel being machined—ψ, the shape of the producing surface, determined by its size and binding, the kinematic scheme of the machining mechanism, determined by the shaping movements of the tool and the wheel, including the constructive movement of the generating surface tool speed relative movement. The specified parameters are independent for the developed shaping model. Thus, the simulation model of shaping, presented in the form of a package of software-implemented algorithms, is a mechanism for controlling the characteristics by means of which it is possible to provide the necessary properties of gear wheels, as well as to synthesize finishing processing methods and tools for their implementation. In particular, high performance processing methods are of considerable interest.

At the present level of technological development, the most promising way to increase productivity is to ensure line contact by synthesizing tools that ensure it is in contact with the wheel.

16.5 Synthesis of Processing Methods and Tools that Provide Line Contact

As a result of the functioning of the model of shaping the gear tooth profile, it is determined that the shape of the working part of the tool teeth has a concavity along the length of the tooth. At the same time, the contact area with the wheel being processed is more than an order of magnitude higher than the contact area at a point touch—Fig. 16.7. Therefore, the implementation of the line contact will dramatically increase the contact area, which means the number of simultaneously working cutting edges. This in turn will provide a significant increase in processing performance. Characteristics of the line contact are the area and location in the space of the engagement field, the length of the contact lines, and the change in the total length of the contact lines, the difference in the total length of the contact lines on the right and left profile within the engagement field.

The area of the engagement field is proportional to the processing performance and tool life, since the area of the engagement field is proportional to the contact area of the tool with the wheel, and therefore proportional to the number of abrasive grains that are simultaneously in contact. Similar to the area of the engagement field,

Fig. 16.7 Tooth shape tool for finishing machining of gear wheels, providing line contact

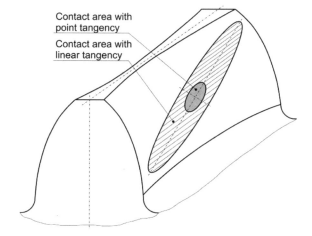

the length of the contact lines is proportional to the machining performance and tool life, since the length of the contact line determines the contour of the engagement field.

One of the factors causing oscillations of the tool and wheel during finishing processing is the periodic change in the stiffness of the gearing [7]. For a line contact, this is caused by a corresponding change in the total length of the contact lines. Thus, reducing the change in the total length of the contact lines leads to stabilization of the movement of the links of the tool-wheel pair, and, consequently, to an increase in tool life, productivity and machining accuracy. The processing of the right and left profiles of the wheel teeth occurs with varying degrees of intensity [2]. This is due to the difference in the efforts and the total length of the contact lines on the opposite profiles. Obviously, reducing the difference in the total length of the contact lines on the opposite profiles will allow to equalize the amount of material removed, and, consequently, to increase the accuracy of processing.

The above considerations make it possible to determine the ways to search for new processing methods and tools. The application of the developed model of geometric contact ensures their synthesis in relation to the line contact of the tool and the wheel being machined.

The most important thing is to increase the productivity of processing for time-consuming methods of finishing processing, for example, gear grinding. Known methods worm globoid grinding [2], providing a significant increase in processing performance. The shaping of a worm tool for the implementation of this method is carried out with the help of a dressing gear. The main disadvantage of these methods is the presence of a theoretically ruled contact only at the end of the processing cycle of each wheel, since the dressing gear has geometrical parameters identical to the corresponding parameters of the wheel being machined. In addition, there is a need for separate processing of each profile with an inevitable reverse for processing of opposite profiles. This leads to loss of processing performance, reduced accuracy and deterioration of the surface layer of the wheels teeth. To eliminate these drawbacks, it is necessary to realize the ruled contact of the globoid grinding wheel with the wheel being machined during the period of the most intensive removal of the allowance, i.e. on rough passes. It is most expedient to carry out the linear contact of the grinding wheel with the gear wheel after reaching a certain kinematic accuracy of the wheel, characterized by the presence of treated sections on all the same tooth profiles of gear wheel. It was established experimentally that such kinematic accuracy of the gear wheel being machined is achieved after removing about a quarter of the total allowance [2, 6]. Processing performance and the state of the surface layer during the finishing treatment of gear wheels can be controlled by changing the contact area of the tool with the wheel. When worm globoid grinding, the contact area can be controlled by changing the geometrical parameters of the dressing wheel relative to the corresponding parameters of the workpiece. In a specific technical solution, this is done by changing the thickness of the tooth of the dressing gear by 0.25–0.75 from the removed allowance—Fig. 16.8.

In accordance with the synthesized method, the worm globoid wheel 1 is introduced into engagement with the wheel 2 being machined. Mutual rotation, radial S_p and axial S feed impart them—Fig. 16.8a. The worm globoid wheel is preliminarily profiled and the dressing gear 3 subsequently restores its cutting ability and accuracy—Fig. 16.8b. The thickness of the dressing gear teeth B exceeds the thickness of the wheel teeth to be processed C by 0.25–0.75 (b) of the total allowance (a) for grinding (A is the surface of the workpiece teeth). When using such a dressing

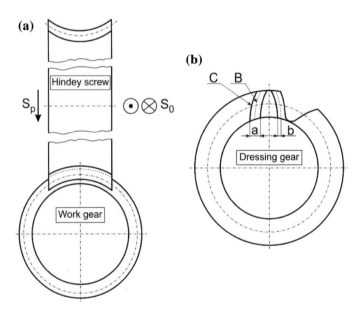

Fig. 16.8 Worm globoid grinding method

(a)

(b)

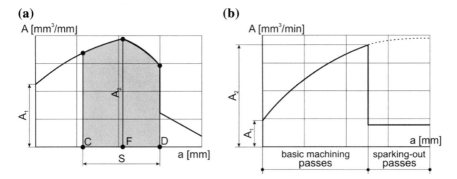

Fig. 16.9 Changes in the performance of the synthesized **a** and traditional **b** method of worm globoid grinding

gear, the performance of gear grinding changes extremely as you remove the stock—Fig. 16.9a. The initial value of productivity—A_1 in this case is much higher than when using the traditional method of grinding—Fig. 16.9b, as long as the contact area in the grinding wheel-gear wheel engagement is much larger, since linear contact is provided on roughing passes at point S. In this case, the performance monotonously increases to the point at which the linear contact is theoretically realized, and then decreases as the contact turns into a zonal one. The optimal implementation range of the line contact is limited to points C and D (on the x-axis) and shaded.

A radical reduction in the change in the total length of contact lines and even the reduction of this difference to zero can be achieved by choosing the shape of the surface of the tool tops. To do this, select a certain value of the contact lines total length (for example, the greatest), and then for different φ_1, the parameter v changes until the total length of the contact lines is the same. For the fixed parameter v, the diameter of the tool teeth tips in each section is determined for given φ_1 by the equation:

$$d_{a0} = d_{b0} \cdot \sqrt{v^2 + 1} \tag{16.18}$$

A fragment of the working part of the diamond gear hone with a constant total length of contact lines and the shape of the working elements of its right and left profile are shown in Fig. 16.10.

The tool consists of a body 1 and a gear ring 2 with a diamond layer 3 applied to it. The active height of the tool tooth profile h_a varies across the width of the gear ring, monotonously decreasing from one butt to the other in opposite directions on different sides of the tooth from the full height h_{amax} to the minimum value h_{amin} determined by the algorithm. Reducing the difference in the total length of the contact lines on the right and left tooth profiles can also be ensured by shifting the right and left working elements relative to each other in the direction of the Z-axis of tool rotation. A fragment of the longitudinal section of the gear hone of this design and its general view is shown in Fig. 16.11.

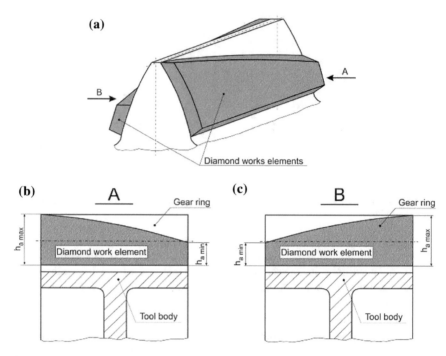

Fig. 16.10 Fragment of the working part of the diamond gear hone teeth with a constant total length of the contact lines

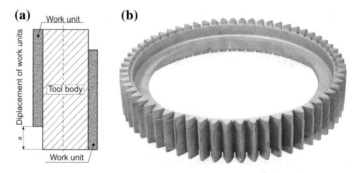

Fig. 16.11 Longitudinal section of the tool tooth (**a**) and a general view of the gear hone (**b**) with a reduced difference in the total length of the contact lines in the unlike profiles

Linear contact of the tool with the gear wheel occurs after a certain period of running in or after dressing. Directly linear contact takes place only in a very short period of time (for example, immediately after dressing) due to tool wear, and in the main contact is quasilinear. To ensure line contact over the entire period of tool life and exclusion of the period of running in of the tool-producing surface, it is necessary to immediately give the necessary shape. To do this, the side surface of the tool body

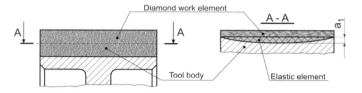

Fig. 16.12 The design of the gear hone, which provides line-contact with the wheel to be machined

teeth is performed as a one-parameter envelope of the producing surfaces family of the wheel teeth in relative motion, and the space between the working element 2 and the body 1 is filled with elastic material 3—Fig. 16.12. In contact with the wheel to be machined, the working elements of the tool will bend under the action of the working force and take on a shape that provides line contact.

Close to ruled contact is also possible when the tool is made of several parts with a common axis of rotation—Fig. 16.13. When loading a gear hone with a radial or circumferential load, its component parts will engage in the entire width of the tool gear ring, realizing a contact close to linear. To increase the total length of contact lines and reduce the magnitude of their change, the shape of the gear hone outer surface is advisable to perform globoid—Fig. 16.13a. A similar result can be obtained by connecting the component parts of the tool with elastic elements with different stiffness—Fig. 16.13b.

The proposed options for processing methods and tools for their implementation provide a linear or quasilinear contact with the processed wheel. When this is achieved the highest performance for the known methods of finishing processing by running.

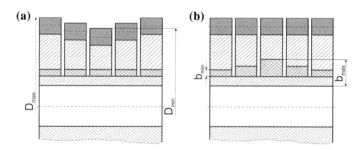

Fig. 16.13 Composite gear hone

16.6 Conclusions

Based on simulation methods, a model has been developed for shaping the profile of the teeth of a machined wheel when specifying a generating surface and relative motion in a general form. The model is part of an information system that describes the relationship of objects of the technological system for finishing processing gears.

Based on the numerical simulation of the finishing process of complex profile geometrically and functionally interconnected surfaces, the analysis of the shaping characteristics of the tooth profile and the analysis of the tool characteristics at the ruled contact with the wheel being processed as objective functions of the developed model are performed.

As a result of the analysis and numerical modeling of the shaping process, finishing processing methods and tools have been synthesized that provide a significant increase in machining productivity and tool life. These processing methods and tools are synthesized using the developed model by numerical simulation of the shaping process. However, in our opinion, the proposed model of forming the workpiece teeth profile allows to synthesize a much larger number of processing methods and tools with the provision of the specified parameters, since it operates with generalized geometric images.

References

1. Litvin, F.L., Fuentes, A.: Gear Geometry and Applied Theory, 2nd edn, p. 800. Cambridge University Press, Cambridge (2004)
2. Storchak, M.: Technological Systems for Finishing Gears, p. 448. Institute for Superhard Materials, Kiev (1994)
3. Krivosheya, A., Danilchenko, J.U., Storchak, M., Pasternak, S.: Design of shaping machine and tooling systems for gear manufacturing. In: Theory and Practice of Gearing and Transmissions, pp. 425–450. Springer, Berlin (2016)
4. Kleene, S.C.: Mathematical Logic, p. 432. Dover (2001)
5. Cooke, D.J., Bez, H.E.: Computer Mathematics, p. 394. Cambridge University Press, Cambridge (1989)
6. Radzevich, S.P.: Theory of Gearing: Kinetics, Geometry and Synthesis, p. 685. Taylor & Francis Group (2013)
7. Babichev, D., Storchak, M.: Synthesis of cylindrical gears with optimum rolling fatigue strength. Prod. Eng. Res. Dev. 9(1), 87–97 (2015)
8. Möhring, H.-C., Kushner, V., Storchak, M., Stehle, M.: Temperature calculation in cutting zones. CIRP Ann. Manufact. Technol. 67(1), 61–64 (2018)
9. Heisel, U., Storchak, M., Eberhard, P., Gaugele, T.: Experimental studies for verification of thermal effects in cutting Production Engineering. Res. Dev. 5(5), 507–515 (2011)

Chapter 17
Design Automation of Cylindrical Gear Manufacturing Processes

Mark M. Kane

Abstract The article shows the urgency of the problem of automation of design of technological processes of manufacturing of cylindrical gears, the possible ways of its solution are considered, the methods of designing the processes of manufacturing of cylindrical gears on the basis of modeling the processes of formation and changes in the parameters of quality of gears at different stages of their manufacture are proposed, the various stages of implementation of these techniques (the choice of routes of processing teeth and base surfaces of gears, the choice of requirements for accuracy and quality of surfaces of teeth and base surfaces of gears at different stages of their processing, the choice of modes of gear cutting by worm cutters) are considered, shows the need for experimental debugging of the designed process and methods of such debugging at different stages of the design process. Implementation of the proposed methods and recommendations with the help of computer technology will automate the design and debugging processes of manufacturing of cylindrical gears, improve the quality of design and gears, as the proposed methods allow to take into account the maximum conditions of manufacture of gears and appropriateness of formation of their quality parameters.

Keywords Indicators of accuracy of teeth of cylindrical gears · The accuracy rate of the base surfaces of the cylindrical gears · The elements of the technological process of processing of gears · The choice of processing routes of the teeth and underlying surfaces of the gears · The choice of the accuracy requirements of the teeth and underlying surfaces of gears for different operations of processing · The choice of modes of gear hobbing

M. M. Kane (✉)
Belorusian National Technical University, Minsk, Republic of Belarus
e-mail: kane_08@mail.ru

© Springer Nature Switzerland AG 2020
V. Goldfarb et al. (eds.), *New Approaches to Gear Design and Production*,
Mechanisms and Machine Science 81,
https://doi.org/10.1007/978-3-030-34945-5_17

17.1 Introduction

Cylindrical gears are among the most mass, complex and critical parts of machines. They largely determine the performance of machines and mechanisms, which are used. The accuracy of the gears is characterized by a large number of parameters, each of which is formed and changed during the processing of the wheels depending on a large number of factors in different ways. These dependences are practically not taken into account by modern computer-aided design of technological processes, so the processes of manufacturing cylindrical gears are designed, as a rule, by experienced technologists in manual mode. At the same time, modern production machines characteristic phenomena of accelerating change machine designs, increase varieties, improve the quality requirements of the machines. In modern engineering, about 85% of all enterprises operate in a diversified (single and serial) production. This leads to a constant increase in the volume of technological design and requirements for its quality, reducing its terms. This increases the relevance of automation of design processes of technological processes of manufacturing machine parts, including gears.

With the help of computer-aided design (CAD) of technological processes (TP), the design of a single TP can be carried out either only on the basis of the description of the design and technological parameters of the object of production (individual design), or on the basis of TP analogues (standard or group). In accordance with this distinguish CAD TP, providing automated synthesis of structures of single processes, and CAD TP, using TP-analogues. Methodology of creation of CAD of TP based on the analogues developed much deeper than the CAD system performing the automated synthesis of single processes.

According to the degree of depth of development of TP using CAD distinguish several levels of computer-aided design TP:

– development of the circuit diagram TP;
– design of technological route of part processing;
– design of technological operations;
– development of control programs for CNC equipment.

In Table 17.1 shows the relative complexity of the various stages of design TP and approximate current level of design automation of these stages [5].

Table 17.1 The ratio of labor intensity of the main stages of individual design of single TP manufacturing parts

Design stage	Share of the total complexity of the design TP, %	Share of the complexity of the work performed in automated mode, from the labor capacity of the stage, %
Technology elaboration route	25…45	Near to 0
Operating-room	50…60	35…45
Technical rate setting	5…10	Near to 0

When automating the design of operating technologies for manufacturing parts of CAD TP are used mainly in solving individual problems of technological design: calculation of allowances and development of the drawing of the workpiece for the adopted method of its production, the calculation of cutting conditions, documentation, etc.

As can be seen from Table 17.1 currently, the CAD system is used primarily to resolve minor technological problems. The greatest difficulty is the choice of routing technology, especially in the manufacture of complex parts. These details include cylindrical gears. In their manufacture, multi-operational processes are used. Existing CAD systems do not allow to perform their complex design and ensure their required quality.

The elaboration of control programs for CNC machines based on the drawing of the part after processing can now be considered fully automated. A significant number of computer-aided programming systems have been created, allowing to obtain high-quality control programs regardless of the used post-processor of the CNC system. Such tasks are solved by integrated systems of automated production-CAD-CAM-systems.

In view of the above, we have set the task of developing a method of designing the basic elements of single technological processes for the manufacture of cylindrical gears and obtaining appropriate recommendations..This technique should have been allowed to undertake the following tasks: selection of processing routes of the teeth and underlying surfaces of the wheels, the choice of the accuracy requirements of the teeth and underlying surfaces in all operations for the manufacture of wheels, the choice of modes of milling cylindrical gear hobbing cutters. At the same time, the results of our studies of the manufacturing processes of cylindrical gears [4, 5] and others were used.

17.1.1   The Method of Selection of the Processing Route of the Teeth of Cylindrical Gears

The choice of the route of treatment of different surfaces of the part is to determine the methods of processing these surfaces and the sequence of their execution. This choice depends mainly on the requirements for the accuracy and quality of the surfaces of the finished part and the workpiece, the nature of changes in these parameters during the manufacture of parts. Traditionally, the surface treatment route is often chosen based on the type of workpiece and the method of final surface treatment by assigning intermediate processing methods that provide a gradual increase in processing accuracy and improve the quality of the surface in question. This is a fairly complex method that requires high qualification of the designer. Its results largely depend on the qualification of the designer, do not contain objective criteria to ensure the design process required accuracy and surface quality of the finished part. To improve the accuracy of the prediction of the results of the technological process of processing of gears, it is necessary to take into account in the design of this process the nature of the operational changes in their quality indicators.

The analysis of technological processes of finishing and hardening of teeth, processing of basic surfaces of cylindrical gears shows that on these operations there can be regular changes in the quality of the wheels, amenable to mathematical modeling. The processes of finishing and hardening of the teeth occur either with free running (chaving, gear rolling, gear honing), or without removing the allowance (chemical-thermal treatment (CTO), gear rolling), or by removing the minimum allowances (gear grinding). When performing such operations, there is a high probability of relationships between the values of quality indicators before and after their execution. When processing the base surfaces of the gears (holes and crown ends), these relationships should also be expected due to the low rigidity of the technological system in these operations.

To study these relationships, 65 batches of cylindrical gears were processed and measured in the production environment and 20 batches of wheels in the laboratory. Studied are the processes of finishing and hardening of teeth as chaving on hard and expandable the mandrel, carburizing and nitrocarburizing, gearhoning, running of teeth of three run-in, geargrinding conical and worm wheels. We touched on the processes underlying surfaces such as drilling, reaming, single, rough and fine broaching, honing bores, rough and finish turning, in face of the crown. The analysis of changes in 11 indicators of accuracy of teeth, 4 parameters of quality of surfaces of teeth (see Table 17.2), 5 indicators of accuracy of basic surfaces (see Table 3.4). Processed cylindrical cogwheels with $m = 2.0-6.0$ mm, $z = 23-51$, with a smooth or splined bearing hole $D_0 = 25 - 70$ mm, 5 steels. The studied gears covered about 90% of the standard sizes of cylindrical gears manufactured in the auto, tractor and machine tool industry.

It was found that for most of the considered indicators of quality and processes of processing gears values of the coefficients of pair correlation between the values of the quality index before (x) and after (y) of the considered operations are within $r_{xy} = 0.28 - 0.9$, the proportion of variance of the quality index after this operation, inherited from the previous operation,—$y_B = 26.5-92.0\%$, the relationship between the values (x) and (y) with sufficient accuracy (the average relative error of the coupling equation ($\varepsilon_{average} = \varepsilon_{av} = 0.5 - 25.0\%$) can be described by a polynomial of the first degree of the form:

$$y = a + bx, \tag{17.1}$$

Exceptions are the processes of gear grinding and stretching of the landing hole.

Based on these results, it can be argued that the processing of teeth and base surfaces of cylindrical gears is a phenomenon of technological heredity. Technological heredity is called the transfer of properties of objects from previous technological operations to subsequent ones.

The presence of technological inheritance of values of parameters of quality of cylindrical gears on the basic operations of their processing gives the basis to consider that such process of production of a wheel which provides for each of the parameters

Table 17.2 The rate of change $K_{ch,i}$ of parameters of quality of a gear ring of cylindrical gears at their production

Characteristics quality	Operations performed sequentially							
	Shaving		CHT		Gearhoning	Running of teeth	Geargrinding	
	Hard mandrel	Expandable mandrel	Nitrocarburizing	Carburizing			Conical wheel	Worm wheel
1	2	3	4	5	6	7	8	9
F''_{ir}	1.25–2.7 (1.9)	1.5–2.9 (2.2)	Fig. 17.2b	Fig. 17.2a	1.05–1.25 (1.15)	1.2–1.4 (1.3)	4–6.3 (5.1)	3.2–6.1 (4.7)
f''_{ir}	1.2–3.6 (2.4)	1.3–3.8 (2.6)	Fig. 17.3b	Fig. 17.3a	1.1–1.35 (1.22)	1.1–1.5 (1.3)	5.2–9.6 (7.4)	2.9–4.2 (3.6)
$F_{\beta r}$	1.4–3.1 (2.2)	1.6–4.5 (3.1)	Fig. 17.4b	Fig. 17.4a	1.05–1.2 (1.12)	1.0–1.15 (1.07)	1.9–7.0 (4.4)	2.1–4.1 (3.1)
F_{rr}	1.4–3.6 (2.5)	–	0.67–0.91 (0.79)	0.65–0.89 (0.77)	–	–	–	–
F_{vwr}	0.3–1.3 (0.8)	0.8–1.4 (1.1)	0.83–0.97 (0.9)	0.77–0.96 (0.85)	–	–	–	–
F'_{ir}	1.1–1.4 (1.25)	–	–	–	–	–	–	–
F_{pr}	1.03–1.2 (1.1)	–	–	–	–	–	–	–
f'_{ir}	1.7–3.4 (2.6)	–	–	–	–	–	–	–
f_{ptr}	1.3–2.1 (1.7)	–	–	–	–	–	–	–
f_{fr}	1.9–3.2 (2.6)	–	–	–	–	–	–	–
f_{pbr}	1.3–1.9 (1.6)	–	–	–	–	–	–	–
R_a 40X	2.1–4.2 (3.2)	–	–	–	–	–	–	–
20XH3A	1.8–4.5 (3.1)	–	0.3–0.95 (0.7)	–	1.05–1.15 (1.1)	1.4–1.6 (1.5)	2.3–2.4 (2.35)	1.8–2.2 (2.0)
25XГТ	1.8–2.2 (2.0)	–	0.8–0.94 (0.87)	–	1.4–1.6 (1.5)	1.3–1.5 (1.4)	2.8–3.0 (2.9)	3.7–4.1 (3.9)
H_μ 40X	1.2–1.6 (1.4)	–	–	–	–	–	–	–
20XH3A	1.1–1.4 (1.25)	–	0.18–0.24 (0.21)	–	1.07–1.2 (1.13)	0.96–0.99 (0.97)	0.85–0.93 (0.89)	0.9–0.94 (0.92)
25XHГТ	1.1–1.3 (1.2)	–	0.1–0.2 (0.15)	–	1.2–1.5 (1.35)	0.86–0.98 (0.93)	0.96–0.98 (0.97)	0.9–0.97 (0.94)
σ_1 40X	(−1.0)–(−1.2) (−1.1)	–	–	–	–	–	–	–

(continued)

Table 17.2 (continued)

Characteristics quality		Operations performed sequentially								
		Shaving		CHT		Gearhoning	Running of teeth	Geargrinding	
		Hard mandrel	Expandable mandrel	Nitrocarburizing	Carburizing			Conical wheel	Worm wheel
1		2	3	4	5	6	7	8	9
σ_1	20XH3A	(−1.1)–(−1.6) (−1.3)	–	0.2–0.65 (0.41)	–	1.3–1.6 (1.15)	0.92–0.97 (0.95)	(−1.5)–(−1.6) (−1.55)	(−1.7)–(−1.9) (−1.8)
	25XГT	(−1.2)–(−4.5) (−2.9)	–	0.1–0.6 (0.35)	–	1.05–2.1 (1.55)	0.96–0.98 (0.97)	(−0.7)–(−0.75) (−0.72)	(−0.81)–(−0.87) (−0.84)
σ_2	40X	1.1–1.3 (1.2)	–	–	–	–	–	–	–
	20XH3A	1.4–1.7 (1.55)	–	0.25–0.35 (0.3)	–	1.2–1.6 (1.4)	0.85–0.97 (0.91)	0.96–1.2 (1.08)	1.03–1.2 (1.1)
	25XГT	1.4–1.8 (1.6)	–	0.15–0.25 (0.2)	–	1.2–1.43 (1.32)	0.97–1.04 (1.0)	0.99–1.2 (1.1)	1.05–1.3 (1.18)

Note 1. F_{ir}'', f_{ir}'', $F_{\beta r}$, F_{ir}', f_{fr}, f_{ir}', F_{pr}, f_{pr}, f_{pkr}, F_{rr}, F_{vwr}—tooth accuracy indicators according to GOST 1643-81

2. R_a, H_μ, σ_1, σ_2—quality parameters of the lateral surface of the teeth (R_a—roughness parameter, μm; H_μ—microhardness of the surface, MPa; σ_1, σ_2—residual stresses of the first and second kind in the surface, MPa)

3. Each filled cell of the table contains three values of the coefficient—minimum, average (in brackets) and maximum, which are selected depending on the condition of the equipment for this operation—bad, average or good

4. The values of the coefficients for the quality parameters F_{ir}'', f_{ir}'', $F_{\beta r}$, $F_{\beta r}$, after the operation of CHT (chemical-heat treatment) are taken from the nomograms of Figs. 17.1, 17.2, 17.3, and 17.4

5. For unfilled cells of the table, experimental data for the corresponding coefficients are currently not available

6. Grade 25XГT (proportions,% C = from 0.22–0.29, Si = 0.17–0.37, Mn = 0.80–1.10, Cr = 1.0–1.3), grade 20XH3A (proportions,% C = 0.17–0.24, Si = 0.17–0.37, Mn = 0.30–0.60 Cr = 0.60–0.90; Ni = 2.75–3.15, grade 40X (proportions,% C = 0.36–0.44, Si = 0.17–0.37, Mn = 0.50–0.80, Cr = 0.80–1.10

7. These indicators of the accuracy of the teeth according to GOST 1643-81 approximately correspond to the content and level of requirements of the same indicators according to DIN 3962

of quality normalized according to the drawing of a ready detail observance of the following condition can be considered acceptable from the technical point of view:

$$K_{ch.tot} \leq \prod_{i}^{m} K_{ch.i},\tag{17.2}$$

Where $K_{ch.tot}$ is he total rate of change of the parameter in the implementation of the technological process; $K_{ch.i}$—coefficient of change of this parameter on the i-th operation; m—the number of operations of the process on which the surface is treated, characterized by this quality parameter.

$$K_{ch.tot} = \delta_{blank}/\delta_{official},\tag{17.3}$$

where\δ_{blank}—tolerance on this parameter of quality in the blank or in the first operation, where is a form surface (e.g., teeth of wheels), which characterizes the present quality parameter; $\delta_{official}$—official tolerance on this parameter according to the drawing of the finished wheel.

$$K_{ch.i} = \delta_{i-1}/\delta_i,\tag{17.4}$$

where δ_{i-1} and δ_i—tolerances for this quality parameter on the preceding $(i-1)$ and given (i) operations that can be provided by the processing methods in question.

The values of the coefficients $K_{ch.i}$ for various parameters of quality of teeth and processes of manufacturing of cylindrical gears are established by us experimentally and are given in Table 17.2.

Taking into account the possibilities of various processes of processing of cylindrical gears, tooth processing operations are selected and for each of the normalized indicators of the quality of the toothed crown, compliance with the condition (17.2) is checked. This takes into account the state of the equipment on which the projected processing process will be implemented (see Table 17.2), and the design of the gear, if the operation of its chemical-heat treatment (CHT).

As technically acceptable are those variants of the route of processing of the toothed crown, in which for all normalized quality indicators the condition (17.2) is observed. From technically acceptable processing routes, providing the required accuracy of the toothed crown and the quality of the surface of the teeth, choose one or two options, based on technical and economic considerations, the condition of the existing equipment, the possibility of acquiring new equipment and other factors.

For the selected options for the route of processing of the teeth, the wheels prepare and master the production and choose one processing process that provides the required quality of the wheels, the required performance at minimum cost.

The degree of change in the accuracy of the teeth in CHT depends not only on the type and conditions of CHT, but also on the design of the gear. Therefore, we carried out a study of the influence of these factors on the values of K_{ch}. Were considered two types of CHT- carburizing and nitrocarburizing. For each type of CHT, the influence of the index of deformation stiffness of the wheel (x_1) and the index of asymmetry of the location of the gear hub relative to its toothed crown (x_2) on the K_{ch} was studied. The x_1 was taken to be the ratio of the diameter of the ring gear and the hole of the wheel d_a/D_0 (see Fig. 17.1). As $x_2(K_{asym})$—the ratio S/b, where S—the distance between the axes of symmetry of the crown and hub in their width, b—the width of the crown, (Fig. 17.1).

d_a/D_0 values varied within 2.3–3.3 K_{asym}—0.2–1.2. The degree of change measures of accuracy of teeth F''_{ir}, f''_{ir}, $F_{\beta r}$, was estimated based on $1/K_{ch.i}$, where $K_{ch.i}$ was determined by the formula (17.4).

The dependence between the studied factors is accepted in the form of

$$1/K_{ch} = V = b_0 + b_1 x_1 + b_2 x_2 \tag{17.5}$$

The coefficients of this equation are calculated based on the results of experiments performed for the processes of cementation and nitrocarburizing of cylindrical gears in non-muffle aggregates. Nomograms are constructed from the obtained equations (Figs. 17.2, 17.3, and 17.4), which allow us to estimate the value of $K_{ch.i}$ for the considered indicators of the accuracy of the gear (F''_{ir}, f''_{ir}, $F_{\beta r}$) and types of CHT, taking into account the constructive and dimensional parameters of the gear.

17.1.2 Method of Selection of Requirements to Accuracy of Cylindrical Gear Teeth on Intermediate Operations of Their Processing

The presence of operational relationships between the parameters of the quality of cylindrical gears on all sequentially performed operations of their processing allows along with single-factor (17.1) to obtain multi-factor models of changes in these indicators of quality in the manufacture of gears:

$$y = f(x_1, x_2, \ldots, x_{n-1}) \tag{17.6}$$

where y-the value of this indicator of quality in the finished part, that is after the last n-th operation; $x_1, x_2, \ldots x_{n-1}$ –values of this indicator of quality in the 1, 2 ,..., $n-1$ operations of the process.

The solution of the problem of optimizing the values of the quality index of the gear on the intermediate operations of its processing is possible by two methods:

(1) sequentially on individual operations of gear processing, starting with the last, n-th operation with the help of dependencies of the equation type (17.1), describing the change of this indicator on each operation of gear processing;
(2) simultaneously for all intermediate operations with the help of type dependencies (17.4), linking the values of a number of quality indicators in the finished

Fig. 17.1 Scheme for
determining the coefficients
of asymmetry and
deformation stiffness of the
gear

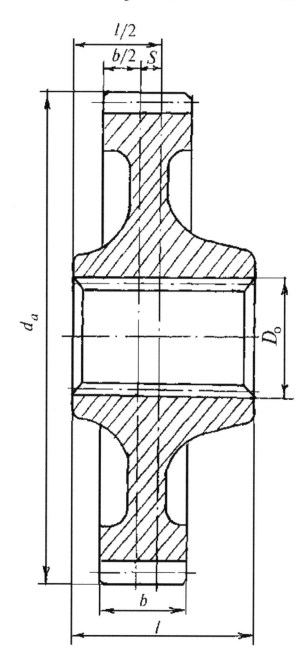

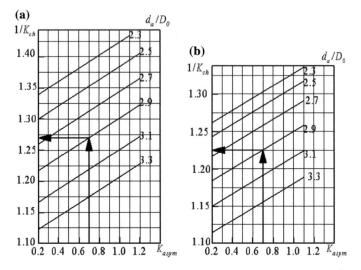

Fig. 17.2 Nomograms to determine the degree of change in F_{ir}'' after CHT ($K_y = \frac{\bar{y}}{\bar{x}}$) depending on the deformation stiffness (d_a/D_0) and asymmetry of the hub relative to the crown of the gear (K_{asym}) during cementation (**a**) and nitrocarburizing (**b**) of cylindrical gears

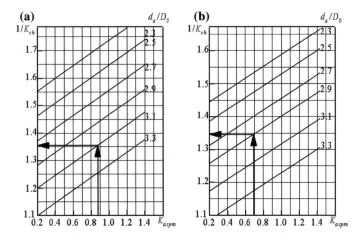

Fig. 17.3 Nomograms to determine the degree of change in f_{ir}'', after CHT (K_{ch}) depending on the deformation stiffness (d_a/D_0) and asymmetry of the hub relative to the crown of the gear (K_{asym}) during cementation (**a**) and nitrocarburizing (**b**) of cylindrical gears

part or after a certain operation with the values of these indicators on all or a number of previous operations.

The second method is more labor-intensive and less accurate in both obtaining multifactor dependencies and optimizing the independent variables included in them. In addition, as shown in [3] for predictor (sequentially interrelated) variables, the

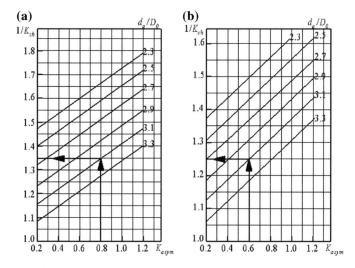

Fig. 17.4 Nomograms to determine the degree of change in $F_{\beta r}$ after CHT (K_{ch}) depending on the deformation stiffness (d_a/D_0) and asymmetry of the hub relative to the veins of the crown of the gear (K_{asym}) during cementation (**a**) and nitrocarburizing (**b**) of cylindrical gears

results of calculations using complex dependencies of type (17.4) are similar to the results of sequential calculations using type (17.1) dependencies.

With this in mind, we have adopted the first method of optimizing the requirements for the quality of cylindrical gears in various operations of their processing.

Currently, when selecting the requirements for the quality parameter of the part on the previous operation of its processing does not take into account the type and nature of the relationship between the values of this parameter before and after the operation in question. This choice is made either on the basis of information about the technological capabilities of the previous operation, or by experimentally determined refinement coefficients ($K_y = \frac{y}{x}$ or the equation of dispersion expansion S_y^2 of the output error [7]. The first method does not take into account the specific processing conditions, the second and third do not provide the accuracy of the optimization requirements for the initial values of the quality parameter, since the coefficient of refinement of the Cu is a random variable and does not reveal the patterns of change in the quality parameter in this operation. The decomposition of the variance S_y^2 assumes a constant accuracy of the regression Eq. (17.1) over the entire range of the argument x, that is a constant level the confidence probability with which the true values of the various characteristics of the statistical model are estimated.

However, as shown in [1], the confidence intervals for the theoretical conditional mean $\bar{y}(x)$ essentially depend on at which point x we construct them. Namely, the further the point x of interest moves away from its mean $\bar{x}$, the less reliable are predictions based on the empirical regression line $\hat{y}(x)$, i.e. the wider the intervals for $\bar{y}(x)$ become.

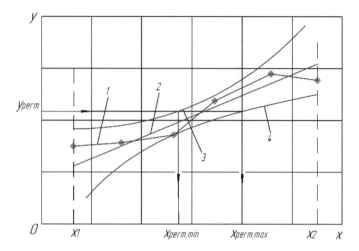

Fig. 17.5 The fundamental calculation pattern of the admissible part quality parameter values in the previous operation of its processing

Taking into account this feature of the linear regression model, the calculation of the maximum permissible initial values of the quality parameter for this operation can be performed using the proposed method, the schematic diagram of which is shown in Fig. 17.5. The initial data for the calculation are:

1. Experimentally established dependence (17.1) between the values of the considered quality parameter before and after this operation.
2. The limit values of x_1 and x_2 that the quality parameter can accept in the preceding operation.
3. The nominal value of the confidence probability P ($P = 0.95; 0.99; 0.9973$, etc.), for which the confidence curves 3 and 4 for the empirical regression line 1 and the averaged line 2 are calculated in the future (see Fig. 17.5).
4. Permissible values of the quality parameter under consideration after this operation. For indicators of accuracy of gears, which are mostly significantly positive values, the maximum permissible value of the v_{per} of this indicator is taken (see Fig. 17.5).

Calculation of the equations of $Y_{max}(x)$ and $Y_{min}(x)$ curves (curves 3 and 4 in Fig. 17.1) limiting the confidence interval for the empirical regression line $y = f(x)$ (line 2 in Fig. 17.2).1), is held in the following sequence:

1. Values are calculated

$$C = \frac{x_1 - \bar{x}}{\sqrt{\sum_{i=1}^{n} (x_i - \bar{x})^2}} \tag{17.7}$$

$$D = \frac{x_2 - \bar{x}}{\sqrt{\sum\limits_{i=1}^{n} (x_i - \bar{x})^2}} \tag{17.8}$$

$$\lambda = \sqrt{\frac{1}{2}\left[1 - \frac{1 + nCD}{\sqrt{(1 + nC^2)(1 + nD^2)}}\right]} \tag{17.9}$$

Here x_i is the value of the considered quality parameter before this operation in the batch of parts for which the regression Eq. (17.1) was calculated; n-volume of this batch of parts.

2. For the values of P, U_v and λ from Table 26 in [1] find $U_v(P, \lambda)$.
3. Write down the equations of the curves $y^{max}(x)$ and $y^{min}(x)$, limiting the confidence zone for the experimental regression line $\hat{y}(x)$:

$$y^{min}(x) = \hat{y}(x) - U_v(P, \lambda)\frac{S}{\sqrt{n}}C_1(x), \tag{17.10}$$

$$y^{max}(x) = \hat{y}(x) + U_v(P, \lambda)\frac{S}{\sqrt{n}}C_1(x), \tag{17.11}$$

where S and $C_1(x)$ are calculated respectively by formulas (17.12) and (17.13):

$$S = \frac{1}{n-2}\sum_{i=1}^{n}\left[y_i - \hat{y}(x_i)\right]^2 = \frac{1}{n-2}\sum_{i=1}^{n}(y_i - a - bx_i)^2, \tag{17.12}$$

$$C_1^2(x) = 1 + \frac{(x - \bar{x})^2}{S_x^2}, \tag{17.13}$$

Here y_i—the values of the considered quality parameter after this operation in the batch of parts for which the regression Eq. (17.1) was calculated; a, b— the coefficients of the regression Eq. (17.1); $\hat{y}(x)$—the value y_i, calculated by Eq. (17.1) for the value x_i ; S_x^2 - the dispersion of the values x_i in this batch of parts.

Possible values of this quality parameter in the preceding operation gdpmp Hopman calculated by substituting in Eqs. (17.10) and (17.11) instead of $y^{min}(x)$ and $y^{max}(x)$ the allowed values of this parameter of quality after the operation in question. For indicators of the accuracy of the gears, as significantly positive values, it is necessary to substitute the value of v_{perm} (see Fig. 17.5). The found values of x_{min} and x_{max}, corresponding to the unknown value of x from Eq. (17.13), will be the required valid initial values of $x_{perm.min}$ and $x_{perm.max}$ of this quality parameter, allowing to provide the required tolerance for this parameter after the operation, taking into account the nature of its change in these processing conditions.

The tolerance for this parameter on the previous operation for alternating errors can be calculated by the formula

$$\delta_x = x_{perm.\max} - x_{perm.\min},\qquad(17.14)$$

and for significantly positive errors-according to the formula

$$\delta_x = x_{perm.\max},\qquad(17.15)$$

Using the proposed method, we obtained recommendations for various accuracy settings of the teeth and underlying surfaces of the wheels, the processes of their treatment for choosing the requirements to the parameters tochnoi gears on zaslonih operations of their processing. These recommendations are given in the State standard of the Republic of Belarus STB 1251-2000 Cylindrical gears. Design methods of manufacturing processes [10]. In Table 17.3 an example of these recommendations for the choice of requirements for the accuracy of the teeth before the operation of stirring on the mandrel is given.

The recommendations given in [10] can be used for the design of technological processes for the manufacture of cylindrical gears, including the selection of accuracy requirements of the teeth and underlying surfaces of the wheels on the interim operations of their processing. Starting with the calculation of the permissible initial

Table 17.3 Recommendations on the choice of performance requirements of precision spur gears before chaving to expandable the mandrel μm (module m over 3.5 to 6.3 mm, crown width b up to 40 mm)

Indicators precision gears	Numerical value (μm) of the accuracy index for the required degree of accuracy according to GOST 1643-81 after shaving							
	6		7		8		9	
	The pitch diameter of the gear, mm							
	To 125	Over 125–400	To 125	Over 125–400	To 125	Over 125–400	To 125	Over 125–400
F''_{ir}	60–108 (84)	84–150 (117)	84–150 (117)	120–216 (168)	106–192 (149)	150–270 (210)	168–300 (234)	210–378 (294)
F_{vwr}	13–18 (16)	22–32 (27)	17–25 (21)	32–46 (39)	22–32 (27)	40–58 (49)	–	–
f''_{ir}	23–67 (45)	26–74 (50)	32–92 (62)	36–104 (70)	46–133 (90)	52–148 (100)	58–166 (112)	65–185 (125)
$F_{\beta r}$	14–38 (26)		17–47 (32)		28–76 (52)		44–118 (81)	

Note 1. Each cell of the table contains 3 values of the accuracy index, allowing to take into account the state of the equipment in this operation. A good state corresponds to the maximum initial value of the accuracy index, a bad state-the minimum the original value, the average condition is the average original value. The equipment must meet the technical requirements of its manufacturer. 2. The degree of accuracy of the gear after stirring corresponds to the minimum value for the three standards of accuracy (kinematic, smooth operation, contact teeth), normalized for the gear after shaving. If these requirements are the same, then the values of F''_{ir} given in the table can be reduced by 1.5–2.0 times, since shaving translates the radial components of the kinematic error into tangential ones

errors for a certain accuracy indicator from the last operation of the toothed crown or the base surface of the wheel and performing it consistently for all previous operations, we obtain rational values of the accuracy of the wheels on all operations of their processing and in the workpiece for the base surfaces of the wheel. To automate the design process using these guidelines, you should use the software at each design stage.

Refining existing recommendations for the specific conditions of the treatment can be performed during the debugging of the designed technological process. For this purpose, under certain conditions [4, 5] it is necessary to process consistently on all the operations under consideration a batch of gears with a volume of 50–100 pieces with measurements of the normalized quality parameter before and after each operation. The data obtained will allow for all the processing operations and quality parameters of the gears to establish the relationship $y = f(x)$ and with their help according to the above method to calculate the requirements for the accuracy of the gears for all intermediate operations, taking into account the operational changes in this indicator during the processing of the wheel in specific processing conditions. The need to use experimentally established relationships $y = f(x)$ to calculate the rational requirements for the accuracy of the gears in the intermediate operations is caused by the following. The values of the accuracy of the wheels before and after the operations, the characteristics of the operational relationships between these indicators depend on many factors (the state of the technological system, allowances, cutting conditions, etc.). Our studies have shown that these values and characteristics vary widely (70–1000%). Therefore, the most accurate description of these relationships in these conditions is possible only by statistical methods based on the results of the Method of selection of requirements to accuracy of base surfaces of cylindrical gears before gear cutting and during intermediate operations of their processing.

17.1.3   Method of Selection of Requirements to Accuracy of Base Surfaces of Cylindrical Gears Before Gear Cutting and on Intermediate Operations of Their Processing

As you know, the main purpose of the base surfaces of the gears (for cylindrical gears is usually the landing hole and the ends of the crown)-to provide the necessary conditions for the process of gear cutting. Based on this, after selecting the requirements for the accuracy of the teeth in gear cutting, it is necessary to choose the requirements for the accuracy of the base surfaces of the wheel, based on the relationship between these indicators of accuracy in gear cutting. We have carried out studies of these relationships and proposed a technique for optimizing the accuracy requirements of the base surfaces of cylindrical gears before gear cutting [5]. This technique makes it possible to determine the optimal values of the beating of the base end of E_T and the diametral gap ΔD_0 in the conjugation of the mandrel of the gear machine with the workpiece, providing the required accuracy of the gear crown in these conditions of gear cutting. The technique also makes it possible to determine the requirements for the accuracy of the Seating surfaces of the device used in gear cutting.

We performed calculations of the optimal values of E_T and ΔD_0 on this technique for all investigated batches of wheels. The results in the form of tables are given in (STB, [10]. In these tables, the valid values for E_T and ΔD_0 given for wide ranges of variation 4 parameters of wheels and 4 degrees of precision wheels according to GOST 1643-81 after gear cutting. An example of such a table for selecting the requirements for the beating of the base end is given in Table 17.4.

Since the collected experimental data cover a large range of sizes of gear wheels, precision bases of the workpieces and the ring gear and the other terms of gear cutting, it has been possible to establish for the considered conditions, the relationship of the maximum allowable under gear cutting values of E_T and ΔD_0 with the geometric dimensions of wheels and the necessary after gear milling or gear shaping precision of the ring gear. We received based on the allowable values of E_T and ΔD_0 such conditions gear cutting as the width of the crown machined wheels b (hereinafter denoted X_1); diameter of tops of teeth $d_a(X_2)$; the ratio of the length of the bore to the diameter of the supporting end $L_{hole}/D_{base}(X_3)$; the degree of accuracy after gear cutting according to GOST 1643-81 (X_4) (if the wheel has a different degree of accuracy according to the norms of the kinematic accuracy and tooth contact, the X_4 should take the lesser of two values; if a DIN or ISO standard is used, for example DIN 3962 or ISO 1328, it is possible to take as X_4 the value of the degree of accuracy according to these standards increased by one compared to the table value (STB, [10], since the tolerances according to GOST 1643-81 are approximately one degree greater than the corresponding DIN or ISO standards); the diameter of the wheel bore Do (X_5). 6 types of regression equations were analyzed and the most adequate dependences were chosen.

For gears of automotive transmissions these dependences have the form:

$$E_T = (82661 + 145514X_3 - 43358X_4 + 2115 \cdot 10^{-2}X_1^2$$
$$+ 160 \cdot 10^{-2}X_2^2 - 118250X_3^2 + 7058X_4^2 - 52 \cdot 10^{-2}X_5^2$$
$$- 46 \cdot 10^{-4}X_2^3 + 15063X_3^3 - 285X_4^3 - 295X_1X_4$$
$$- 278X_2X_3 + 412X_3X_5 + 15X_2X_4) \cdot 10^{-6} \text{ mm}, \qquad (17.16)$$

$$\Delta D_o = (1810689 - 118650X_5 + 83765X_3 + 2354X_5^2$$
$$+ 88X_4^3 - 1439 \cdot 10^{-2}X_5^3 - 665X_1X_3$$
$$- 12630X_3X_4 + 154X_4X_5 + 452 \cdot 10^{-3}X_2^2) \cdot 10^{-6} \text{ mm}, \qquad (17.17)$$

These are based on or derived from their recommendations in the form of tables, which are listed in the developed world the standard of the Republic of Belarus (STB, [10], can be used for pre-selecting the tolerance band of the base surfaces of the workpieces cylindrical gears $m = 3.5-5$ mm before their hob cutter gear milling

Table 17.4 Permissible values of the run out of the base end of E_T (μm) of cylindrical gears by cutting teeth

Characteristics gears			The required degree of accuracy of the gear according to GOST 1643-81 after gear milling or shaping															
			6				7				8				9			
			The pitch diameter of the gear, mm															
b, mm	$\dfrac{L_{hole}}{D_{base}}$	D_{hole}, mm	From 50 to 85	Over 85–125	Over 125–165	Over 165–200	From 50 to 85	Over 85–125	Over 125–165	Over 165–200	From 50 to 85	Over 85–125	Over 125–165	Over 165–200	From 50 to 85	Over 85–125	Over 125–165	Over 165–200
To 20	From 0.20 to 0.35	From 30 to 50	19	25	33	41	27	34	43	52	35	43	53	63	46	54	64	74
		From 50 to 80	20	26	34	42	28	35	44	53	36	44	54	64	47	55	65	75
	From 0.35 to 0.70	From 30 to 50	32	36	42	48	40	45	52	59	48	54	62	70	59	65	73	81
		From 50 to 80	34	38	44	50	42	47	54	61	50	56	64	72	61	67	75	83
	From 0.70 to 1.05	From 30 to 50	44	46	48	51	52	54	58	62	60	63	68	73	71	74	79	84
		From 50 to 80	48	49	52	55	56	58	62	66	64	67	72	77	75	78	83	88
Over 20 to 30	From 0.20 to 0.35	From 30 to 50	12	18	26	34	17	24	33	42	22	30	40	50	30	38	48	58
		From 50 to 80	13	19	27	35	18	25	34	43	23	31	41	51	31	39	49	59
	From 0.35 to 0.70	From 30 to 50	25	29	35	41	30	35	42	49	35	42	49	57	43	49	57	65

(continued)

Table 17.4 (continued)

Characteristics gears

b, mm	$\frac{L_{hole}}{D_{base}}$	D_{hole}, mm	The required degree of accuracy of the gear according to GOST 1643-81 after gear milling or shaping																
			6				7				8				9			
			The pitch diameter of the gear, mm															
			From 50 to 85	Over 85–125	Over 125–165	Over 165–200	From 50 to 85	Over 85–125	Over 125–165	Over 165–200	From 50 to 85	Over 85–125	Over 125–165	Over 165–200	From 50 to 85	Over 85–125	Over 125–165	Over 165–200
		From 50 to 80	27	31	37	43	32	37	44	51	37	43	51	59	45	51	59	67
	From 0.70 to 1.05	From 30 to 50	37	39	41	44	42	44	48	52	47	50	55	60	55	58	63	68
		From 50 to 80	41	42	45	48	46	48	52	56	51	54	59	64	59	62	67	72
Over 30 to 40	From 0.20 to 0.35	From 30 to 50	9	15	23	31	11	18	27	36	13	21	31	41	18	26	36	46
		From 50 to 80	10	16	24	32	12	19	28	37	14	22	32	42	19	27	37	47
	From 0.35 до 0.70	From 30 to 50	22	26	32	38	24	29	36	43	26	32	40	48	31	37	45	53
		From 50 to 80	24	28	34	40	26	31	38	45	28	34	42	50	33	39	47	55
	From 0.70 to 1.05	From 30 to 50	34	36	38	41	36	38	42	46	38	41	46	51	43	46	51	56
		From 50 to 80	38	39	42	45	40	42	46	50	42	45	50	55	47	50	55	80

and hobbing or shaping. At the stage of debugging the technological process, it is necessary to process a batch of parts with a volume of about 50 pieces on operations gear cutting, of measure to gear cutting in each blank features precision bases, and after gear cutting-normalized according to the working drawing details the indicators of the kinematic accuracy and tooth contact. These indicators of the accuracy of the teeth to the greatest extent depend on the accuracy of the base surfaces of the gears during gear cutting [5]. After calculating the relationships between these parameters, it is necessary to calculate the optimal requirements for these conditions of tooth processing for the accuracy of the base surfaces of the wheel blanks during gear cutting according to the method given in [5].

Next, you need to select the routes of treatment of the base surfaces before and after gear cutting and the requirements for the accuracy of the base surfaces in the intermediate operations of their processing.

When choosing the routes of treatment of the basic surfaces of the gears before and after gear cutting, dependencies (17.2)–(17.4) are used. When calculating $K_{ch.tot}$ about the formula (17.3) as δ_{blank} for the treatment of the base surfaces of the wheels to gear cutting accepted tolerances for runout of the end face of the crown and the diameter of the hole in the billet wheel, for machining the base surfaces after gear cutting the same tolerances for gear cutting wheels. As $\delta_{official}$ for the first stage of processing of basic surfaces tolerances are taken on the beating of the ends of the crown and the diameter of the bore of the gear wheel when gear cutting, for the second stage of processing –tolerances on the same accuracy of the base surfaces for the finished gear.

When choosing the routes of treatment of the base surfaces of the wheels before and after gear cutting, we can use the data obtained on the values of the coefficients of change in the accuracy of the base surfaces to ISM on various operations of their processing (Tables 17.5 and 17.6) (STB, [10]. The method of solving these problems is similar to the method of choosing the route of processing the teeth of cylindrical gears, described above in paragraph 17.1.2.

Then, you must select the accuracy requirements for the base surfaces of the gears in the intermediate machining operations of these surfaces. At the design stage of the technological process of manufacturing the gear to solve this problem can be used the recommendations given in (STB, [10]. When debugging the process—you can use the technique described in paragraph 17.1.2.

17.1.4   The Choice of Best Cutting Conditions by Hobbing of Cylindrical Gears

We have carried out studies of the influence of gear milling modes of cylindrical gears by worm mills (cutting speed V and feed S) on a number of indicators of accuracy of teeth and quality parameters of their side surfaces (Y). Interrelations of the considered parameters of quality of teeth with V and S at gear milling with

Table 17.5 Coefficients of change $K_{ch.i}$ quality parameters of the bore holes cylindrical gears in their manufacture

Base hole quality settings	Sequentially executed operations						CHT		Shot blasting	Mandrel
	The drilling of the workpiece	Countersinking billets	Broaching a single	Broaching finish	Fine boring	Honing or grinding	Nitrocarburizing	Carburizing		
Diameter holes D_0	3.5–5.1 (4.3)	3.8–5.9 (4.85)	3.7–5.6 (4.65)	1.3–1.7 (1.5)	2.3–2.7 (2.5)	2.1–2.8 (2.45)	0.1–0.37 (0.3)	0.08–0.35 (0.28)	1.0–1.07 (1.03)	2.5–5.2 (3.8)
Out-of-round holes H_{or}	1.7–1.9 (1.8)	1.9–2.2 (2.05)	3.5–8.6 (6.0)	1.1–2.1 (1.6)	1.1–1.6 (1.35)	1.5–2.1 (1.8)	0.2–0.7 (0.45)	0.15–0.6 (0.4)	1.0–1.05 (1.02)	2.3–4.9 (3.6)
Non-cylindrical holes H_{nc}	2.1–2.8 (2.45)	2.5–3.6 (3.0)	3.6–5.5 (4.5)	1.1–1.9 (1.5)	1.3–1.9 (1.6)	1.4–2.0 (1.7)	0.3–0.7 (0.5)	0.2–0.66 (0.45)	1.0–1.03 (1.01)	2.1–4.7 (3.4)

Table 17.6 Coefficients of change $K_{ch.i}$ quality parameters of end faces of the ring gear • cylindrical gear wheels during their production

Characteristic precision ends	Sequentially executed operations					
	Draft turning	Semi-finishing turning	Finishing turning	CHT		Grinding
				Nitrocarburizing	Carburizing	
Beat end E_T	2.5–6.2 (4.3)	1.7–4.7 (3.2)	1.5–3.8 2.6	0.8–0.9 (0.85)	0.7–0.85 (0.77)	2.2–4.1 (3.15)
Non-parallelism of ends, E_1	2.3–5.9 (4.1)	1.5–4.5 (3.1)	1.3–3.5 (2.4)	0.85–0.95 (0.9)	0.75–0.9 (0.82)	2.1–3.8 (2.95)

sufficient accuracy ($\varepsilon_{av} = 5.2–21.6\%$) can be described by polynomials of the first or second degree:

$$Y = b_0 + b_1 V + b_2 S, \tag{17.18}$$

$$Y = b_0 + b_1 V + b_2 S + b_{12} V S + b_{11} V^2 + b_{22} S^2, \tag{17.19}$$

The dependence (17.18) is true for measures of accuracy of teeth F''_{ir}, F_{rr}, f''_{ir}, $F_{\beta r}$, and the parameter of surface roughness of teeth R_a, μm, the dependence (17.19)—for the quality parameters of the surface of the teeth H_μ(surface microhardness, MPa), σ_1 and σ_2 (residual voltage 1 and 2 of the births in the surface of the teeth, MPa).

As a result of these studies, we have established:

1. Feed S when gear milling with a worm cutter has 11–28 times greater impact on the accuracy of the teeth and up to 10 times-on the quality parameters of the surface of the teeth than the cutting speed V.
2. The dimensions of the gears and gear cutting equipment do not have a significant impact on the relationship between the accuracy of the teeth with V and S, the material of the wheels affects the relationship of the quality parameters of the surfaces of the teeth with V and S.
3. For 3 materials of gear wheels are obtained depending on the considered quality parameters of gears from the milling of the V and S [2, 4]. The dependence obtained for gears made of steel 40X, 25ХГТ, 20ХН3А, m=3,5–6.3 mm, d up to 400 mm in the processing of their standard hobbing cutters with teeth mostly of steel P6M5. Based on these dependencies is made up of table values is achievable in terms of accuracy and quality parameters of the surface of the gear teeth when the gear milling and hobbing and worm milling for different cutting conditions (Table 17.7 and 17.8).

These data, along with the recommendations of reference books on cutting modes (see, for example, [6] can be used in the design of technological processes for the manufacture of gears for the selection of gear milling modes. The following sequence of actions can be taken:

1. According to the guide [6], the modes of gear milling are established, providing mainly rational tool life.

Table 17.7 Achievable value in terms of accuracy cylindrical gear wheels in different modes of milling (material-steel 25XГT, 20XH3A, $m = 3.5 - 6.3$mm, $d = 120 - 400$mm)

Cutting conditions		Indicators of accuracy cylindrical gears			
V, m/min	S_0, mm/speed	F_{rr}, mm	F_{ir}'', mm	f_{ir}'', mm	$F_{\beta r}$, mm
25–30	1.5	0.100	0.120	0.040	0.019
	2.5	0.110	0.130	0.044	0.028
	3.5	0.120	0.140	0.048	0.037
30–35	1.5	0.110	0.130	0.044	0.028
	2.5	0.120	0.140	0.048	0.037
	3.5	0.130	0.150	0.052	0.046
35–40	1.5	0.120	0.140	0.048	0.037
	2.5	0.130	0.150	0.052	0.046
	3.5	0.140	0.160	0.056	0.055
40–45	1.5	0.130	0.150	0.052	0.046
	2.5	0.140	0.160	0.056	0.055
	3.5	0.150	0.170	0.060	0.064
45–50	1.5	0.140	0.160	0.056	0.055
	2.5	0.150	0.170	0.060	0.064
	3.5	0.160	0.180	0.064	0.073

2. According to Tables 17.7 and 17.8, cutting conditions are set to ensure the required values of accuracy and surface quality of the teeth of the machined wheels.
3. From the found values of the cutting conditions (feed S and cutting speed V), smaller values are selected to ensure both rational tool life, and the requirements for accuracy and quality of the tooth surface established earlier (see paragraph 17.1.2).

When debugging the technological process of gear milling of cylindrical gears, the proposed method of determining the cutting modes can be used, which comprehensively ensure the efficiency of the process and the conditions of its normal implementation in these production conditions. To use this technique, it is necessary to determine the relationship of the cutting modes V and S with the main quality parameters for this gear with the use of experimental planning.

The established relationships between the characteristics of the quality of cylindrical gears and gear milling modes allow for the considered intervals of mode change to solve the problems of their complex optimization, as well as to predict the achievable parameters of the quality of the wheels for certain modes of gear milling.

The task of complex optimization of cutting conditions consists in the determination of such values which would ensure maximum effectiveness in one sense or another cutting process in the given specific operating conditions for fulfillment of all requirements of the technological process. This problem belongs to the class of parametric optimization problems in the development of technological processes. In its solution, operational models are widely used, which reflect the requirements for

Table 17.8 Achievable values of the quality parameters of the surface of the teeth of cylindrical gears at different modes of gear milling

Cutting conditions		The quality parameters of the surface of the teeth												
V, m/min	S, mm/min	R_a, μm			H_μ, kgf/mm²			σ_1, MPa			σ_2, MPa			
		40X	25ХГТ	20XH3A	40X	25ХГТ	20XH3A	40X	25ХГТ	20XH3A	40X	25ХГТ	20XH3A
20–25	2.5	3.1	4.65	2.7	294.0	191.5	186.2	– 66.0	– 92.0	– 79.4	468.8	237.2	293.4
	5.0	3.8	5.25	3.5	302.0	202.1	196.4	– 94.8	– 94.8	– 158.8	495.4	270.5	335.7
	10.0	5.3	6.65	5.1	308.2	224.2	218.9	– 184.7	– 242.7	– 206.3	536.0	315.4	371.0
25–30	2.5	3.4	5.0	2.9	279.8	186.3	181.7	251.6	625.4	590.5	445.6	231.0	282.2
	5.0	4.1	5.6	3.7	284.8	197.2	191.0	148.8	505.5	468.8	467.5	261.0	303.4
	10.0	5.6	7.0	5.3	288.6	2201	215.0	25.4	202.7	166.4	498.6	288.0	335.2
30–35	2.5	3.7	5.3	3.2	269.3	181.5	177.5	720.4	908.6	870.5	438.0	225.0	274.6
	5.0	4.4	5.9	4.1	274.0	192.0	187.7	602.1	792.4	746.7	454.2	235.1	287.8
	10.0	5.9	7.3	5.6	278.6	214.3	210.1	293.8	501.0	448.7	503.3	281.3	332.6
35–40	2.5	4.0	5.6	3.4	263.2	176.6	173.2	291.5	334.3	290.2	405.5	195.2	240.1
	5.0	4.7	6.2	4.1	267.7	186.5	183.8	106.3	265.5	221.4	430.6	220.3	265.3
	10.0	6.2	7.6	5.8	273.0	209.4	207.2	– 132.2	– 105.6	– 92.3	464.8	255.5	292.4
40–45	2.5	4.3	6.0	4.6	261.7	171.0	169.1	– 316.5	– 415.5	– 388.2	395.6	177.5	225.6
	5.0	5.0	7.0	5.3	265.9	180.8	179.5	– 502.0	– 532.3	– 421.1	215.2	215.2	249.2
	10.0	6.5	8.0	7.2	272.0	205.1	203.2	– 718.6	– 781.3	– 749.2	454.9	242.4	281.5

the desired design solution, and the factors that must be taken into account when making this decision.

In technological design operational models are recorded in the following form:

$$\left.\begin{array}{l} F(x_1, x_2, \ldots, x_n) \to \min(\max) \\ g_j(x_1, x_2, \ldots, x_n) \le b_j, \quad j = \overline{1, m} \\ a_{1i} \le x_i \le a_{2i}; \quad i = \overline{1, n}, \end{array}\right\}, \qquad (17.20)$$

where $x_1, x_2, \ldots, x_n$—optimized process parameters that can take values from the set of $[a_{1i}, a_{2i}]$ real numbers; $F(x)$—the target function to be optimized (it must take the minimum or maximum value under the conditions imposed on the function $g_j(x_1, x_2, \ldots, x_n)$, and the values x_i, b_j are the given real functions.

Very often when optimizing cutting conditions as $F(x_1, x_2, \ldots, x_n)$ the cost share of C_p operation depending on the cutting conditions is taken

$$C_p = t_o E_p + W_k / n_1,$$

where t_0—the main time of operation; E_p—the cost of 1 min of the machine and the worker; W_k—the sum of all costs for the period of tool life; n_1—the number of blanks processed during the period of tool life.

Since

$$t_0 = \frac{L_{wor.str} z}{s_0 n \varepsilon q} \quad \text{and} \quad R = \frac{T S_0 n \varepsilon q}{L_{wor.str} z},$$

that

$$C_p = \frac{L_{wor.str} z}{S_0 n \varepsilon q} \left(E_p + \frac{W_k}{T} \right),$$

where $L_{wor.str}$—the length of the working stroke of the cutter or table of hobbing machine, mm; z—the number of teeth of the gear being cut, S_0—feed per revolution of the workpiece; ε—the number of visits of the cutter; q—the number of simultaneously processed parts; n—the number of revolutions of the cutter; T—tool life of gear-milling cutter.

The minimization of the C_p value should be achieved when certain restrictions are imposed on the most important under these conditions, the quality parameters of the gears (for gears of tractor transmissions, such parameters can be $F_{\beta r}, R_a, H_\mu$), cutting power N, torque M_{tor} from the maximum cutting force, limit values n and S_0 for this machine, wear h on the back face of the cutter during T of its operation. These parameters can be calculated by formulas.

Cutting power:

$$N = K_p m^{1.7} S_0^{0.9} V / D_{hob},$$

where m—the module of the cut wheel, mm; D_{hob}—cutter diameter; $K_r = 0.12$ for steel and 0.06 for cast iron.

Power consumed by electric motor of machine:

$$P_{mach} = N/\eta,$$

where $\eta = 0.4 - 0.5$—coefficient of performance gear-cutting machine.

The power of P_{mach} shall not exceed the actual power of the electric motor of the $P_{mach.akt}$ machine.

Torque of maximum cutting forces on the spindle of the hob

$$M_{tor.hob} = P_z R_{hob} = K_p m^{x_p} S_0^{y_p} R_{hob},$$

where the values of K_p, x_p, y_p coefficients can be taken from [8], Table 11); R_{hob}—radius of the cutter.

Torque from the maximum cutting force on the spindle of the item:

$$M_{tor.it} = P_z R_{it} = K_p m^{x_p} S_0^{y_p} r,$$

where r is the diameter of the dividing circle of the cut wheel.

Values of $M_{tor.hob}$ and $M_{tor.it}$ should not exceed the corresponding limit values on the passport of the gear hobbing machine.

Wear h on the back face of the cutter during time T of its operation is equal

$$h = 1.36 \cdot 10^{-6} \cdot T^{0.79} S_0^{1.34} V^{2.3} m^{0.5} z^{-0.25}.$$

The value of h should not exceed h_{permis} for the accepted period of resistance of the cutter T. The value of h_{permis} can be taken from the reference literature.

Thus, the operating model of the gear milling process of cylindrical gears for the optimization of cutting modes S and $V(n)$ can be written as follows:

$$\left.\begin{array}{l} L_{wor.str}z\left(E_p + \frac{W_k}{T}\right)/S_0 n\varepsilon q \to \min \\ a_0 + a_1 V + a_2 S \leq f_i''; b_0 + b_1 V + b_2 S \leq F_\beta; c_0 + c_1 V + c_2 S \leq R_a \\ H_{\mu min} \leq d_0 + d_1 V + d_2 S \leq H_{\mu max} \\ K_p m^{1.7} S_0^{0.9} V/D_{hob}\eta < N_{mach} \\ K_p' m^{x_p} S_0^{y_p} R_{hob} < M_{tor.hob.permis}; K_p'' m^{x_p'} S_0^{y_p'} R_{it} < M_{tor.it.permis} \\ 1.36 \cdot 10^{-6} \cdot T^{0.79} S_0^{1.34} V^{2.3} m^{0.5} z^{-0.25} < h_{permis} \\ n_{min} \leq n \leq n_{max} \\ S_{min} \leq S \leq S_{max} \end{array}\right\} \quad (17.21)$$

The solution of this problem is possible with the help of linear programming methods.

If the restriction for H_μ is described using the dependence of the form (17.19), then for the solution of the system (17.21) it is necessary to use methods of nonlinear programming, for example, gradient methods. In both cases, the solution of the system (17.21) can be found graphically.

17.2 Conclusions

Tasks of automation of design of technological processes of production of cylindrical gears are very actual and difficult. The relevance of these tasks is caused by the mass production of cylindrical gears, a large amount of work on the design of their production processes, a low degree of automation of these works. The complexity of the automation of the design of technological processes for the manufacture of cylindrical gears is explained by the multi-operational nature of these processes, the diversity and low study of the relationships between the characteristics of the quality of gears and the conditions of their manufacture.

The author on the basis of the researches of the majority of processes of processing of basic surfaces and teeth of cylindrical gears with the module $m = 3.5$–6.0 mm, the dividing diameter to 400 mm having 5–12° of accuracy according to GOST 1643-81, offers methods of performance of various stages of design of processes of production of cylindrical gears and the corresponding recommendations are given. These techniques can be realized by means of computer technologies and represent scientific and methodical base of automation of performance of design of processes of production of cylindrical gears.

The proposed methods are based on the statistical analysis of the processes of formation of the quality parameters of the teeth when they are cut by a worm cutter, as well as changes in the quality parameters of the teeth and base surfaces on all operations of their processing. These methods allow to provide the required quality of gears with a given probability, taking into account the basic processing conditions.

In all the design stages of technological processes of manufacturing cylindrical gears (election of routes of the treatment of the teeth and underlying surfaces of the wheels, the choice of the accuracy requirements on these surfaces in all operations of processing, the choice of the accuracy requirements of the underlying surfaces of the wheel and cutting in hobbing) provides methods and recommendations for the design and debugging of technological processes. The need to debug technological processes can improve the accuracy of technological design and more accurately take into account the actual conditions of production. As our studies have shown, the spread of values of quality indicators of gears and characteristics of their relationships with different factors for different production conditions can be in the range of 70–1000%. Therefore, when debugging the process, it is necessary to clarify these values and relationships for the actual production conditions. Data processing and analysis during debugging can also be performed using computer technology.

References

1. Aivazyan, S.A.: Statistical Study of Dependencies. Metallurgy (1968) (in Russian)
2. Algin, V.B., Antonyuk, V.E., Kane, M.M.: Gears and Transmissions in Belarus: Design, Technology, Evaluation of Properties. Belarusian science, Minsk (2017). (in Russian)
3. Draper, N., Smith, G.: Applied Regression Analysis. Finance and Statistics (1986) (in Russian)
4. Kane, M.: (2016). Quality control of spur gears on the basis of simulating their production processes. In: Litvin, F.L., Goldfarb, V., Barmina, N. (eds.) Theory and Practice of Gearing and Transmissions, vol. 34, pp. 393–403. Springer (2016)
5. Kane, M.: Optimization of requirements for accuracy of base surfaces for spur and helical gears at their tooth cutting. In: Goldfarb, V., Trubachev, E., Barmina, N. (eds.) Advanced Gear Engineering Mechanisms and Machine Science, vol. 51, pp. 345–364. Springer (2018)
6. Korchemkin, D.A.: Cutting of Metals. Handbook, Niitavtoprom (1995). (in Russian)
7. Lukomsky, Ya. I.: Correlation Theory and Its Application to the Analysis of Production. 2rd edn. State Statistical Office (1961) (in Russian)
8. Medveditskov, S.N.: High-performance Gear Cutting Cutters. Mashinostroenie, Moscow (1981). (in Russian)
9. Starzhinsky, V.E., Kane, M.M.: Production Technology and Quality Provision of Gears and Transmissions. Manual. SPb., Profession (2007) (in Russian)
10. STB 1251-2000: State Standard of the Republic of Belarus. Cylindrical Gears. Methods of Design of Technological Processes of Manufacturing. States Standard, Minsk (2000) (in Russian)

Chapter 18
Advanced Methods of Simulation of Gears with Teeth Outlined by Low-Order Surfaces

Boris P. Timofeev, Mikhail Yu. Sachkov and Alexander V. Kovalevich

Abstract The present paper describes the advanced methods of engineering calculation, design, simulation, analysis and production of parts and units of mechanisms and machines. In particular, special attention is paid to methods of full cycle computer-aided design on the basis of the mathematical software MathCAD (by PTC), SolidWorks CAD system (by Dassault Systèmes) and additive technologies. The objects under consideration are non-conjugated toothed and rod-toothed gears. The paper presents the example of engineering design and the cycle of development of specimens and experimental models beginning from mathematical simulation of kinematics of gear meshing. The considered combination of software programs and design approaches will allow for swift developing of study guides or applied tutorials for training of engineers. The full cycle design is also the advantageous tendency in industry due to the reduction of the design period, and to the possibility to apply the advanced production technologies such as additive production.

Keywords Gears · Non-conjugated gears · Position error · Error of gear ratio · Full cycle design · Computer-aided design

18.1 Introduction

Time reduction for design and manufacture of products is the fixed requirement of the reality. This problem is successfully solved by applying the existing software programs of mathematical modeling, computer-aided design systems and additive technologies.

B. P. Timofeev · M. Yu. Sachkov · A. V. Kovalevich (✉)
ITMO University Saint, P.O. Box 197101, Saint-Petersburg, Russia
e-mail: kovaevich@mail.ru

B. P. Timofeev
e-mail: timofeev@mail.ifmo.ru

M. Yu. Sachkov
e-mail: urie2006@yandex.com

© Springer Nature Switzerland AG 2020   413
V. Goldfarb et al. (eds.), *New Approaches to Gear Design and Production*,
Mechanisms and Machine Science 81,
https://doi.org/10.1007/978-3-030-34945-5_18

The necessity of joint application of different software programs is primarily explained by requirements of instrument and mechanical engineering enterprises. In market competition conditions the enhancement of the enterprise operation at stages of the design, production preparation and final product manufacture implies, as a rule, implementation of administration and technical systems that provide the control of all information about the product (Product Data Management). Application of such systems allows for simplifying the document flow between subdivisions of enterprises and implementing the already existing process of the series-parallel design electronically.

In turn, it is the most interesting to fulfill the full cycle design method or WAVE methods where the virtual model of the product and its modifications are implemented.

The present research touches upon the process of full cycle design by the example of non-conjugated gears and their modifications.

The first part of the work presents the results of mathematical modeling of gears with linear contact by means of MathCAD. The mathematical model implement the calculation of geometrical parameters of the gear with the consequent calculation of values of kinematic parameters of meshing. The condition of fulfillment of this calculation is the absence of the edge contact.

In the second part the results of mathematical modeling are compiled into the CAD system to develop the solid model and verify the kinematic parameters.

The third part shows the results of experimental modeling of the developed gears with application of additive technologies.

18.2 Mathematical Model of Gearwheels

The object of mathematical modeling is the non-conjugated gear. The driving gearwheel is the toothed pinion which is meshed with the pin wheel [14].

Let us start with the example of the gearing with the theoretical (nominal— i_{n21}) gear ratio equal to one: $i_{n21} = z_2/z_1 = 1$. This gear is necessary for numerical evaluation of the character of variation of kinematic parameters of meshing with different gear ratios. We will use the geometric calculation similar to that applied for calculation of the involute gearing [7].

When radii (ρ) of teeth and pins are taken to be equal, the analog for the gear module m can be determined from two formulas for the gear circumferential pitch p. The first one is $p = 4 \cdot \rho + c' \cdot m$, where c' is the backlash factor and the second one is $p = \pi \cdot m$. The considered value of the circumferential pitch is related to the analog of the pitch diameter by the formula $D = z \cdot m$. These expressions are valid at the pitch contact point of the non-conjugated gear. The tooth addendum is taken to be equal to ρ. The tooth dedendum is taken as $h_f = 1.2 * \rho$. The value ρ in the considered analysis is the summarized parameter of mathematical modeling.

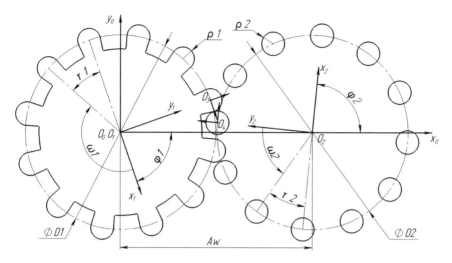

Fig. 18.1 Meshing of gearwheels and coordinate systems

Geometrical parameters and arrangement of coordinate systems of the mathematical model for meshing of the pinion and pin gearwheel are shown in Fig. 18.1.

The mathematical model represents the system of trigonometric equations and it allows for calculating the values of kinematic parameters of meshing such as the position error function, the error function of the gear ratio and the transverse contact ratio. By means of these criteria we can evaluate the cyclic error of the gear, and the character and value of impacts at reconjugation [7].

The initial condition of the contact point in the meshing was determined at the profile of the driving gearwheel tooth and pin. Coordinates were reduced to the fixed coordinate system by means of the transition matrices. The resulting system of equations of the equality of radii vectors at the contact point and unit normal vectors assigns the position function $(\varphi_2(\varphi_1))$ implicitly.

The error of the position function is determined as $\Delta\phi_2(\phi_1) = \phi_2(\phi_1) - \phi_1 \cdot i_{n21}$ where φ_1 is the angle of the driving gearwheel rotation, i_{n21} is the nominal gear ratio; $\varphi_2(\varphi_1)$ is the function of the driven gearwheel position.

The instant gear ratio is the derivative position function of the angle of the driving gearwheel rotation φ_1: $i_{21}(\phi_1) = \frac{d\phi_2(\phi_1)}{d\phi_1}$.

In turn, the error of the gear ratio function deviated from the nominal value is determined as $\Delta i_{21}(\phi_1) = i_{21}(\phi_1) - i_{n21}(\phi_1)$.

Figure 18.2 presents the diagrams of the position error and gear ratio error at reconjugation of three pairs of teeth in the gear with radii of teeth and pins $\rho = 5$, the gear ratio $i_{12} = 1$, and the tooth number $z = 12$.

The mathematical modeling of the gearing and analysis of functions of the position error and the error of the gear ratio proves that within the process of meshing of one pair of teeth the considered functions do not have points that characterize

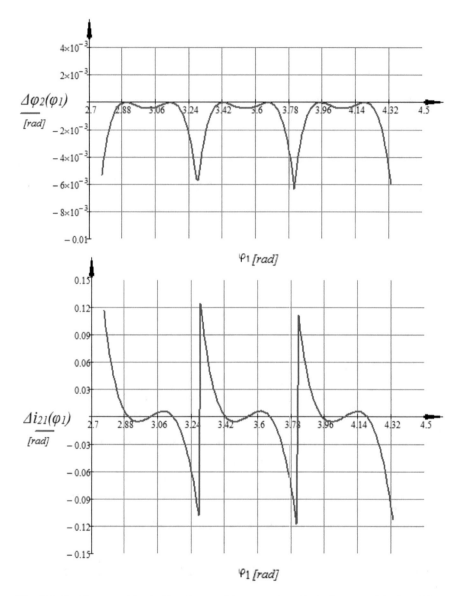

Fig. 18.2 Position error ($\Delta\varphi_2(\varphi_1)$) and error of the gear ratio ($\Delta i_{21}(\varphi_1)$) at reconjugation of three pairs of teeth

"impacts". The position function and the function of the gear ratio are differentiable and continuous within the angular pitch. However, at reconjugation of neighboring pairs of teeth the impact is noticed. Let us optimize the gear in order to reduce the negative effects to minimum.

The presented results of mathematical modeling characterize the gear at the stage of its design and the computation time for the considered example does not exceed one minute at the personal computer with the processor AMD_Ryzen_3_1200_Quad-Core under the system Windows 8.1.

As compared to the gear design software "Shafts and mechanical gears 3D" ("Compas 3D") the mathematical model illustrates visually the "inner processes of meshing".

Analyzing the meshing by means of the mathematical model allows for comparing the kinematic characteristics of groups of gears at the stage of design.

Let us consider the influence of simultaneous variation of radii of the tooth and pin with keeping their equality.

When increasing the tooth radius and, therefore, the angle α, the gear smoothness is improved due to reduction of the difference of boundary values of the error of the gear ratio (Fig. 18.3). However, the position error is increased. And when decreasing the tooth radius, there is the function jump which implies the transition from the meshing phase "circumference-circumference" to the phase "circumference-straight line". It means that the contact line will move to the tooth root passing through the line of conjugation of two surfaces. The curvature of the surface on the line of surface conjugation is instantly changed from the definite value ρ to the infinity which is the negative factor.

Figure 18.3 shows the families of curves characterizing the kinematic parameters of the gear when changing the tooth radius and the pin radius from 7 mm (bold dashed line) to 4 mm (bold dot and dash line) with the pitch 0.5 mm.

The mathematical model of the meshing allows not only for varying the geometrical parameters of gearwheels but for "scaling" the meshing in order to improve the gear characteristics. However, variation of geometric parameters did not lead to solving the problems of the accuracy and smoothness of motion transmission. In order to improve the meshing characteristics fundamentally, it is necessary to change the approach to upgrading the gear elements. In this connection, the combination of the described above gearwheels is considered.

Let us arrange the described above elementary gear consisting of the gearwheel and pin wheel in n rows by shifting each next row at the angle $\gamma = \tau/n$ where τ is the angular pitch of the gearwheel equal to $\tau = 2\pi/z$ and n is the number of rows. It will result in a multi-row (pseudo helical) gear with the linear contact in which the axial components of the load are completely absent when transmitting the motion (Fig. 18.4).

Figure 18.4 presents the example of the gearwheel and pin wheel with six rows ($n = 6$). The axial offset of rows is the same.

Figure 18.5 shows that due to transition of the meshing from one row to the following rows (layers) of the gearwheel the decrease in the amplitude of the function of the position error ($\Delta\varphi_2(\varphi_1)$) is achieved. The accuracy of meshing is thus increased.

The diagram $\Delta\varphi_2(\varphi_1)$ (Fig. 18.5) shows that the amplitude of the function Δy_{max} $\approx 3 \times 10^{-4}$ rad. We can determine for comparison the analog of the parameter of the involute gear—the error of the cyclic frequency. For the multi-row gear the analog is $f_{zz0} = 12.4$ mcm. This value is a little smaller than the value of the sixth degree

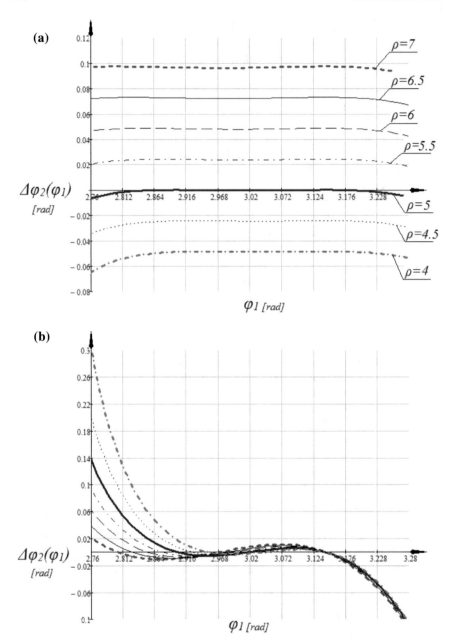

Fig. 18.3 Position errors (**a**) and errors of the gear ratio (**b**) within variation of parameters

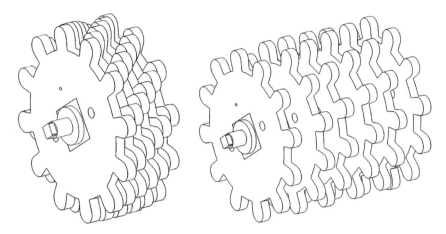

Fig. 18.4 Multi-row version of the gearwheel

of accuracy regulated by smoothness of the involute gear in accordance with the Russian State Standard GOST 1643-81 for similar parameters of the module and tooth number [10]. Evidently, with increase in the number of rows of the multi-row gear the value of the amplitude of the function of the gear ratio error will be decreased.

Figure 18.5 shows the specific jumps of the function $\Delta i_{21}(\varphi_1)$ at reconjugation. However, unlike the single-row meshing the angle at which one pair is contacting is equal to $\gamma = \frac{\tau}{n} = \frac{2 \cdot \pi}{z \cdot n}$, thus decreasing the maximum value of the jump of the instant gear ratio from 0.24 for the single-row meshing to 0.02 for the multi-row meshing with six rows. Noises and wearing out of gearwheels at operation are expected to be decreased.

In case of the sufficiency of kinematic parameters of meshing obtained by analysis of the mathematical model, the designing engineer makes a decision on the developments of the three-dimensional solid model.

18.3 Solid Modeling of Gearwheels

Mathematical modeling and development of kinematic parameters of the researched gear was carried out by MathCAD software program (developed by Parametric Technology Corporation—PTC) with application of matrix methods [7, 8, 13]. This software is convenient and common. Matrices of mutual arrangement of radii vectors and unit normal vectors are assigned directly in the client area in order to make up the system of trigonometric equations with the consequent solution by Runge-Kutta method.

Application of computer-aided computation systems simplifies significantly the process of developing functional relations with the consequent analysis and visualization of the obtained results. The products of PTC company also allow for performing

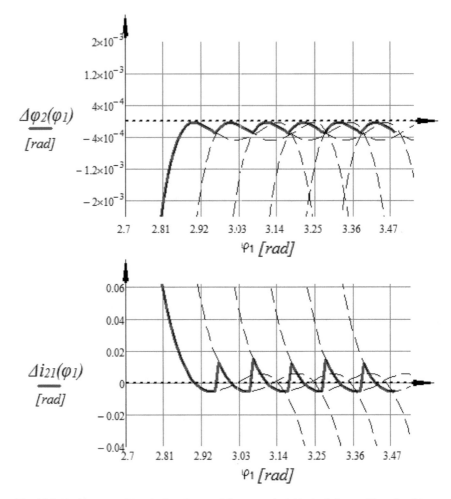

Fig. 18.5 Position error ($\Delta\varphi_2(\varphi_1)$) and error of the gear ratio ($\Delta i_{21}(\varphi_1)$) for meshing of multi-row gearwheels

complex design procedures by uniting, for instance, MathCAD and Creo [2, 3, 6]. The mathematical apparatus of these programs is very similar, thus allowing to use their integration for computer-aided design of parts and units on the basis of the calculation file. The end-to-end transfer of parameters of the mathematical model is also possible in CAD systems developed by other program builders (Fig. 18.6).

Figure 18.4 presented the solid model of the multi-row gear composed of elementary plane gearwheels considered above with geometrical parameters in accordance with Table 18.1 for the gear ratio $i_{n12} = 1$. This solid model was developed by SolidWorks software (Dassault Systèmes company).

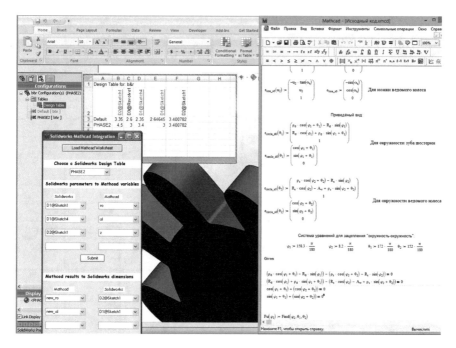

Fig. 18.6 Transfer of values of variables from MathCAD to SolidWorks

The solid model provides the computer-aided computation of the values of the mass, the tensor of moments of inertia in axes of the global or local coordinate systems. Therefore, SolidWorks system calculates the values necessary for simulation of kinematic parameters of the gear, including the consideration of dynamic processes.

On the basis of the performed calculations for the multi-row gear the computer-aided development of the three-dimensional model of single-layer gearwheels and multi-row layout versions was carried out. Modeling was performed by means of the SolidWorks computer-aided design software. This program was also used to perform physical simulation by the software suite Motion and Simulation (Fig. 18.7).

Modeling of the multi-row gearing is well agreed with the obtained diagrams (Fig. 18.8).

In Fig. 18.8 the straight line corresponds to the angular velocity of the pinion (ω_1); and the broken line is for the angular velocity of the pin wheel (ω_2). Here, the error of the velocity is the same as of the gear ratio ~2% for the driven gearwheel. Therefore, the diagram obtained by means of the dynamic modeling of the gear corresponds to the diagram obtained by mathematical modeling (Fig. 18.5, $\Delta i_{21}(\varphi_1)$).

Having determined the values of preferable layout versions of the gear for solving the assigned applied problem already at the design stage (in the real time mode), it is possible to specify the range of the required materials, standard parts and to estimate the necessary working area of the manufacturing equipment.

Table 18.1 Numerical values of the main geometrical and kinematic parameters of the pinion and pin gearwheel for different gear ratios

Parameter	Designation	Values for different gear ratios		
Nominal gear ratio	i_{n12}	1	2.5	3.17
Pinion tooth number	Z_1	12	12	12
Gearwheel tooth number	Z_2	12	30	38
Angular pitch of the pinion (°)	τ_1	30	30	30
Angular pitch of the gearwheel (°)	τ_2	30	12	9.47
Rounding radius of teeth and radius of pins (mm)	ρ	5	5	5
Module (mm)	m	6.9	6.9	6.9
Reference diameter of the pinion (mm)	D_1	83	83	83
Reference diameter of the gearwheel (mm)	D_2	83	207.5	262.8
Cone angle of the tooth at the center of the gearwheel (°)	α	13.84	5.5	4.35
Interaxial distance (mm)	A_w	83	291	346
Amplitude of the position error for meshing of one pair (rad.)	Δy_{max}	6×10^{-3}	8×10^{-3}	5×10^{-3}
Cyclic error of the pinion tooth frequency (mm)	f_{zzo}	0.249	0.332	0.208
Drop in gear ratios for one meshing pair	Δi_{max}	0.232	0.189	0.13

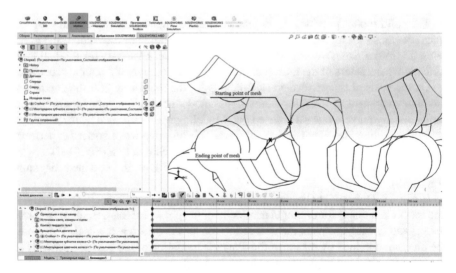

Fig. 18.7 Dynamic modeling of the gear (the contact zone is marked)

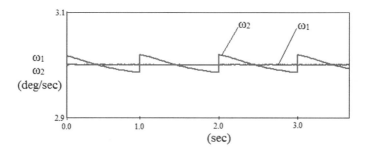

Fig. 18.8 Angular velocities of the driving and driven gearwheels versus time

Nowadays the most actively developing niche within the manufacturing equipment is the area of additive technologies. Their high level of connectivity with advanced CAD systems and versatility allows for manufacturing the products of various geometrical shapes. The verified solid model is transferred by CAD means to the program software of a printer for 3D printing.

18.4 Production of Experiments Models of Gears with Application of Additive Technologies

Application of additive technologies allows for simplifying the process of manufacturing preparation of production. Positioning of the part in the working area of the printer is possible at the developer's computer with sending the assignment for the printer nearby or via internet.

The possibility of complex shape printing and material economy combined with the mobility allows for making one-of-a-kind products within the shortest time with account of the digital end-to-end design.

Quick prototyping of the experimental model of the gear was carried out for the mathematical and solid models described above. The result of the previous stage was the development of the three-dimensional model by SolidWorks. The information of the model can be written in many common formats such as .sldprt, .stl, .step and other.

There is the variety of 3D printing technologies applying different physical principles and materials [1, 4]. All elements of the experimental model were printed by the 3D printer with FDM (Fused Deposition Modeling) technology using the ABS (acrylonitrile butadiene styrene) plastic (Fig. 18.9). This material was chosen mainly due to its high mechanical properties (as compared to plastics available for everyday printing) and ease of machining.

The end-to-end design was used to obtain gearwheels that form cylindrical and bevel gears [9, 11] (Fig. 18.9).

Fig. 18.9 Examples of gearwheels made by 3D printer

In case of interaction of bevel gearwheels with the second-order operating flanks it is possible to obtain easily the gear ratios higher than 6. For commonly applied bevel gears with straight barrel-type teeth the gear ratios are $i = 1, \ldots, 4$; and with circular teeth they are up to 6. In turn, the methods of production of involute bevel gearwheels of conjugated gears (spherical involute surface) are still absent [5, 10, 12]. Providing the gear ratios above 5 for small numbers of teeth decreases the values of geometrical contact ratios to critical ones.

In order to produce rods of the rod-toothed gearwheels it is the most reasonable to use the ready round section assortment. Cylindrical pins by the Standard GOST 3128-70 can be applied as the rods. Diameters and lengths of standard products available at stocks can be compared with the mathematical model within the end-to-end design even at the stage of the primary calculation and changed in order to provide the assigned interaxial distance.

Experimental modeling by methods of end-to-end design was carried out in the students university laboratory for gearwheels with the second order operating flanks (Fig. 18.10). The total time of development and production of the experimental model did not exceed two working days of one expert.

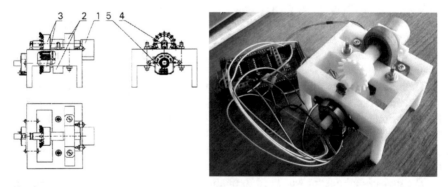

Fig. 18.10 Laboratory test stand. 1—step motor 28BYJ-48 (China), 2—bearings ISB-B000 (China), 3—rod-toothed gearwheels, 4—incremental encoder EC10E1 220501 (ALPS company, Japan), 5—incremental encoder AMT103-V (CUI Inc., USA)

The presented above version of gearwheels is described in the Russian Patent N146159 [9].

18.5 Conclusions

The result of the presented work is the production of experimental models and stands for non-conjugated gears and rod-toothed gears by means of the end-to-end computer-aided design. The joint operation of software programs MathCAD, SolidWorks and 3D printing is optimized.

Mathematical modeling of gears is carried out, the analysis of variation of geometrical parameters is presented. The computer-aided development of the three-dimensional model, the analysis of parameters of meshing by CAD tools and experimental modeling of gearwheels and the stand (FDM printing) are performed.

The results are obtained by application of principles of the end-to-end modeling of products and are visually demonstrated by the example of the non-conjugated gear consisting of kinematic pairs with the second-order surfaces.

The obtained results can be useful for on-the-fly development of auxiliary mechanisms such as the gearbox and the system of motion transmission in order to optimize the engineering skills and skills at the stage of education, to make stands with pre-assigned parameters, and to perform functional experimental modeling and prototyping of full-scale projects.

The complex of solutions presented in this manuscript can be upgraded by implementation of production by additive technologies on the basis of metals, such as LENS (Laser Engineered Net Shaping), SLS (Selective Laser Sintering), SLM (Selective Laser Melting), DMD (Direct Metal Deposition) and other.

References

1. Additive technologies. Information. TorTec Ltd. http://tornado.co.com/, 23 May 2019
2. Creo Parametric 5.0.4.0. Interactive reference. http://support.ptc.com/help/creo/creo_pma/russian/index.html, 23 May 2019
3. Gavrilov, P.Ya., Brezgin, V.I.: Integration of calculation and design subsystems when developing the equipments of steam turbine plants. In: Proceedings of the 3rd Scientific Technical Conference of Young Researchers of the Ural Energy Institute, pp. 143–145 (in Russian) (2018)
4. Ilyushchenko, A.F., Savich, V.V.: Powder metallurgy is one of the first additive technologies. In: Proceedings of the Scientific Technical Conference "Additive Technologies, Materials and Structures", Grodno, GrGU, pp. 20–30 (in Russian) (2016)
5. Kabatov, N.F., Lopato, G.A.: Bevel Gearwheels with Circular Teeth. Mashinostroenie, Moscow (in Russian) (1966)
6. Kulikov, D.D.: Electronic textbook on "CAD systems for manufacturing processes (Part 1)". Saint-Petersburg, SPb ITMO. https://de.ifmo.ru/bk_netra/start.php?bn=4, 20 May 2019 (in Russian) (2010)
7. Litvin, F.L.: Theory of Gearing. Nauka, Moscow (in Russian) (1968)

8. Litvin, F.L.: Design of Mechanisms and Parts of Instruments. Mashinostroenie, Moscow (in Russian) (1973)
9. Russian Patent N 146159, МПК F16H 55/10 F15H 55/17. Gearwheel for transmission of the rotational motion. Timofeev, B.P., Sachkov, M.Yu. (in Russian) (2014)
10. Russian State Standard GOST 1643-81. Basic regulations of interchangeability. Spur and helical gears. Tolerances. Moscow, IPK Izdatelstvo standartov (in Russian) (1981)
11. Sachkov, M.Yu.: Synthesis of rod toothed gears. Ph.D. Thesis, Saint-Petersburg (in Russian) (2015)
12. Segal, M.G.: Types of Localized Contact of Bevel and Hypoid Gears. Moscow, Mashinovedenie, vol. 1, pp. 56–63 (in Russian) (1970)
13. Sheveleva, G.I.: Theory of Generation and Contact of Moving Bodies. Mosstankin, Moscow (in Russian) (1999)
14. Timofeev, B.P., Ponomarenko, M.Yu., Kovalevich, A.V.: Approximated gears with the sectional linear contact. J. Izvestiya VUZov. Priborostroenie. **61**(2), 135–140 (in Russian) (2018)

Chapter 19
New Approach to Computer-Aided Design of Gearbox Systems: Models and Algorithms

Olga V. Malina

Abstract The paper considers the design procedure as the process of structural synthesis. Classification factors proposed in the paper describe essentially different models of the design system; they also allow for stating the limits of applying the pointed models in order to develop computer-aided design systems for medium complexity products of mechanical engineering involving spiroid gearboxes. The new approach to the development of the computer-aided design system for spiroid schemes implies the development of a model implementing the synthesis of a possible layout version by choosing from the set of possible ones by means of application of the exhaustive search algorithms at the set of all features describing the considered class of objects. The pointed approach allows the user to assign the initial design data by values of any subset of features (parameters and criteria) describing the class of objects; and the system will supplement the definition of values of all the rest features. The peculiarity of implementation of exhaustive search algorithms at the set of high power elements (and the power of features describing the class of spiroid gearboxes is considerable) is the high probability of the effect of the "curse of dimensionality" when the computer power is not enough to solve the design engineering assignment. The paper considers the models of the synthesis process as the optimized exhaustive search algorithm with each iteration involving the stages of synthesis and analysis where the knowledge on the existing relations within the subject area are mainly applied as the tool for analysis of the obtained versions.

Keywords Spiroid gearbox design · Model of structural synthesis · Exhaustive search algorithms · Sets of features · Forbidden figures

O. V. Malina (✉)
Kalashnikov Izhevsk State Technical University, Studencheskaya Str. 7, 426069 Izhevsk, Russia
e-mail: malina_0705@mail.ru

© Springer Nature Switzerland AG 2020
V. Goldfarb et al. (eds.), *New Approaches to Gear Design and Production*,
Mechanisms and Machine Science 81,
https://doi.org/10.1007/978-3-030-34945-5_19

19.1 Introduction

Current trends of development of new layouts are impossible without consideration of, first, the existing practice of synthesis of new design solutions and, second, the possibility of computerization of the pointed process.

Both described above factors are of essential influence on the structure of the design process and the quality of obtained layouts.

The pointed factors are definitely interrelated, since only the existing practice formalized as algorithms can become the foundation for computerization of the design processes. However, the lack of the pointed experience or, vice versa, the appearance of the new experience can become the serious obstacle for computerization of the design process.

This contradiction is even more obvious with the increasing complexity of the object to be designed automatically.

Gearbox systems are related to medium complexity objects basing on the number of their component parts and their variety.

Similar to any other mechanical engineering object, the process of gearbox design requires the solution of a number of problems through a number of stages from the technical assignment up to the final layout meeting the requirements of production and economic efficiency.

In less details these stages can be divided into the following groups:

– development of the structure: choice of parts which are components of the future product;
– calculations: engineering design, manufacturing and economic;
– preparation of engineering design documentation (engineering drawings, 3D models, etc.).

It is evident that the pointed stages do not follow one after another, they are divided into steps which further comprise the process of synthesis. So, for example, the choice of the shaft radius (the stage of choosing a part) is accompanied by strength and rigidity analyses, and the choice of the shape and dimensions of the casing may require the development of a 3D model to evaluate the gearbox assembling ability.

The sequence of steps is determined by the chosen ideology of computerization of the structural synthesis process including, of course, the engineering design process.

19.2 Classification Factors of the Design Process

The ideology is determined by combination of four classification factors. Let us consider the pointed classifications.

The first classification K_1 is related to the statement of the design assignment:

(A) achieving the best design version that meets the requirements of the technical assignment;

(B) achieving the design version that meets the requirements of the technical assignment.

Unlike the second version, the first one requires the statement of the optimization assignment: what should be considered the best version and what will be the next steps if the pointed requirements contradict each other.

The second classification K_2 considers the basic approach to obtaining the assigned layout that meets the requirements of the technical assignment.

In accordance with this classification all design processes can be divided into:

(A) processes of direct synthesis;
(B) processes of choosing from the set.

The process of direct synthesis implies the presence of the algorithm that allows to obtain the layout that meets the requirements of the technical assignment.

As for complex objects, implementation of the pointed approach may face the problem of inconsistency of initial data or the problem of the absence of the end-to-end algorithm of the structural synthesis.

The process of choice requires forming the set of possible versions with the consequent choice of the sought one.

The main difficulty is the formation of the set of possible versions.

The third classification K_3 is related to definition of the base function of the system and the role of the technical assignment within the computer-aided design process.

The classical approach to the design process implies that the technical assignment is the number of criteria to be agreed with the new design solution; and the base function of the design system is the determination of the number of parameters describing the layout.

Implementation of automatization of such an approach is related to three essential problems:

1. The absence of the algorithm that transforms the assigned set of criteria into the sought set of parameters.
2. The variety of statements of the assignment when the consumer puts different sets of criteria as initial data and, therefore, he defines the value of not all criteria.
3. The absence of the clear definition of the set of criteria and the set of parameters (overall dimensions can become both the criterion and the parameter depending on the statement of the assignment).

The approach proposed by the author is alternative to the classical one; and it implies the absence of division of the set of characteristics describing the object into criteria and parameters. Within this approach a complete set of characteristics of the object (its features) is generated. The consumer defines values of the subset of characteristics as the technical assignment; and the base function of the system is to complete the definition of the set, that is, to find possible values of all the rest features.

The forth classification K_4 is determined by the base algorithm of the synthesis process.

In the first case, the framework of the synthesis is the algorithms of the subject area formalized as calculation relations, empiric curves, tables, etc.

In the second case, it is the search algorithms at the set of possible values of all characteristics (parameters and criteria) excluding impossible combinations. In this case, the subject algorithms become the framework of analysis of already developed layouts.

The given above classifications seem to be rather trivial from the first sight, however, their joint consideration allows for obtaining the set of methodological models of the process of structural synthesis.

The set of methodological models can be described by the formula:

$M = K_1 \times K_2 \times K_3 \times K_4$, where K_i is the ith classification assigned by the set of its values $K_i = \{k_{i1}, k_{i2}\}$.

The power of such a set is equal to 16, since each classification has two values.

Therefore, the described analysis resulted in 16 methodological models of the synthesis process. Evidently, they have different degrees of implantation ability that in turn depends on:

– the complexity of the synthesis object;
– the level of formal characterization of the design process;
– the level of coverage of the design process by formal characterization.

It is obvious that for simple objects with an unambiguously defined set of values of criteria and parameters with all stages of development being formalized, it is easy to develop the system of the computer-aided design synthesis that allows for finding the best version of the layout by implementing the classical approach by means of the direct synthesis, applying the calculation relations of the subject area as algorithms of the synthesis.

As for the complex objects, the development of a computer-aided design system can also involve the model that allows for finding the best version of the layout by implementing the classical approach by means of the direct synthesis, applying the calculation relations of the subject area, if there is a complete formal characterization of the synthesis process for each set of initial criteria.

Similar to the majority of mechanical engineering products of the medium and high degree of complexity, implementation of the pointed model for gearboxes and gearbox systems is impossible nowadays.

This explains the composition and the composed function of the existing and commonly spread CAD systems.

The pointed systems allow for executing individual calculations, develop 3D and 2D models, plot the technical documentation (including the graphic one), develop and store the base of graphic primitives. However, the process of structural synthesis (that is, the choice of design elements and their mating) is implemented by a human expert; and the quality of the technical solution is determined by his experience and intuition.

Analysis of all the obtained methodological models showed that the model that implements the synthesis of a possible version of the layout by selecting from the set of possible versions by means of search algorithms at the set of all features describing the considered class of objects is the universal one.

Having chosen the pointed model as the base one, let us study what is necessary for its functioning and which problems should be solved to make its functioning possible.

19.3   Shortly About Features and Forbidden Figures

The information base of the system that implements the proposed model contains groups of initial data: the set of features, describing the class of objects [8] and the set of forbidden figures fixing the possible reasons of impracticality of the version [2, 3].

The feature is the question the answer to which in fact forms one characteristics of the synthesis object. For instance, the feature "Number of bearing supports of the worm unit for a spiroid gearbox" can have the following values "One", "Two", "Three". A specific gearbox can have only one value of the feature in description of its layout, for instance, "Number of bearing supports of the worm unit for a spiroid gearbox" is "Two".

Form mathematical point of view, the feature is the set the elements of which are its values:

$p_i = \{a_{ij}\}$, where p_i is the ith feature in the description of the layout, a_{ij} is the jth value of the ith feature.

All the set of features is divided into two subsets: obligatory and non-obligatory.

The feature is the obligatory one if one of its values is present in the description of each layout of the object of the assigned class. For instance, the feature "Material of a spiroid gearwheel" with values "Bronze", "Steel", "Plastic" is obligatory, since each layout of a spiroid gearbox comprises a spiroid gearwheel, and the gearwheel is made of a certain material.

Non-obligatory feature is the feature with an undefined value within certain versions of layouts of the group of objects.

For example, there is the feature "Number of bearings of the third support of the gearwheel unit" in the description of the class of objects, while the structure of the object comprises only two supports of the gearwheel unit.

The set of values of non-obligatory features comprises the empty value that denotes the absence of the feature in the design solution:

$p_i = \{a_{ij}, nil\}$, where p_i is the ith feature in the description of the layout, a_{ij} is the jth value of the ith feature, nil is the empty value of the feature.

The empty value gets the feature describing the non-obligatory layout part as the element.

Introduction of empty values into the set of values of the feature allows for obtaining the combination of features describing the layout of the product or its certain part by executing the Cartesian multiplication of features describing the class of objects.

In this case, in the final set of versions each value of each feature will be present at least in one correct structure.

Let us consider the example. Two features are assigned: p_1—"Number of supports of the gearwheel unit" with values a_{11}—"One", a_{12}—"Two", a_{13}—"Three" and the feature p_2—"Number of bearings of the third support" with values a_{21}—"One", a_{22}—"Two", a_{23}—"Nil".

Having multiplied two features, we get the following sets:

N	Description of the version		Correctness
	"Number of supports of the gearwheel unit—one"	"Number of bearings of the third support—one"	Incorrect
	"Number of supports of the gearwheel unit—two"	"Number of bearings of the third support—one"	Incorrect
	"Number of supports of the gearwheel unit—three"	"Number of bearings of the third support—one"	
	"Number of supports of the gearwheel unit—one"	"Number of bearings of the third support—two"	Incorrect
	"Number of supports of the gearwheel unit—two"	"Number of bearings of the third support—two"	Incorrect
	"Number of supports of the gearwheel unit—three"	"Number of bearings of the third support—two"	
	"Number of supports of the gearwheel unit—one"	"Number of bearings of the third support—nil"	
	"Number of supports of the gearwheel unit—two"	"Number of bearings of the third support—nil"	
	"Number of supports of the gearwheel unit—three"	"Number of bearings of the third support—nil"	Incorrect

Let us continue only with correct versions.

Having multiplied two features, we get the following sets:

N	Description of the version		Values of features
	"Number of supports of the gearwheel unit—three"	"Number of bearings of the third support—one"	a_{13}, a_{21}
	"Number of supports of the gearwheel unit—three"	"Number of bearings of the third support—two"	a_{13}, a_{22}
	"Number of supports of the gearwheel unit—one"	"Number of bearings of the third support—nil"	a_{11}, a_{23}
	"Number of supports of the gearwheel unit—two"	"Number of bearings of the third support—nil"	a_{12}, a_{23}

The presented example shows that all values of initial features are involved into development of new structures.

If the second feature did not have the nil value, the values a_{11} and a_{12} would not be present in the resulting structures, which is the loss of the meaning data.

When considering the example, we deleted incorrect versions after multiplication which are called the forbidden ones in the theory of characterization analysis that will be applied further for the development of the synthesis algorithm. The reasons of appearance of forbidden versions are forbidden figures—impossible combinations of values of different features.

Since we multiplied only two features, our forbidden versions are the forbidden figures.

In the real design process, in order to get the description of structures that correspond to the layout of the product, one should multiply all features that describe the class of the designed objects and the power of description is equal to the number of features of the class of objects. Within the obtained set of structures the part of versions will turn out to be forbidden due to the presence of forbidden figures that can have very low power.

Forbidden figures are the second element of the information support of the system developed on the bases of the proposed model [1, 2, 4].

Forbidden figures are the reason of impracticability of the version. In accordance with the method of generation and influence on the synthesis process, forbidden figures are divided into empiric and functional.

An empiric forbidden figure is the impossible combination of values of features obtained, as a rule, by the expert survey. The specific feature of such figures is that the empiric figure is the randomized set that can be represented by an absolutely symmetric function depending on implementation of the synthesis process. That is, the forbidden figure $\{a_{ik}, a_{jl}\}$ can be represented by the function $a_{ik} = f(a_{jl})$ or the function $a_{jl} = f(a_{ik})$, and the forbidden figure $\{a_{ik}, a_{jl}, a_{nt}\}$ can be represented by functions $a_{ik} = f(a_{jl}, a_{nt})$, $a_{jl} = f(a_{ik}, a_{nt})$ or $a_{nt} = f(a_{ik}, a_{jl})$.

A functionally forbidden figure is the set of features inter-related with each other by the functional relation $p_i = f(\{p_k\})$, where $\{p_k\}$ is the set of arguments of the function, and p_i is its value.

By substituting all the possible sets of values of features p_k into the function, we get the corresponding values of the feature p_i. As the result of the pointed manipulations all the rest sets of values of features can be considered as forbidden ones.

19.4 Model of the Class of Designed Objects and the Process of Synthesis

With two described above sets (features and forbidden figures) available, one can formally represent the set of possible versions as $V = \prod_{i=1}^{N} p_i - z$, where V is the set of possible versions, N is the number of features p_i describing the class of the designed objects, and Z is the set of forbidden versions.

The obtained formula shows the basic idea of the proposed approach:

- generation of the set of the sought structures is the process consisting of two stages: synthesis and analysis;
- the framework of synthesis is the search algorithm implemented by Cartesian multiplication of the set of features describing the class of objects;
- the procedure of synthesis is the exclusion of sets that contain the forbidden figures representing the formalized knowledge of the subject area.

The pointed structure of the design process allows for making the intermittent conclusions:

1. Stages that constitute the model of the process are equivalent, that is why, the quality of preparation of data used at these stages should be maximally complete and absolutely correct;
2. Despite of the search algorithm of the synthesis, the necessary condition of the proficient execution of the stage of analysis is the correctly presented information on inter-relations fixed within the subject area. That is why, any research allowing to obtain new knowledge, relations and influences of features on each other are important for implementation and promotion of the proposed methodology.
3. Despite of the apparent simplicity, application of this formula as the model of the process of synthesis is not possible, since the combinatorial search implementing the synthesis stage leads to informational blow-up due to avalanche increase in the data volume; the test of intermediate sets for the presence of forbidden figures in them after multiplication of the next feature will consume the unreasonable time.

The latter problem can be overcome by optimization of the process of structural synthesis. The optimized parameters are the capacity of the used memory and the time of solving the problem of structural synthesis.

The idea which is the basis of the proposed optimization method and algorithm consists, firstly, in the consequent execution of synthesis stages (multiplication by the next feature) and analysis (when intermediate versions are checked only for those forbidden figures that could potentially occur by that instant), secondly, in such ordering of multiplied features that the maximum number of sets containing forbidden figures could be excluded at early stages of synthesis when the variety as a set still has the low power and, therefore, is not demanding for storage in memory.

The performed research showed that exactly the functional forbidden figures are of the key influence on the order of multiplication of features.

The process of synthesis can be presented by the following number of steps:

1. multiplication of features that are components of functional forbidden figures with resulting in allowed sets of values of features;
2. multiplication of the obtained variety of sets by features with their values being present in empiric forbidden figures that apart from values of multiplied features

contain a part of values of features that are components of functional forbidden figures;

3. checking of the obtained sets for the presence of such forbidden figures there and exclusion of forbidden versions;

4. multiplication of the variety of the obtained sets by all the rest features of the set of features that describe the class of objects.

For example:

Let us assume that the class of objects is described by the following set of features:

$$p_1 = \{a_{11}, a_{12}\}$$
$$p_2 = \{a_{21}, a_{22}, a_{23}\}$$
$$p_3 = \{a_{31}, a_{32}\}$$
$$p_4 = \{a_{41}, a_{42}\}$$
$$p_5 = \{a_{51}, a_{52}\}$$
$$p_6 = \{a_{61}, a_{62}\}$$

The set of forbidden figures is as follows:
Functionally forbidden figures are

$z_1 = \{p_1, p_2, p_3\}$, where $p_1 = f(p_2, p_3)$,
$z_2 = \{p_2, p_4\}$, where $p_2 = f(p_4)$;

Empirically forbidden figures are

$$z_3 = \{a_{11}, a_{51}\}, z_4 = \{a_{21}, a_{52}\}, z_5 = \{a_{31}, a_{51}\}, z_6 = \{a_{51}, a_{62}\}.$$

Since first of all we have to multiply functionally forbidden figures, it is obvious that the order of their multiplication should be determined. Since p_2 is the argument in one case and the value in the other case, all possible values of the feature p_2 are to be firstly determined for the set of values of the feature p_4. Let us assume that implementation of the function $p_2 = f(p_4)$ resulted in the following sets: $\{(a_{21}, a_{41}), (a_{22}, a_{41}), (a_{23}, a_{42})\}$.

Let us calculate the value of the function $p_1 = f(p_2, p_3)$ for all possible sets of values of features p_2 and p_3. We assume that the following sets will be obtained: $\{(a_{11}, a_{21}, a_{31}, a_{41}), (a_{12}, a_{22}, a_{31}, a_{41}), (a_{21}, a_{23}, a_{32}, a_{42})\}$.

In order to determine the order of the next multiplication, the set of empirically forbidden figures should be divided by subsets: figures involving the outcomes of features that compose functional figures and other forbidden figures.

Forbidden figures $z_3 = \{a_{11}, a_{51}\}, z_4 = \{a_{21}, a_{52}\}, z_5 = \{a_{31}, a_{51}\}$ are related to the first type. Values of the feature p_5 are components of the pointed forbidden figures. It means that the previously obtained set should be multiplied by the set of values of this feature:

$$\{(a_{11}, a_{21}, a_{31}, a_{41}), (a_{12}, a_{22}, a_{31}, a_{41}), (a_{21}, a_{23}, a_{32}, a_{42})\} \times \{a_{51}, a_{52}\}$$
$$= \{(a_{11}, a_{21}, a_{31}, a_{41}, a_{51}), (a_{12}, a_{22}, a_{31}, a_{41}, a_{51}), (a_{21}, a_{23}, a_{32}, a_{42}, a_{51}),$$
$$(a_{11}, a_{21}, a_{31}, a_{41}, a_{52}), (a_{12}, a_{22}, a_{31}, a_{41}, a_{52}), (a_{21}, a_{23}, a_{32}, a_{42}, a_{52})\}$$

Let us exclude the versions containing forbidden figures $z_3 = \{a_{11}, a_{51}\}$, $z_4 = \{a_{21}, a_{52}\}$, $z_5 = \{a_{31}, a_{51}\}$.

We get:

$$\{(\cancel{a_{11}, a_{21}, a_{31}, a_{41}, a_{51}}), (a_{12}, a_{22}, \cancel{a_{31}, a_{41}, a_{51}}), (a_{21}, a_{23}, a_{32}, a_{42}, a_{51}), (\cancel{a_{11}, a_{21}, a_{31},}$$
$$\cancel{a_{41}, a_{52}}), (a_{12}, a_{22}, a_{31}, a_{41}, a_{52}), (a_{21}, a_{23}, a_{32}, a_{42}, a_{52})\} = \{(a_{21}, a_{23}, a_{32}, a_{42}, a_{51}),$$
$$(a_{12}, a_{22}, a_{31}, a_{41}, a_{52}), (a_{21}, a_{23}, a_{32}, a_{42}, a_{52})\}.$$

Let us multiply the obtained sets by the remained feature $p_6 = \{a_{61}, a_{62}\}$ and we get:

$$\{(a_{21}, a_{23}, a_{32}, a_{42}, a_{51}), (a_{12}, a_{22}, a_{31}, a_{41}, a_{52}), (a_{21}, a_{23}, a_{32}, a_{42}, a_{52})\} \times \{a_{61}, a_{62}\}$$
$$= \{(a_{21}, a_{23}, a_{32}, a_{42}, a_{51}, a_{61}), (a_{12}, a_{22}, a_{31}, a_{41}, a_{52}, a_{61}), (a_{21}, a_{23}, a_{32}, a_{42}, a_{52}, a_{61}),$$
$$(a_{21}, a_{23}, a_{32}, a_{42}, a_{51}, a_{62}), (a_{12}, a_{22}, a_{31}, a_{41}, a_{52}, a_{62}), (a_{21}, a_{23}, a_{32}, a_{42}, a_{52}, a_{62})\}.$$

Having deleted the versions that contain the forbidden figure $z_6 = \{a_{51}, a_{62}\}$, we get the sought variety of possible sets:

$$\{(a_{21}, a_{23}, a_{32}, a_{42}, a_{51}, a_{61}), (a_{12}, a_{22}, a_{31}, a_{41}, a_{52}, a_{61}), (a_{21}, a_{23}, a_{32}, a_{42},$$
$$a_{52}, a_{61}), (\cancel{a_{21}, a_{23}, a_{32}, a_{42}, a_{51}, a_{62}}), (a_{12}, a_{22}, a_{31}, a_{41}, a_{52}, a_{62}), (a_{21}, a_{23}, a_{32}, a_{42},$$
$$a_{52}, a_{62})\} =$$
$$\{(a_{21}, a_{23}, a_{32}, a_{42}, a_{51}, a_{61}), (a_{12}, a_{22}, a_{31}, a_{41}, a_{52}, a_{61}), (a_{21}, a_{23}, a_{32}, a_{42}, a_{52},$$
$$a_{61}), (a_{12}, a_{22}, a_{31}, a_{41}, a_{52}, a_{62}), (a_{21}, a_{23}, a_{32}, a_{42}, a_{52}, a_{62})\}$$

Analysis of the proposed algorithm allows for stating the practicability of the proposed optimization method:

1. Multiplication of features takes turns with analysis for the presence of intermediate versions of forbidden figures.
2. Checking for each forbidden figure is performed only once.

However, practical application of the pointed algorithm faces the difficulty related to imposing the restrictions of the technical assignment imposed on the design process.

What is the technical assignment exactly for this approach?

The technical assignment is the specification of values of individual features.

Therefore, if the designer determined the value of a certain feature within the initial assignment, all the rest values automatically turn into forbidden figures that are called relative.

Let us assume that the user stated the values of features a_{11}, a_{41} and a_{52} within the technical assignment. Therefore, values a_{42}, a_{51} and a_{12} became forbidden. How will execution of the design process be changed?

At the first stage of execution of the function $p_2 = f(p_4)$ the following sets have been obtained: $\{(a_{21}, a_{41}), (a_{22}, a_{41})\}$.

When calculating the second function $p_1 = f(p_2, p_3)$ we get the following sets:

$$\{(a_{11}, a_{21}, a_{31}, a_{41})\}$$

Then the obtained set is multiplied by the value a_{52}, since a_{51} is forbidden: $\{(a_{11}, a_{21}, a_{31}, a_{41}, a_{52})\}$. The obtained set contains the forbidden figure $z_4 = \{a_{21}, a_{52}\}$, therefore, the obtained version is forbidden.

Having obtained the empty set of versions, we can finalize that the designer incorrectly stated the requirements of the technical assignment.

Another problem appears when the designer assigns the value of the feature that is the value of the functional forbidden figure as the initial one. For instance, the designer defined the value of the feature $p_2 - a_{21}$ in the technical assignment. Then the functional forbidden figure $p_2 = f(p_4)$ starts operating as empiric. All possible combinations (as it was mentioned above, they are $\{(a_{21}, a_{41}), (a_{22}, a_{41}), (a_{23}, a_{42})\}$) are generated with its help, and then all versions where the value of the feature p_2 is not equal to a_{21} are deleted. Therefore, we get the variety of intermediate sets $\{(a_{21}, a_{41})\}$.

19.5 Conclusion

It is obvious that the proposed algorithm does not lead to the change in the power of the final set of versions of structural solutions that correspond to the requirements of the technical assignment. It solves the problem of optimization of the required time and memory capacity for storing the intermediate versions within the structural synthesis which the design process is totally related to.

Implementation of the pointed algorithm became possible due to the solution of a number of mathematical problems [4–7] which is itself an interesting scientific result that can be discussed by the author in other manuscripts.

References

1. Gorbatov, V.A., Demyanov, V.F., Kuliev, G.B.: Automation of design of complex logic structures. Energy, Moscow (1978)
2. Gorbatov, V.A.: Characterization. Calculation of semantics. Artificial intelligence. In: Proceedings of 13th All-Union Symposium on Computer-Aided Logic Control, pp. 3–7. Moscow (1990)

3. Malina, O.V.: Analysis of the set of forbidden figures of the structural synthesis of objects and processes. J. Inform. Math. **1**(3):138–143 (2003) (Moscow, Publication of Phys.-Math. Lit.)
4. Malina, O.V., Morozov, S.A.: Analysis of the set of allowed figures and their place in the model of the structural synthesis of complex objects. J. Inform. Math. **1**(6), 174–183 (2007)
5. Malina, O.V., Valeev, O.F.: Model of the process of structural synthesis of objects developed at discrete structures and features of its implementation. J. Bull. Kalashnikov ISTU **2**(57), 24–26 (2013)
6. Malina, O.V., Valeev, O.F.: Mathematical support of computer-aided design of gearbox systems. In: Proceedings of International Symposium on Theory and Practice of Gearing, pp. 426–431. ISTU Publ, Izhevsk (2014)
7. Malina, O.V., Zarifullina, E.G.: Computer-aided design of mechanical engineering products—a new approach and new assignments. J. Intel. Syst. Prod. **3**(27), 32–34 (2015)
8. Malina, O.V.: Problems of developing the model of class of objects in intelligent CAD of gearbox systems. In: Goldfarb, V., Trubachev, E., Barmina. (eds.) Advanced Gear Engineering Mechanisms and Machine Science, vol. 51, pp. 393–418. Springer (2018)

Chapter 20
New Approach to Computer-Aided Design of Gearbox Systems: Conception and Development of Information Support

Olga V. Malina and Maksim N. Mokrousov

Abstract The new approach to automation of the design process consists in application of search algorithms as the means of synthesis of new structural solutions for the set of existing data and knowledge of spiroid gearboxes. The quality of gathering the data and formalizing the knowledge about the class of considered objects is crucial for the consequent quality of design solutions implemented by the system. Informational support of such a system represents a generalized graph model representing all functional elements of the class of spiroid gearboxes and the classifier of the pointed functional elements that allows for indentifying any gearbox or its module. Being the elements of searching, sets of features of the classifier are involved not only into synthesis but also into analysis within forbidden figures—the reasons of impracticability of the layout. Once the information support is developed, it should be supplemented with new data during operation of the system. Searching and filtration of new data are a serious assignment that also requires automation. And the automation should be available at all stages of operation: when stating the search query, when searching and filtrating the obtained information.

Keywords Generalized model of the class of objects · Forbidden figures · Classifier · Information search and filtration · Automation

20.1 Introduction

Automation of intelligent human activity, in particular, automation of the process of solving the engineering problems at the absence of end-to-end design algorithms, is a serious issue that requires implementation of a new vision on the process of structural synthesis involving the design of gearboxes and gearbox systems.

Attempts to develop a computer-aided design system for spiroid gearboxes with functional relation of the subject area as its algorithmic basis turned to be useless,

O. V. Malina (✉) · M. N. Mokrousov
Kalashnikov Izhevsk State Technical University, Studencheskaya str. 7, 426069 Izhevsk, Russia
e-mail: malina_0705@mail.ru

© Springer Nature Switzerland AG 2020
V. Goldfarb et al. (eds.), *New Approaches to Gear Design and Production*,
Mechanisms and Machine Science 81,
https://doi.org/10.1007/978-3-030-34945-5_20

since nowadays there is no end-to-end algorithm that allows for making a structure of a new gearbox that meets the requirements of the technical assignment.

The authors propose the way out of the pointed problem by applying a principally different approach to implementation of the process of structural synthesis. The mentioned system will allow for scanning the variety of layouts that meet the requirements of the technical assignment by searching of modules (combinations of values of parameters and criteria that describe the class of objects). The searching process is in turn subdivided into two stages: synthesis (obtaining the combination from the initial variety of modules) and analysis (check of practicability of the obtained combination of modules).

The following experiment can demonstrate the implementation of the pointed approach: several toy cars are disassembled into parts and several children are offered to create a number of new cars with the existing parts, each child will take the necessary set for a new car: a body, wheels, axles for wheels, a steering wheel. The selection of elements for a set is in fact random (a child does not have experience in car design). During the design process the child can face the difficulty that the chosen parts can not be assembled with each other, for instance: the steering wheel is not housed in the body or the axle for wheels is greater than the hole in the wheel. Having excluded such versions, the child will get new toy cars.

The given example shows the possibility of implementation of the proposed approach. Moreover, some researchers in the field of philosophy of scientific activity state that unlike cause-and-effect algorithms of conscience, searching algorithms are the base algorithms of sub-conscience that operate with the variety of data accumulated by humans. Exactly this consciousness of a new, previously unknown set of known data obtained by searching is perceived by a person as a new idea—"Eureka!!!"

What information is necessary to implement the proposed approach and how should it be organized to avoid the information explosion within the searching process?

It is evident from analysis of the considered above example that the information support of systems that implement the searching algorithm should have two components: the set of modules for searching and the set of rules to check the practicability of obtained versions.

Thus, let us consider the process of development of information support of computer-aided design systems functioning on the base of searching algorithms.

20.2 Development of Generalized Model of Class of Objects

The first problem that requires solution is the problem of obtaining the set of modules describing the layout.

Development of the pointed set requires development of the generalized model of the class of designed objects. So how is the pointed model developed?

The expert designing engineer is offered to choose several existing layouts of the object and make their description. The description is performed in two stages. The first stage is the decomposition of the structure into functional modules in accordance with the technique of decomposition accepted in the engineering practice: the product is decomposed into units, units into sub-units, sub-units into assemblies, assemblies into parts, parts into surfaces. As the result, the tree-type graph structure will be obtained, with its vertexes being the functional elements and its branches being the relations that denote the "occurrence" operation [5].

The basic principles of implementation of this stage are the principles of: decomposition (the correct choice of the number of levels); analysis and synthesis (reverse assembly is possible); sufficiency (the reasonable level of specification—it is unnecessary to decompose standard parts, for example, bearings, into elements) (Fig. 20.1).

The second stage is the description of each functional element. At this stage the emergence principle should be strictly followed—the characteristics belongs to the functional vertex only when it can not be used to characterize the components of this vertex. For example, the gear ratio of the gearbox characterizes the gearbox as a whole, since it can not be the characteristics of its units or individual parts, while the number of spiroid gearwheel teeth will be related to the gearwheel rim rather than the gearwheel as a whole.

It is also essential to follow the principle of sufficiency (it makes no sense pointing characteristics that do not influence the structure of the product, for instance, the color of the gearbox). In fact, the variety of gearbox characteristics is the set of all criteria, parameters and other features that are important for the designing engineer.

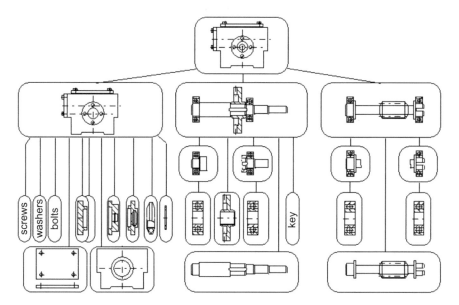

Fig. 20.1 Decomposition of the gearbox into functional modules

As the result of performing the pointed stages for several known versions of the layout we get their several model representations. Their generalization will allow for speaking about the set of modules for searching. Methods of description of objects and development of the generalized model are considered earlier in [6].

As the result of generalization we get the generalized functional graph model that consists of inter-related functional elements and the classifier of the class of objects where the characteristics are united into features and their corresponding values. The graph model includes all possible layout functional elements of objects of the considered class independently if they are obligatory for all versions of the layout or not. The vertex of the graph model has two basis attributes: the feature of obligation (if the vertex is not obligatory, then there is the feature "Presence of the pointed vertex" in the classificator of the parent vertex; and all features of the current vertex have NIL value in the set of values) and the reference to the part of the classifier that describes this vertex.

The classifier represents the set of features with each feature having the set of values (the fragment is shown in Table 20.1).

The obtained generalized model is not final. Prior to its inclusion into the information basis of the system it will be supplemented with data on the gearboxes dropped out of the primary query. All types of data are to be extended: the set of functional vertexes, sets of features, and the set of values of features. Let us consider the changes of the model after the mentioned supplementation omitting the mechanism of obtaining new data.

20.3 Extension of Generalized Model

Let us consider the extension of the set of values of the feature. It should be noted that there are two types of values of the feature: simple and generating. The values of features are simple if their appearance in the classifier does not lead to any changes of the generalized model. For example, if the feature "Method of assembly" is supplemented with the new value "Press fit", then the graph model will not be changed; and the feature "Method of assembly" of the classifier will be supplemented with the value "Press fit", thus increasing the power of the feature. If the feature "Method of assembly" is supplemented with the value "By key", then other qualitative changes will take place alongside the supplementation of the feature by this value. It is evident that another child functional vertex "Key" will appear in the graph model at the apex that characterizes the feature "Type of assembly". This vertex is not obligatory and it should be described further in the classifier, that is, its features and all possible values should be further defined. Therefore, it is obvious that the value "Press fit" of the feature "Type of assembly" is simple and the value "By key" is the generating one. The given example shows that one feature can have different types of values. The value is the generating one if the name or reference to the functional element is present in its statement. Values of features that are quantitative characteristics (for instance, "Number of supports of gearwheel unit") are also the generating ones.

Table 20.1   Fragment of the classificator of the class of spiroid gearboxes

Function vertex	Feature	Values of features
Gearbox	Gear ratio	60
		40
		65
		Other
	Overall dimension	100
		150
		Other
Support 1	Number of bearings	2
		Other
Bearing 1	Type of bearing	N
		B
		Other
Bearing 2	Type of bearing	M
		C
		Other
Support 2	Number of bearings	1
		Other
Gearwheel unit	Presence of the 3rd support	Yes
		No
Support 3	Number of bearings	1
		Other
		NILL
	Presence of bearing	Yes
		No
Assembly of shaft and gearwheel	Method of assembly of shaft and gearwheel	Press fit
		By key
		Other
	Presence of key	Yes
		No
Shaft of gearwheel unit	Length	L
		Other
Gearwheel	Type of gearwheel	Solid
		Assembled
Gearwheel hub	Diameter	D1
		D2
		NILL

(continued)

Table 20.1 (continued)

Function vertex	Feature	Values of features
		Other
Assembled gearwheel hub	Diameter	D3
		NILL
		Other

If this feature had two values "One" and "Two" and after supplementation of the model the value "Three" appeared, then the functional vertex "Gearwheel unit" will get the additional non-obligatory child functional vertex "The third support" that should be decomposed into non-obligatory functional modules; and the vertex "The third support" itself and all its components should be described by features with the corresponding values. Each newly added features will have NIL value in the set of values.

Extension of the set of features is not so unambiguous as extension of the set of their values. In practice, it is not the feature that is stated as the question, but another characteristics that can not be represented within the existing set of features. Exactly in this case the feature and the outcome appear. Similar to outcomes, the features can be simple and generating. Simple features have values that do not extend the set of functional vertexes. However, features can be stated that will essentially extend the pointed set and upgrade the model for example, in the initial model the gearwheel is always solid. That is why, the feature of the gearwheel structure was not revealed. When the expert includes the characteristics "Structural type of gearwheel - solid", the pseudo-feature "Structural type of gearwheel" with the value "Solid" will appear which will be primarily characterize the gearwheel. Consideration of the pointed pseudo-feature will allow for obtaining the additional value "Assembled gearwheel", the pseudo-feature will turn into the meaningful feature, the feature "Type of assembly" will appear in the gearwheel structure; and connecting functional elements will appear in the gearwheel structure which are non-obligatory in their nature and which should also be described by the set of features.

The model is subjected to even more complex modification, when characteristics appear that extend the subject area. For example, the expert introduced the characteristics "Single-stage" into the gearbox description. Since the generalized model has been developed only for single-stage gearboxes, this new characteristic will turn into the pseudo-feature "Number of stages" (which does not take part in the synthesis process) with the only value "1"; and the intermediate level (the functional vertex "The first stage") will appear in the gearbox structure. However, if other values of the feature "Number of stages" appear in the process of analysis (for example, "2"), then the pseudo-feature will turn into the quantitative feature; and the new value of the feature will lead to appearance of a new functional non-obligatory vertex "The second stage" which will require decomposition and description.

When new functional vertexes appear, different versions of modification of the generalized model are also possible.

If the functional vertex added to the model is not an alternative to some vertex previously existing in the model, then it is included into the model along with its description. As a rule, such vertexes are non-obligatory, since they are added into the a priori correct model. The example of such a vertex can be the structural element "The third support of the gearwheel unit", "The second stage of the gearbox", "The second bearing of the first support of the gearwheel unit".

Sometimes the situation occurs when the functional vertex introduced into the model becomes the alternative to the existing one. For instance, if the existing generalized model provided only key connection of the motor and gearbox and the worm unit comprised the obligatory vertex "Key", then introduction of the spline joint will result in a new functional vertex "Spline surface of the worm shaft" into the structure of the generalized model which in fact will be alternative to the vertex "Key". Supplementation of the model with new vertexes will among other things change the character of the functional vertex "Key". It will stop to be obligatory; and the set of values of all features describing this vertex will be supplemented with the value "NIL".

The generalized model obtained by this means for the class of objects consisting of the generalized graph of functional vertexes and the classifier is the first element of the informational support of the considered system involved into the searching process at the stage of synthesis.

20.4 Set of Forbidden Figures: Structure and Classification

The second element necessary to implement both the stage of synthesis and the stage of analysis is the set of forbidden figures [3, 4].

The forbidden figures is the reason of an impracticable structure. The typical example of the forbidden figure is the combination of values of features "Output shaft to the left" and "Blank left cap of the casing unit" or "Gearwheel diameter 50 mm" and "Overall dimensions of the casing 20×25".

In the given example the forbidden figure is the impossible combination of values of different features within one set describing the structure.

Versions of structures containing forbidden figures become forbidden, that is, impracticable.

Formation of the set of forbidden figures is the complex problem that requires the thorough and qualitative analysis of the subject area.

First of all, it should be noted, that forbidden figures can be absolute and relative.

Absolutely forbidden figures are combinations of values of features which are always forbidden. Previously given examples describe exactly forbidden figures.

Relatively forbidden figures appear due to imposing restrictions on the synthesis process by stating the technical assignment. When stating the technical assignment, the customer in fact denotes possible (acceptable) values of certain features describing the class of objects, thus making all the rest values of specified features forbidden.

Formation of the set of relatively forbidden figures allows for further eliminating all absolutely forbidden figures with relatively forbidden figures at the stage of analysis.

In accordance with their influence on the process of structural synthesis, forbidden figures can be functional and empiric.

Functional forbidden figures are relations of the subject area which relate a certain number of features that describe the class of objects. Features-arguments and the feature-value are singled out from the set of the pointed features. By substituting all possible sets of features-arguments into the pointed function, we get the value of the feature-value for each set. Therefore, other combinations become impossible—forbidden figures. Application of functional forbidden figures allows for using them as the means of synthesis, saving the system from the necessity to generate a priori impossible versions. Functionally forbidden figures can be applied in this way if the value of the feature, which is the value of the function, is not assigned by the user as the initial data or if it is not determined by the system as the fixed value. In these cases the functionally forbidden figure turns into the means of analysis: for the variety of possible sets of initial data the value of the function is calculated which is compared with the assigned one. If the calculation result and the assigned value do not coincide, the figure is declared to be forbidden and withdrawn from consideration.

Empiric forbidden figures are the typical means of analysis. An empiric forbidden figure is the set of values of features the combination of which is impossible. The forbidden figure is the reason for formation of a forbidden version. Any forbidden figure is the forbidden version, however, not every forbidden version is the forbidden figure. The forbidden version is the forbidden figure only when no subset of values of features of lower power is the forbidden figure. For example, combination of values of three features ("Material of gearwheel rim - steel", "Material of worm - steel", "Lubrication - absent") is the forbidden figure, since the pointed set is the forbidden version; and any of its subsets {"Material of gearwheel rim - steel", "Material of worm - steel"}, {"Material of gearwheel rim - steel", "Lubrication - absent"} and {"Material of worm - steel", "Lubrication - absent"} are possible combinations. In order to get forbidden figures it is necessary to consider all combinations of values of all features and assess them. For this purpose, outcomes of all features are united into one set which is Cartesian multiplied by itself. At the first multiplication all combinations of values of features with the power 2 will be obtained. Each set of the obtained variety is assessed from practicability point of view. If it is impracticable, the set is declared to be the forbidden figure and excluded from the intermediate variety. Then the obtained variety is multiplied by the initial one; and those forbidden versions are excluded from the resulting variety of sets that contain previously revealed forbidden figures and the remained sets are checked for practicability. The number of intermediate multiplications should theoretically be equal to the number of features describing the class of objects, however, practical experience shows that empiric figures with the power above 3 are not present in reality. It also makes no sense analyzing the practicability of sets of values of features included into one functionally forbidden figure. Basing on the algorithm of the synthesis process, functionally forbidden figures take part in the synthesis process earlier, so, all impossible combinations of their values will be already excluded by the time of analysis of empirically forbidden figures.

Similar to the set of features, the set of forbidden figures is not once developed and unchangeable.

The set of forbidden figures can be changed if a new functional relation describing the dependence of features in the subject area appeared after the research process. Appearance of the functionally forbidden figure allows for excluding all empiric forbidden figures from consideration that were developed in the set of values of the pointed features.

The set of forbidden figures can be changed if a certain combination of values of features changed its status and became allowed or otherwise.

However, the pointed set is subjected to maximum changes when the set of features is supplemented or the set of values extended within the existing set of features. In order to recover the correctness of the set of empiric forbidden figures, it is necessary to Cartesian multiply the set of new values by the set of all values of all features and perform the described above procedure of revealing the forbidden figures.

Therefore, we determined that the informational support represents the generalized model of the class of objects (uniting the graph model of functional modules of the object and the classifier representing the set of features describing each functional module and their values) and the set of absolutely forbidden figures.

Also we showed that the pointed informational support is modified both within the process of the primary accumulation of data, and in the process of functioning of the system due to appearance of new knowledge in the subject area which comprises the objects to be designed by the system of structural synthesis.

20.5 Actualization of Informational Support

The question of actualization of the informational support of the system and the search of new useful information is still an open issue.

Till recent times this part has been trusted to a human expert, however, the traditional manual technique of processing the increasing document flow does not provide necessary conditions for the effective solution of the problem of extending the informational support of software systems; and rapidly developing means of linguistic analysis of implementation of the target search of information (including Internet) allow for proposing the possibility of automation of this process.

The problem of informational search is specially urgent due to constantly increasing information volume. To overcome this difficulty, it is necessary to solve a number of accompanying issues, including: the development of informational searching languages, the development of quality syntax analyzers of structured (html, xml) and partially structured texts (technical specifications of devices, description of parameters and properties of objects), the development of structures to describe and store the knowledge on objects, processes and phenomena within the subject areas and also the methods of searching, filtration and comparison of these concepts.

The problem of extension of the informational support includes two inter-related issues:

(1) search and filtration of new information;
(2) introducing of new information without breaking of the existing structure integrity.

The ways of solving the second issue have been considered above.

Let us consider in details the issue of searching and filtration of new information.

20.6 Methods of Computer-Aided Search and Filtration of Information

The problem of searching and filtration of data is related to the problem of informational search which can be solved by methods of computer-aided processing of texts.

The first of them are the methods of semantic analysis of natural language texts (NL texts). Algorithms of their functioning can be divided into three large groups.

1. Algorithms based on rules; they imply operation with different types of structured sets of data (ontology, dictionaries) where the versions of application of all lexical units are described.
2. Algorithms based on static methods; they imply computer training and application of the statistical approach to analysis of texts. Basing on the probability of appearance of a lexical unit in texts used for the training, its value in the considered text is determined.
3. Mixed-type algorithms; they imply the joint application of dictionaries and statistical methods. It is preferable to apply this approach for texts analyzed within the predetermined subject area.

The semantic analysis is, as a rule, the final stage of processing the text by the system of informational search. In most systems the division of analysis into four stages is used, each of them applying the results of the previous ones and impossible without them [1].

1. Tokenization is the division of the text into sentences and sentences into words.
2. Morphological and morpheme analysis is the obtaining of morphological information for each word (kind, number, case, declension and oth.) and definition of its morpheme composition.
3. Syntax analysis (parsing) is the process of comparison of the linear sequence of lexemes of the NL text with the formal grammar of the language in order to get the structural representation (usual as the tree) [2].

4. Semantic analysis is the determination of meaning relations between elements of the text. The semantic analysis is in turn performed in three stages.

   (1) extraction of semantics of each word resulting in the development of a semantic structure where all possible values of the word in this context or probabilities of appearance of these values are determined;
   (2) extraction of semantics for each sentence when an attempt is made to eliminate the uncertainty of recognition of the essence of each word and generate the semantic kernel of each sentence and the text as a whole;
   (3) displaying the result of analysis and representation of the initial text as the structured set of data in the form depending on the aims of further research (a table, graph model, classifier, etc.).

Methods of semantic search directly depend on the method used for the semantic analysis of the text.

The second group of methods of analysis and search can be called formally statistic. These methods include formal grammars (finite-state automations), regular expressions, methods based on language models and statistics. Finite-state automations are often used in practice for analysis of texts, development of syntactic and lexical analyzers with their specific syntactic structure. Regular expressions represent a specialized language for searching and execution of other operations with lines of the initial text. The language of regular expressions is based on application of metasymbols to form the rule of searching, that is, the line-the sample (pattern or template). Along the sample line the alternative line is sent to the input of the analyzer; it contains information for replacing the found coincidences and it can also have special symbols [9].

Regular expressions are a rather powerful and commonly spread means of computer-aided analysis of the text. In practice regular expressions are used by the majority of text editors to search the correspondences in the text and replace the sequences.

Statistical methods of search are more often based ***on the vector model of informational search*** where the document is considered as the unordered set of terms. Terms in informational search are lexically independent structures of the text, more often words and fragments of the text like *2010, II-5, Tien Shan*, etc.

To order the output of results of the search system operation due to its correspondence to the query, it is necessary to introduce weights of correspondence of documents to the query which should be determined on the basis of words included into the query.

There are different methods to determine the weight of the term in the document—the "importance" of the word for identification of this text. For example, one can simply calculate the number of usage of the term in the document, the so-called frequency of the term—the more often the word is met in the document, the higher its weight will be.

All terms met in documents of the processed collection can be ordered. If weights of all terms for a certain document are written out in the order (including those that are absent in this document), we get the vector which can be the representation of this document in the vector space. Similar to dimensionality of the space, the dimensionality of this vector is equal to the number of different terms in the whole collection; and it is the same for all documents.

More formally it is

$$dj = \left(w_{1j}, w_{2j}, \ldots, w_{nj}\right)$$

where d_j is the vector representation of the jth document, w_{ij} is the weight of the ith term in the jth document, n is the total number of different terms in all documents of the collection.

Having such a representation of all documents available, one can determine, for example, the distance between points of the space and thus solve the problem of similarity of documents—the closer the points are to each other, the more similar the corresponding documents are. In case of search of the document in accordance with the query, the query is also represented as the vector of the same space; and the correspondence of documents to the query can be determined.

In order to determine the vector model, it is necessary to point the method of calculation of the term weight in the document. There are several standard methods of determination of the function of weighting:

- Boolean weight is equal to 1 if the term is met in the document and 0 otherwise;
- tf (term frequency)—the weight is determined as the function of the number of inclusions of the term in the document;
- tf-idf (term frequency—inverse document frequency)—the weight is determined as the product of the function of the number of inclusions of the term in the document and the function of the value inverse to the number of documents of the collection where this term is met.

The more often the term is met in the document, the greater weight it has. But the more often the term is met in the collection of documents, the less weight it has.

Therefore, the correspondence of the query to the document is measured by a specific number; and all documents can be orderly put out by the search system in accordance with this number.

The advantage of the vector model of information search is that the model provides a simple model for the ordered putout of results of informational system operation. The specific method of determining the weights of words in the document can be changed here depending on the problem to be solved and the operating collection. The drawback of this approach is the assumption of independency of words in the text which contradicts with the statement that the text uses the set of words which are inter-related by their meaning.

In 1977 Robertson and Sparck-Jones reasoned and implemented the probability model. The relevance in this model is considered as the probability of the fact that this document can be interesting to the user. The presence of already existing initial set of relevant documents is implied here, the documents being chosen by the user or obtained automatically within a certain simplified assumption. The probability to become relevant for every next document is determined basing on the relation of the term frequency in the relevant set and in the rest part of the collection.

The document can be defined as the set of words without account of the word frequency in the document. One can also represent the set as a usual Boolean vector $D = \{d_1, \ldots, d_n\}$ where n is the number of all terms and d_i can take values from the set $\{0, 1\}$. The query can be defined as the set of words.

The correspondence of the document to the query will be determined as follows: let us assume that for each fixed query Q_k we have distributions of probabilities for all documents "to be relevant" and "to be irrelevant" to the query Q_k. It is correspondingly denoted as $P(R|Q_k, D)$ and $P(\overline{R}|Q_k, D)$. Then the correspondence function will be the ratio of these two values:

$$\frac{P(R|Q_k, D)}{P(\overline{R}|Q_k, D)}.$$

The third group of methods can be described as computer training—the group of methods of artificial intelligence, the feature of which is not the direct solution of the problem, but the solution in accordance with the results of training based on the experience of solving the similar problems [7]. As a rule, the basis of methods of computer training are the means of mathematical statistics, optimization methods, methods of numerical solutions, theory of probability, etc.

Classes of computer training are commonly divided into two large parts:

1. Inductive training (training by precedents) is based on searching the empiric regularities in data.
2. Deductive training implies formalization of information for experts in the subject area and collecting them as the knowledge base.

The second class of methods is commonly associated with expert systems, in this connection, the inductive training is rendered with computer training.

Trees of solutions are the method of representation of rules in the hierarchic consequent structure where the only node that gives the solution corresponds to each object. The rule is defined here as the production "*IF… THEN…*".

Other names of this method are given to it in accordance with the problems to be solved—classification tree or regressive tree.

The decision tree is the means of the decision support which can be applied for analysis of statistic data and development of forecasting models. The tree structure represents "leaves" and "branches". The objective function depends on attributes that are written in branches of the decision tree, the values of the objective function are written in "leaves"; and attributes of different cases are written in the rest nodes. In order to classify a new case, one should go down the tree to the leaf and give the corresponding value. The objective is to develop a model which forecasts the value of the objective variable on the basis of several variables at the input.

Within the whole variety of methods of informational search and processing of natural languages for every specific assignment, it is necessary to choose methods of solving the problems very thoroughly. The criteria of choice can be the structures of processed texts, their lexical and semantic coherence and complexity, specifics of the subject area, information capacity, the presence of preliminary parsed corpuses for training of algorithms and many other.

Nowadays, probability statistical methods of search are commonly applied due to their relative computational simplicity. Such an approach allows the searching computers for developing statistical models of documents representing numerical characteristics of lexical units and texts as a whole. Advanced DBMS successfully solve the problems of storing and processing of data having such a format. Comparison of such models within the search takes less computer resources but the more statistical data has been accumulated in a specific subject area, the higher the relevance of searching results is for such an approach.

Classical text methods of searching that include algorithms of precise and fuzzy search, regular expressions, algorithms of full text search that consider morphology word inflexions and transpositions are, as rule, applied in search modules that are components of the informational system.

Semantic methods of search that consider the meaning of documents and represent the texts of documents as complex by complete semantic models found their application in extremely specialized systems that solve the problems within a definite area of knowledge. Application of thesauruses and ontologies is relevant for such methods.

In accordance to [2], from the point of view of informational search a natural language has a number of drawbacks.

1. The presence of words in the natural language that have little semantic charge (prepositions, conjunctions) and words and word combinations that can be neglected when transmitting the data. In this case we deal with the redundancy of a natural language; and this problem is successfully solved by humans, in particular, when composing the text of a telegram.

2. The presence of synonyms which humans use to make their speech more expressive, flexible, to achieve the accuracy of expression of thoughts and to pay attention to slight details. However, the presence of synonyms in the text reduces the completeness of the informational search. Synonyms can be expressed here by complex word combinations: *laser and optical quantum generator, computer and electronic computing machine,* etc. When developing the informational-search languages and systems it is important to unite synonymic structures into synonymic rows—classes of conditional equivalence.

3. The presence of polysemanticity in a natural language when one and the same word can have the variety of meanings; and homonymy, that is, words which are equally pronounced and written but have different meanings. Polysemic words are different from homonymic ones by keeping the integrity of meanings of words within different contexts ("address" is the housing of a person and the location of information in the computer). Uneliminated homonymy and polysemanticity are the source of the "informational noise".

Terms are the basis of the vocabulary of the informational search language (ISL), since they are the main carriers of the scientific technical information in texts of documents and queries. The knowledge of mechanisms of appearance and functioning of terms in scientific technical texts allows for increasing the quality of preparation of search images of documents and search prescriptions, thus advancing the procedure of indexing.

Depending on language means of expression, all terms can be divided into two groups: language and non-language.

Language terms comprise terms-words, terms-word combinations and terms-sentences. Non-language terms are represented by special signs—graphic symbols.

Experimental investigations of terminology show that the most typical models of term generation are the following:

(1) adjective+noun (for instance, rare metals, refractory alloy, planetary gear);
(2) noun+noun (for instance, data process, inductor coil);
(3) adjective+adjective+noun (for instance, integral informational systems, arc flame lamps).

Similar to words of a natural language, there are definite paradigmatic relations among lexical units of any ISL, the relations being explained by the presence of logical connections between subjects and phenomena designated by these words. Paradigmatic relations are: identity, hierarchy, derivation, contiguity, contrast and multiplicity. Without account of paradigmatic relations it is impossible to execute correctly the informational search by the query, make the quality indexing, annotation or summarization of the document, and to prepare the literature review. Establishment of paradigmatic relations between lexical units play an important role in the development of an ISL; and methods of their establishment can be conditionally divided into two groups: logically intuitive and formalized. The logically intuitive method of establishment of paradigmatic relations is aimed at revealing the essential meaning connections between concepts by analysis of real scientific technical texts, addressing the encyclopedias, terminology dictionaries, summarizing reference books in the specific area or by addressing experts of this specific area which implies correction of preplanned paradigmatic relations. In practice, all three methods of revealing the paradigmatic relations are usually applied.

The formalized method is based on the means of "recognition" of paradigmatic relations in the text: lexical (usage of verbs, adverbs, parenthetic words, prepositions, etc.); punctuation (usage of colon, dash, brackets, etc.); application of schemes, drawings; application of different fonts (italics, letter spacing).

Development of any ISL comprises five main stages.

1. Selection of lexical units—an important stage in the development of any ISL, since any language is created by lexical units exactly. The "key words" that carry the greatest semantic charge are usually selected into the composition of the lexicon of an ISL. The key word is a full word or word combination which is the carrier of the essential information from the point of view of informational search.

2. After lexical units (key words, terms) have been selected, the problem of their uniform record—standardization—arises. Standardization of the lexicon is implemented by two operations: (1) representation of key words in the unified grammar form (morphological level of lexicon standardization); (2) elimination of synonymy and polysemanticity (semantic level).

3. At the stage of lexicon systematization it is necessary to systemize, order all selected and recorded in the standard form lexical units, to establish meaning relations between them and to generate classes of close meaning words. At this stage an important issue is the establishment of paradigmatic relations between lexical units.

4. The stage of development of classification schemes of concepts is necessary for denoting the boundaries of the concept, determination of the denotation, contents and structure of the concept and for establishment of its relations with other concepts.

5. Organizational formalization of the lexicon implies its graphic and sign actualization as classifiers, subject chapters, descriptors, etc.

Therefore, the correctly developed structure of the ISL allows not only to organize reasonably the process of informational search, but also to perform the comparative analysis of already existing ISL and to execute the substantiated choice of the ISL to solve the definite number of problems.

The search can be determined by two assignments:

(1) search by the query of a user generated in the natural language form or in formal expressions;

(2) search of new subjects, processes and their properties that are not present in the initial knowledge base.

Both assignments have similar algorithms and depend on the aim of search and the form of query.

The search can be performed within the preliminary prepared array of documents of a specific theme, and within unsorted arrays. In the first case, more relevant results of search should be expected, since the unsorted array of documents can contain homonymic words of the query.

The text of the search prescription should be stated for the search computer with the maximum degree of specificity, compactness, consistency and completeness. The higher quality the search query has, the higher is the probability of obtaining the relevant results.

Let us consider the versions of search queries (texts of search prescriptions).

1. Word listing. For example, *gearbox worm*. The search computer will consider such a query as the united conjunction. Usually such a list is composed in the order of descending the validity of words or from the higher class of meanings to the lower one, without application of prepositions and punctuation marks. In this case, the search system should look for documents that contain these words with account of their normal forms by applying the algorithms of complete and partial coincidence search, and also the full text search and probability statistical

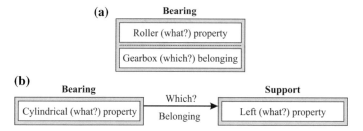

Fig. 20.2   Models of the search query

models of search, including metrics of fuzzy texts search [8]. This search can be supplemented with databases of synonyms, homonyms, and paradigmatic classifiers. Analogs of such a search are commonly known search services by Google, Yandex, Mail, etc.

2. A simple sentence in which the main word and secondary parts can be singled out. For example, (a) *roller bearing of the gearbox* or (b) *cylindrical bearing of the left support.* For this query it is necessary to develop a formal model with pointing the key word and its attributes. Secondary words should be considered by a search computer as specifying, increasing the accuracy and completeness of return. The model of the search query for the pointed examples is shown in Fig. 20.2a, b.

   There can be any type of text representation of the model. In the ideal case, in order to perform the search, it is necessary to present the texts of documents in the same formal form and perform the search by comparison of models with account of inter-relations of words in different sentences.

   Such an approach allows to neglect within the search the syntactic relations between words in the text, but to handle formal models having the metalinguistic representation. It is easier to translate these models into other languages and perform the search without account of the language of the text of the document.

3. A simple sentence that contains the verbal noun, participle, adverb, that is, the description of an action or process. These queries are similar to those described in p. 2, but they are different only by the presence of "words-actions" and their attributes. It should be noted that queries containing participles, participle phrases and subordinate clauses reduce the time efficiency of the search process, since it is necessary to develop more complex formal schemes of the search prescription and compare greater number of elements of such formal models. In general case, the model of the search query can look like a classic semantic scheme of the text where objects are the network nodes and actions are relations between them. Another version of the model implies the development of the network with objects-nodes and processes-nodes; and relations between these nodes are the questions: *where? what? what for? for what purpose?* etc. For example, *bearing with rotation to one direction* or *bearing rotating to one direction.* Formally, it is one and the same text, but in the second case the participle is used which can be simplified at the stage of morphological analysis of the text prior to the

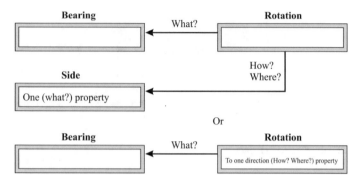

Fig. 20.3 Versions of the formal model

development of the semantic scheme of the sentence to the verb *rotate* or verbal noun *rotation*. That is, both versions of texts of queries will be simplified to the sentence *bearing rotation to one direction*. Versions of the formal model are shown in Fig. 20.3.

20.7 Settings of Search and Analysis of Results

Any search system should have a variety of settings for the searching process to help a user guide himself within the search results. These settings can have standard instructions such as: search by the precise or partial coincidence, search in documents having a definite format, sizes, location, languages, focusing on a specific theme, etc. The semantic type of results to be obtained can also be specified in settings:

– new object;
– new property of the object;
– new principle of action or cancellation;
– new property of the action;
– new composition;
– new name and definition of the essence;
– and so on.

Moreover, the user can point the part of the existing knowledge base for which and by means of which the search is performed, and also refer the search result: the part of the structure of the part of characteristics.

Developers of the informational search language will need to state the corresponding markers in formal semantic models for each of these settings; these markers will allow for referring the found essence and its attributes with each type of markers.

No doubt, the search results must be subjected to the expert assessment, that is why, it is necessary to provide the mechanism of ranking and sifting of search results. The easiest and common method of ranking of search results is the application of statistical metrics tf-idf, that is, the application of mathematical and semantic methods of calculation of similarity of the found results and the search query. Behavioral and user-defined factors of the search based on the previously performed search should be considered last of all, since specialized search systems require an individual approach to each fact of the information search.

20.8 Process of Search for New Knowledge Within the Development of Informational Support

The search of new information is rather consuming for both the person executing the search and the software-hardware system used for the search. The preferable version of the search is its starting with a certain periodicity independently of the expert, when the computer resources are free.

The version of the manual search when the expert himself states the text of the query by typing it on the keyboard or applying specialized design kits based on the knowledge of the existing classifier has been described above. In this case the expert performing the search is also responsible for the quality of the search query and the adequacy of interpretation of the search results.

The authors propose to use the version of automatic (or automized, semi-automized) search which consists of three stages:

(1) automatic generation of the search query;
(2) search on the basis of the generated search query;
(3) automatic filtration of the search results, their analysis by the expert.

Let us consider each stage in more details.

1. Automatic generation of the search query.

The important stage that precedes the informational search is the generation of the text of the search prescription which defines the successful search process.

This stage implies automatic generation of the text of the search prescription with application of knowledge of the existing classifier and rules of forming the text of the query. Depending on the result to be found, the corresponding rules of query c the search system:

(1) when searching the value of the feature of the functional element, it is necessary to perform integration of the text of the search prescription from the node designating the functional element to the node-feature: *gearbox-casing-material* or *key-type*. In the first case, all possible values of *materials of the gearbox casing* are truncated from the knowledge base, in the second one—all *types of keys*.
(2) when searching the feature of the functional element, it is necessary to perform integration of the text of the search prescription from the node designating the functional element to the nodes-values of the feature thus eliminating the node feature: *gearbox-casing-steel* or *key-prismatic, segment, cylindrical, wedge, tangent.*

(3) when searching <u>functional elements</u>, it is necessary to generate the text of the search prescription basing on characteristics of functional elements, that is, from the node designating the functional elements (eliminating it) to nodes-values of the feature: *material–steel* or *type—prismatic, segment, cylindrical, wedge, tangent.*

The pointed versions of search queries are purposed for the expedient search the results of which will be filtrated basing on orientation of queries. If it is necessary to perform the extensive search, for instance, of functional elements and their characteristics, the searching query can contain only names of functional elements of the classifier.

It should be noted that due to the presence of homonymy and polysemanticity in natural languages, the search query should not consist of separate words that have more than one version of interpretation. For example, the search query *worm* can give results related to zoology, game theme, types of mechanical gears and so on. However, the search query *worm of gearbox* confines the area of search, and the search results will be related to gearbox systems in question. At the same time, the search query *key* is allowable in the proposed system, since this term is unambiguous determining the machine part.

2. Search on the basis of the generated search queries

Within the standard internet search the results of the query (text documents, manuscripts) will consist of those documents which contain words of the search prescription.

At this stage the search is executed, as a rule, with application of probability statistical methods, methods of explicit and fuzzy search, and full text algorithms of comparison. This group of methods has the least computational complexity in the informational search and it allows for obtaining the search results rather quickly.

Then it is necessary to single out the fragments from texts, that are mostly similar to the text of the query, perform their syntactic analysis, define the composition of sentences and compare with those elements of the classifier which are present in the knowledge base. At this stage it is reasonable to apply methods of syntactic analysis, formation of syntactic models of sentences and the text which can be used for the more effective automatic filtration of search results.

3. Automatic filtration of the search results, their analysis by the expert.

The automatic filtration of the search results implies selecting those fragments of the text that correspond to orientation of the search query generated at the first stage.

When searching the <u>feature</u> and <u>value</u> of the feature of the functional element, it is necessary to select and present that part of the text to the expert which is syntactically close to the text of the search prescription. This syntactic closeness can be quantitatively determined in the algorithm of filtration either with pointing the number of sentences before and after revealing the closeness with the query, or applying stylistic marks: *consists of, is, are related to,* etc.

When filtrating the search results within the search of <u>values</u> of features, it is necessary to compare the search results with those values that are already present in the classifier but to present to the expert only those vales that are absent in the classifier. The similar procedure should be done when searching the <u>features</u>.

Search results in accordance with the first two versions of queries are supposed to contain more relevant search results that are related to the subject area interesting for an expert, since they comprise references to functional elements. Information should be presented to the expert in fragments with the possibility to examine the source.

When searching the <u>functional elements</u>, when only characteristics are pointed without relation to specific objects, the search results can cover any subject areas. In this case, the search results are documents as a whole, references to manuscripts, records of issue-related blogs, etc. It is necessary to show to the expert the headings of such documents and manuscripts with fragments of the text in which the words of the search query have been found.

Then the expert looks through the sought results, analyzes them and, if necessary, adds new knowledge into the initial knowledge base of the classifier. It is also reasonable to give the expert the possibility to mark those results of the search that, in his opinion, do not contain new information and are not interesting in their contents for the stated assignment. These documents should not be involved into the search again. Moreover, other documents that contain the texts of "forbidden" documents as a whole or in fragments, should not be included into the selection for other search queries. Filtration of "forbidden" documents should be executed in the background, unnoticed for the expert, with application of algorithms of the full text search and algorithms of calculating the level of similarity of big fragments of the text. These algorithms operate effectively on the basis of methods of a vector model of the informational search.

For the effective search of new knowledge, especially for that performed in the semi-automatic mode, the stage of preliminary preparation of data is necessary. At this stage, firstly, the work with dictionaries and reference books of the natural language is performed; and secondly, the expert preparation of data is executed. Dictionaries, in particular, of definitions and synonyms, are necessary to reveal homonymous and polysemic words applied in the classifier, so that polysemantic search prescriptions could be excluded at the stage of generation of the search query.

Spelling dictionaries are necessary for standardization of the lexicon, that is, for putting the words to their normal forms for further search, and to correct grammar and orthographic errors of the manual input. The expert preparation of data implies specification and correction of data by an expert in this subject area, and the preliminary setting of the search.

20.9   Conclusion

Therefore, the proposed system has three large blocks: the block of preparation of knowledge on the subject area, the black of the main functional of the system, that includes generation of the search query, execution of the search and filtration of results; and the block of extension of the informational support performed by the expert.

References

1. Bolshakova, E.I., Klyshinskiy, E.S., Lande, D.V., Noskov, A.A., Peskova, O.V., Yagunova E.V.: Computer-Aided Processing of Texts by a Natural Language and Computer Linguistics. Teaching Guide, Moscow, MIEM (2011) (in Russian)
2. Gendina, N.I.: Linguistic Support of Computer-Aided Library Systems. Alma-Ata, Nauka (1991) (in Russian)
3. Gorbatov, V.A.: Characterization. Calculation of semantics. Artificial intelligence. In: Proceedings of 13th All-Union Symposium on "Computer-Aided Logic Control", Moscow, pp. 3–7 (1990) (in Russian)
4. Malina, O.V.: Analysis of the set of forbidden figures of the structural synthesis of objects and processes. J. Inf. Math. **1**(3), Moscow, Publ. Phys.-Math. Lit., 138–143 (2003)
5. Malina, O.V., Zarifullina, E.G.: Class of object graph model as dataware of structural synthesis system. In: Zawislak, S., Rysiński, J. (eds.) Graph-Based Modelling in Engineering, vol. 42, pp. 211–222. Springer (2017)
6. Malina, O.V.: Problems of developing the model of class of objects in intelligent cad of gearbox systems. In Goldfarb, V., Trubachev, E., Barmina, N. (eds.) Advanced Gear Engineering, vol. 51, pp. 393–418. Mechanisms and Machine Science Springer (2018)
7. Nedelko, V.M.: Fundamentals of statistic methods of computer training. Teaching Guide. Novosibirsk, NSTU (2010). http://www.iprbookshop.ru/45418.html. Accessed 5 June 2019 (in Russian)
8. Smetanin, N.: Fuzzy search in the text and dictionary. HABR (2019). https://habr.com/ru/post/114997. Accessed 1 August 2019 (in Russian)
9. Smirnov, I.V., Shelmanov, A.O.: "Semantic syntactic analysis of natural languages. Part 1. Review of methods of syntactic and semantic analysis of texts. J. Artif. Intell. Decis. Mak. 1 (2013) (in Russian)

Chapter 21
Improvement of Methods of Design and Analysis of Load-Carrying Capacity of Case-Hardened Cageless Bearing Units for Power Drives of Mobile Machines

Efim I. Tesker

Abstract Improvement of layouts of power drives for different types of machines is impossible without common application of new technical solutions aimed at the increase in the technical level and operation characteristics of heavy-loaded rapidly wearing units and parts that determine the life-time of a power drive as a whole by their normal operation. Peculiar conditions are specific for operation of machines comprising non-standard cageless bearing heavy-loaded supports. Being reasonably applied in layouts of units and aggregates, cageless bearings allow for increasing essentially the mass and overall dimensions of drives. It should be noted at the same time that the drawback of cageless supports is their sensitivity to stress concentration reducing the gear reliability. In this connection, in order to provide the required level of reliability and reduce the influence of stress concentration, it is necessary to provide higher accuracy of geometry of contacting surfaces within the production process which increases the cost of products. Therefore, stress concentration on local areas of contacting surfaces is the main reason of part fractures and early breakdowns of drives.

Keywords Loading · Surface contact strength · Depth contact strength · Hardened layer · Cageless bearing supports · Characteristics of case-hardened layer · Layer thickness · Equivalent shear stresses · Minimum safety factor · Subsurface cracks

In order to increase the load-carrying capacity of drives, the most efficient trend of development of designs with high performance in mechanical engineering is the application of progressive methods of generation of surface layers of parts with high quality characteristics [1–4]. Basing on research results, one can state that the greatest effect is achieved by application of hardening techniques focused on using highly concentrated energy flows of laser radiation.

Thorough investigations of properties of laser-hardened layers showed that laser modification of metallic surfaces is a highly effective hardening method for parts of

---

E. I. Tesker (✉)
Volgograd State Technical University, Volgograd, Russia
e-mail: agromash-vlg@rambler.ru

© Springer Nature Switzerland AG 2020
V. Goldfarb et al. (eds.), *New Approaches to Gear Design and Production*,
Mechanisms and Machine Science 81,
https://doi.org/10.1007/978-3-030-34945-5_21

power gears operating under multi-factor action (dynamic loads, aggressive medium, etc.).

The present work continues the number of investigations on the contact strength of case-hardened parts of various types of drives subjected to the hardening laser machining. The main research stage is the study of the resistivity to contact fractures, the kinetics and mechanism of development of in-depth fractures for both the hardened layer itself and the adjacent part of the loaded power contact.

First of all, we are interested in studying the mechanisms of generation and properties of the structurally non-homogeneous layer with its strength characteristics varying with the distance from the contact surface depthward the material.

Besides the high interest of scientists to the problem under consideration [5–8] there are no system investigations of conditions for the contact fracture of operating surfaces of laser-hardened parts. Only basing on the results of such investigations one can optimize the properties of surface layers and develop the valid calculation methods for the assessment of the contact strength of contacting parts.

The feature of laser machining of structural steels is the generation of a microstructure of the fine-needle (crypto-needle) martensite with high hardness.

Figure 21.1 shows the results of studying the regularities of distribution of microhardness in the surface layer obtained by laser machining of medium carbon steel.

It follows from analysis of microstructure and phase state of structural steel after laser machining that steel properties vary with the distance from the hardened surfaces depthward the main metal (core).

It is seen that after laser machining the structurally non-homogeneous layer can consist of several zones. The surface layer with the thickness up to 0.2 mm has the highest hardness (HV $\geq$ 10,000 MPa). This layer possesses the highest wearing resistance. Then there is the layer with the hardness HV $\geq$ 6000 MPa. Properties of this layer are optimal from the point of view of resistance to depth contact fractures. Deeper to the core there is the layer with the decreased hardness which represents the transient zone (heat affected zone).

It is evident that similar to chemical heat treatment, in laser machining under unfavorable conditions and non-stationary loading the sites of subsurface contact fractures can appear both in the hardened layer and in the main metal or at the boundary of the main metal and the core.

The obtained results of metallographic investigations prove the main advantages of laser machining as compared to other known methods of hardening surface machining.

Investigations showed that along with high strength the surface layer generated by laser radiation has the necessary viscosity thus increasing the resistivity to development of elastoplastic deformations and fatigue breakage of the hardened layer. This statement is shown in Fig. 21.2 which presents the results of impact resistance testing of the surface laser layer.

Testing results indicate the considerable increase in strength characteristics within the zone of laser machining; it allows for recommending the laser modification for the parts operating under multi-factor actions including impact loads.

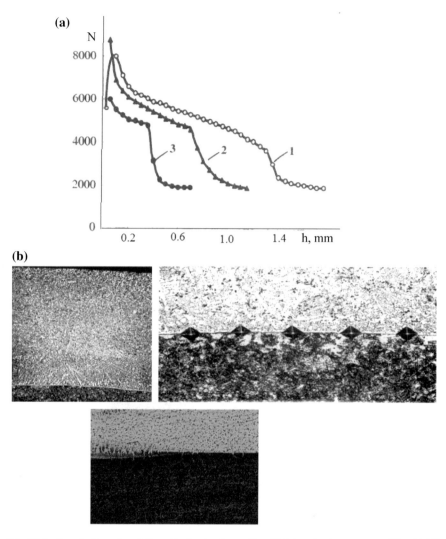

Fig. 21.1 Regularities of variation of microhardness (Fig. 21.1a) and microstructure (Fig. 21.1b) after laser machining of medium carbon steel. Curves 1, 2 and 3 correspond to different modes of laser machining

Optimization of laser machining modes in each specific case requires the application of valid methods of analysis of load carrying capacity for surfaces of hardened parts operating at contact friction interaction. Investigations results are given below that became the basis for the development of methods of calculating the characteristics of laser machined surface layers under the action of contact loads.

Analysis of operation data shows that the main type of fracture of operating surfaces of bearing parts is the progressive depth contact fractures (Fig. 21.3).

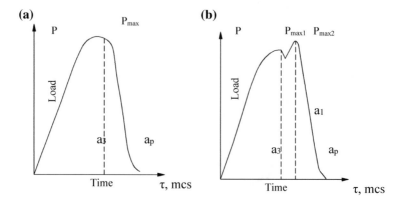

Fig. 21.2 Regularities of development of hardened layer fractures at impact loading: **a** solid hardening (martempering), **b** fracture at the presence of the case hardened layer

Fig. 21.3 Progressive depth contact fractures: **a** semi-axis of the cageless bearing; **b** bearing cage

The solution of the problem of determining the location of a risky zone of the laser surface layer has been found in accordance with investigations of the stress strain state with application of the contact problem of the elasticity theory at the initial contact of conjugating rolling parts.

Figure 21.4 shows the scheme of the stress state for case-hardened gears [1]. The scheme (Fig. 21.4) can be used to analyze stresses in the contact of laser hardened parts of cageless bearings.

One can accept that maximum equivalent contact stresses at cyclic loads are acting at the distance $0.5b$ from the surface of contact; and maximum contact shearing stresses at static loads τ_{max} are acting under the surface at the depth $z = 0.786b$. These stresses can lead to punching shear of the laser hardened layer under overloads, geometry distortion and breakdown of the bearing unit.

In accordance with the solution of the contact problem of the elasticity theory, stress components are determined by the condition:

Fig. 21.4 Scheme of
stresses at contact of two
cylinders under static and
cylindrical contact loads

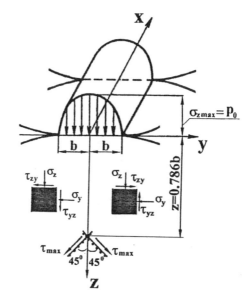

$$\frac{\sigma_z - \sigma_y}{2} = -p_0 \frac{z}{b}\left(1 - \frac{z}{\sqrt{b^2 + z^2}}\right);$$

$$\frac{\sigma_z - \sigma_x}{2} = -p_0 \frac{1}{2b}\left(2\mu(z - \sqrt{b^2 + z^2}) + \frac{b^2}{\sqrt{b^2 + z^2}}\right);$$

$$\frac{\sigma_y - \sigma_x}{2} = -p_0 \frac{1}{2b}\left(2(1 - \mu)(\sqrt{b^2 + z^2} - z) - \frac{b^2}{\sqrt{b^2 + z^2}}\right). \tag{21.1}$$

After calculation of stress components by formula [1] it is established that the
most risky surfaces with contact fractures are the contacting surface and the zone
of action of maximum equivalent stresses at a definite distance from the surface of
contact where the safety factor of the depth contact strength is $n = n_{min}$.

In order to investigate the conditions of contact interaction of heavy-loaded parts
of rolling supports with laser modification of contacting surfaces and to develop on
their basis new methods of contact strength analysis of laser layers, the generalized
mathematical model has been developed that allows for describing the joint influence
of two basic conditions of contact loading – variation of acting stresses and properties
of the metal surface layer within the force contact. The model is based on the fact
that the load carrying capacity of contacting surfaces of bearing parts is determined
by the value of contact stresses and by laws of variation of material properties with
the depth of the laser hardened layer. In accordance with the model, the basic design
conditions that determine the load carrying capacity are as follows:

at cyclic loading

$$\tau'_{eq}(z) < \tau_K(z); \tag{21.2}$$

at static loading

$$\tau_{eq}(z) < \tau_S(z),\qquad(21.3)$$

where τ'_{eq}, τ_{eq} are equivalent contact tangent stresses; τ_κ, τ_s, are critical endurance points within the risky zone of the laser layer determined by the coordinate z that represents the distance from the surface to the zone where primary fractures can appear.

Results of research by means of the proposed model are based on the following scientifically substantiated assumptions and experimentally proved facts.

1. Depth contact fractures of the laser layer are caused by the features of the stressed state, structural non-homogeneity and distribution of physical and mechanical properties within the most loaded zone of the layer.
2. The stressed state at any point of the contact zone is mathematically described by the basic relations for stresses which can be obtained from equations of the contact problem of the elasticity theory.
3. For all schemes of contact loading of the surface layer the resistance to the contact fracture and, therefore, the contact fatigue limits are related to the surface layer hardness by the linear dependence. Regularities describing hardness variations along depth can be represented as the corresponding functions depending on the coordinate z practically for all applied types of hardening processing of parts (chemical thermal treatment, laser modification, etc.).
4. At static loading the contact strength of the layer is determined by the value of tangent stresses τ_{max} which act on the areas located at the angle 45° to the central axis of symmetry z.
5. At cyclic loading the depth contact fatigue strength is determined by the action of variable sign tangent stresses τ_y at the boundary of the contact area ($y \approx \pm b$) at the depth $z = 0.5b$ on planes arranged parallel and perpendicular to the contact surface (see Fig. 21.4).

Therefore, in accordance with the developed model, the loading at any point of the material within the zone of force contact located at the depth z is in general case characterized not by the absolute value of the equivalent tangent stress τ_{eq} but by the strength safety factor n, its value varying along the thickness of the hardened layer together with stresses and properties of the hardened layer; that is, the safety factor is the continuous function $n(z)$.

Therefore, unlike the known design methods for the surface contact strength of structural steels, the proposed model considers the basic distinctive feature of contact interaction of case-hardened machine parts which specifies that the material resistance to contact loads and equivalent stresses that cause contact fractures are varying simultaneously with the distance from the surface of contact (that is, along the layer thickness). When analyzing the depth strength, the safety factor is not the constant value, that is $h = f(z)$.

Fig. 21.5 Distribution of hardness H(z) in case-hardened parts

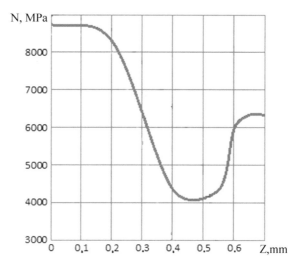

Analysis by means of the pointed model showed that depth fatigue cracks will first of all appear in zones located at the distance z_{min}. Within this zone the function $n(z)$ have the least value. In accordance with the criteria of the depth contact strength obtained when solving the contact problems, functional relations $n(z)$ has been derived which describe the regularities of variation of the safety factor variation along the depth of the laser hardened layer and which have the following form:
    at cyclic loads

$$n_K(z) = \frac{\tau_K(z)}{\tau'_{eq}(z)};$$ (21.4)

at static loads

$$n_s(z) = \frac{\tau_s(z)}{\tau_{eq}(z)}.$$ (21.5)

Therefore, the application of the model gives the possibility to control optimal characteristics of the hardened layer which in turn allows for either stating the required properties of the surface layer, or searching for the ways of reducing the loading of the contact zone.

In accordance with the model, in order to find the location of the risky zone of the laser layer $n = n_{min}$, it is necessary to know not only the stress components that determine the stress strain state and their variations within limits of the laser hardened layer, but also the laws of material property (hardness) distribution in the layer.

Figure 21.5 shows the typical curve describing the laws of hardness variation within the hardened layer.

The function $H(z)$ can be represented as the equation:

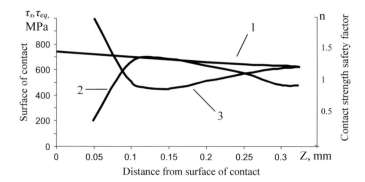

Fig. 21.6 Regularities of variation of allowable stresses [1], functions of equivalent stresses [2] and safety factors [3] in the laser hardened layer

$$H = \frac{H_n}{\left(\frac{H_n}{H_c} - 1\right)\left(\frac{z}{\delta_c}\right)^4 + 1}. \tag{21.6}$$

where H_n is the hardness on the surface of the laser layer; H_c is the core hardness (basic metal); δ_c is the thickness of the laser layer.

In contact of two cylinders, in accordance with the model of contact interaction of the laser hardened metal at static loads (overloads) the plastic strains can first of all appear in the zone of contact with the least value of the safety factor $n_{s\ min}$. The coordinate $z = z_{min}$ that defines the location of this risky zone of the laser layer can be determined by the equation:

$$\frac{d}{dz} n_s(z) = \frac{d}{dz}\left(\frac{\tau_s(z)}{\tau_{eq\max}(z)}\right) = 0. \tag{21.7}$$

Figure 21.6 shows the example of functions of stresses and variation of the safety factor for the case of contact loading of cylindrical surfaces of the cageless bearing. Calculations are performed by means of the model. It is evident that the risky zone is located in the hardened layer at the depth $0.1 - 0.2$ mm.

$$\frac{d}{dz} n(z) = \frac{d}{dz}\left(\frac{\tau_K(z)}{\tau'_{eq}(z)}\right) = 0. \tag{21.8}$$

The analysis proved that at cyclic contact loading of the laser layer, fatigue depth fracture also appear first of all in the zone $z = z_{min}$. The coordinate that determines the location of this zone can be obtained by the equation:

$$\frac{d}{dz}\left(\frac{cH}{\left[\left(\frac{H_n}{H_c}-1\right)\left(\frac{z}{\delta_c}\right)^4+1\right](\tau_{zy}+k\sigma_y)}\right)=0.\qquad(21.9)$$

The performed theoretical and experimental investigations allowed for developing the scientific technical base for calculations and optimization of properties and characteristics of laser hardened machine parts that operate under high contact loads.

When designing and analyzing, it is recommended to state the regularities of hardness distribution in the surface layer, the type of functions $H(z)$, the effective thickness H^{ef} of the layer δ_y^{ef} and the relation δ_y^{ef}/δ_e as one of the main characteristics of properties of the laser surface hardened layer.

In order to calculate the values δ_y^{ef} and H^{ef}, first of all it is necessary to find the location of the zone with the minimum safety factor for the depth contact strength (the coordinate $z=z_{min}$).

If the risky zone is located within the hardened layer, the following condition should be fulfilled in the zone of action of the maximum equivalent stresses:

$$\delta^{ef}>0.8b\qquad(21.10)$$

or

$$\delta^{ef}>1.386\times10^{-5}\sigma_n.\qquad(21.11)$$

The effective thickness of the laser layer can be determined by the formula:

$$\delta_y^{ef}=1.386\times10^{-5}\sigma_{\mathcal{H}}[n]\rho_{red}.\qquad(21.12)$$

In order to prevent the depth fractures at the boundary of the laser hardened layer and the basic metal, it is necessary to provide the definite relation between the total δ_e and effective δ_y^{ef} thicknesses of the laser hardened layer. For laser machining this relation is as follows:

$$k_n\frac{\delta_e}{\delta_y^{ef}}=\sqrt{\frac{[n]\sigma_n-4.1Hc}{0.85Hc}}.\qquad(21.13)$$

The necessary hardness H^{ef} of the laser hardened layer at the thickness δ_y^{ef} is calculated by the equation describing the hardness distribution $H(z)$ along the thickness of the hardened layer for specific conditions of contact loading:

$$H^{ef}=\frac{H_n}{\left(\frac{H_{surflas}}{H_{core}}-1\right)\cdot\left(\frac{\delta_y^{ef}}{\delta_c}\right)+1}.\qquad(21.14)$$

The value H^{ef} should meet the following condition:

$$H^{ef} \geq 0.22\sigma_{\mathcal{H}}[n] \tag{21.15}$$

Investigation results show that the increase in load carrying capacity of case hardened rolling bearings can be provided by manufacturing methods. Basing on the investigation results the optimal characteristics of laser hardening of cageless bearing parts has been determined. The design model has been used to find the required characteristics of the laser hardened steel in accordance with the depth contact strength. The values H^{ef} and δ_y^{ef} determined exactly by the condition of the depth contact strength allow for choosing purposefully the material of parts and effective methods of hardening.

Basing on investigation results, the following characteristics of the laser hardened layer are recommended for laser hardened heavy-loaded parts:

$$H^{ef} \geq 8500 \, MPa; \; \delta_y^{ef} > 0.3; \; [n]_{min} = 1.6. \tag{21.16}$$

21.1 Conclusions

1. The analysis of loading has been made and factors have been determined that influence the operation performance of parts hardened by the laser.
2. The mathematical model has been developed that describes the conditions of contact loading of parts with surface hardening obtained by laser modification.
3. New methods of analysis have been developed for the depth contact strength of laser hardened parts at cyclic and static types of loading.
4. It has been established that in order to provide the required safety factors of the depth strength, to prevent contact fractures and to increase the life time of gears, the thickness of the laser hardened layer with high hardness (HV $\geq$ 8000 MPa) should be not less than 0.5 mm. Herewith, the optimal relation between the total and effective thicknesses of hardened layers should be provided.

References

1. Airapertov, E.L.: State and Prospects of Development of Methods of Calculation of Gear Loading and Strength. Guidance materials, Izhevsk, ISTU (in Russian) (2000)
2. Berestnev, O.V., Soliterman, Y.L., Goman, A.M.: Reliability and life-time design of transmissions of mobile machines. In: Proceedings of the BITU "Modern Methods of Machine Design, Engineering and Production Techniques", vol. 2, No. 1, pp. 312–316 (in Russian) (2002)
3. Birger, I.A., Shorr, B.F., Iosilevich, G.B.: Strength Analysis of Machine Parts. Reference book, Moscow, Mashinostroenie (in Russian) (1993)
4. Petrusevich, A.I.: Contact Strength of Machine Parts. Mashinostroenie (in Russian), Moscow (1969)

5. Pinegin, S.V.: Contact Strength and Rolling Resistance. Mashinostroenie (in Russian), Moscow (2011)
6. Tesker, E.I.: Investigation of regularities of wearing of friction surfaces modified by radiation of CO_2-laser. In: Proceedings of the XXXVI Conference of Department of Machine Parts and Machine Design, Brno, Czech Republic, Technical University of Brno, vol. 2, pp. 177–180 (1995)
7. Tesker, E.I.: Modern Methods of Analysis and Increase in Load Carrying Capacity of Surface Hardened gears of Transmissions and Drives. Mashinostroenie (in Russian), Moscow (2011)
8. Tesker, E.I., Tesker, S.E., Guryev, V.A., Koryakin, A.F.: Progressive methods of increase in operation properties of base parts of compressive equipment. In: Proceedings of the International Symposium. "Consumers, Producers of Compressive Equipment", Saint-Petersburg, pp. 7–11 (in Russian) (2004)

Chapter 22
Vibration Analysis of Gearboxes

Claudia Aide González-Cruz and Marco Ceccarelli

Abstract This chapter presents the application of different signal processing tools for characterization of the dynamic behavior of gear transmission systems. The working principle and design characteristics for simple and planetary gear transmissions are presented in order to highlight the most important dynamic characteristics to be considered. The experimental requirements for testing are discussed. Some signal processing tools are introduced as useful to extract the dynamic characteristics of the system from vibration signals. Finally, the case of study of a planetary gear transmission is presented. The gearbox is analyzed by means of the presented signal processing tools in order to show the powerful the feasibility and efficiency to extract the transient responses and nonlinear behavior for a performance characterization of gear design.

22.1 Introduction

Gear transmission is one of the most used mechanisms to transmit motion and torque in a broad areas of systems, such as manufacturing, aeronautic, automotive, wind energy and robotics automation to mention a few. However, it has some drawbacks, mainly noise and vibration that are caused by internal and external excitation sources, such as the varying gear mesh stiffness, transmission errors, variations on the load and fluctuations on the operational speed. The reduction of the mesh variation has been widely studied since mesh fluctuation is one of the main dynamic self-excitation sources that induce vibrations, dynamic overload and transmission error [1–3].

The inherent nonlinear features of the multiple components of a gear transmission system produce a complex dynamic performance of the whole system. Considerable

C. A. González-Cruz
Universidad Autónoma de Querétaro, 74010 Querétaro, Mexico

M. Ceccarelli (✉)
University of Rome Tor Vergata, Department of Industrial Engineering, Laboratory of Robot Mechatronics, 00133 Rome, Italy
e-mail: marco.ceccarelli@uniroma2.it

© Springer Nature Switzerland AG 2020
V. Goldfarb et al. (eds.), *New Approaches to Gear Design and Production*,
Mechanisms and Machine Science 81,
https://doi.org/10.1007/978-3-030-34945-5_22

attention has been dedicated on the study of wave transmission and energy dissipation through the gear transmission elements. For example, Xiao et al. [4] study the transmission of vibration from the gear to the housing through multiple interfaces on which there is a path transfer. They define a vibration transmission ratio as the ratio between the acceleration amplitude of the n-th component and the acceleration amplitude of the gear to evaluate the energy dissipation. Results shown that the interfaces inner race-outer race of bearings and gear-shaft dissipate about the 60% and 40% of the total energy. Gonzalez et al. [5] investigate the dynamic synchronization of a gearbox as function of the housing stiffness, since the housing is the coupling element that enable the propagation of the vibrations among the gearbox elements. They find that the synchronization among the gears and bearings increases 40% the cyclic loading. Guo et al. [6] analyze the wave propagation from the meshing gear teeth through the shafts and bearings to the housing and surrounding environment. The developed vibro-acoustic model shown capability to estimate the vibration and noise propagation of a spur gearbox. Furthermore, the numerical results shown that the housing flexibility has an important role on the bearings forces instead of the gear dynamics where it has a small impact.

The complexity of the dynamic behavior of gear transmissions is higher in the case of planetary gears due to the multiple planets gearing. Planetary gears are used for the substantial advantages that they offer, such as large weight-to-load ratio, small volume design and high efficiency [7]. Many models have been developed in order to identify the modal proprieties of planetary gears and failure identification. Furthermore, experimental studies have been done to improve the models and to have a better understanding of the time varying parameters. Like in the following examples from a wide literature.

Liu et al. [8] introduce an elastic-discrete model to study the effects of the sliding friction and an elastic continuum ring gear on the time varying mesh stiffness of a two-stage planetary gear system. The effect of the elasticity of ring gear on the modal properties of the TSPG is studied through the variation of the planet number, ring bending stiffness and sliding friction coefficient in the dynamic model. The natural frequencies calculated by the elastic-discrete model are compared with those of the lumped parameter model. The effect of sliding friction coefficient on the natural frequencies is also analyzed. Li et al. [9] developed a set of vibration signal models to represent the modulation sidebands induced by a single planet gear mesh and multiple planet gears meshes on the dynamic response of a planetary gear train with floating sun gear. The models are based on signal transfer path functions and measuring-direction projection functions of the gear mesh force. The influence of floating sun gear on some specific frequency components is obtained and it is used as a feature indicator to distinguish whether the planetary gear train is assembled with a floating sun gear, or whether the planetary gear train without a floating sun gear exists some distributed defects. Numerical results are consistent with the analysis results of experimental data from a single stage planetary gear train and a wind turbine planetary gearbox. Kong et al. [10] propose a method for fault diagnosis of wind turbine planetary ring gear based on the evaluation of the significance level of meshing frequency modulation and an adaptive empirical wavelet transform framework. The meshing modulation regions are determined by means of a Fourier spectrum segmentation. Then, a series of empirical wavelet filters are designed an applied to decompose the vibration signal into several modes. Finally, the failure identification is achieved by an envelope spectrum analysis. Experimental results shown superiority in the fault diagnosis compared with other methodologies such as empirical mode decomposition and Spectral Kurtosis techniques.

Li et al. [11] present a method for failure identification in planetary gears based on variational mode decomposition, power spectral entropy and deep neural network. First, VMD is used to decompose the signal into narrow-band components with independent centre frequencies. Then, PSE is used as feature extraction tool to quantify the size and distribution of the amplitudes of the side band in the components. Finally, a DNN is used to execute a feature reduction and to classify failures. Experimental results shown satisfactory performance of the algorithm. Mbarek et al. [12] identify experimental and numerically the modal properties of a planetary gear transmission by means of the frequency response function. Impact tests were developed on back-to-back planetary gears with different load conditions and under mesh stiffness fluctuation. Experimental results were compared with the numerical results from a 6-DOF lumped parameter model having a deviation about 2–3%.

Park et al. [13] propose a method, positive energy residual (PER), for fault detection of planetary gears. It uses the wavelet transform (WT) and the Gaussian process (GP) to remove the variability of the signals while extracting the faulty signals. It is highlighted that the method does not require angular information and it has low sensitivity for fluctuations of speed. Numerical and experimental results show the performance of the method for advanced stage fault detection. Later, Park et al. [14] propose a variance of energy residual (VER) method for planetary gear fault detection under variable-speed conditions. The method is based on the computation of the short time Fourier transform (STFT) and it does not need angular information. Numerical and experimental results shown that the VER method have better fault sensitivity and take less computation time than other methods such as the wavelet transform. However, since the method lack of angular information the effects of the variable-speed conditions on the dynamics of the planetary gears can be missed.

Tsai et al. [15] propose an approach to analyze the gear meshing in planetary gear trains with a floating sun gear. The proposed model is based on the exact involute gear geometry and it take in consideration the influence of number of planets, the backlash and the assembly/manufacturing errors. The results helps to define the suitable tolerance for the movable area and assemblability of the sun gear, the backlash and other components of the gear train. Zghal et al. [16] developed a 2-D lumped parameter model of one stage planetary gearbox is proposed in order to analyze the modulation phenomenon caused by the time varying gear meshing and position of the planet gears. Vibration signals at the transducer location are expressed by the sum of all vibrations induced from each component and influenced by the time varying path. The model results are validated experimentally. Shen et al. [17] propose a modified Time Varying Mesh Stiffness (TVMS) model to evaluate the influence of the tooth wear on the TVMS. Results prove that the tooth wear only have effect on the bending stiffness, shear stiffness, and axial compressive stiffness because they depend of the tooth profile and tooth thickness. Furthermore, the TVMS decreases conforming to the tooth wear depth.

This chapter deals with the experimental characterization of the dynamics of gearboxes. The content of this chapter is as follows. Section 22.2 presents a review with the main problems and requirements in the design of gearboxes, including gear design issues and the technical requirements and considerations during the design of a test bed for gearboxes experimentation. Section 22.3 describe the experimental tools and procedures for the signal processing of the data. Section 22.4 presents the case of study of a 2-DOF planetary gearbox designed at LARM[2]. Section 22.5 presents the results of the analysis and finally, Sect. 22.6 presents the conclusion.

22.2 Problems and Requirements in the Design of Gearboxes

Nowadays, the weight and design optimization of machines is a major concern in the industry. Especially in gearboxes, since they are one of the most used mechanical transmission systems in heavy industry.

The intrinsic nonlinear characteristics of the gear transmission elements make that gearboxes have high failure and high downtime per failure rates among other components in different machines, for example wind turbines. The combination of internal and external excitation sources generates operation conditions that can exceed the design limitations of the gearbox elements. Consequently, conditions such as, fatigue in shafts, large variations of torques and excessive stress on gears and bearings are common [18, 19].

A schematic diagram of the design process of gearboxes is shown in Fig. 22.1. The first stage of the process is the definition of the design requirements, such as operational load and speed, size, material, etc. (1). Then, the design, conceptual and geometrical, is developed (2). The design stage includes the definition of the transmission typology, kind of gears, dimensioning of the gearbox elements, etc. Later, a numerical simulation is required in order to evaluate the static and dynamic response of the gearbox (3). The accuracy of the numerical results depends of the simplification of the model, the numerical methods used and the load and input excitation considered. According to the numerical response, a design optimization stage is necessary to improve the characteristics of the system (4). Once no changes are necessary in the design, the prototype can be manufactured (5) using the appropriated machining process, special treatments and assembly specification to achieve specific mechanical properties. The prototype is experimental tested (6) in laboratory and/or in-service (field). In this stage, supervision systems have an important role in the evaluation of the gearbox under real operation conditions. So that, design improvements and future innovations can be proposed. Finally, in case of successful testing, the final product is delivered (7).

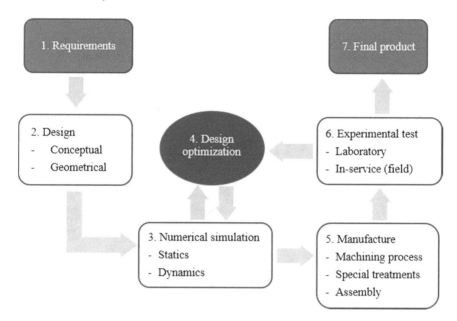

Fig. 22.1 Schematic diagram of the design process of gearboxes

There are different challenges in the design process of gearboxes. Mainly the reduction of the nonlinear behavior caused for the nonlinear characteristics of its elements and the development of experimental testing tools that enable a more accurate analysis of the gearbox dynamics. These topics are discussed in the following subsections.

22.2.1   Gear Design Issues

The intrinsic existence of gear transmission error, friction between components and stiffness fluctuation in a gear transmission system can produce vibrations, noise and dynamic overload. Furthermore, the operating conditions have a strong influence in the dynamic transmission error, even in absence of contact loss of the teeth. Hence, there is a constant interest in the development of design optimization methods, which allow the achievement of the growing demand for efficiency, high-strength and high precision on gear design [20, 21].

Figure 22.2 shows the main internal and external excitation sources in gearboxes. The interaction of them during the system operation can amplify the nonlinear behavior of the whole system. Therefore, there is the constant interest to improve the design and manufacturing process of gears in order to reduce the noise and avoid failures in the machinery.

Fig. 22.2 Classification of the main excitation sources in gear transmission systems

Many authors have been worked on the improvement of the parameters design of a gearbox by means of the analysis, measurement and reduction of the static and dynamic transmission error [22–24]. The variation of the contact ratio has been studied by means of the modification of the addendum and the depth of the teeth [25]. Marafona et al. [26] develop an algorithm for designing gears, which considers safety factor and meshing efficiency to generate the design parameters of a gear. Results show a potential reduction of noise, vibration and dynamic overload on the gears.

22.2.2 Test Bed Design

There is a continuous technological challenge for measuring of the transient response of gearboxes, since this contains the fault signatures that characterize the incipient faults. Therefore, it is necessary to have sensors and data acquisition systems that ensure the accuracy and reliability of the data, as well as, testing tools that enable the repeatability and reproducibility of the experiments. However, it is important to take into account that the quality of the signals is also function of the sensors location and the synchronization of the data acquisition systems.

Vibrations have been the most common used variable for condition monitoring and failure identification in gearboxes, as well as, acoustic emission, temperature and electricity consumption [7]. However, additional sensors are essential, such as the encoders and others are necessary for supervising of specific components. For example, torque and displacement sensors to analyze the power transmission [27], oil quality sensors to monitoring the health life of bearings [28] and strain gauges to measure the gear mesh stiffness [29]. Nevertheless, it is important to consider that the use of extra sensors in the test-bed can increase the complexity of both, the assembly and the maintenance. Therefore, non-invasive techniques based on the analysis of electrical signatures are been developed [30].

There are many considerations for test bed designing for gearboxes [31]. Figure 22.3 shows a scheme with the major subsystems that integrate the test bed. The structure includes the mechanical structure, which acts like the base support and all the devices that are involved in the test. The instrumentation includes all the sensors to measure the physical variables of the system under test and those from the auxiliary equipment. The supervision system comprises the panel view with graphical interfaces and signal processing tools to supervise the development of the test,

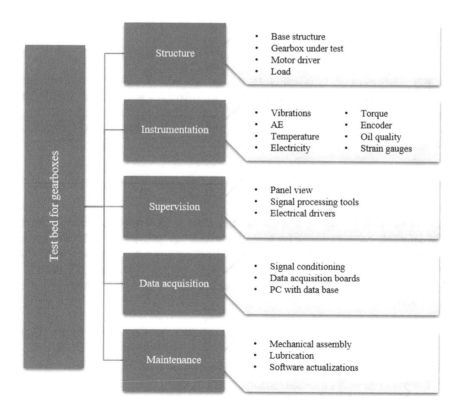

Fig. 22.3 A scheme of a test bed configuration for gearboxes

as well as, the electrical drivers for the devices control. The data acquisition system includes the signal conditioning stages, devices for the data acquisition and the PC for information storage. Finally, the maintenance is essential to keep the test bed in optimal operating conditions. It involves the supervision of the mechanical assembly, lubrication and software actualization.

22.3 Experimental Tools and Procedures

Gear transmission systems are often subject to hybrid failures. This condition increases the challenge in the earlier and localized detection of failures. There are several methodologies for condition monitoring developed in the last decades. Many of them deals with the identification the transient response of the failure signatures.

In this chapter there are included the time-frequency analysis by means of the continuous wavelet transform and phase diagram. A brief definition of each analysis tool is presented in the next subsections.

22.3.1 Wavelet Transform

Wavelet transform (WT) is a time-frequency tool widely used to reveal transient spectral information of a time domain signal $x(t)$. The WT of $x(t)$ is obtained by the correlation between x(t) and a set of functions obtained by scaling and translating a fundamental wavelet function ψ, also called mother wavelet. WT is expressed as follows [32]

$$WT(s, \tau) = \langle x(t), \psi_{a,b}(t) \rangle = \frac{1}{\sqrt{s}} \int_{-\infty}^{+\infty} x(t) \psi^* \left(\frac{t - \tau}{s} \right) dt \qquad (22.1)$$

where s and τ are the scaling and translating factors, respectively, and the super index * indicates de complex conjugate of ψ. The similarity between the wavelet function and the nonstationary contents in the signal determines the successfulness of this analytical tool. In this work, the Morlet wavelet is used as mother wavelet since it is able to detect the impulses that characterize transitory signals [33]. Morlet wavelet is defined as

$$\psi(t) = \pi^{-1/4} \left(e^{-j\omega_0 t} - e^{-j\omega_0^2/2} \right) e^{-t^2/2} \qquad (22.2)$$

where ω_0 is the central frequency of the Morlet wavelet.

The powerful of WT for fault identification in rotating machinery has been demonstrated in many scientific works. For example, Chen et al. [34] make a review to analyze the inner product operation of WT and to present a summary of the major developments of WT in the fault diagnosis field. Teng et al. [35] use the complex Gaussian wavelet to extract the weak features of bearing faults from vibrations signals of a wind turbine gearbox.

22.3.2   Phase Portrait

Phase portrait describes the stable and unstable steady states of a system by means of a graphical representation of the energy variations in a space composed by the position q and momentum p. The phase portrait shows the trajectory of the system for each set of initial conditions, such as the limit cycles, attractors and repellors. The trajectory followed by the system can be determined from the equation of Hamilton as follows [36]

$$d\mathbf{q}/dt = \partial H[\mathbf{p}(t), \mathbf{q}(t)]/\partial \mathbf{p}, \tag{22.3}$$

$$d\mathbf{p}/dt = -\partial H[\mathbf{p}(t), \mathbf{q}(t)]/\partial \mathbf{q} \tag{22.4}$$

where the dimensionality of the vectors $\mathbf{p}$ and $\mathbf{q}$ corresponds at the number of degree of freedom of the system. Assuming that the Hamiltonian has no explicit time dependence, $H = H(\mathbf{q}, \mathbf{p})$, the Hamilton equation shows that, as $\mathbf{p}$ and $\mathbf{q}$ vary, the value of $H(q(t), p(t))$, remain constant

$$\frac{dH}{dt} = \frac{d\mathbf{q}}{dt} \cdot \frac{\partial H}{d\mathbf{q}} + \frac{d\mathbf{p}}{dt} \cdot \frac{\partial H}{d\mathbf{p}} = 0 \tag{22.5}$$

Then, for a time-independent system, the energy is conserved $E = H(\mathbf{p}, \mathbf{q}) = c$. So that, the dynamic behavior of the system can be studied from the graphical representation of $\mathbf{p}$ and $\mathbf{q}$ in the phase portrait.

22.4   Case of Study

22.4.1   Two-DOF Planetary Gearbox

This section presents the analysis of a 2-DOFs planetary gearbox designed at LARM[2] [37]. Figure 22.4 shows a schematic diagram of the testing setup. It consists basically of the 2-DOF planetary gearbox driven by an AC motor and a varying load provided by a flywheel. Vibrations and torques are measured on the bearings of the input and output shafts. The angular velocity of the output shaft is measured by an encoder Two data acquisition systems from NI are used to acquire data and storage it in a PC.

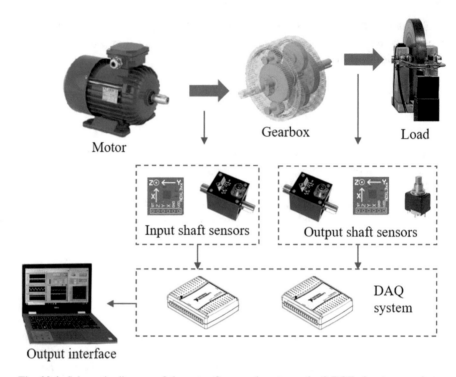

Motor   Gearbox   Load

Input shaft sensors   Output shaft sensors

DAQ system

Output interface

Fig. 22.4 Schematic diagram of the setup for experiments on the 2-DOF planetary gearbox at LARM[2]

A schematic diagram of the 2-DOF planetary gearbox is shown in Fig. 22.5. The input stage comprises a sun gear S_1, a couple of planet gears $P_{2,3}$, a ring gear R_1 and an input carrier C_1. Similar, the output stage comprises a sun gear S_2, a couple of planet gears $P_{4,5}$, a ring gear R_2 and an output carrier C_2.

Fig. 22.5 Schematic diagram of the 2-DOF planetary gearbox designed at LARM[2]. Input stage: S_1—sun gear, $P_{2,3}$—planet gears, R_1—ring gear, C_1—input carrier; Output stage: S_2—sun gear, $P_{4,5}$—planet gears, R_2—ring gear, C_2—output carrier

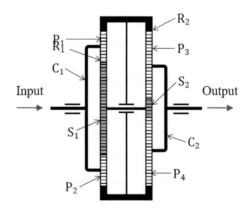

The planetary gearbox operates differentially as function of the friction forces between gears and the load torque [38]. The first DOF operates when the friction forces between gears are higher than the load force generated by the applied torque on the output carrier, so that the gears of the closed chain $P_{1,2}$-$R_{1,2}$-$P_{3,4}$-S_2-S_1 are locked between them. Thus, the angular displacement of the output carrier is equal to that in the input carrier. Other case, when the load increases until it overcomes the friction forces, the second DOF actuates. The large load locks the output carrier, so that, the output planets rotate on their own axis and the ring gear becomes the output link with an inverted rotation to the input carrier.

22.4.2 Experiments

The features of the test bed at LARM[2] allow to test the complete operation conditions of planetary gear prototypes by varying the operating conditions of speed and load. In the particular case of the 2-DOF planetary gear prototype in Fig. 22.5, tests are developed with varying operating conditions of speed and load. Then, the transmission efficiency and the dynamic response of the gear transmission can be evaluated, especially during the activation of the second DOF.

The first test at constant speed and varying load is carried out as follows: (1) $\omega_n = 120$ rpm is set as operating speed; (2) the systems is turned on and the first DOF starts to operate; (3) it is allowed that the output shaft reaches de nominal speed ω_n and it is held for few seconds; (4) the load is increased, then, the second DOF is actuated and it operates for few seconds; (5) the load is decreased; (6) steps 3–5 are repeated four times; (7) it is allowed that the output shaft reaches the nominal speed before turn off the system.

The test at varying load and varying speed is carried are out as follows: (1) the output shaft is preload by means of the brake; (2) the input speed is smoothly increased until $\omega_n = 120$ rpm, while, the second DOF is actuated; (3) the load is decreased, then, the first DOF is actuated; (4) it is allowed that the output shaft reaches the nominal speed; (5) the load is increased and it is allows that the second-DOF operates for few seconds; (6) the operational speed is smoothly increased $\Delta = 60$ rpm; (7) steps 3–6 are repeated until reach $\omega_n = 300$ rpm; (8) steps 3–6 are repeated but now decreasing the nominal speed of $\Delta = 60$ rpm each time until reach $\omega_n = 120$ rpm; (9) the load is decreased and it is allowed that the first DOF operates; (10) the initial load condition is applied meanwhile, the operating speed is smoothly deceased until $\omega_n = 0$. The experimental data are stored in a PC during the tests and later, they are post-processed by using the analysis techniques that ate described in the next section.

22.5   Results

22.5.1   Test at Constant Speed and Varying Load

The measured signals from the test at constant speed and varying load conditions are presented in Fig. 22.6. The test has been carried out with a nominal speed of the motor drive $\omega_n = 190$ rpm while the variation of the load is produced by means of the load braking in order to increase the load torque.

Figure 22.6a shows the acquired torques in the input and output shafts. The operation of the system starts with the low load condition having the inertia of the flywheel. Therefore, the first DOF of the planetary gear transmission starts to operate and the rotation of the input shaft is directly transmitted to the output shaft, as it can be seen in Fig. 22.6b. Once the moment of inertial of the load is overcome, the torques decrease until they reach their nominal value at the nominal operating speed. Then, when the load is increased, the output carrier is braked ($\omega_C = 0$) and the torques present a suddenly increment, which is managed by the actuation of the second DOF. The torques achieve new nominal values as the operation of the second DOF reach the nominal conditions. Finally, when the load decreases, the first DOF is reactivated and the torque values decrease until they reach again their corresponding nominal

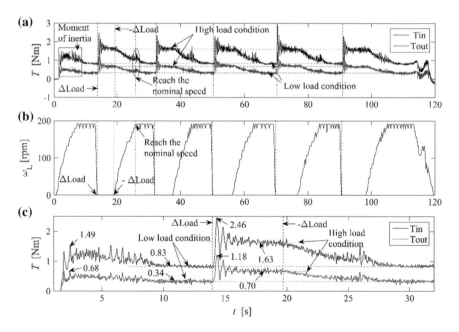

Fig. 22.6 Experimental results from the tests at the nominal speed $\omega_n = 190$ rpm and varying load: **a** dynamic torques in the input and output shafts, **b** angular velocity of the output shaft, **b** zoom-in of the torque signals for $t = 1$ to 32 s

values (5). Small disturbances appear in the torque signals when the nominal speed is reached during the reactivation of the first DOF.

It is assumed that these disturbances are caused due to the backlash in the gearbox design. The behavior of the torque signals proves the importance of the second DOF of the planetary gear transmission in order to avoid excessive torques which can damage the components of the system.

A zoom-in in the torque signals is shown in Fig. 22.6c. It can be noted that the nominal values of the torques during the operation of the first DOF are $T_{in} = 0.83$ Nm, $T_{out} = 0.34$ Nm and the nominal values during the operation of the second DOF are $T_{in} = 1.63$ Nm, $T_{out} = 0.70$ Nm. An estimation of the friction torque gives $T_f = 0.83$ Nm. The efficiency is computed and it found $\eta = 39.9\%$.

22.5.2   Test at Varying Speed and Varying Load

Time domain signals of the angular speed, torques and radial vibrations of the gear transmission during the test at varying speed and varying load are presented in Fig. 22.7. The system is started with a preload that is generated by means of the load braking. Therefore, the output shaft does not rotate while, the amplitude of the torques at the beginning of the tests are higher than those in the previous test. The development of the test can be seen through Figs. 22.7, as follows.

The motor drive is turned on and its operational speed is smoothly increased until $\omega_n = 120$ rpm. Since there is a high preload condition, the amplitude of the torques increases continuously until the torque that is produced by the load and the internal friction forces is overcome ($t \approx 20$ s). Immediately, the second DOF of the planetary gear transmission is activated and the input shaft starts to rotate driven by the motor, while the output shaft remains braked. At the same time, the amplitude of the torques decreases and the vibrations increase suddenly in both, input and output shafts. With the second DOF operating, the speed of the input shaft is still increased in order to change the speed condition. It can be seen that as the operating speed increases, the input shaft torque increases and the output shaft torque decreases. Then, when the nominal speed is reached ($t \approx 30$ s), the input torque decreases and the output torque increases until they reach their nominal values ($T_{in} = 1.64$ Nm, $T_{out} = 0.78$ Nm). Vibrations continue increasing until the nominal speed is reached. The higher amplitudes of the input shaft vibrations than those of the output shaft, are because the output shaft remains locked during the actuation of the second DOF.

The first DOF is actuated by means of the load decrement at t $\approx$ 45 s, so that, the rotation of the input shaft is directly transmitted to the output shaft. Under this operating condition, the input torque increases due to the moment of inertia produced for the load, but it decreases as the nominal speed is achieved ($t \approx 53$ s). On the other hand, the output torque and the vibrations in both shafts decrease until they reach the stable operating condition. The second DOF is actuated by means of the load increment ($t \approx 62$ s). When the load is increases, the output torque increases suddenly and this decreases as the second DOF reaches the nominal operating condition.

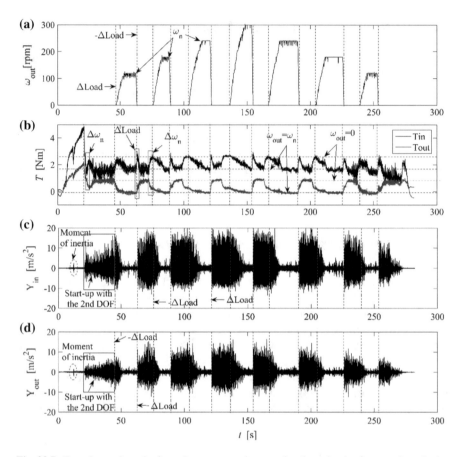

Fig. 22.7 Experimental results from the test at varying speed and varying load: **a** angular velocity in the output shaft; **b** dynamic torques; **c** radial vibrations on the input shaft; **d** radial vibrations on the output shaft

On the other hand, the input torque and vibrations increases until the nominal operating condition is reached. The output torque and vibrations increase when the load is increased and they reach the stable operating condition as the output link of the second DOF reaches its nominal speed.

The variation of the load and operating speed is repeated different times in order to characterize the dynamic behavior of the planetary gear transmission, mainly during the actuation of the second DOF. Steps 3 and 4 are repeated, increasing each time the operating speed in $\Delta \omega n = 60$ rpm until $\omega n = 300$ rpm is reached. Then, steps 3 and 4 are repeated but now decreasing the operating speed until reach $\omega n = 120$ rpm. Finally, the load is increased and the nominal speed is smoothly decreased until turn-down the system.

The behavior of the torques in Fig. 22.7 shows the effectiveness of the second DOF of the planetary gear transmission in order to avoid excessive load in the gear transmission elements. It can be seen that the nominal values of the torques do not depend of the operating speed.

The radial vibrations are analyzed with the continuous wavelet transform in order to characterize the dynamic behavior of the planetary gear transmission during the operation of the second DOF. The continuous wavelet transform is computed by means of AutoSIGNAL™ software and the results are displayed graphically in time-frequency spectrograms. The time-frequency spectrogram for the radial vibration signals are shown in Fig. 22.8. It can be noted that the system shows a strong nonlinear behavior since the spectrograms present frequency discontinuities and amplitude variations. Furthermore, the frequency content increases at high operating speed.

It is assumed that the higher frequency amplitudes in the input shaft spectrogram than those in the output shaft spectrogram are because the output shaft is braked during the operation with the second DOF.

The power spectral density $P(f)$ of the vibration signals from the continuous wavelet transform results is presented in Fig. 22.9. It shows the variation of the frequency content in the vibration signals across time. The amplitude of $P(f)$ increases

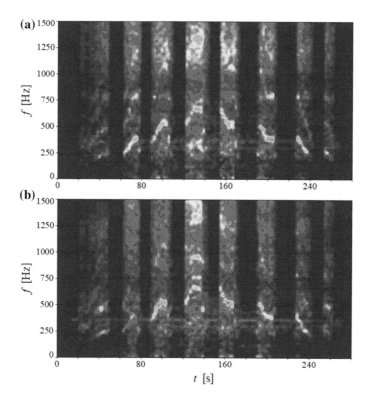

Fig. 22.8 Time-frequency spectrograms from the continuous wavelet transform of the radial vibrations in Fig. 22.7 on: **a** input shaft, **b** output shaft

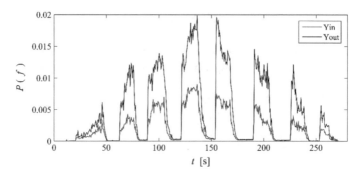

Fig. 22.9 Power spectral density $P(f)$ of the radial vibration from the time-frequency spectrograms in Fig. 22.8 (Y_{in}—radial vibrations on the input shaft, Y_{out}—radial vibrations on the output shaft)

as the operating speed increases and vice versa, $P(f)$ decreases as the operating speed decreases.

The radial vibrations on the input and output shafts are analyzed with the CWT in a short time period. This analysis provides a further characterization of the dynamic behavior of the planetary gear transmission during the operation of the second DOF. Figure 22.10 shows the spectrograms for a test in ramp-up from 120 rpm to 190 rpm and a ramp-down from 190 rpm to 120 rpm. In both cases, the system has a high load condition in order to actuate the second DOF.

It can be seen in the spectrograms in Fig. 22.10 that the amplitude of the frequencies increases as the operating speed increases, such as the nonlinearity and complexity of the spectrograms is higher during the ramp-up than during the ramp-down process.

The power spectral density from the time-frequency spectrums are presented in Fig. 22.11. The evolution of the frequency density across time shows that the system presents similar performance during the actuation of the second DOF, even when it works under varying operating conditions. The complexity of the system can be noted since the frequency content and the frequencies density change through time.

The phase diagrams are built for the same data in Fig. 22.10 in order to analyze the stability of the gearbox during the operation of the second DOF.

Figure 22.12 shows the phase diagrams of the input and output shafts during the activation of the second DOF for the test at $\omega_{in} = 190$ rpm. It can be seen that the system is unstable since the loops in the diagrams for both shafts oscillate between two atractors. Figure 22.13 shows the phase diagrams for both shafts when the second DOF is operating at the nominal conditions. In this case the system is stable since the loops in the phase diagrams have one attractor.

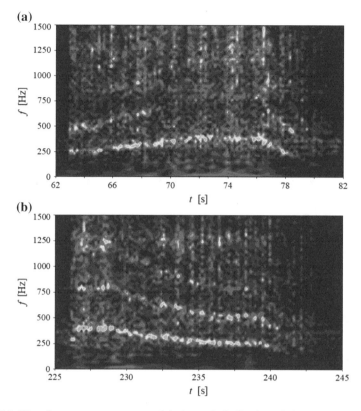

Fig. 22.10 Time-frequency spectrograms of the input shaft vibrations during different steps of the second DOF actuation: **a** ramp up from 120 to 190 rpm, **b** ramp-down from 190 to 120 rpm

22.6 Conclusions

This chapter presents a review of the issues and requirements on the design of gearboxes, such as, an experimental characterization of a two DOF planetary gear prototype under varying operating conditions. The design characteristics of the planetary gear transmission allow a differential operation as function of the internal friction forces and the load. Therefore, in order to characterize the dynamic behavior and the usefulness of the second DOF, two main tests are carried out in a test bench at LARM[2]; namely the first test is made at constant operating speed and varying load and the second one at varying load and varying operating speed. The variation of the load enables the actuation of the second DOF of the planetary gear transmission, meanwhile, the variation of the speed during the operation of the second DOF is made to study the dynamic behavior and dynamic synchronization of the gear transmission. Torques and vibrations are acquired and they are analyzed by means of the continuous wavelet transform and statistical descriptors.

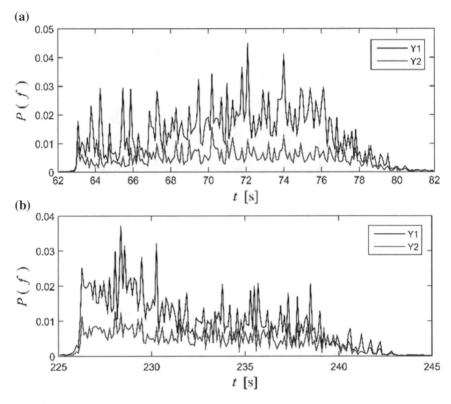

Fig. 22.11 Power spectral density $P(f)$ of the vibrations on the input shaft $(-Y_{in})$ and the output shaft $(-Y_{out})$, during different steps of the second DOF actuation: **a** ramp up from 120 to 190 rpm, **b** ramp-down from 190 to 120 rpm

The results shows that the efficiency of the torque transmission is DOF, $\eta \approx 40$ when the first DOF is active and they prove the effectiveness of the second DOF in the design of the planetary gear transmission to avoid excessive stress in the elements of the system. However, it is important to note that the ring gear, which acts like the output ring during the operation of the second DOF, does not transmit the torque to another system, since it does not coupled to another transmission torque element. The strong nonlinear behavior of the system is appreciated through the time-frequency spectrograms that are computed for the radial vibrations by means of the continuous wavelet transform and through the variation in the density of the frequency content.

The experimental characterization of the planetary gearbox shows its useful to drive low torques, since it helps to avoid excessive stresses in the elements of the gearbox and it does not need an actuator to exchange between the operation of the first and the second DOF.

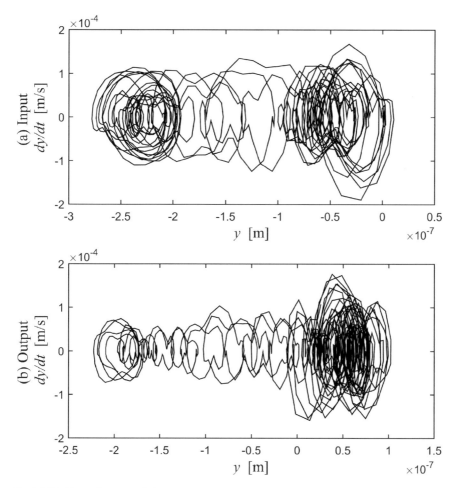

Fig. 22.12 Phase diagrams of the system test at 180 rpm during the actuation of the second DOF: **a** input shaft, **b** output shaft

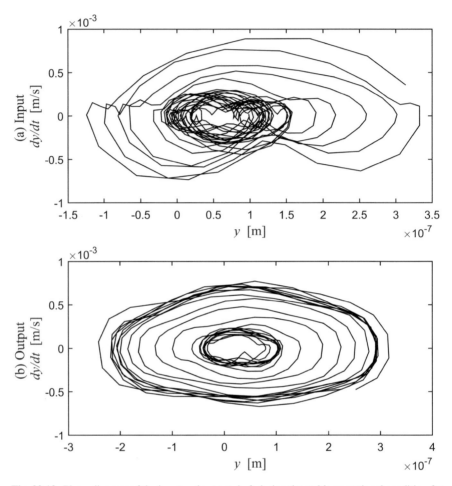

Fig. 22.13 Phase diagram of the input and output shaft during the stable operational condition for the test at 190 rpm

Acknowledgements The first author gratefully acknowledges the Mexican Government Foundation CONACYT for the fellowship for her postdoctoral research at LARM[2]: Laboratory of Robot Mechatronics in Rome University Tor Vergata in the academic years 2017–2018 and 2018–2019.

References

1. Guangjian, W., Lin, C., Li, Y., Shuaidong, Z.: Research on the dynamic transmission error of a spur gear pair with eccentricities by finite element method. Mech. Mach. Theory **109**, 1–13 (2017)
2. Luo, Y., Baddour, N., Liang, M.: Effects of gear center distance variation on time varying mesh stiffness of a spur gear pair. Eng. Fail. Anal. **75**, 37–53 (2017)

3. Liang, X., Zuo, M.J., Feng, Z.: Dynamic modeling of gearbox faults: a review. Mech. Syst. Signal Process. **98**, 852–876 (2018)
4. Xiao, H., Zhou, X., Liu, J., Shao, Y.: Vibration transmission and energy dissipation through the gear-shaft-bearing-housing system subjected to impulse force on gear. Measurement **102**, 64–79 (2017)
5. González-Cruz, C.A., Jáuregui-Correa, J.C., Domínguez-González, A., Lozano-Guzmán, A.: Effect of the coupling strength on the nonlinear synchronization of a single-stage gear transmission. Nonlinear Dyn. **85**, 123–140 (2015)
6. Guo, Y., Eritenel, T., Ericson, T.M., Parker, R.G.: Vibro-acoustic propagation of gear dynamics in a gear-bearing-housing system. J. Sound Vib. **333**(22), 5762–5785 (2014)
7. Wang, T., Han, Q., Chu, F., Feng, Z.: Vibration based condition monitoring and fault diagnosis of wind turbine planetary gearbox: a review. Mech. Syst. Signal Process. **126**, 662–685 (2019)
8. Liu, W., Shuai, Z., Guo, Y., Wang, D.: Modal properties of a two-stage planetary gear system with sliding friction and elastic continuum ring gear. Mech. Mach. Theory **135**, 251–270 (2019)
9. Li, Y., Ding, K., He, G., Yang, X.: Vibration modulation sidebands mechanisms of equally-spaced planetary gear train with a floating sun gear. Mech. Syst. Signal Process. **129**, 70–90 (2019)
10. Kong, Y., Wang, T., Chu, F.: Meshing frequency modulation assisted empirical wavelet transform for fault diagnosis of wind turbine planetary ring gear. Renew. Energy **132**, 1373–1388 (2019)
11. Li, Y., Cheng, G., Liu, C., Chen, X.: Study on planetary gear fault diagnosis based on variational mode decomposition and deep neural networks. Measurement **130**, 94–104 (2018)
12. Mbarek, A., Hammami, A., Del Rincon, A.F., Chaari, F., Rueda, F.V., Haddar, M.: Effect of load and meshing stiffness variation on modal properties of planetary gear. Appl. Acoust. (2017)
13. Park, J., Hamadache, M., Ha, J.M., Kim, Y., Na, K., Youn, B.D.: A positive energy residual (PER) based planetary gear fault detection method under variable speed conditions. Mech. Syst. Signal Process. **117**, 347–360 (2019)
14. Park, J., Kim, Y., Na, K., Youn, B.D.: Variance of energy residual (VER): an efficient method for planetary gear fault detection under variable-speed conditions. J. Sound Vib. **453**, 253–267 (2019)
15. Tsai, S.J., Huang, G.L., Ye, S.Y.: Gear meshing analysis of planetary gear sets with a floating sun gear. Mech. Mach. Theory **84**, 145–163 (2015)
16. Zghal, B., Graja, O., Dziedziech, K., Chaari, F., Jablonski, A., Barszcz, T., Haddar, M.: A new modeling of planetary gear set to predict modulation phenomenon. Mech. Syst. Signal Process. **127**, 234–261 (2019)
17. Shen, Z., Qiao, B., Yang, L., Luo, W., Chen, X.: Evaluating the influence of tooth surface wear on TVMS of planetary gear set. Mech. Mach. Theory **136**, 206–223 (2019)
18. Igba, J., Alemzadeh, K., Durugbo, C., Henningsen, K.: Performance assessment of wind turbine gearboxes using in-service data: current approaches and future trends. Renew. Sustain. Energy Rev. **50**, 144–159 (2015)
19. Li, Z., Jiang, Y., Hu, C., Peng, Z.: Recent progress on decoupling diagnosis of hybrid failures in gear transmission systems using vibration sensor signal: a review. Measurement **90**, 4–19 (2016)
20. Abderazek, H., Ferhat, D., Ivana, A.: Adaptive mixed differential evolution algorithm for bi-objective tooth profile spur gear optimization. Int. J. Adv. Manuf. Technol. **90**(5–8), 2063–2073 (2017)
21. Pedersen, N.L.: Reducing bending stress in external spur gears by redesign of the standard cutting tool. Struct. Multidiscip. Optim. **38**(3), 215–227 (2009)
22. Palermo, A., Britte, L., Janssens, K., Mundo, D., Desmet, W.: The measurement of gear transmission error as an NVH indicator: theoretical discussion and industrial application via low-cost digital encoders to an all-electric vehicle gearbox. Mech. Syst. Signal Process. **110**, 368–389 (2018)
23. Xun, C., Long, X., Hua, H.: Effects of random tooth profile errors on the dynamic behaviors of planetary gears. J. Sound Vib. **415**, 91–110 (2018)

24. Dai, X., Cooley, C.G., Parker, R.G.: Dynamic tooth root strains and experimental correlations in spur gear pairs. Mech. Mach. Theory **101**, 60–74 (2016)
25. Bozca, M.: Transmission error model-based optimization of the geometric design parameters of an automotive transmission gearbox to reduce gear-rattle noise. Appl. Acoust. **130**, 247–259 (2018)
26. Marafona, J.D., Marques, P.M., Martins, R.C., Seabra, J.H.: Towards constant mesh stiffness helical gears: the influence of integer overlap ratios. Mech. Mach. Theory **136**, 141–161 (2019)
27. Marques, P.M., Camacho, R., Martins, R.C., Seabra, J.H.: Efficiency of a planetary multiplier gearbox: influence of operating conditions and gear oil formulation. Tribol. Int. **92**, 272–280 (2015)
28. Coronado, D., Kupferschmidt, C.: Assessment and validation of oil sensor systems for on-line oil condition monitoring of wind turbine gearboxes. Procedia Technol. **15**, 747–754 (2014)
29. Raghuwanshi, N.K., Parey, A.: Experimental measurement of gear mesh stiffness of cracked spur gear by strain gauge technique. Measurement **86**, 266–275 (2016)
30. Kia, S.H., Henao, H., Capolino, G.A.: Trends in gear fault detection using electrical signature analysis in induction machine-based systems. In: 2015 IEEE Workshop on Electrical Machines Design, Control and Diagnosis (WEMDCD), pp. 297–303. IEEE (2015)
31. González-Cruz, C.A., Ceccarelli, M., Alimehmeti, M., Jáuregui-Correa, J.C.: Design and experience of a test-bed for gearboxes. In: IFToMM World Congress on Mechanism and Machine Science, pp. 967–976. Springer, Cham (2019)
32. Zhang, Y., Lu, W., Chu, F.: Planet gear fault localization for wind turbine gearbox using acoustic emission signals. Renew. Energy **109**, 449–460 (2017)
33. Kong, Y., Wang, T., Li, Z., Chu, F.: Fault feature extraction of planet gear in wind turbine gearbox based on spectral kurtosis and time wavelet energy spectrum. Front. Mech. Eng. **12**(3), 406–419 (2017)
34. Chen, J., Li, Z., Pan, J., Chen, G., Zi, Y., Yuan, J., Chen, B., He, Z.: Wavelet transform based on inner product in fault diagnosis of rotating machinery: a review. Mech. Syst. Signal Process. **70**, 1–35 (2016)
35. Teng, W., Ding, X., Zhang, X., Liu, Y., Ma, Z.: Multi-fault detection and failure analysis of wind turbine gearbox using complex wavelet transform. Renew. Energy **93**, 591–598 (2016)
36. Ott, E.: Chaos in Dynamical Systems. Cambridge University Press (2002)
37. Balbayev, G., Ceccarelli, M., Carbone, G.: Design and numerical characterization of a new planetary transmission. Int. J. Innov. Technol. Res. **2**(1), 735–739 (2014)
38. González-Cruz, C.A., Ceccarelli, M.: Experimental characterization of the coupling stage of a two-stage planetary gearbox in variable operational conditions. Machines **7**(2), 45 (2019)

Chapter 23
Selection of Vibration Norms and Systems Structures When Designing Means of Monitoring Units with Gear Transmissions

Oleg B. Skvorcov

Abstract The features of vibration monitoring of gears used to form models of vibration state, damage diagnostics, prediction of operational failures and emergency protection are considered. Recommendations on the transition from monitoring individual vibration parameters to complex synchronous collection of information about all the basic parameters of spatial vibration: acceleration, speed and movement in the form of averaged and extreme values for short and long realizations are proposed. In addition to monitoring the intensity of vibration, it was proposed to normalize the threshold tolerances for vibration and the duration and frequency composition of the oscillations, which allows forming the vibration rationing taking into account the test results on the cyclic (vibration) strength of the used structural elements and materials. This structure of collecting primary information allows us to simultaneously solve the problems of diagnosing and forecasting in the extended frequency and dynamic range and to obtain high response speed and high reliability when implementing the emergency protection functions of monitoring systems implemented using the proposed budgetary structural solutions of the industrial Internet of things. The proposed structural solutions allow adaptively changing the structure of the system for collecting data on vibration processes by changing the number of control channels with expanding the dynamic range for both low-frequency and high-frequency vibrations with full use of the dynamic range of modern vibration control sensors. The proposed solutions can be applied in the study of both individual gears, and in the development of models of units, including both gears and bearing support elements and drives, as well as in the creation of systems for continuous monitoring and protection of such systems.

O. B. Skvorcov (✉)
Structural Acoustics Laboratory, Department of Theoretical and Applied Acoustics, Mechanical Engineering Research Institute of the Russian Academy of Sciences (IMASH RAN), 4, M. Kharitonyevskiy Pereulok, 101990 Moscow, Russian Federation
e-mail: skv@balansmash.ru

Development of Electronic Systems, Scientific and Technical Center, Zavod Balansirovochnykh Mashin Limited Leability Company, Varshavskoye Shosse, d. 46, 115230 Moscow, Russian Federation

© Springer Nature Switzerland AG 2020
V. Goldfarb et al. (eds.), *New Approaches to Gear Design and Production*,
Mechanisms and Machine Science 81,
https://doi.org/10.1007/978-3-030-34945-5_23

Keywords Machines · Gears · Reducers · Multipliers · Vibration · Strength ·
Monitoring · Diagnostics · Emergency protection · Rotation of coordinate axes

23.1 Introduction

Solving the tasks of emergency protection, monitoring, diagnostics of defects and
predicting the residual life by the vibration parameters of the gear units assumes
the existence of a model of the vibration state of such a unit. Such a model may be
analytical, which is not always possible due to the complex nature of the processes and
their dependencies on operating modes and individual characteristics of the control
object. Models are also possible in the form of sets of numerical indicators defined,
for example, in tabular form, as is often done in normative documents on vibration
control. In any of these representations, the initial data for constructing a model and
for comparing it with a real object during operation are the results of monitoring
vibration parameters. The volume and accuracy of such results largely determine the
practical utility of the model of the vibrational state of the gear mechanism.

In the development, testing and operation of complex mechanical equipment,
including gears, the use of vibration diagnostics methods is one of the most effective means of assessing their condition. The use of vibration diagnostics is the basis
for the implementation of maintenance and repair strategies based on actual condition. As the industrial Internet of Things (IIoT) is implemented, these methods are
extended both to large and expensive units, as well as to more and more simple and
budget mechanisms and equipment. A feature of the vibration nodes with gears is
the generation of vibrations in a wide range of frequencies. Along with vibrations of
circulating frequencies, their harmonics and subharmonics, oscillations are formed
with combinations of teeth gearing frequencies, as well as shock processes with frequencies 400–500 times larger than typical circulating ones [2]. Vibrations with a
broadband spectrum can not only serve as an effective diagnostic sign of the condition of the equipment, but also cause damage to the equipment elements. This leads
to the need to use new methods for assessing the vibration state [6]. The harmful
effects of vibration processes are associated with the cyclic strength of materials,
including the processes of ultrahigh-cycle (gigacycle) fatigue. There are contradictory approaches to assessing such strength under conditions of vibration, as assessed
by displacement, velocity, and acceleration. Accounting for these parameters, measured on equipment support elements, can be used in analyzing the results obtained
when testing the cyclic strength of materials under dual-frequency loading [13] to
justify the strong influence of relatively weak high-frequency vibrations. The study
of the effect of broadband vibration on structural elements to ensure the solution
of problems of vibration diagnostics and emergency protection is associated with
the possibility of increasing the information content of such multiparameter measurements [5]. Practical implementation of such measurements, including taking

into account the spatial nature of vibration processes and implementation possibilities in a low-budget version, allows performing precise measurements, diagnostics, prediction and emergency protection using MEMs and IIoT technologies.

23.2 Gear Vibration Monitoring

Gears are a common node of various rotary assemblies, and their operation depends on phenomena typical of rotary equipment, for example, the presence of unbalances of rotating elements.

The complexity of the problems associated with balancing rotor elements is limited by the fact that the information content of the processes generated in this process is limited only by the components corresponding to the circulating frequency. Gears, like many other rotary assemblies, are characterized by the presence of complex vibration processes, the spectra and temporal characteristics can change during operation, for example, when operating modes change. The existing regulatory framework for monitoring vibration of rotary equipment is based on current estimates of the intensity of vibration, as a rule, one of the characteristics of vibration signals, most often the mean square value of the vibration velocity in ISO 10816-3.

The random and time-varying nature of the vibration processes in this case is proposed to be assessed by the maximum of the resulting estimates in ISO/TR 19201-2013.

Monitoring estimates of the intensity of vibration of a rotor equipment at a given point in time is explained by the fact that the purpose of such monitoring was to obtain diagnostics on the current level of vibration without taking into account the processes of vibration changes and their effect on fatigue strength.

Vibration processes in machines are in fact not only a tool for diagnosing a condition, but can also be the cause of the origin and development of defects and are the main cause of damage associated with the cyclic strength of materials and elements [6].

The development of the regulatory framework of standards relating to the testing of machines and equipment, as well as structural materials for cyclic strength, showed the need to use criteria for evaluating vibration as a process in time. Vibration as a process in time is considered as the basis of vibration monitoring of the effect of vibration on a person. The introduction of criteria for exposure time and characteristics of changes in the intensity of vibration over time to the standards relating to the vibration monitoring of rotary equipment would harmonize such standards with the regulatory framework relating to vibration tests. Vibration monitoring as a time process is essentially reflected in the analysis of vibration trends as the basis for diagnostics and prediction in ISO 13373-3-2015 and ISO 13381-1-2015.

The features of vibration processes in gear units are the presence of components associated with rotation like rotors, the presence of significant components related to collisions during operation when the teeth are touched, and friction processes generating random vibration processes. The structure of the vibration signals is complicated by the fact that the design of gearboxes is usually include rolling bearing units, which also generate vibrations of a complex nature. The intensity of vibration of these elements significantly depends on changes in loads and modes of operation, and its measurement in most cases is carried out by sensors installed on stationary elements of the unit, which are transmitted by different, often quite remote from the place of its generation, by propagation along structural elements.

The control of the vibration state of aggregates with gears is considered in such regulatory documents as API standard 677 and ISO 8579-2-1993-02.

The API677 standard on vibration monitoring is largely focused on monitoring vibration movements using eddy current probe meters, which can effectively monitor vibration only in the low frequency range.

The ISO 8579-2 standard is based on measuring the vibration of rotary equipment in accordance with the basic standards of ISO 2954, 7919, 10816 groups in terms of vibration intensity level in terms of displacement and speed. Informative for diagnostics, the frequency range of vibration of gears corresponds, first of all, to the gearing frequencies and their harmonics. The processes of mechanical contact of teeth are more of a shock than a harmonic oscillation, and the energy released during this process, which is spent on local heating and deformation, is primarily associated with high-frequency oscillations. To assess the effect of such a vibration, it is more efficient to perform measurement and analysis of acceleration values. Such measurements are often used in diagnostic measurements of gears with respect to vibration parameters, but it is precisely significant local accelerations of the surface layers of gear elements that can cause the formation and development of defects, and not just their sign.

In addition to these two main standards relating to measuring the intensity of vibration of gears, vibration control is reflected in other regulatory documents relating to gear units in GOST 16162-85. The level of vibration on the supporting elements in the frequency range of 20–8000 Hz is according to the presented method. Vibration control is performed at a load level of at least 40%.

Essentially, these methods are similar to acoustic control methods in ISO 8579-1-2002. In this case, the measurement points of acoustic signals using microphones are selected at angular and equidistant (middle) points of the polyhedron describing a gear unit. This is the peculiarity of the application of acoustic methods, which for each measuring channel form integral (averaged over spatial directions) estimates. Vibration measurements, with the desired location of sensors in the immediate vicinity of possible sources, allow to localize the source of vibration. Such a difference allows us to recommend acoustic methods for general assessments of the level of radiated mechanical vibrational energy, and vibrational ones for analyzing the state of individual elements. Obtaining an overall assessment may be more effective in solving the problem of emergency protection, and the localization of sources of oscillations is more convenient in solving problems of diagnosing individual parts of a gear.

The standard dependencies of maximum and average radiation levels on the magnitude of the load, reverse speed and geometric characteristics given in this standard are based on a set of measurements for a certain set of types of specific gearboxes and can hardly be taken as recommendations for setting threshold values for diagnostic algorithms and emergency protection.

In the regulatory documentation and in the publications, not completely unambiguously defined terms are used, such as "vibroactivity" or "vibro-loading". This is due to the complexity of the analysis of vibration processes and the desire to apply in this case the approaches of evaluating the entire unit or node as a whole integrally generating oscillatory processes, as is done when assessing the state of such equipment based on its noise characteristics. Such methods are also widely used in the diagnosis of gear units. The difference in methods of assessing the state of the emitted noise and vibration is largely due to the difference in spatial resolution capabilities in multidimensional measurements. If multicomponent accelerometers providing high spatial resolution are low cost and are widely used [11], obtaining similar capabilities using acoustic methods with typical spatial geometric dimensions is difficult. An example of the use of the term "vibration load" in relation to the analysis of the state of gears can serve. In vibration measurements, the vibration load or vibration overload A is usually understood to be an indicator defined as the ratio of the acceleration a acting at the measurement point a to the earth acceleration g.

$$A = a/g$$

This value characterizes the locally acting forces, which can be accompanied by the appearance of damage with high acceleration and/or their long-term exposure. Unfortunately, often the possible influence of vibration effects on the strength of the elements of toothed assemblies is underestimated, or not taken into account at all, although they are fundamental principles of cyclic strength. Even when solving the problems of ensuring the durability of modern gearing, the contribution questions.

The effect of high-frequency vibrations is accompanied by a drastic decrease in cyclic strength, although the corresponding movements are extremely small compared to simultaneously operating low-frequency vibrations, which cause corresponding low-frequency deformations that have a noticeable magnitude.

This may be due to the local influence of large dynamic sig from the impact of high-frequency vibrations that do not manifest themselves outside this local region or their action for neighboring local regions is compensated for by the difference in the oscillation phases for such neighboring local regions.

Other features of the effect of high-frequency vibrations on the fatigue strength of gearing elements are their random nature and a large number of loading cycles recruited during operation over a relatively short time interval. Vibroacoustic diagnostics of gears operating under conditions of significant mechanical loads, shows a random nature, changes over time (Fig. 23.1).

The analogy between the dependences of the boundary values of mechanical stresses and linear accelerations of structural elements on the number of loading cycles was noted in [1].

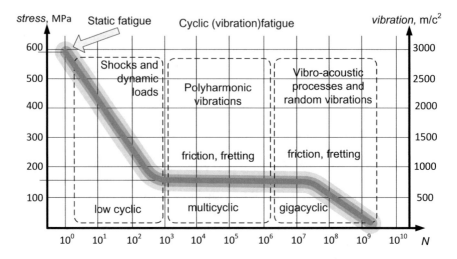

Fig. 23.1 S–N diagram and assessment of the impact of vibration effects, estimated by the magnitude of the accelerations

The phenomenon of gigacycle fatigue discovered and described in subsequent publications shows that the contribution of high-frequency vibrations can be quite significant even at their relatively small values.

23.3   Limits of Vibration of Gears

For vertical motors and orthogonal gears, thrust bearings must be at the non-drive end and must limit axial displacement to 125 μm in ISO 13709-2009.

The magnitude of the axial vibration should not exceed 60 μm with a peripheral speed of up to 20 m/s in API standard 677. Radial vibrations in this case, when the size of the shaft journal is less than 305 mm, it should not exceed 6.5 μm and for large shafts 10 μm. Axial vibration with up to 15 μ values of permissible unbalance of shafts:

$$UA = 6,350 \, W/N,$$

where

W is the static load;
N working speed $\min^{-1}$

Rationing of vibration intensity levels on the fixed elements of the gear unit design in API standard 677 is performed for the frequency range from 10 Hz to 10 kHz both according to the general level, and for individual frequency components according

Table 23.1 Vibration rates on fixed elements API standard 677

	Velocity RMS (mm/c)	Acceleration PEAK (g)
Frequency (KHz)	0.01–2.5	2.5–10
All frequencies	5	8
Discret frequency	4	

Table 23.2 Norms of the limit levels of the RMS intensity intensity of vibration of the body elements of gear units GOST R ISO 20283-4-2017

Zone	RMS velocity (mm/c)
A	<4.5
B	4.5–11.2
C	>11.2

to mean square values and peak estimates of speed and acceleration. Standards for vibration of the gearbox housing are shown in Table 23.1.

Depending on the class of equipment, typical norms of average values in GOST R ISO 20283-4-2017 of vibration velocity, presented in Table 23.2, are also applied to estimates of the intensity of vibration on non-rotating frequent gearbox cases.

The previously mentioned basic standard for measuring vibration of gear units ISO 8579-2-1993 provides recommendations for measuring the levels of displacements in the frequency range 0–500 Hz and the acceleration levels on fixed elements in the frequency range 10–10000 Hz. In this case, the amount of movement of the shaft should not exceed 25% of the maximum allowable or 6 μ. It is allowed to evaluate both relative and absolute movements of the shafts. The absolute movements of the shafts are obtained by subtracting the measured relative displacements and measurements at the place of installation of the sensors, taking into account the phase relationships of the signals.

Measurement of sensor movement can be performed using low-frequency accelerometers, followed by double integration. As a double integration link, it is advisable to use a low-pass filter with multi-loop feedback on an inverting amplifier and a cut-off frequency below the lower cut-off frequency of the measurement. Such a scheme makes it possible to obtain a displacement signal without a substantial phase shift with respect to the signal from the proximity meter relative shaft movements.

The measured parameters are the RMS of the velocity and range of motion. Vibration estimates of speed at frequencies above several kiloherz have low information content due to the rapid attenuation of the signal at higher frequencies during integration. Therefore, when estimating vibration on fixed parts, it is more correct to use acceleration estimates without using signal integration.

Depending on the class, the DR permissible displacement levels are constant from 31.5 to 200 μm (the dimension in millimeters shown in the figure seems to be a typo) in the frequency range from 0 to 50 Hz, and at higher frequencies the values decrease with a slope of 10 dB per decade. For the VR velocity, the permissible norms are constant depending on the class in the range from 3.15 to 20 mm/s in the frequency range from 45 to 1590 Hz and below and above this frequency fall with a slope of 14 dB per decade. The choice of the indicated numerical values is essentially not supported by any scientific substantiation, as well as the numerical values indicated in the current standards for the assessment of the vibration norms of rotary equipment, developed on the basis of ISO 2372. In such standards, the numerical indicators of the norms of vibration laid down in the original versions for clearly subjective reasons are repeated in newer versions instead of introducing norms based, for example, on objective measurements of the vibration strength of structural materials. In the ISO 8579-2-1993 standard, the DR and VR standards are used to classify the vibration threshold values of gear equipment depending on its output power. The standard also provides a list of factors affecting the vibration characteristics and related to the magnitude and changes of the load, features of the assembly, angular and linear movements, speed of rotation. These factors essentially characterize the gear unit itself and the changes in the vibration state associated with them are essential for the diagnosis and emergency protection of the node. In addition, factors related to the influence of the engine, as well as vibrations from nearby equipment, can affect the measurement of vibration. Such equipment is often of the same type with controlled and separation of vibration processes in them can be a significant problem. The reverse vibration transmitted to the body, for example, due to the presence of imbalances, as well as its second harmonic, associated with misalignment and ovality can be transferred to the supporting element at low loads in support bearings very weakly, and at high loads their contribution to the overall level, as well as high frequency vibrations associated with jagged frequencies may be decisive. The application for such separation of methods associated with independent sequential launches with subsequent separation of processes, taking into account the phase mismatch of oscillations [7, 10], is difficult because the vibration signals in the gear units are complex and the phase difference for them is difficult to implement. To separate such effects, you can use the methods of correlation analysis. Calculating autocorrelation and cross-correlation functions in this case may be useful for such a separation. In addition, estimates of the correlation functions allow us to estimate the time during which previously occurring processes become unrelated to the current ones. Evaluation of such a time allows you to get recommendations on the choice of the time of formation of one set of measurement results.

In most cases, for such an assessment, it is possible to use an estimate of the delay time at which the envelope of the autocorrelation function becomes lower than the noise level of the corresponding measuring channel.

In addition, the previously listed individual features of the vibration processes for a specific instance of a gear unit imply the need to obtain a set of statistical estimates for it. This is possible if the controlled vibration processes meet the criteria for ergodicity.

23.4 Vibration Resistance and Monitoring

The issues of calculating the strength of gears using the methods of finite element analysis and analytical methods presented in the standards of the ISO 6336 series are considered in GOST 21354-75. Such a comparison shows a satisfactory agreement between the results of the simulations obtained by these methods.

The strength of the gear elements is determined by possible damage during static loads, as well as cyclic loads. Cyclic loads arise as dynamic forces associated with the rotation of the rotor elements and with the shock dynamic loads on the gear elements in the area of the gears. There are regulatory documents containing methods for calculating static and low-cycle loads in gears. An example of such a document is the standard ISO 6336, which presents a methodology for the assessment of bending strength. The impact of cyclic loading leads to significantly lower estimates of tensile strength. A particularly significant decrease in strength is observed under high-cycle and gigacycle loads. High values of current frequencies make the issues of multi-cycle strength damage relevant, since such a number of cycles can accumulate in a relatively short time. Percussion processes during teeth engagement can be generated by different teeth, but the frequency of such gears is proportional to the product of the circulating frequency and the number of teeth, which frequencies turn out to be large and the shock vibrations generated by this have repetition frequencies that suggest gigacyle fatigue of materials. The decrease in strength with such fatigue is noticeable compared with the static strength of materials, and the amplitude of vibration overloads is such that damage development processes can become one of the leading factors determining reliability. High-frequency vibrations at the same time can act in local areas and not be transmitted to the hull bearing fixed structural elements. In these cases, such vibrations do not manifest as noticeable deformations of the supporting elements, but their effect on the reduction of strength can be quite significant and decisive compared with low-frequency loads, causing noticeable deformations [13]. Such a significant decrease in strength when exposed to high-frequency vibrations can be explained by the high-cycle nature of the decrease in strength and the large dynamic forces occurring in the range of such high-frequency vibrations, although the corresponding dynamic frequency movements are negligible.

23.5 Features of Vibration Gear Mechanisms. Frequency and Dynamic Ranges

The vibration of gear elements is usually characterized by a complex spectral composition and changes in spectral composition with time. The complex spectrum measured on the fixed parts of the support elements of the gear unit contains circulating frequencies and their harmonics, with the number of circulating frequencies f_{Ni} corresponding to the number of shafts on which the wheels and gears of the gear train are mounted. As a rule, these revolving frequencies do not match and correlate as a given gear ratio of a gear.

The gear ratio for standard gearboxes can reach 4,000 [15]. The second group of frequency components is determined by the number of tooth loading cycles per revolution or, the resonant speed of the gear, the frequency of f_{Ci} gearing [12].

In this case, S/N is the ratio of the working and resonant frequencies in the zone of the main resonance [12]. The third group of frequency components is determined by the resonant frequencies of the rotating gear, and such frequencies can correspond to both linear resonances and resonances associated with torsional oscillations.

The processes of collision and friction in the gearing elements generate broadband spectral components. Additional frequency components of vibration can be associated with the operation of equipment in which gears are used, for example, pressure pulsation frequencies or various cavitation processes in elements of pumping equipment. These components are usually the nature of random fluctuations.

In addition, many of the components, depending on mode changes, for example, due to variable loading forces, are modulated, which complicates the analysis of the received vibration data.

The combination of these states must be taken into account in the models of processes described analytically and used in the construction of monitoring systems.

Thus, the vibration signals and their generating processes usually cover a very wide frequency range from low-frequency to high-frequency oscillations. Measuring in such a wide frequency range is in itself a rather complicated technical task. The complexity of such measurements lies in the fact that in the analysis of low-frequency vibrations they use data on vibrational displacements measured by specialized low-frequency sensors [9].

When measuring high-frequency components, the level of signals proportional to displacements and generated by vibration sensors becomes lower than the level of intrinsic noise and during such measurements acceleration signals are usually measured, such as those generated by accelerometers. The components of displacements and accelerations are recalculated among themselves with coefficients proportional to the square of the frequency ratio. With a large difference in frequencies, typical for vibration of nodes with gears, reaches tens and hundreds of millions. In practice, the measurement is performed in a relatively narrow frequency band according to one of the parameters of displacement, velocity, or acceleration. Most often, the monitoring equipment is focused on obtaining speed estimates, which allows us to adequately estimate the vibrational state in only the mid-frequency range. Therefore, such modern monitoring systems are focused on diagnosing a narrow circle of possible defects and do not allow timely detection of all possible defects. An example of the construction of broadband vibration monitoring systems shows a significant distinction between the scope of the possible use of displacement meters, speed and acceleration [14].

23.6   Repeatability Interval and Ergodicity of Processes

Even instances of the same type of gear units can have significant differences in the vibration signals generated during their operation. This is explained not only by the differences in the modes of operation and loading, but also by the individual features of the processes when the teeth engage and the wear that occurs. The accumulation of statistical information necessary for the reasonable establishment of vibration standards should take into account such features. The limited possibilities of parallel monitoring of a large number of similar objects can be replaced by obtaining statistical data on each of the monitored objects for a long time. The theoretical rationale for this possibility is the application of the Berghoff-Khinchin theorem, according to which the statistics of estimates for the set of objects is equivalent to statistics for averaging data on a sequence of realizations over time if the observed processes are ergodic. For the ergodicity of the process, it is necessary to evaluate the constancy of estimates of the expectation and variance, and also to show that the autocorrelation function of the process was independent of time and, with an increase in the delay value, tended to zero. In most cases, the processes of vibration meet these criteria. Monitoring the processes of vibration using continuous monitoring systems in the process of collecting information allows us to estimate the average values of their basic parameters and their constancy in time under identical operating conditions. According to the data obtained, it is also possible to estimate the constancy of the variance or the standard deviation. The recordings of long-term implementations of vibration processes obtained for solving problems of diagnostics also allow us to estimate the nature of the autocorrelation functions of vibration processes and their convergence to zero values. For such an estimate, it is convenient to control the convergence to zero of the functions of the envelope of the autocorrelation function. In most cases, the vibration-controlled processes monitored by the monitoring system, as well as the processes of change in their characteristics over time, such as the values of the mean square estimates, can be considered ergodic [4].

The resulting estimates of time intervals beyond which the autocorrelation function becomes close to zero can be used to select the duration of the generated implementations used in diagnosing the monitoring object. The accumulation and analysis of statistics of data on such long-term implementations allows us to make a reasonable conclusion about changes in the vibration state over time. These results can be used to predict the behavior of equipment based on the results of vibration monitoring, taking into account the time-integrated vibration effects, and not be limited to the current assessment of the vibration intensity. On the other hand, the correlation of vibration processes over short time intervals, due to the previously mentioned theorem, makes it possible to use such data when creating algorithms for generating emergency protection by vibration with increased reliability [3, 8]. In order to do this, the vibration monitoring data obtained from the multicomponent sensors of each of the vibration control points of the gear unit can be used, the installation of which is regulated by the regulatory documents on the vibration control mechanisms with gears.

23.7 Structural Methods for Improving the Metrological Characteristics of Monitoring Systems. Methods for Expanding the Frequency and Dynamic Ranges

Practical issues of constructing vibration monitoring systems of gear equipment, oriented to work in the industrial Internet environment, should provide a very wide frequency and dynamic range of control of vibration signals. This allows you to exclude the loss of data that can be observed either at very short time intervals or occur rarely during implementations. The expansion of the dynamic range allows you to eliminate the loss of information about low-level vibrations, which may be important in solving diagnostic problems. Large dynamic range is also important to prevent possible overloads of the channel during the formation of the assessment. Such overloads can lead to loss of data reliability and, which is very significant, to false alarms of emergency protection systems or to the omission of a moment of rapid development of damage. In the latter case, the emergency protection system will not provide reliable protection of the monitored unit from an accident. Since non-contact proxymeters do not allow to obtain a large dynamic range (typical measurement range from 10 μm to several millimeters) and have a relatively narrow frequency range (from 0 to 200 to 500 Hz) used for monitoring vibration of gear units with limited individual tasks.

As noted in ISO 8579-2-1993, the control of high-frequency vibrations is usually performed using acceleration sensors, moreover, integrating matching devices are used to obtain speed (movement) signals from such signals. The standard notes that special attention should be paid to reducing the influence of low-frequency noise. The effect of such noise can be reduced by performing integration directly in the input circuits of the matching amplifier, which is usually used as charge amplifiers. Such circuit solutions provide high gain at low frequencies and reduce the likelihood of overload of the matching amplifier by the high-frequency components of the acceleration signal [7].

Measurement of displacements with non-contact vibration sensors (eddy current detectors) is practically limited to frequencies of several hundred hertz and such sensors are unsuitable for vibrations associated with jagged gears ISO 8579-2-1993-02, although they are used successfully to determine imbalances, displacements and deviations from roundness. The lower threshold of measurement with contactless sensors is usually limited to 10 μ, although, as previously noted, some of the norms require monitoring of vibration values of the order of 6 μ. Considering the above, the use of contactless sensors cannot replace the need to use equipment with gears for accelerometers, which have a wider dynamic measuring range, for measuring vibrations. The performance of measurements at low levels of measurable quantities requires efficient algorithms for extracting information against the background of relatively large noise and interference. In this case, one can apply correlation algorithms for extracting useful data against the background of noise and compensating for the influence of the latter on the measurement results.

Fig. 23.2 Block diagram of
the vibration data collection
module during monitoring

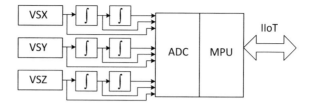

An example of a block diagram of the module for collecting vibration signals from a three-component sensor (components (VSX, VSY and VSZ) on an analog-digital converter ADC and IIoT MPU—microcontroller with parallel data inputs from sensors and analog integrators), marked with an integral sign is shown in Fig. 23.2.

23.8   Diagnosis Gears

The main provisions for performing diagnostics of 4 nodes with gears using vibration measurements are presented in the annexes to the standard GOST R 50891-96. The standard provides recommendations for the diagnosis and monitoring of the status of gears and their bearing units based on the measurement of vibrations caused by shock processes. In particular, it is recommended to use shock vibration level control in the immediate vicinity of bearing assemblies. Such monitoring is performed in order to prevent the sudden development of damage to the gear train or bearing. In this case, piezoelectric accelerometers with built-in or external charge amplifiers with a measuring frequency range of at least 20 kHz are used as measuring equipment. Frequency analysis and the recommended synchronous accumulation of signals are quite simply implemented in modern nodes of digital signal processing. As an auxiliary device in such systems use one or several sensors of a phase mark. Getting the binding to different frequencies, which is typical for gears, use frequency converters of pulses in the form of dividers and frequency multipliers of signals from phase mark sensors.

Such a conversion can be implemented using built-in internal timers embedded in digital signal processing units, as well as the use of specialized hardware multipliers and frequency dividers, starting from such as 1901 by Brüel and Kjer. The use of the phase locked loop in the latter, whose operation is based on the use of low-pass filters, limits the possible speed of response to changes in the frequency of revolutions. If the mode of operation of a gear train changes rapidly, it is possible to use specialized frequency multiplication schemes with a high response speed.

To solve diagnostics problems, they use estimates of the level of vibration at the frequency of re-conjugation of the teeth of the driving and driven gear wheels, as well as at the frequencies of their harmonics. They use synchronous accumulation, analysis of the envelope and its spectrum, control the spectra with the use of FFT and wavelet transform and perform harmonic analysis, calculate cepstrs, kurtosis and other parameters of vibration processes. Such analysis in real time can be performed by modern processors built into the measuring equipment.

The manifestation of incipient defects by the methods of vibration diagnostics requires observation during long time intervals reaching several months. The use of diagnostic methods for vibration and other parameters (acoustic noise control, temperature monitoring, etc.) allows you to identify many defects and proceed to maintenance and repair based on an assessment of the actual condition of the equipment. Diagnostic methods cannot guarantee 100% detection of all defects. A number of defects can develop very quickly and the diagnostic control system based on the collection and analysis of long-term implementations may not provide timely response to the appearance of a dangerous failure. The functions of emergency protection of equipment, including rotor assemblies with gears, are usually performed by the system of emergency protection of the unit according to vibration parameters. If the diagnostic system is usually focused on obtaining data on relatively small changes in the early stages of defect manifestation, which requires collecting long implementations and, as a rule, using long-term averaging with a practically unlimited signal acquisition time, the emergency control system should provide automatic response to rapid changes in the vibrational state with reaction time in units or fractions of a second. In this case, besides the response speed, the emergency protection system usually requires a higher accuracy of operation. Such a system should, as far as possible, eliminate both false alarms and the transmission of actual emergency situations. One possible solution would be to use a redundant emergency protection system. Such requirements are in many respects opposite to the requirements for a diagnostic vibration monitoring system.

It should also be noted that systems with gears are often operated in non-stationary modes (with changes in revolutions, loads, etc.), and for diagnosing such modes are undesirable. The observation time of the minimum cycle with the repetition of the state of the gear transmission can be quite large and correspond to a large number of revolutions.

Satisfying the contradictory requirements of building a vibration monitoring system without using an excessive number of sensors and data acquisition equipment is possible by ensuring the use of multi-channel synchronous parallel data acquisition by data acquisition devices and conversion to digital formats that are adapted to a specific task. Since solving the problems of diagnostics requires the use of vibration measurements in three orthogonal directions, it is possible to use three-component vibration sensors in the system with programmatically extracting vibration information with precision accuracy for diagnosing the gear unit. Simultaneously, vibration data from these three channels to confirm rapid changes for the organization of emergency protection. This can be provided by matrix transformations of vectors.

23.9   3-Axials Vibration Gear Mechanisms

In the standard for measuring the vibration of gear units in ISO 8579-2-1993-02, it is recommended to change the vibration of shafts in several orthogonal directions. It is recommended to measure the vibration of body elements and bearing supports in three orthogonal directions. Measurement of spatial vibration by single-component and multi-component sensors is a common task in monitoring the vibration of complex rotor assemblies [16].

To implement the diagnostics, the position of the sensitivity axes of the three-component sensor is orthonormal to eliminate their mutual influence vector and the sensitivity axes are orthonormal. At the same time, the measurement results by channels describe the vibrational behavior in the transverse, vertical, and axial directions most completely independently.

To organize emergency protection of the sensitivity axis, matrix transformation of the vector of signals m is arranged so as to ensure equal contribution of each of these independent components, i.e. get close in importance signals. This transformation is illustrated by the vector diagram shown in Fig. 23.3.

An example of using a three-component accelerometer for collecting data on spatial vibration when solving the problem of vibration diagnostics and extracting three components from the vector of signals of such a sensor, which can be used for mutual confirmation of changes. Such changes are correlated with each other and can be used to implement crash protection. In this case, the cost of hardware implementation is minimized.

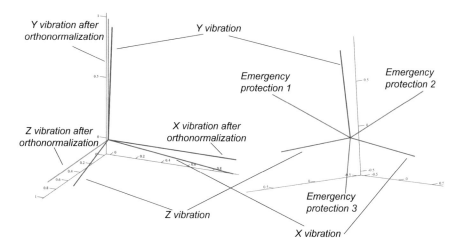

Fig. 23.3 Co-ordination of the sensitivity axes when monitoring spatial vibration with a three-component sensor for diagnostics (left) and anti-emergency protection (right)

23.10   Conclusion

The operation of mechanical units with gears is characterized by a complex combination of vibration processes. Control over each of these processes, which can be either periodic or non-periodic, is necessary for the correct assessment of the conformity of a real object with a model of its behavior during operation. Performing continuous vibration monitoring of complex gear units allows you to create a set of statistically valid data, taking into account the characteristics of each of the monitored units. This is possible taking into account the ergodicity of the processes of vibration in such rotary units. Continuous monitoring using the technologies of the industrial Internet of things allows you to collect data on the history of vibration effects and to supplement monitoring of vibration intensity parameters with data that allows you to build models that take into account vibration strength in high-cycle and gigacycle fatigue conditions of materials and gear components.

The spatial nature of the vibration processes in the control points, which is necessary in accordance with the current regulatory recommendations, allows for obtain reliable data by eliminating measurement errors. In addition to obtaining high-resolution high-precision data, such measurements make it possible to simultaneously implement emergency vibration protection algorithms with high response speed and increased accuracy of operation. In this case, the operation of a three-component sensor is equivalent to a redundant control of three independent sensors. Such opportunities are provided by matrix transformations of the vector signals of spatial vibration. Providing a wide dynamic and frequency ranges to provide broadband vibration monitoring in the presence of budget constraints is provided by parallel multichannel data collection from various channels with switched links between them through elements that perform analog integration of vibration signals.

Referenes

1. Lenk, A., Rehnitz, J.: VEB Verlag Technik. Berlin (1974)
2. Luo, H.: Physics-based data analysis for wind turbine condition monitoring. Clean Energy **1**(1), 4–22 (2017)
3. Pravotorova, E.A., Skvorcov, O.B.: Compensation of the degradation of the parameters of a multicomponent vibration sensor by comparing the statistical characteristics of the signals. In: Scientific papers of the IV International Scientific Conference Basic Research and Innovative Technologies in Mechanical Engineering M. Publishing house « Spektr », pp. 212–214 (2015)
4. Pravotorova, E.A., Skvortsov, O.B.: Modelling of vibration tests of winding elements of power electric equipment. J. Mach. Manuf. Reliab. **44**(5), 103–110 (2015)
5. Radchik, I.I. et al.: Patent RU 2644620 (2016)
6. Skvorcov, O.B.: Prospects for the development of the regulatory framework and the expansion of vibration monitoring of rotary equipment. Power Stations **8**, 46–53 (2017)
7. Skvorcov, O.B.: Vibration safety of large power units. In: Security Problems of Complex Systems: Materials of the XXVI International Conference. Publishing house IPU RAS, pp. 310–313 (2018)

8. Skvorcov, O.B., Pravotorova, E.A.: Redundant mode of operation of a three-component vibration sensor. In: Dynamics and Structural Strength of Aerohydroelastic Systems. Proceedings of the Fourth All-Russian Scientific and Technical Conference, M., Publishing House IMASH RAS, pp. 50–51 (2017)
9. Skvortsov, O.B.: Measurements of low frequency vibration. Sens. Syst. **4**, 16–21 (2017)
10. Skvortsov, O.B.: Analysis of vibrational signals in solving problems of balancing rotors. Autom. Modern Technol. **2**, 60–66 (2018)
11. Skvortsov, O.B.: The accuracy limiting at measuring spatial quantities with orthonormalization. Autom. Modern Technol. **5**, 217–223 (2019)
12. Starzhinsky, V.E., Antonyuk V.E. et al.: A dictionary of gears in Russian-English-German-French. Gomel., 242, (2011)
13. Terentyev, V.F., Oksogoev, A.A.: The Cyclic Strength of Metallic Materials, p. 61. Publishing house NGTU, Novosibirsk (2001)
14. Trunin, E.S., Skvortsov, O.B.: Operational monitoring of the technical condition of hydroelectric plants. Power Technol. Eng. **44**(4), 314–321 (2010)
15. Tsehnovich, L.I., Petrichenko, I.P.: Atlas of Gear Designs. Graduate School, p. 151 (1990)
16. Tumer, I.Y., Huff, E.M.: Analysis of triaxial vibration data for health monitoring of helicopter gearboxes. J. Vib. Acoust. **125**(1), 19 (2003)

Chapter 24
Curvature Interference Characteristic of ZC1 Worm Gear

Qingxiang Meng and Yaping Zhao

Abstract The main objective of this paper is to establish the theory of the curvature interference characteristic of a ZC1 worm gear based on the meshing theory for gearing. Some fundamental and important results are obtained, for example the tooth surface equations, the meshing function, and the curvature interference limit function of the ZC1 worm pair. A method to judge the existence of the curvature interference limit line is put forward. Then the curvature interference limit line of ZC1 worm gear is determined by solving a system of nonlinear equations. The numerical example is provided for validation and verification and the result shows that there is one significative curvature interference limit line on the worm gear tooth surface, which generally does not enter into the conjugate area of the worm pair. The conjugate area of the worm pair on the worm gear tooth surface is located on the non-curvature-interference side of the curvature interference limit line. Consequently, the undercutting of the worm gear seldom takes place. When the modification coefficient of worm gear and the radius of circular arc profile of the grinding wheel are less and the shaft angle of the worm pair is larger, the curvature interference limit line on the worm gear tooth surface may be closer to the intersection point of the conjugate line of the worm addendum and the conjugate line of the meshing limit line of the worm pair, namely, this area has the greatest potential hazards to be undercut.

Keywords ZC1 worm pair · Curvature interference characteristic · Tooth surface · System of nonlinear equations

Q. Meng · Y. Zhao (✉)
College of Mechanical Engineering and Automation, Northeastern University, Shenyang 110819, China
e-mail: zhyp_neu@163.com

Q. Meng
e-mail: mengqingxiang_neu@163.com

© Springer Nature Switzerland AG 2020
V. Goldfarb et al. (eds.), *New Approaches to Gear Design and Production*,
Mechanisms and Machine Science 81,
https://doi.org/10.1007/978-3-030-34945-5_24

513

24.1 Introduction

The so-called ZC1 worm pair as illustrated in (Fig. 24.1) was invented by Gustave
Niemann in the 1930s [3, 6] and its worm was originally called as the Niemann worm
[1]. In regard to the ZC1 worm pair, its meshing theory was continuously studied by
Litvin [5], Wu [8], Wang [9] and so on. Not only that, the meshing limit line of the
worm pair has also been studied in literature [11] by Zhao. However, the curvature
interference characteristic of ZC1 worm gear was not paid enough attention to. From
the meshing theory for gearing [2, 7], the curvature interference limit line on the
enveloped surface, i.e. the worm gear tooth surface, is the boundary between the
undercutting area and the non-undercutting area. As a result, the appearance of the
curvature interference limit line on the worm gear tooth surface will lead the tooth
surface to be undercut, and the ineffective part will appear on the tooth surface. In the
existing literature [4, 8], although some researchers drew the curvature interference
limit line on the worm gear tooth surface, the computing method concerning the
curvature interference limit line and its detailed numerical results were not provided.
Therefore, the curvature interference characteristic of the ZC1 worm gear needs to
be investigated systematically.

   In this paper, the theory of solving the curvature interference limit line for a
ZC1 worm gear is fully established. The numerical example investigation is also
implemented.

24.2 Generation of ZC1 Worm Pair and Its Characteristic
   Parameters

24.2.1 Geometry of Toroidal Grinding Wheel

As shown in Fig. 24.2, the unit vector $\vec{k}_a$ of a coordinate system $\sigma_a \left\{ O_a; \vec{i}_a, \vec{j}_a, \vec{k}_a \right\}$ is
along the axis of the toroidal grinding wheel. The intersecting line of an axial section
of the grinding wheel and its generating torus is a segment of circular arc with the

Fig. 24.2 Toroidal grinding wheel in σ_a

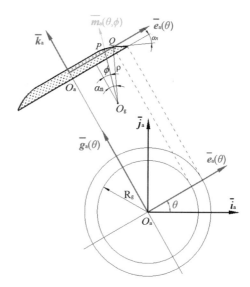

radius ρ. Therefore, based on the sphere vector function [2], it is easy to obtain the equations for the generating surface, Σ_a, of the toroidal grinding wheel and its unit normal vector, $\vec{n}_a$, in σ_a as follows

$$(\vec{r}_a)_a = \overrightarrow{O_a O_g} + \rho \vec{m}_a(\theta, \phi) = x_a \vec{i}_a + y_a \vec{j}_a + z_a \vec{k}_a, \qquad (24.1)$$

$$(\vec{n}_a)_a = \frac{\frac{\partial(\vec{r}_a)_a}{\partial\theta} \times \frac{\partial(\vec{r}_a)_a}{\partial\phi}}{\left|\frac{\partial(\vec{r}_a)_a}{\partial\theta} \times \frac{\partial(\vec{r}_a)_a}{\partial\phi}\right|} = -\vec{m}_a(\theta, \phi) = -\sin\phi\cos\theta\,\vec{i}_a - \sin\phi\sin\theta\,\vec{j}_a - \cos\phi\,\vec{k}_a,$$

$$(24.2)$$

where $x_a = \rho\sin\phi\cos\theta + (R_g - \rho\sin\alpha_n)\cos\theta$, $y_a = \rho\sin\phi\sin\theta + (R_g - \rho\sin\alpha_n)\sin\theta$, and $z_a = \rho(\cos\phi - \cos\alpha_n)$. Here, the symbols R_g and α_n are the nominal radius and the shape angle of the grinding wheel, respectively.

Obviously, the direction of $\vec{n}_a$ in Eq. (24.2) is from the space to the inside of the grinding wheel.

Moreover, based on the unit normal vector $\vec{n}_a$, it can be established a moving principal frame $\sigma_p\{P; \vec{g}_1, \vec{g}_2, \vec{n}_a\}$ at an arbitrary point on the generating toroidal surface Σ_a. The two principal directions, $\vec{g}_1$ and $\vec{g}_2$, of Σ_a can be figured out as below

$$(\vec{g}_1)_a = \frac{\frac{\partial(\vec{r}_a)_a}{\partial\theta}}{\left|\frac{\partial(\vec{r}_a)_a}{\partial\theta}\right|} = \vec{g}_a(\theta) = -\sin\theta\,\vec{i}_a + \cos\theta\,\vec{j}_a, \qquad (24.3)$$

$$(\vec{g}_2)_a = (\vec{n})_a \times (\vec{g}_1)_a = \vec{n}_a(\theta, \phi) = \cos\phi\cos\theta\,\vec{i}_a + \cos\phi\sin\theta\,\vec{j}_a - \sin\phi\,\vec{k}_a. \qquad (24.4)$$

Based on this, the principal curvatures of Σ_a along $\vec{g}_1$ and $\vec{g}_2$ can be respectively worked out as

$$k_1 = \frac{\sin\phi}{\rho\sin\phi - \rho\sin\alpha_n + R_g}, \quad k_2 = \frac{1}{\rho}. \tag{24.5}$$

24.2.2   Cutting Meshing of ZC1 Worm

During grinding the ZC1 worm, the unit vector $\vec{k}_{o1}$ of a static coordinate system $\sigma_{o1}\{O_1; \vec{i}_{o1}, \vec{j}_{o1}, \vec{k}_{o1}\}$ is along the axis of the worm rough as shown in Fig. 24.3. The original point O_1 is at the middle point of the thread length of the worm. The crossing angle between the worm axis $\vec{k}_{o1}$ and the grinding wheel axis $\vec{k}_a$ is the lead angle, γ, of the worm on its pitch cylinder as shown in Fig. 24.3b. The unit vector $\vec{i}_a$ is along the common perpendicular of the two vectors $\vec{k}_a$ and $\vec{k}_{o1}$, and is parallel to the unit vector $\vec{i}_{o1}$. A rotating coordinate system $\sigma_1\{O_1; \vec{i}_1, \vec{j}_1, \vec{k}_1\}$ is used to indicate the current position of the worm, whose rotating angle relative to its initial position is φ. At initial position, the coordinate system σ_a locates at the black position, i.e. the point O_a coincides with the point O_d. The distance from the point O_1 to the point O_d is the operating center distance a_d during grinding ZC1 worm as shown in Fig. 24.3b. At current position, i.e. the red position of $\sigma_a\{O_a; \vec{i}_a, \vec{j}_a, \vec{k}_a\}$, the displacement distance of the coordinate system σ_a along the axis of the worm

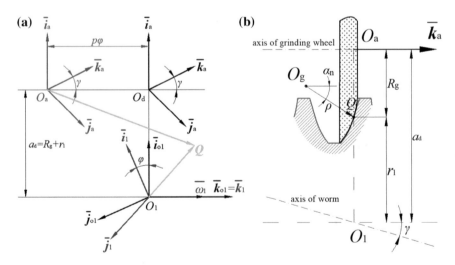

Fig. 24.3 **a** Coordinate systems for grinding ZC1 worm. **b** Relative position between grinding wheel and ZC1 worm

is $\left|\overrightarrow{O_d O_a}\right|$, and $\left|\overrightarrow{O_d O_a}\right| = p\varphi$. Herein the symbol p is the spiral parameter of the worm.

Without loss of generality, it can be assumed that the angle velocity vector $\vec{\omega}$ of the worm rough rotation around its axis $\vec{k}_{o1}$ is $1 \text{ rad}/\text{s}$. Meanwhile, the angle velocity vector $\vec{\omega}_d$ of the grinding wheel is 0. The relative angle velocity vector $\vec{\omega}_{d1}$ between the grinding wheel and the worm rough can thus be obtained as follows

$$(\vec{\omega}_{d1})_a = (\vec{\omega}_d)_a - (\vec{\omega})_a = (\vec{\omega}_d)_a - R\left[\vec{i}_a, -\gamma\right](\vec{\omega})_{o1} = -\sin\gamma\,\vec{j}_a - \cos\gamma\,\vec{k}_a. \tag{24.6}$$

The relative velocity vector between the grinding wheel and the worm rough can also be acquired by means of the meshing theory for gearing [2] as follows

$$\left(\vec{V}_{d1}\right)_a = (\vec{\omega}_{d1})_a \times (\vec{r}_a)_a - (\vec{\omega})_a \times \left(\overrightarrow{O_1 O_a}\right)_a + \frac{d\left(\overrightarrow{O_1 O_a}\right)_a}{d\varphi} = V_{d1}^{(x)}\vec{i}_a + V_{d1}^{(y)}\vec{j}_a + V_{d1}^{(z)}\vec{k}_a, \tag{24.7}$$

where $\left(\overrightarrow{O_1 O_a}\right)_a = \left(\overrightarrow{O_1 O_d}\right)_a + \left(\overrightarrow{O_d O_a}\right)_a = a_d\vec{i}_a - p\varphi\sin\gamma\,\vec{j}_a - p\varphi\cos\gamma\,\vec{k}_a$, $V_{d1}^{(x)} = y_a\cos\gamma - z_a\sin\gamma$, $V_{d1}^{(y)} = -(x_a + a_d)\cos\gamma - p\sin\gamma$, and $V_{d1}^{(z)} = (x_a + a_d)\sin\gamma - p\cos\gamma$.

According to the meshing theory for gearing [2], the meshing function, Φ_d, of the cutting meshing of the ZC1 worm can be attained as

$$\Phi_d = \Phi_d(\theta, \phi) = (\vec{n})_a \cdot \left(\vec{V}_{d1}\right)_a = A_d\sin\phi - B_d\cos\phi, \tag{24.8}$$

where $A_d = A_d(\theta) = (p\sin\gamma + a_d\cos\gamma)\sin\theta - \rho\cos\alpha_n\sin\gamma\cos\theta$ and $B_d = B_d(\theta) = \left(R_g - \rho\sin\alpha_n\right)\sin\gamma\cos\theta + a_d\sin\gamma - p\cos\gamma$.

Letting $\Phi_d = 0$ in Eq. (24.8), in σ_{o1}, the equations for the surface of action during grinding the ZC1 worm rough and its unit normal vector can be respectively obtained by coordinate transformation as below

$$(\vec{r}_1)_{o1} = R\left[\vec{i}_{o1}, \gamma\right](\vec{r}_a)_a + \left(\overrightarrow{O_1 O_a}\right)_{o1} = x_{o1}\vec{i}_{o1} + y_{o1}\vec{j}_{o1} + z_{o1}\vec{k}_{o1}, \quad \Phi_d(\theta, \phi) = 0, \tag{24.9}$$

$$(\vec{n})_{o1} = R\left[\vec{i}_{o1}, \gamma\right](\vec{n})_a = n_x\vec{i}_{o1} + n_y\vec{j}_{o1} + n_z\vec{k}_{o1}, \quad \Phi_d(\theta, \phi) = 0, \tag{24.10}$$

where $x_{o1} = x_a + a_d$, $y_{o1} = y_a\cos\gamma - z_a\sin\gamma$, $z_{o1} = y_a\sin\gamma + z_a\cos\gamma - p\varphi$, $n_x = -\cos\theta\sin\phi$, $n_y = -\cos\gamma\sin\theta\sin\phi + \sin\gamma\cos\phi$, and $n_z = -\sin\gamma\sin\theta\sin\phi - \cos\gamma\cos\phi$.

Via coordinate transformation, from Eq. (24.9), the equation for the ZC1 worm helicoid Σ_1 can be figured out in σ_1 as

$$(\vec{r}_1)_1 = R\left[\vec{k}_1, -\varphi\right](\vec{r}_1)_{o1} = x_1\vec{i}_1 + y_1\vec{j}_1 + z_{o1}\vec{k}_1, \quad \Phi_d(\theta, \phi)v0, \quad (24.11)$$

where $x_1 = x_{o1}\cos\varphi + y_{o1}\sin\varphi$ and $y_1 = -x_{o1}\sin\varphi + y_{o1}\cos\varphi$.

By definition [7], from Eq. (24.8), it can be figured out the meshing limit function, $\Phi_{d\varphi}$, of the cutting meshing of the ZC1 worm as

$$\Phi_{d\varphi} = \frac{\partial \Phi_d}{\partial \varphi} = 0. \quad (24.12)$$

Based on the meshing theory for gearing [2], the normal vector, $\overrightarrow{N}_d$, of the instantaneous contact line between the grinding wheel and the ZC1 worm rough can be obtained from Eqs. (24.3), (24.4), (24.6) and (24.7) as follows

$$\left(\vec{N}_d\right)_a = \lambda_d\left(\vec{g}_1\right)_a + \mu_d\left(\vec{g}_2\right)_a, \quad (24.13)$$

where $\lambda_d = k_1\left(\vec{V}_{d1}\right)_a \cdot \left(\vec{g}_1\right)_a + (\vec{\omega}_{d1})_a \cdot \left(\vec{g}_2\right)_a$ and $\mu_d = k_2\left(\vec{V}_{d1}\right)_a \cdot \left(\vec{g}_2\right)_a + \sin\gamma\cos\theta$.

Moreover, the results of the preceding dot products $\left(\vec{V}_{d1}\right)_a \cdot \left(\vec{g}_1\right)_a$, $\left(\vec{V}_{d1}\right)_a \cdot \left(\vec{g}_2\right)_a$, and $(\vec{\omega}_{d1})_a \cdot \left(\vec{g}_2\right)_a$ are

$$\left(\vec{V}_{d1}\right)_a \cdot \left(\vec{g}_1\right)_a = -V_{d1}^{(x)}\sin\theta + V_{d1}^{(y)}\cos\theta, \quad \left(\vec{V}_{d1}\right)_a \cdot \left(\vec{g}_2\right)_a = V_{d1}^{(x)}\cos\theta\cos\phi + V_{d1}^{(y)}\sin\theta\cos\phi - V_{d1}^{(z)}\sin\phi, \quad (24.14)$$

$$\left(\vec{\omega}_{d1}\right)_a \cdot \left(\vec{g}_2\right)_a = -\sin\gamma\sin\theta\cos\phi + \cos\gamma\sin\phi. \quad (24.15)$$

According to Eqs. (24.3), (24.4), (24.7) and (24.13), the curvature interference limit function, Ψ_d, of the cutting meshing of the ZC1 worm can be determined as

$$\Psi_d = \Psi_d(\theta, \phi) = \lambda_d\left(\vec{V}_{d1}\right)_a \cdot \left(\vec{g}_1\right)_a + \mu_d\left(\vec{V}_{d1}\right)_a \cdot \left(\vec{g}_2\right)_a. \quad (24.16)$$

In accordance with the meshing theory for gearing [2], the curvature parameters of Σ_1, including the normal curvatures, $k_\xi^{(1)}$ and $k_\eta^{(1)}$, of Σ_1 along $\vec{g}_1$ and $\vec{g}_2$, and its geodesic torsion $\tau_\xi^{(1)}$ along $\vec{g}_1$ can be respectively acquired as follows

$$k_\xi = k_1 - \frac{\lambda_d^2}{\Psi_d}, \quad k_\eta = k_2 - \frac{\mu_d^2}{\Psi_d}, \quad \tau_\xi = -\frac{\lambda_d\mu_d}{\Psi_d}. \quad (24.17)$$

24.2.3 Meshing of ZC1 Worm Pair

The cutting meshing of the worm gear is same as the working meshing of the ZC1 worm pair, so that we will not distinguish these two courses in this paper.

As shown in Fig. 24.4, the unit vector $\vec{k}_{o2}$ of a static coordinate system $\sigma_{o2}\left\{O_2; \vec{i}_{o2}, \vec{j}_{o2}, \vec{k}_{o2}\right\}$ is along the center line of the worm gear. The positions and roles of the static coordinate system $\sigma_{o1}\left\{O_1; \vec{i}_{o1}, \vec{j}_{o1}, \vec{k}_{o1}\right\}$ and the rotating coordinate system $\sigma_1\left\{O_1; \vec{i}_1, \vec{j}_1, \vec{k}_1\right\}$ are same as those of the cutting meshing of the ZC1 worm. The two unit vectors $\vec{i}_{o1}$ and $\vec{i}_{o2}$ are all along the common perpendicular of the center lines of the worm and the worm gear. The shaft angle Σ of the worm pair is equal to $180° - \angle(\omega_1, \omega_2)$. A rotating coordinate system $\sigma_2\left\{O_2; \vec{i}_2, \vec{j}_2, \vec{k}_{22}\right\}$ is used to indicate the current position of the worm gear. At current position, the rotation angles of the worm and worm gear are φ_1 and φ_2, respectively, and $\varphi_2 = \varphi_1/i_{12}$. Here, the symbol i_{12} is the drive ratio of the worm pair.

When the ZC1 worm rotates around $\vec{k}_{o1}$, its helicoid can form a surface family, $\{\Sigma_1\}$, with one parameter in σ_{o1}. The equations for the surface family $\{\Sigma_1\}$ and its unit normal vector can be respectively attained from Eqs. (24.11) and (24.10) as

$$(\vec{r}_1^*)_{o1} = R\left[\vec{k}_{o1}, \varphi_1\right](\vec{r}_1)_1 = R\left[\vec{k}_{o1}, \varphi_1 - \varphi\right](\vec{r}_1)_{o1} = x_{o1}^*\vec{i}_{o1} + y_{o1}^*\vec{j}_{o1} + z_{o1}\vec{k}_{o1}, \ \Phi_d(\theta, \phi) = 0, \tag{24.18}$$

$$(\vec{n}^*)_{o1} = R\left[\vec{k}_{o1}, \varphi_1 - \varphi\right](\vec{n})_{o1}, \ \Phi_d(\theta, \phi) = 0, \tag{24.19}$$

where $x_{o1}^* = x_{o1}\cos(\varphi_1-\varphi) - y_{o1}\sin(\varphi_1-\varphi)$ and $y_{o1}^* = x_{o1}\sin(\varphi_1-\varphi)+y_{o1}\cos(\varphi_1-\varphi)$.

Without loss of generality, the angular velocity vector $\vec{\omega}_1$ of the worm can also be assumed as $|\vec{\omega}_1| = 1\text{rad}/\text{s}$ during the meshing of the worm pair. Accordingly, the angular velocity of the worm gear is $|\vec{\omega}_2| = 1/i_{12}\text{rad}/\text{s}$. As a result, the relative angular velocity vector and the relative velocity vector of the ZC1 worm pair can be respectively figured out in σ_{o1} as follows

Fig. 24.4 Coordinate systems for meshing of ZC1 worm pair

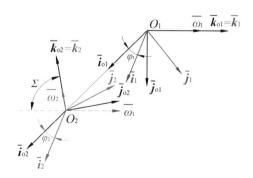

$$(\vec{\omega}_{12})_{o1} = (\vec{\omega}_1)_{o1} - (\vec{\omega}_2)_{o1} = \frac{1}{i_{12}}\left[\sin \Sigma \vec{j}_{o1} + (i_{12} + \cos \Sigma)\vec{k}_{o1}\right], \qquad (24.20)$$

$$\left(\vec{V}_{12}\right)_{o1} = (\vec{\omega}_{12})_{o1} \times \left(\vec{r}_1^*\right)_{o1} - (\vec{\omega}_2)_{o1} \times \left(\overrightarrow{O_2O_1}\right)_{o1} + \frac{d}{d\varphi}\left(\overrightarrow{O_2O_1}\right)_{o1} = \frac{1}{i_{12}}\left(V_{12}^{(x)}\vec{i}_{o1} + V_{12}^{(y)}\vec{j}_{o1} + V_{12}^{(z)}\vec{k}_{o1}\right),$$
$$(24.21)$$

where $V_{12}^{(x)} = z_{o1}\sin\Sigma - y_{o1}^*(i_{12} + \cos\Sigma)$, $V_{12}^{(y)} = x_{o1}^*(i_{12} + \cos\Sigma) - a\cos\Sigma$, and $V_{12}^{(z)} = (a - x_{o1}^*)\sin\Sigma$.

According to the meshing theory for gearing [2], the meshing function of the cutting meshing of the ZC1 worm gear can be attained as

$$\Phi = \left(\vec{n}^*\right)_{o1} \cdot \left(\vec{V}_{12}\right)_{o1} = \frac{1}{i_{12}}[A\sin(\varphi_1 - \varphi) + B\cos(\varphi_1 - \varphi) + C], \qquad (24.22)$$

where $A = n_y p\varphi \sin\Sigma + \frac{b_A}{\cos\theta}$, $B = -n_x p\varphi \sin\Sigma + b_B$, $C = n_z[a\sin\Sigma - p(i_{12} + \cos\Sigma)]$, $b_A = \frac{\sin\theta}{\sin\gamma}(a_d n_y + pn_z)\sin\Sigma - n_x a\cos\Sigma\cos\theta$, and $b_B = \left(-pn_z\cot\gamma + \frac{a_d}{\sin\gamma}\sin\theta\sin\phi\right)\sin\Sigma - n_y a\cos\Sigma$.

Based on Eqs. (24.18) and (24.22), it is easy to obtain the equation of the worm gear tooth surface, Σ_2, in σ_2 as

$$\begin{cases} (\vec{r}_2)_2 = R\left[\vec{k}_2, -\frac{\varphi_1}{i_{12}}\right]\left[R\left[\vec{i}_{o2}, \pi + \Sigma\right](\vec{r}_1^*)_{o1} + \left(\overrightarrow{O_2O_1}\right)_{o2}\right] = x_2\vec{i}_2 + y_2\vec{j}_2 - z_2\vec{k}_2, \\ \Phi_d(\theta, \phi) = 0 \\ \Phi(\theta, \phi, \varphi, \varphi_1) = 0 \end{cases}$$
$$(24.23)$$

where $x_2 = \left(x_{o1}^* - a\right)\cos\frac{\varphi_1}{i_{12}} + \left(-y_{o1}^*\cos\Sigma + z_{o1}\sin\Sigma\right)\sin\frac{\varphi_1}{i_{12}}$, $y_2 = -\left(x_{o1}^* - a\right)\sin\frac{\varphi_1}{i_{12}} + \left(-y_{o1}^*\cos\Sigma + z_{o1}\sin\Sigma\right)\cos\frac{\varphi_1}{i_{12}}$, and $z_2 = y_{o1}^*\sin\Sigma + z_{o4}\cos\Sigma$.

From Eq. (24.23), it can be figured out the meshing limit function, $\Phi_{\varphi 1}$, of the cutting meshing of the ZC1 worm gear according to its definition [2] as

$$\Phi_{\varphi 1} = \frac{\partial\Phi}{\partial\varphi_1} = \frac{1}{i_{12}}[A\cos(\varphi_1 - \varphi) - B\sin(\varphi_1 - \varphi)]. \qquad (24.24)$$

After coordinate transformation from σ_a to σ_{o1}, the two unit vectors $\vec{g}_1$ and $\vec{g}_2$ can be respectively represented in σ_{o1} as follows

$$\left(\vec{\alpha}_\xi\right)_{o1} = R\left[\vec{k}_{o1}, \varphi_1 - \varphi\right]R\left[\vec{i}_{o1}, \gamma\right](\vec{g}_1)_a = \alpha_{\xi x}\vec{i}_{o1} + \alpha_{\xi y}\vec{j}_{o1} + \sin\gamma\cos\theta\vec{k}_{o1},$$
$$(24.25)$$

$$\left(\vec{\alpha}_\eta\right)_{o1} = R\left[\vec{k}_{o1}, \varphi_1 - \varphi\right]R\left[\vec{i}_{o1}, \gamma\right](\vec{g}_2)_a = \alpha_{\eta x}\vec{i}_{o1} + \alpha_{\eta y}\vec{j}_{o1} + \alpha_{\eta z}\vec{k}_{o1}, \qquad (24.26)$$

where $\alpha_{\xi x} = -\sin\theta\cos(\varphi_1 - \varphi) - \cos\gamma\cos\theta\sin(\varphi_1 - \varphi)$, $\alpha_{\xi y} = -\sin\theta\sin(\varphi_1 - \varphi) + \cos\gamma\cos\theta\cos(\varphi_1 - \varphi)$, $\alpha_{\eta x} = \cos\theta\cos\phi\cos(\varphi_1 - \varphi) -$

$(\cos\gamma \sin\theta \cos\phi + \sin\gamma \sin\phi)\sin(\varphi_1 - \varphi),\alpha_{ny} \quad = \quad \cos\theta \cos\phi \sin(\varphi_1 - \varphi) +$
$(\cos\gamma \sin\theta \cos\phi + \sin\gamma \sin\phi)\cos(\varphi_1 - \varphi),$ and $\alpha_{nz} \quad = \quad \sin\gamma \sin\theta \cos\phi -$
$\cos\gamma \sin\phi.$

By means of the method proposed in literature [10], the normal vector, $\vec{N}_{12}$, of the instantaneous contact line between the ZC1 worm and the worm gear can be obtained from Eqs. (24.17), (24.20), (24.21), (24.25), and (24.26) as follows

$$(\vec{N})_{o1} = N_\xi (\vec{\alpha}_\xi)_{o1} + N_\eta (\vec{\alpha}_\eta)_{o1}, \tag{24.27}$$

wherein

$$N_\xi = k_\xi (\vec{V}_{12})_{o1} \cdot (\vec{\alpha}_\xi)_{o1} + \tau_\xi (\vec{V}_{12})_{o1} \cdot (\vec{\alpha}_\eta)_{o1} + (\vec{\omega}_{12})_{o1} \cdot (\vec{\alpha}_\eta)_{o1}, \tag{24.28}$$

$$N_\eta = \tau_\xi (\vec{V}_{12})_{o1} \cdot (\vec{\alpha}_\xi)_{o1} + k_\eta (\vec{V}_{12})_{o1} \cdot (\vec{\alpha}_\eta)_{o1} - (\vec{\omega}_{12})_{o1} \cdot (\vec{\alpha}_\xi)_{o1}. \tag{24.29}$$

According to Eqs. (24.21), (24.24), (24.25), (24.26), (24.28) and (24.29), the curvature interference limit function, Ψ, of the cutting meshing of the ZC1 worm gear can be determined as

$$\Psi = N_\xi (\vec{V}_{12})_{o1} \cdot (\vec{\alpha}_\xi)_{o1} + N_\eta (\vec{V}_{12})_{o1} \cdot (\vec{\alpha}_\eta)_{o1} + \Phi_{f\,\varphi 1}. \tag{24.30}$$

24.3 Calculating Method of Curvature Interference Limit Line

For the propose of determining a typical point P_ψ located on the curvature interference limit line, it can be assumed that the symbol L_ψ denotes the axial position parameter of this typical point P_ψ along the axial line of the ZC1 worm gear. That is to say, $z_2 = -L_\psi$. As a result, a system of nonlinear equations with four unknowns ϕ, θ, φ, and φ_1 to determine the typical point P_ψ can be established from Eqs. (24.16), (24.23) and (24.30) as follows

$$\begin{cases} y_{o1}^* \sin\Sigma + z_{o1} \cos\Sigma = -L_\psi \\ \Phi_d(\theta, \phi) = 0 \\ \Phi(\theta, \phi, \varphi, \varphi_1) = [A\sin(\varphi_1 - \varphi) + B\cos(\varphi_1 - \varphi) + C]/i_{12} = 0 \\ \Psi(\theta, \phi, \varphi, \varphi_1) = 0 \end{cases} \tag{24.31}$$

Solving System (24.31) directly is very complicated because the judgement of its solution existence and the supply of its iterative initial value are extremely difficult to implement. Therefore, we should eliminate some variables in System (24.31) in order to solve it rapidly and accurately. The specific steps are:

From the second expression of System (24.31), the variable ϕ can be expressed by the variable θ as below

$$\phi(\theta) = \arctan\left(\frac{B_d}{A_d}\right). \tag{24.32}$$

Using Eq. (24.32) to substitute all the variable ϕ in System (24.31), the variable ϕ will be eliminated in System (24.31).

By combining the first and third expressions of System (24.31), it can be obtained a binary nonlinear equation group with the unknowns $\sin(\varphi_1 - \varphi)$ and $\cos(\varphi_1 - \varphi)$ as follows

$$\begin{cases} x_{o1} \sin(\varphi_1 - \varphi) + y_{o1} \cos(\varphi_1 - \varphi) = L_\psi^* \\ A \sin(\varphi_1 - \varphi) + B \cos(\varphi_1 - \varphi) = -C \end{cases}, \tag{24.33}$$

where $L_\psi^* = -z_{o1} \cot \sum - L_\psi \csc \sum$.

The solutions of System (24.33) are

$$\sin(\varphi_1 - \varphi) = \frac{E_S}{D}, \quad \cos(\varphi_1 - \varphi) = \frac{E_C}{D}, \tag{24.34}$$

where $D = x_{o1} B \cos\theta - y_{o1}\left(n_y p \sin \sum \varphi \cos\theta + b_A\right)$, $E_S = \left(L_\psi^* B + y_{o1} C\right) \cos\theta$, and $E_C = -x_{o1} C \cos\theta - L_\psi^*\left(n_y p \sin \sum \varphi \cos\theta + b_A\right)$. Moreover, the variables in the components D, E_S, and E_C are θ and φ.

Substituting Eq. (24.34) into the components $\alpha_{\xi x}, \alpha_{\xi y}, \alpha_{\eta x}$, and $\alpha_{\eta y}$ of Eqs. (24.25) and (24.26) leads up to

$$\alpha_{\xi x}(\theta, \varphi) = \frac{\hat{\alpha}_{\xi x}}{D}, \quad \alpha_{\xi y}(\theta, \varphi) = \frac{\hat{\alpha}_{\xi y}}{D}, \quad \alpha_{\eta x}(\theta, \varphi) = \frac{\hat{\alpha}_{\eta x}}{D}, \quad \alpha_{\eta y}(\theta, \varphi) = \frac{\hat{\alpha}_{\eta y}}{D}, \tag{24.35}$$

where $\hat{\alpha}_{\xi x} = -E_C \sin\theta - E_S \cos\gamma \cos\theta$, $\hat{\alpha}_{\xi y} = -E_S \sin\theta + E_C \cos\gamma \cos\theta$, $\hat{\alpha}_{\eta x} = E_C \cos\phi \cos\theta - E_S(\cos\gamma \cos\phi \sin\theta + \sin\gamma \sin\phi)$, and $\hat{\alpha}_{\eta y} = E_S \cos\phi \cos\theta + E_C(\cos\gamma \cos\phi \sin\theta + \sin\gamma \sin\phi)$.

Substituting Eq. (24.34) into the components x_{o1}^* and y_{o1}^* of Eq. (24.18) and then substituting the obtained results into the components $V_{12}^{(x)}$, $V_{12}^{(y)}$, and $V_{12}^{(z)}$ of Eq. (24.21) result in

$$V_{12}^{(x)}(\theta, \varphi) = \frac{\hat{V}_{12}^{(x)}}{D}, \quad V_{12}^{(y)}(\theta, \varphi) = \frac{\hat{V}_{12}^{(y)}}{D}, \quad V_{12}^{(z)}(\theta, \varphi) = \frac{\hat{V}_{12}^{(z)}}{D}, \tag{24.36}$$

where $\hat{V}_{12}^{(x)} = z_{o1} D \sin \sum - (x_{o1} E_S + y_{o1} E_C)\left(i_{12} + \cos \sum\right)$, $\hat{V}_{12}^{(y)} = (x_{o1} E_C - y_{o1} E_S)\left(i_{12} + \cos \sum\right) - aD \cos \sum$, and $\hat{V}_{12}^{(z)} = (aD - x_{o1} E_C + y_{o1} E_S) \sin \sum$.

Based on Eqs. (24.20), (24.35), and (24.36), the dot products $\left(\vec{V}_{12}\right)_{\text{o1}} \cdot \left(\vec{\alpha}_{\xi}\right)_{\text{o1}}$, $\left(\vec{V}_{12}\right)_{\text{o1}} \cdot \left(\vec{\alpha}_{\eta}\right)_{\text{o1}}$, $(\vec{\omega}_{12})_{\text{o1}} \cdot \left(\vec{\alpha}_{\xi}\right)_{\text{o1}}$, and $(\vec{\omega}_{12})_{\text{o1}} \cdot \left(\vec{\alpha}_{\eta}\right)_{\text{o1}}$ in Eqs. (24.28) and (24.29) can be respectively expressed by the two variables θ and φ as

$$\left(\vec{V}_{12}\right)_{\text{o1}} \cdot \left(\vec{\alpha}_{\xi}\right)_{\text{o1}} = \frac{V_{\xi}}{i_{12}D^2}, \ \left(\vec{V}_{12}\right)_{\text{o1}} \cdot \left(\vec{\alpha}_{\eta}\right)_{\text{o1}} = \frac{V_{\eta}}{i_{12}D^2}, \ (\vec{\omega}_{12})_{\text{o1}} \cdot \left(\vec{\alpha}_{\xi}\right)_{\text{o1}} = \frac{\omega_{\xi}}{i_{12}D}, \ (\vec{\omega}_{12})_{\text{o1}} \cdot \left(\vec{\alpha}_{\eta}\right)_{\text{o1}} = \frac{\omega_{\eta}}{i_{12}D}.$$

$$(24.37)$$

where $V_{\xi} = \hat{V}_{12}^{(x)} \hat{\alpha}_{\xi x} + \hat{V}_{12}^{(y)} \hat{\alpha}_{\xi y} + \hat{V}_{12}^{(z)} D \sin \gamma \cos \theta$, $V_{\eta} = \hat{V}_{12}^{(x)} \hat{\alpha}_{\xi x} + \hat{V}_{12}^{(y)} \hat{\alpha}_{\xi y} + \hat{V}_{12}^{(z)} D\alpha_{\eta z}, \omega_{\xi} = \hat{\alpha}_{\xi y} \sin \sum + (i_{12} + \cos \sum) D \sin \gamma \cos \theta$, and $\omega_{\eta} = \hat{\alpha}_{\eta y} \sin \sum + (i_{12} + \cos \sum) D\alpha_{\eta z}$.

Then substituting Eqs. (24.17) and (24.37) into Eqs. (24.28) and (24.29), the components N_{ξ} and N_{η} can also be expressed by the two variables θ and φ as follows

$$N_{\xi}(\theta, \varphi) = \frac{\hat{N}_{\xi}}{i_{12} \Psi_{\text{d}} D^2}, \ N_{\eta}(\theta, \varphi) = \frac{\hat{N}_{\eta}}{i_{12} \Psi_{\text{d}} D^2}, \tag{24.38}$$

where $\hat{N}_{\xi} = V_{\xi}\left(k_1 \Psi_{\text{d}} - \lambda_{\text{d}}^2\right) - V_{\eta} \lambda_{\text{d}} \mu_{\text{d}} + \omega_{\eta} \Psi_{\text{d}} D$, and $\hat{N}_{\eta} = -V_{\xi} \lambda_{\text{d}} \mu_{\text{d}} + V_{\eta}\left(k_2 \Psi_{\text{d}} - \mu_{\text{d}}^2\right) - \omega_{\xi} \Psi_{\text{d}} D$.

Finally, substituting Eq. (24.34) into Eq. (24.24), and then substituting the acquired outcome and Eqs. (24.37–24.38) into Eq. (24.30) give rise to

$$\Psi = \frac{\left(\hat{N}_{\xi} V_{\xi} + \hat{N}_{\eta} V_{\eta}\right) \cos \theta + i_{12} D^3 \Psi_{\text{d}} \hat{\Phi}_{\varphi 1}}{i_{12}^2 D^4 \Psi_{\text{d}} \cos \theta} = \frac{f_1(\theta, \varphi)}{i_{12}^2 D^4 \Psi_{\text{d}} \cos \theta} = 0. \tag{24.39}$$

where $\hat{\Phi}_{\varphi 1} = E_C \left(n_{\text{y}} p \sin \sum_{12} \varphi \cos \theta + b_A\right) - E_S B \cos \theta$.

In Eq. (24.39), $f_1(\theta, \varphi) = 0$ is a binary nonlinear equation group with the two unknowns θ and φ. On the other hand, another binary nonlinear equation group $f_2(\theta, \varphi) = 0$ with the unknowns θ and φ can also be determined from the two trigonometric equations of Eq. (24.34). As a result, the system of nonlinear equations with the two unknowns θ and φ for determining the typical point P_{Ψ} can be obtained as

$$\begin{cases} f_1(\theta, \varphi) = \left(\hat{N}_{\xi} V_{\xi} + \hat{N}_{\eta} V_{\eta}\right) \cos \theta + i_{12} D^3 \Psi_{\text{d}} \hat{\Phi}_{\varphi 1} = 0 \\ f_2(\theta, \varphi) = E_S^2 + E_C^2 - D^2 = 0 \end{cases}. \tag{24.40}$$

Essentially, System (24.40) is the equivalent form of System (24.31), i.e. the solutions of System (24.40) are also the solutions of System (24.31). According to the geometric construction approach proposed in literature [12], the curves of $f_1(\theta, \varphi) = 0$ and $f_2(\theta, \varphi) = 0$ can be drawn in a figure. If the intersection points between these two curves exist over the given solving intervals of θ and φ, the solutions of System (24.40), i.e. the solutions of System (24.31), will exist over its given solved domain. Not only that, the values of θ and φ at preceding intersection points can also be used as the initial values to solve System (24.40) iteratively. Consequently, the existence issue of the solution and the supply of the iterative initial values for System (24.40) are all settled.

After solving System (24.40) iteratively, the values of the variables θ and φ can be obtained. Based on this, the values of the variables ϕ and φ_1 can be respectively obtained from Eqs. (24.32) and (24.34). Then the typical point P_ψ can be determined. After determining a series of typical points by changing the value of L_ψ, the curvature interference limit line will be determined by means of the interpolation method.

24.4 Numerical Example

24.4.1 Main Parameters of ZC1 Worm Pair

The main design parameters of the ZC1 worm pair discussed in this paper are supplied in Table 24.1. The main technological parameters of the ZC1 worm pair are listed in Table 24.2.

Table 24.1 Design parameters of ZC1 worm pair

Center distance of worm pair	a/mm	200
Drive ratio of worm pair	i_{12}	15.5
Number of worm thread	Z_1	2
Module of worm pair	m/mm	10
Shaft angle of worm pair	$\Sigma\,/^\circ$	90
Helix parameter	$p = mZ_1/2/\text{mm}$	10
Tooth number of worm gear	$Z_2 = i_{12}Z_1$	31
Modification coefficient of worm gear	x_2	0.7
Thread length of worm	$L_w \approx 2.5\,m\sqrt{Z_2 + 1}/\text{mm}$	141
Diametral quotient of worm	$q = 2a/m - Z_2 - 2x_2$	7.6
Reference radius of worm	$r_1 = qm/2/\text{mm}$	38
Lead angle of worm	$\gamma = \arctan Z_1 / q\,/^\circ$	14.7436
Addendum of worm	$h_{a1} = m/\text{mm}$	10
Dedendum of worm	$h_{f1} = 1.16m/\text{mm}$	11.6
Addendum of worm gear	$h_{a2} = (1 + x_2)m/\text{mm}$	17
Dedendum of worm gear	$h_{f2} = (1.16 - x_2)m/\text{mm}$	4.6
Reference radius of worm gear	$r_2 = mZ_2/2\,/\text{mm}$	155
Tooth width of worm gear	$b_2 \approx 2m(0.5 + \sqrt{q + 1})/\text{mm}$	69

Table 24.2 Technological parameters of ZC1 worm pair

Nominal radius of grinding wheel	R_g/mm	120
Radius of circular arc profile	$\rho = 5.5$ m/mm	55
Shape angle of the grinding wheel	α_n/$^\circ$	22
Operating center distance during grinding ZC1 worm	$a_d = R_g + r_1$/mm	158

Table 24.3 Computational results of curvature interference limit line

Typical point	①	②	③	④	⑤	⑥	⑦
L_ψ /mm	-45	-30	-15	0	15	30	45
ϕ/$^\circ$	6.3018	8.1912	10.0697	9.7573	8.0564	9.8797	5.8220
θ/$^\circ$	163.2482	170.3129	173.5157	173.0954	169.9983	164.8477	158.8694
φ/$^\circ$	92.2356	114.8290	164.6947	200.0864	232.7520	259.9235	274.1619
φ_1/$^\circ$	110.6309	130.3963	169.2347	546.8011	558.9538	566.0629	563.3797
z_{01}/mm	-4.8128	-12.119	-22.5579	-28.4919	-32.5408	-34.8111	-34.6207
$\sqrt{x_{01}^2 + y_{01}^2}$/mm	63.7435	54.8594	50.9095	51.4476	55.2391	61.6362	69.8633
z_2/mm	-45	-30	-15	0	15	30	45
$\sqrt{x_2^2 + y_2^2}$/mm	154.928	154.546	153.0223	151.2601	150.399	150.2458	150.5932

24.4.2 *Computational Results of Curvature Interference Limit Line*

In this paper, in order to determine the curvature interference limit line, seven typical points are selected on the curvature interference limit line in accordance with the position of the worm gear in the coordinate system σ_2. The values of L_ψ at these typical points are listed in Table 24.3.

The typical point with $L_\psi = -30$ mm is used as an instance to explain the determination process of a typical point. The curves of $f_1(\theta, \varphi) = 0$ and $f_2(\theta, \varphi) = 0$ at this typical point are drawn in Fig. 24.5. The angle θ is the abscissa of Fig. 24.5 and the angle φ is its ordinate. In addition, for the propose of avoiding the appearance of all the meaningless curvature interference limit lines, the solving intervals of θ and φ in Fig. 24.5 are reasonably determined as $[150^\circ, 180^\circ]$ and $[-300^\circ, 300^\circ]$ based on the computation of the conjugated zone of the ZC1 worm pair [11].

Obviously, there is one intersection point between these two curves over the whole solved region as shown in Fig. 24.5. This means that System (24.40) has a solution over the given solved region and there is one curvature interference limit line on the worm gear tooth surface Σ_2. Based on this figure, the point $(\theta, \varphi) = (2.9725, 2)$, whose location is close to the foregoing intersection point, can be utilized as the initial guess to solve System (24.40) iteratively. Then the typical point aforesaid above can be ascertained.

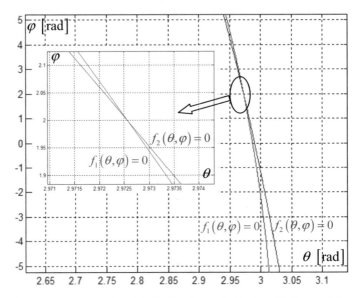

Fig. 24.5 Curves of $f_1(\theta, \varphi) = 0$ and $f_2(\theta, \varphi) = 0$ at typical point with $L_\psi = -30$ mm

Moreover, all the typical points on the curvature interference limit line are also similarly determined by modifying the value of L_ψ. The obtained computational results for all the typical points are also supplied in Table 24.3.

For the sake of revealing the position relation of the curvature interference limit line with regard to the worm pair tooth surfaces. The projection drawings of the curvature interference limit line and its conjugate line are drawn in the axial sections of the worm gear and worm in Figs. 24.6 and 24.7, respectively.

Fig. 24.6 Curvature interference limit line in axial section of worm gear

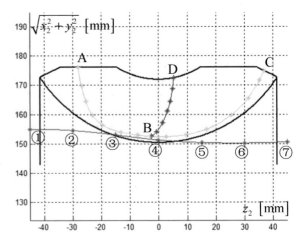

Fig. 24.7 Conjugate line of
curvature interference limit
line in axial section of worm

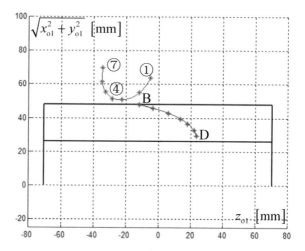

In Fig. 24.6, the blue line BD is the conjugate line of the meshing limit line
of the worm pair on the worm gear tooth surface [11, 12]. The cyan line ABC is
the conjugate line of the worm addendum on the worm gear tooth surface, i.e. the
boundary line of the conjugate zone on the worm gear tooth surface. The red line is
the curvature interference limit line of the worm gear, and the points ①–⑦ are the
typical points of the curvature interference limit line. In Fig. 24.7, the blue line BD
is the meshing limit line of the worm pair on the worm helicoid. The red line is the
conjugate line of the curvature interference limit line of the worm gear on the worm
helicoid.

These two figures show that the curvature interference limit line and its conjugate
line are all located at the outside of the conjugate zone of the worm pair. Moreover,
from Eq. (24.30), the value of Ψ at point B can be obtained as -0.0695. This means
that the conjugate zone of the worm pair is located on the non-curvature-interference
side of the curvature interference limit line because the direction of the normal vector
of the grinding wheel is from the space to its inside, i.e. the undercutting of the worm
gear does not take place.

On the other hand, a lot of computing results indicate that the curvature interference limit line on the worm gear tooth surface generally does not enter into the
conjugate zone of the worm pair. Nevertheless, if the modification coefficient of
worm gear and the radius of circular arc profile of the grinding wheel are less and
the shaft angle of the worm pair is larger, the curvature interference limit line on the
worm gear tooth surface may be closer to the intersection point of the conjugate line
of the worm addendum and the conjugate line of the meshing limit line of the worm
pair. Accordingly, the modification coefficient of worm gear and the radius of circular arc profile of the grinding wheel need to be deeply focused on when designing a
ZC1 worm pair.

24.5 Conclusions

The theory of the curvature interference characteristic of a ZC1 worm gear is well established on the basis of the meshing theory for gearing. The tooth surface equations of the worm pair are well obtained based on the forming principle of the worm pair. The meshing function, the curvature interference limit function of the ZC1 worm pair and so on, are also obtained.

By means of the elimination and the geometric construction, a method to judge the existence of the curvature interference limit line is put forward. Then the curvature interference limit line of ZC1 worm gear is determined by solving a system of nonlinear equations.

The numerical results show that there is one significative curvature interference limit line on the worm gear tooth surface, which generally does not enter into the conjugate area of the worm pair. The conjugate area of the worm pair on the worm gear tooth surface is located on the non-curvature-interference side of the curvature interference limit line. Consequently, the undercutting of the worm gear seldom takes place.

When the modification coefficient of worm gear and the radius of circular arc profile of the grinding wheel are less and the shaft angle of the worm pair is larger, the curvature interference limit line on the worm gear tooth surface will move to the intersection point of the conjugate line of the worm addendum and the conjugate line of the meshing limit line of the worm pair, namely, this area has the greatest potential hazards to be undercut.

Acknowledgements This study was funded by the Open Fund of the Key Laboratory for Metallurgical Equipment and Control of Ministry of Education in Wuhan University of Science and Technology (2018B05) and the National Natural Science Foundation of China (51475083).

References

1. Crosher, W.: Design and Application of the Worm Gear. ASME Press, New York (2002)
2. Dong, X.: Foundation of Meshing Theory for Gear Drives. China Machine Press, Beijing (1989). (in Chinese)
3. Dudas, I.: The Theory and Practice of Worm Gear Drives. Penton Press, London (2000)
4. Hu, L.: Principle and Application of Space Meshing. Coal Industry Press, Beijing (1986). (in Chinese)
5. Litvin, F.L.: Meshing Principle for Gear Drives. Shanghai Science and Technology Press, Shanghai (1964). (in Chinese)
6. Litvin, F.L.: Development of Gear Technology and Theory of Gearing. NASA Reference Publication, Cleveland (1997)
7. Litvin, F.L., Fuentes A.: Gear Geometry and Applied Theory, second edn. Cambridge University Press
8. Wu, H., et al.: Design of Worm Drive, vol. 1. Mechanical Industry Press, Beijing (1986). (in Chinese)

9. Wang, S.: Circular Arc Cylindrical Worm Drive. Tianjin University Press, Tianjin (1991). (in Chinese)
10. Zhao, Y., Zhang, Y.: Computing method for induced curvature parameters based on normal vector of instantaneous contact line and its application to Hindley worm pair. Adv. Mech. Eng. **9**, 1–15 (2017)
11. Zhao, Y., Sun, X.: On meshing limit line of ZC1 worm pair. In: European Conference on Mechanism Science 2018, pp. 292–298. Aachen, Germany (2018a)
12. Zhao, Y., Kong, X.: Meshing principle of conical surface enveloping spiroid drive. Mech. Mach. Theory **123**, 1–26 (2018b)